Imre J. Rudas, János Fodor, and Janusz Kacprzyk (Eds.)

Towards Intelligent Engineering and Information Technology

T0191658

Studies in Computational Intelligence, Volume 243

Editor-in-Chief

Prof. Janusz Kacprzyk
Systems Research Institute
Polish Academy of Sciences
ul. Newelska 6
01-447 Warsaw
Poland
E-mail: kacprzyk@ibspan.waw.pl

Imre J. Rudas, János Fodor, and Janusz Kacprzyk (Eds.)

Towards Intelligent Engineering and Information Technology

 Springer

Prof. Dr. Imre J. Rudas
Budapest Tech
John von Neumann Fac.
Informatics
Dept. Intelligent Engineering
Systems
Bécsi út 96/B
Budapest 1034
Hungary
E-mail: rudas@ bmf.hu

Prof. Dr. Janusz Kacprzyk
PAN Warszawa
Systems Research Instiute
Newelska 6
01-447 Warszawa
Poland
E-mail: kacprzyk@ibspan.waw.pl

Prof. Dr. János Fodor
Budapest Tech
John von Neumann Fac.
Informatics
Dept. Intelligent Engineering
Systems
Bécsi út 96/B
Budapest 1034
Hungary
E-mail: fodor@ bmf.hu

ISBN 978-3-642-26926-4   e-ISBN 978-3-642-03737-5

DOI 10.1007/978-3-642-03737-5

Studies in Computational Intelligence   ISSN 1860-949X

ⓒ 2009 Springer-Verlag Berlin Heidelberg
Softcover reprint of the hardcover 1st edition 2009

This work is subject to copyright. All rights are reserved, whether the whole or part
of the material is concerned, specifically the rights of translation, reprinting, reuse
of illustrations, recitation, broadcasting, reproduction on microfilm or in any other
way, and storage in data banks. Duplication of this publication or parts thereof is
permitted only under the provisions of the German Copyright Law of September 9,
1965, in its current version, and permission for use must always be obtained from
Springer. Violations are liable to prosecution under the German Copyright Law.

The use of general descriptive names, registered names, trademarks, etc. in this
publication does not imply, even in the absence of a specific statement, that such
names are exempt from the relevant protective laws and regulations and therefore
free for general use.

Typeset & Cover Design: Scientific Publishing Services Pvt. Ltd., Chennai, India.

Printed in acid-free paper

9 8 7 6 5 4 3 2 1

springer.com

Foreword

This year marks the 130th anniversary of the foundation of Budapest Tech, the largest polytechnical institution in Hungary. The Institute's motto of 'Knowledge and innovation in support of the economy' has been the guiding principle of all its activities. For 130 years of existence through its predecessors, Budapest Tech's educational excellence has remained a primary goal. The institution has been implemented a comprehensive academic program with emphasis on competitive, flexible and high-level training, scientific research and development, which complies with the changes of economic and social life and permanently renews. Budapest Tech offers a focused, technologically-based education on a wide range for about 12000 students in a stimulating environment at five faculties and two educational centers.

Among the many activities undertaken by Budapest Tech towards the achievement of its goals, publication of conference proceedings played an important role. This activity includes international and national-level seminars, symposia, conferences and workshops, highlighting the emerging areas of ongoing frontline research in various fields of intelligent engineering systems. As an outcome of these events and results achieved, the Institute is bringing out a book under the title Towards Intelligent Engineering and Information Technology, published by Springer.

Chapters in the present volume have been written by eminent scientists from different parts of the world and from Budapest Tech, dealing with challenging problems for efficient modeling of intelligent systems. The reader can find different aspects, characteristics and methodologies of computational intelligence with real life applications. I believe that the state-of-the art studies presented in this book will be very useful to a broad readership. Thanks to the authors for their excellent contributions and to the editors for their valuable work and effort bringing out this volume.

June 2, 2009

József Pálinkás
President
Hungarian Academy of Sciences

Preface

An intelligent system is a system that emulates some aspects of intelligence exhibited by nature. These include learning, adaptability, improving efficiency, and management of uncertain and imprecise information. Intelligent systems are developed for handling problems in which algorithmic solutions require "almost infinite" amount of computation, or even more: there exist no algorithms for solving them.

Extracting from diverse dictionaries, intelligence means the ability to comprehend; to understand and profit from experience. There are, of course, other meanings such as ability to acquire and retain knowledge; mental ability; the ability to respond quickly and successfully to a new situation; etc.

Within such a framework, intelligent engineering systems try to replicate fundamental abilities of humans and nature in order to achieve sufficient progress in solving complex problems. In an ideal case multi-disciplinary applications of different modern engineering disciplines can result in synergistic effects. Information technology and computer modeling are the underlying tools that play a major role at any stages of developing intelligent systems.

The present edited book is a collection of 51 chapters written by internationally recognized experts and well-known colleagues of the fields. Chapters contribute to diverse facets of intelligent engineering and information technology. The volume is organized in six parts according to the main subjects, and in harmony with the fields of interest of the jubilarian institution. In this landmark year Budapest Tech celebrates its past and looks to its future in research through this volume.

The first part of the book is devoted to theoretical issues. This includes pseudo-analysis in engineering decisions; a study of possibilistic counterparts of well-known probabilistic concepts and results; recent results in matrix perturbation theory; an outline of computationalism in a non-traditional context; a study of measuring voting power; parametric classes of digital fuzzy conjunctions for hardware implementation of fuzzy systems.

The second part of the book concerns with different aspects of control. This includes a study of quantum control systems; a model-based control design of integrated vehicle systems; design and applications of cerebellar model articulation controller; an iterative feedback tuning in linear and fuzzy control systems; model-based design issues in fuzzy logic control; situational control, modeling and diagnostics of large scale systems; a hybrid approach in power electronics and motion control.

The third part addresses issues of robotics, including mobile mini robots for space applications, motion control for formation of mobile robots in environment with obstacles, robot systems for play in education and therapy of disabled children, an overview of recent advances in intelligent robots at J. Stefan Institute, new trends in robotic reinforcement learning, different points of view on building an intelligent robot, autonomous locomotion of humanoid robots in presence of mobile and immobile obstacles, and an autonomous advertising mobile robot for exhibitions developed at Budapest Tech.

The fourth part presents approaches and results on information technology. This includes high-level specification of games; new trends in non-volatile semiconductor memories; biometric motion identification based on motion capture; through wall tracking of moving targets by M-Sequence UWB radar; studies on advanced industrial communications; logical consequences in partial knowledge bases; distributed detection system of security intrusions based on partially ordered events and patterns; the theory and practice of wireless sensors networks; image processing using polylinear functions on HOSVD basis; intelligent short text assessment in eMax system; determination the basic network algorithms with gains; the role of RFID in development of intelligent human environment.

The fifth part of the book consists of papers on machines, materials and manufacturing. This includes the history and challenges of mechanism and machine science within IFToMM community; a service orientated architecture for holonic manufacturing control; sensitivity of power spectral density analysis for measuring conditions; numerical prediction of friction, wear, heat generation and lubrication in case of sliding rubber components; design of a linear scale calibration machine; engineering objective driven product lifecycle management with enhanced human control; in-situ investigation of the growth of low-dimensional structures; monitoring of heat pumps; the nozzle's impact on the quality of fabric on the pneumatic weaving machine.

The last part of the book is devoted to intelligent systems and complex processes, including a comprehensive evaluation of the efficiency of an integrated biogas, trigen, PV and greenhouse plant using digraph theory; an approximation of a modified traveling salesman problem using bacterial memetic algorithms; an approach towards intelligent systems with incremental learning ability; a systems engineering approach to sustainable energy supply; fuzziness of rule outputs by the DB operators-based control problems; observer based iterative fuzzy and neural network model inversion for measurement and control applications, the role of the cyclodextins in analytical chemistry; an intelligent GIS-based route/site selection plan of a metro-rail network.

The editors are grateful to the authors for their excellent work. Thanks are also due to Ms. Anikó Szakál for her editorial assistance and sincere effort in bringing out the volume nicely in time.

June 22, 2009   I.J. Rudas
                                                                         J. Fodor
                                                                      J. Kacprzyk
                                                             Budapest and Warsaw

Contents

Part I
Theoretical Issues

Pseudo-analysis in Engineering Decision Making

Endre Pap

Department of Mathematics and Informatics
Trg Dositeja Obradovica 4, 21 000 Novi Sad, Serbia
pape@eunet.yu, pap@im.ns.ac.yu

Abstract. There is presented probabilistic von Neumann-Morgenstern type approach to engineering design. Further generalization of the utility theory, using pseudo-analysis, first based on possibility theory, and second, as a common generalization through hybridization of the both preceding approaches are given. In modeling uncertainty in engineering design it is very useful the fuzzy system approach, which involves further real operations as aggregation functions.

1 Introduction

Design is a process where the human intellect with creativity produce useful artifacts, and which involves pure and applied sciences, but also behavioral and social sciences. Engineering design is recognized as a decision-making process at the core [23, 24, 35, 49, 56]. Engineering design conducted with incomplete and imperfect information, yet most traditional design approaches treat the design problem are deterministic. The proposed research is to develop tools for decision making under risk and uncertainty and apply the tools to engineering design.

The approach with probability has three main elements [24]: identification of the options, determination of expectations on each option and the expression of values. The main decision rule is: the preferred decision is the option whose expectation has the highest value. Classical decision theory [32] separates expectations and values - a common mistake is to make them equal. The decision making involve options, expectations and values.

The advantage of the pseudo-analysis [38, 40], as a generalization of the classical real analysis, based on a semiring structure (see [18, 31]) on a real interval $[a, b] \subset [-\infty, \infty]$, is the fact that it coveres as one theory and so with unified methods equations (usually nonlinear), and models with uncertainty (not only with probability) from many different fields (system theory, optimization, control theory, differential equations, difference equations, decision making, etc.).

I.J. Rudas et al. (Eds.): Towards Intelligent Engineering & Information Tech., SCI 243, pp. 3–16.
springerlink.com   © Springer-Verlag Berlin Heidelberg 2009

An engineering decision cannot be done in the absence of human values, whereas problems in the science are solved in the absence of options and values. The purpose of values in decision making is to rank order alternatives. This ranking is managed by a preference relation, which is connected by the usual real order relation through the utility function.

In Section 2 we present probabilistic von Neumann-Morgenstern type approach to engineering design. Section 3 contains further generalization of the utility theory, first based on possibility theory, and second, as a common generalization through hybridization of the both preceding approaches. In modeling uncertainty in engineering design is very useful the fuzzy system approach, presented in Section 4, and which involves further real operations as aggregation functions, presented in Section 5.

2   Probabilistic Approach Based on von Neumann–Morgenstern Theory

We start with an axiomatic approach to engineering design which guarantees a rational treatment of all information that the designer uses for the design and enables a rational decision making [23, 24, 57]. We shall use the symbol $\succ$ for the relation "is preferred to", the symbol $\sim$ for "is indifferent to" and the symbol $\succeq$ for "is preferred or indifferent to".

Axiom 1: *The axiom of deterministic making.* Given a defined set of options from which to choose, each with a known and deterministic outcome, the decision maker's preferred choice is that option whose outcome is most desired.

Axiom 2: *Ordering of alternatives.* Preference and indifference orderings hold between any two outcomes, and they are transitive.

Axiom 3: *Reduction of compound lotteries.* Any compound lottery is indifferent to a simple lottery with the same outcomes and associated probabilities.

Axiom 4: *Continuity.* Given outcomes of a lottery ordered by preference from A_1 through A_r, there exists a number u_i such that each outcome A_i is indifferent to a lottery containing only A_1 and A_r.

With mathematical symbols

$$A_i \sim [u_i A_1, (1 - u_i) A_r] = \hat{A}_i,$$

where $\hat{A}_i$ is the lottery.

Axiom 5: *Substitutability.* In any lottery L, $\hat{A}_i$ is substitutable for A_i.

Axiom 6: *Transitivity.* Preference and indifference among lotteries are transitive relations.

Axiom 7: *Monotonicity.* A lottery $[pA_1,(1-p)A_r]$ is preferred or indifferent to a lottery $[p'A_1,(1-p')A_r]$ if and only if $p \geq p'$.

Axiom 8: *Reality of engineering design.* All engineering designs are selected from among the set of potential designs that are explicitly considered.

Let us mathematically formalize the preceding axioms. Let p be a simple probability measure on $X = \{x_1,...,x_n\}$, thus $p = (p(x_1), p(x_2)..., p(x_n))$, where $p(x_i)$ are probabilities of outcome $x_i \in X$ occurring, i.e., $p(x_i) \geq 0$ for all $i = 1, 2..., n$, and $\sum_{i=1}^{n} p(x_i) = 1$. Define $\mathbf{P}(X)$ as the set of simple probability measures on X. A particular lottery p is a point in $\mathbf{P}(X)$. A compound lottery (mixture) is an operation defined on $\mathbf{P}(X)$ which combines two probability distributions p and p' into a new one, denoted $V(p, p'; \alpha, \beta)$, with $\alpha, \beta \in [0,1]$ and $\alpha + \beta = 1$, and it is defined by

$$V(p,p';\alpha,\beta) = \alpha p + \beta p'.$$

Note that $V(p, p'; \alpha, \beta) \in \mathbf{P}(X)$. Let $\preceq f$ be a binary relation over $\mathbf{P}(X)$, where $p \preceq fq$ means that lottery q is "preferred to or equivalent to" lottery p.

The preceding system of axioms corresponds to the following utility axioms:

NM1 $\mathbf{P}(X)$ is equipped with a complete preordering structure $\preceq f$.

NM2   (Continuity): For $p \prec q \prec r$ there exists α such that
$$q \sim V(p,r;\alpha,1-\alpha).$$

NM3   (Independence): $p \sim q$ implies
$$V(p,r;\alpha,1-\alpha) \sim V(q,r;\alpha,1-\alpha) \quad (r \in \mathbf{P}(X), \alpha \in [0,1]).$$

NM4   (Convexity): For $p \prec q$ we have
$$p \prec V(p,q;\alpha,1-\alpha) \prec q \quad (\alpha \in]0,1[.$$

The theorem below shows that the preference ordering on set of states which satisfies the proposed axioms can always be represented by a utility function.

Theorem 1 (Representation Theorem ([57], 1944)). *A preference ordering relation $\preceq f$ on $\mathbf{P}(X)$ satisfies axioms NM1, NM2, NM3 and NM4 if and only if, there is a real-valued function $U: \mathbf{P}(X) \to \mathbb{R}$ such that*

(i) U represents $\preceq$, i.e., for $p,q \in \mathbf{P}(X)$ holds $p \preceq fq$ if and only if
$$U(p) \preceq U(q);$$

(ii) U is affine, i.e., for every $p,q \in \mathbf{P}(X)$ and every $\alpha \in]0,1[$ we have
$$U(\alpha p+(1-\alpha)q) = \alpha U(p)+(1-\alpha)U(q).$$
Moreover, U is unique up to a linear transformation.

As consequence of these axioms there were deduced the following three important theorems [23, 24].

Theorem 2 (The expected utility theorem.). *Given a pair of options, each with a range of possible outcomes and associated probabilities of occurrence, that is, two lotteries, the preferred choice is the option (or lottery) that has the highest expected utility.*

Theorem 3 (The substitution theorem.). *A decision maker is indifferent between a lottery L and a certainty outcome whose utility is equal to the expected utility of the lottery.*

The person who has a transitive preference relation usually is called rational, in the opposite case he is irrational. The famous Arrow's Impossibility Theorem [3, 23] states that a group consisting only of rational individuals need not exhibit transitive preferences.

Theorem 4 (Arrow's Impossibility Theorem.). *Groups consisting of rational people are not necessarily rational.*

3   Generalization of the Probabilistic Approach

Pseudo-analysis is based on the semiring structure on the real interval $[a,b] \subseteq [-\infty,\infty]$, see [38, 40]. For some engineering applications see [4, 43]. In this paper we restrict ourselves on the special case, operations on the interval $[0,1]$ (see [28]) and therefore on special non-additive measures on so called pseudo-additive (decomposable) measures (see [28, 38, 40]).

Definition 1. A triangular conorm (t-conorm for short) is a binary operation on the unit interval $[0,1]$, i.e., a function $S : [0,1]^2 \to [0,1]$ such that for all $x, y, z \in [0,1]$ the following four axioms are satisfied:

(S1)   *Commutativity* $S(x,y) = S(y,x)$,

(S2)   *Associativity* $S(x,S(y,z)) = S(S(x,y),z)$,

(S3)   *Monotonicity* $S(x,y) \le S(x,z)$ whenever $y \le z$,

(S4)   *Boundary Condition* $S(x,0) = x$.

If S is a t-conorm, then its dual t-norm $T : [0,1]^2 \to [0,1]$ is given by
$$T(x,y) = 1 - S(1-x, 1-y).$$

Definition 2. A t-norm T is restricted distributive over a t-conorm S if for all $x, y, z \in [0,1]$ we have

(RD) $T(x,S(y,z))=S(T(x,y),T(x,z))$,

whenever $S(y,z) < 1$.

The complete characterization of the pair (S,T) satisfying condition (RD) is given in [28].

A mapping $m : 2^X \rightarrow [0,1]$ is called a pseudo-additive measure (S -measure), if $m(\varnothing) = 0, m(X) = 1$ and if for all $A, B \in 2^X$ with $A \cap B = \varnothing$ we have $m(A \cup B) = S(m(A), m(B))$, see [10, 28, 38]. Important example is the maxitive measure, i.e., max-measure, where $m(A) = \sup_{x \in A} \pi(x)$.

We present now the possibilistic approach to the utility theory [13]. The belief state about which situation in X is the actual one is supposed to be represented by a possibility distribution π. A possibility distribution π defined on X takes its values on a valuation scale V, where V is supposed to be linearly ordered. V is assumed to be bounded and we take $\sup(V) = 1$ and $\inf(V) = 0$. Define $\mathbf{Pi}(X)$ as set of consistent possibility distributions over X, i.e.,

$$\mathbf{Pi}(X) = \{\pi : X \rightarrow V \mid \exists x \in X \pi(x) = 1\}.$$

The possibilistic mixture is an operation defined on $\mathbf{Pi}(X)$ which combines two possibility distributions π and π' into a new one, denoted $P(\pi, \pi'; \alpha, \beta)$, with $\alpha, \beta \in V$ and $\max(\alpha, \beta) = 1$, given by

$$P(\pi, \pi'; \alpha, \beta) = \max(\min(\alpha, \pi), \min(\beta, \pi')).$$

Let $\sqsubseteq$ be a binary relation over $P(\pi, \pi'; \alpha, \beta)$. Hence, we can write $\pi \sqsubseteq \pi'$ to indicate that possibilistic lottery π' is "preferred to or equivalent to" lottery π.

The proposed axiom systems for the possibilistic optimistic utility is

Pos 1 $\mathbf{Pi}(X)$ is equipped with a complete preordering structure $\sqsubseteq$.

Pos 2   (Continuity) For every $\pi \in \mathbf{Pi}(X)$ there exist $\lambda \in V$ such that $\pi \sim P(\overline{\pi}, \underline{\pi}; \lambda, 1)$, where $\overline{\pi}$ and $\underline{\pi}$ are a maximal and a minimal element of $\mathbf{Pi}(X)$ w.r.t. $\sqsubseteq$, respectively.

Pos 3   (Independence) $\pi \sim \pi'$ implies $P(\pi, \pi''; \lambda, \mu) \sim P(\pi', \pi''; \lambda, \mu)$, for every $\pi'' \in \mathbf{Pi}(X)$ and every $\lambda, \mu \in V$.

Pos 4   (Uncertainty prone): $\pi \leq \pi'$ implies $\pi \sqsubseteq \pi'$.

The set of axioms Pos1, Pos2, Pos3 and Pos4 characterize the preference ordering induced by an optimistic utility.

Theorem 5 (Representation Theorem ([13], 1998)). *A preference ordering relation $\sqsubseteq f$ on $\mathbf{Pi}(X)$ satisfies axioms Pos1, Pos2, Pos3 and Pos4 if and only if, there exist*

(i)   *a linearly ordered utility scale* U, *with* $\inf(U) = 0$ *and* $\sup(U) = 1$;

(ii)   *a preference function* $u : X \to U$ *such that* $u^{-1}(1) \neq \varnothing \neq u^{-1}(0)$, *and*

(iii)   *an onto ordered preserving function* $h : V \to U$ *such that*
$$h(0)=0,\ h(1)=1,$$

in such a way that it holds: $\pi \sqsubseteq \pi'$ *if and only if* $\pi \lhd_u \pi'$, *where* $\lhd_u$
is the ordering on $\mathbf{Pi}(X)$ *induced by the qualitative utility*
$$QU(\pi) = \max_{x \in X} \min\left(h(\pi(x)), u(x) \right).$$

We present now the hybrid probabilistic-possibilistic utility theory [14, 15]. In order to generalize stated sets of axioms for utility theory, we denote $X = \{x_1, x_2, ..., x_n\}$ set of outcomes, $\Delta(X)$ the set of all S-measures defined on X.

Definition 3. A hybrid mixture operation which combines two S-measures m and m' into a new one, denoted $M(m, m'; \alpha, \beta)$, for $a \in [0,1]$ and that (α, β) belongs to with $\Phi_{S,a} = \{\alpha, \beta \,|\, \alpha, \beta \in \,]0,1[\,, \alpha + \beta = 1 + a$ or $\min(\alpha, \beta), \leq a, \max((\alpha, \beta) = 1 \ \}$

is given by
$$M(m, m'; \alpha, \beta) = S(T(\alpha, m), T(\beta, m')),$$

where (S, T) is a pair of continuous t-conorm and t-norm, respectively, which satisfy the property of restricted distributivity (RD).

We propose the following set of axioms for a preference relation $\preceq_h$ defined over $\Delta(X)$ to represent optimistic utility

H1 $\Delta(X)$ is equipped with a complete preordering structure $\preceq_h$ (i.e., $\preceq_h$ is reflexive, transitive and complete).

H2   (Continuity) If $m \preceq_h m' \preceq_h m''$ then we have

(i)   for $m, m', m'' > a$ there exists $\alpha \in \,]a,1[$ such that
$$m' \sim_h M(m, m''; 1 + a - \alpha, \alpha);$$

(ii)   there exists $\alpha \in \,]0, a]$ such that $m' \sim_h M(m, m''; 1, \alpha)$.

H3   (Independence) For all $m, m', m'' \in \Delta(X)$ and for all $\alpha, \beta \in \Phi_{S,a}$ we have that $m' \preceq_h m''$ is equivalent with $M(m', m; \alpha, \beta) \preceq_h M(m'', m; \alpha, \beta)$.

H4   (Uncertainty prone)

(i)   if $m, m' > a$ then $m \preceq_h m'$ implies

$m \preceq_h M(m, m'; \alpha, 1 + a - \alpha) \preceq_h m'$ for $\alpha \in]a, 1[$;

(ii)   otherwise $m < m'$ implies $m \preceq_h m'$.

Now, we define a function of optimistic utility for all $m \in \Delta(X)$ in the following way

$$U(m) = S_{x_i \in X} T(m(x_i), u(x_i)),$$

where $u : X \to U$ is a preference function that assigns to each consequence of X a preference level of U, such that $u^{-1}(1) \neq \varnothing \neq u^{-1}(0)$.

Remark 1. It is interesting to note that U preserves the hybrid mixture in the sense that

$$U(M(m, m'; \alpha, \beta)) = S(T(\alpha, U(m)), T(\beta, U(m')))$$
$$= M(U(m), U(m'); \alpha, \beta).$$

Theorem 6 (Representation Theorem - Optimistic Utility, [44]). *Let $\Delta(X)$ be the set of all S-measures defined on 2^X, and $\preceq_h$ a binary preference relation on $\Delta(X)$. Then the relation $\preceq_h$ satisfies the set of axioms H1, H2, H3, H4 if and only if there exist*

(i)   a linearly ordered utility scale U, with $\inf(U) = 0$ and $\sup(U) = 1$;

(ii)   a preference function $u : X \to [0, 1]$,

such that $m \preceq_h m'$ if and only if $mf \sqsubseteq_h m'$, where $f \sqsubseteq_h$ is the ordering in $\Delta(X)$ induced by the optimistic utility function given by

$$U(m) = S(T(m(x), u(x)),$$

where (S, T) is a pair of continuous t-conorm and t-norm, respectively, which satisfy the condition (RD).

4   Fuzzy Systems

A decision is made under the risk if the only available knowledge related the outcome states is the probability distribution. This can be used in the optimization of the utility function. If the knowledge about the probabilities of the outcome is unknown, then the decision have to be made under uncertainty. In engineering design one of the most critical problems is the preliminary design decision when the design is imprecise and most costly [2, 37, 59]. In such situations, fuzzy decision making can be used to handle this vagueness. This fuzziness can be modeled in different ways: fuzzy sets (membership function) [37] (The Method of Imprecision), [62], fuzzy measures (Choquet and Sugeno integrals) [21, 38, 58]. There are other design methodologies as optimization tools (linear, nonlinear, integer programming, multi-objective optimization - e.g., with weighted sum technique [36]), probability methods [54].

An overall evaluation of design alternatives have two parts: their partial evaluation and the importance of the criteria taken into account. The first step consists in the determination of fuzzy sets representing partial evaluation of the alternatives. Since there are many different judgments with respect to the expression of the suitability of variety of alternatives there is need for some methods for this purpose. The Analytical Hierarchy Process [48] and some other matrix methods as [45, 46] are very convenient tools for that purpose. At the second step all partial information is aggregated into a final rating. In engineering design the preliminary design decisions are very important although the design description is still imprecise. Fuzzy design methods are convenient for representing and manipulating design imprecision [25, 37, 62]. By the Method of Imprecision [37, 50] constraints can be imprecise permitting to choose preferences over a range of values. This method was specially developed for engineering design and implies that the trade-off combination functions (aggregation operators) have to satisfy the boundary conditions, monotonicity, continuity, annihilation and idempotency. Then it follows by [37, 51] that any weigthed quasi-linear mean that satisfies the annihilation property is design-appropriate. A weighted aggregation function which continuous, strictly monotonic, idempotent and bisymmetrical has the representation ([1, 17])

$$\mathbf{M}^{f}_{\omega_1,\ldots,\omega_n}(x_1,\ldots,x_n) = f^{-1}\left(\frac{\omega_1 f(x_1) + \cdots + \omega_n f(x_n)}{\omega_1 + \cdots + \omega_n}\right),$$

where f is a strictly monotone continuous function. By this representation it is possible to construct special convenient families of aggregation functions ([17, 21, 30, 37]). Drakopoulos [9] proved that probabilities have a higher representational power than fuzzy sets (with respect to max-min) and possibilities (special fuzzy measure) for finite domain, but at the cost of higher computational complexity and reduced computational efficiency (they have equal representational power when their domains are infinite).

5 Aggregation in Engineering Design

The aggregation of incoming data plays a key role in applications of several intelligent systems. The aggregation functions (operators) form a fundamental part of multi-criteria decision making, engineering design, expert systems, pattern recognition, neural networks, fuzzy controllers, genetic algorithms, etc. ([17, 20, 28, 37, 61]).

We restrict ourselves to the inputs and outputs from the unit interval $[0,1]$. Note that the case of any other closed interval is the question of rescaling only.

Definition 4. An *aggregation function* $\mathbf{A}$ is a non-decreasing mapping

$$\mathbf{A}: \bigcup_{n \in \mathbb{N}} [0,1]^n \to [0,1]$$

fulfilling the following conditions

(i) $\quad 0 \leq x_i \leq y_i ?1, i = 1, \leq, n$ imply $\mathbf{A}(x_1, ..., x_n) \leq \mathbf{A}(y_1, ..., y_n)$;

(ii) $\quad \mathbf{A}(x) = x$ for all $x \in [0,1]$;

(iii) $\quad \mathbf{A}(0,...,0) = 0$ and $\mathbf{A}(1,...,1) = 1$.

Property (i) in Definition 4 is *the monotonicity* and properties (ii) and (iii) are *the boundary conditions*. Each aggregation function $\mathbf{A}$ can be represented as a system $(\mathbf{A}_n)_{n \in \mathbb{N}}$ of n-ary operators $\mathbf{A}_n, n \in \mathbb{N}$, on the unit interval, where $\mathbf{A}_1$ is the identity operator on $[0,1]$ and each $\mathbf{A}_n, n \geq 2$, is non-decreasing and $\mathbf{A}_n(0,...,0) = 0, \mathbf{A}_n(1,...,1) = 1$.

Depending on the field of application, several additional properties can be required and/or examined, such as commutativity, associativity, continuity, idempotency, compensation, cancellativity, etc. Note, for example, that the associativity of an aggregation function $\mathbf{A}$ means that the binary function $\mathbf{A}_2$ is associative and its corresponding n-ary extensions (for $n > 2$) are just the relevant n-ary operators $\mathbf{A}_n$. Therefore, an associative aggregation function $\mathbf{A}$ is fully determined by $\mathbf{A}_2$. If $\mathbf{A}$ is an aggregation function, then the operator $\mathbf{DA}: \bigcup_{n \in \mathbb{N}} [0,1]^n \to [0,1]$ defined by

$$\mathbf{DA}(x_1, ..., x_n) = 1 - \mathbf{A}(1 - x_1, ..., 1 - x_n)$$

is called the *dual operator* of $\mathbf{A}$. $\mathbf{DA}$ is also an aggregation function.

Fuzzy design methods are convenient for representing and manipulating design imprecision [37, 62]. The Method of Imprecision [37] was specially developed for engineering design and implies that the trade-off combination functions (aggregation operators) have to satisfy the boundary conditions, monotonicity, continuity, annihilation, and idempotency, where annihilation means that if one argument (when the preference for any one attribute of the design sinks to zero) of the aggregation operator is zero then the value of the aggregation operator (the overall preference of the design) is zero. We remark that if the weights $\{\omega_i\}$ in a weighted aggregation function are given with respect to a ratio scale, then ω_i are not uniquely determined, since any other system of weights $\{\omega_i'\}$ with $\omega_i' = C\omega_i$ for a positive rational number C is convenient, e.g., $\omega_i' = \frac{\omega_i}{\sum_i \omega_i}$. Specially important cases are $f(x) = x, f(x) = \log x$ and $f(x) = x^s$. The last case is interesting since it generates a parameterized family of aggregation functions. For $s > 0$, the annihilation property fails, but it can be handled in the practical engineering design by assuming that preferences smaller than some small ε are not relevant for the designer [51]. There are also design-appropriate aggregation functions which are not weighted quasi-arithmetical means, that is, they are not strictly monotone.

Starting from a given t-norm and/or t-conorm, several useful operations on $[0,1]$ can be introduced. The conditions (i) - (iii) required for an aggregation operator $\mathbf{A}$ are the genuine properties of triangular norms and conorms.

From the practical application point of view, there are suggestions to use the special aggregation functions, so-called *compensatory operators,* in order to model intersection and union in many-valued logic. The main goal of compensatory operators is to model an aggregation of incoming values. If two values are aggregate by a t-norm, then there is no compensation between low and high values. On the other hand, a t-conorm based aggregation provides the full compensation. None of the above cases covers the real decision making. To avoid such inaccuracies, in [63] suggested two kinds of so-called compensatory operators, see [33]. The first of them was γ-operator, $\Gamma : [0,1] \to [0,1], \ \gamma \in [0,1], \ n \geqslant 2$

$$\Gamma_\gamma(x_1,...,x_n) = \left(\prod_{i=1}^n x_i\right)^{1-\gamma} \left(1 - \prod_{i=1}^n (1-x_i)\right)^\gamma.$$

Here parameter γ indicates the degree of compensation. Note that γ-operators are a special class of exponential compensatory operators [28]. For a given t-norm T t-conorm S (not necessarily dual to T) and parameter γ indicating the degree of compensation, the exponential compensatory operator $E_{T,S,\gamma} : [0,1]^n \to [0,1], \ n \geq 2,$ is defined by

$$E_{T,S,\gamma}(x_1,...,x_n) = (T(x_1,...,x_n))^{1-\gamma}(S(x_1,...,x_n))^\gamma.$$

It is obvious that γ-operator is based an $T_\mathbf{P}$ $S_\mathbf{P}$, $\Gamma_\gamma = E_{T_\mathbf{P},S_\mathbf{P},\gamma}$. Further note that $E_{T,S,\gamma}$ is a logarithmic convex combination of T and S and up to the case when $\gamma \in \{0,1\}$ it is non-associative. Another class of compensatory operators proposed by [63, 64] are so-called convex-linear compensatory operators.

It was proposed an *associative class of compensatory operators* in [27]. The degree of compensation is ruled by two parameters, namely by the neutral element e and the compensation factor k. Let T be a given strict t-norm with additive generator $f, f(\frac{1}{2}) = 1$, and let S be a given strict t-conorm with an additive generator $g, g(\frac{1}{2}) = 1$. For a given $e \in]0,1[, \ k \in]0,\infty[$, we define an associative compensatory operator

$$C(T, S, e, k) = C : [0, 1]^2 \setminus \{(0, 1), (1, 0)\} \rightarrow [0, 1]$$

by

$$C(x, y) = h^{-1}(h(x) + h(y)),$$

where $h : [0, 1] \rightarrow [-\infty, \infty]$ is a strictly increasing bijection such that

$$h(x) = \begin{cases} kf(\frac{x}{e}) & \text{if } x \in [0, e] \\ g(\frac{x-e}{1-e}) & \text{if } x \in]e, 1]. \end{cases}$$

Engineering decision making need more general mathematical models, which involve also non-additive measures. Previously used additive probability measures could not model some situations, e.g., the Ellsberg Paradox, see [21]. For the non-additive set function (measure) m defined on a σ-algebra $\mathcal{A}$ of subsets of a set X (for finite X it is usually taken $\mathcal{A} = 2^X$, the family of all subsets), the difference $m(A \cup B) - m(B)$ depends on B and can be interpreted as the effect of A joining B, [21, 38, 52, 53, 58]. A monotone set function m with $m(\varnothing) = 0$ is usually called fuzzy measure. More than the contribution of the extension principle of the fuzzy sets [55], fuzzy connectives [16, 21, 27, 63] and fuzzy measures are important in the problem of the modeling of the behavior of decision makers. Utility theory [17, 21, 47] deals with preference relations describing the decision behavior, and as the basis of decision theory, is well axiomatically based on the fuzzy measures and Choquet integral [17, 20, 22, 38]. The Choquet integral approach is generalized in many directions ([28, 34, 38]). As the mapping, the fuzzy integral is defined by a set of 2^n (for n elements basic set X) parameters and a t-conorm system. The word "identification" has the origin in the system theory and is preferred to the word "learning", though the algorithms for finding the appropriate fuzzy measure could be the learning samples minimizing certain criterion. Unknown measure to be identified can be regarded as the part of the parameter identification [19, 20, 21].

Conclusion

We have given a short overview of some basic facts from the theory of pseudo-analysis, mostly related to pseudo-operations. As a generalization of von Neumann and Morgestern utility theory, using pseudo-analysis, there are presented approach based on possibility theory, and as a common generalization through hybridization of the both approaches is given. We modeled uncertainty in engineering design with fuzzy systems, which involves more general real operations: aggregation functions [20]. We remark that S-measures and corresponding integrals have the advantage that for n elements of the basic set X they require only n parameters.

References

[1] Aczél, J.: Lectures on Functional Equations and their Applications. Academic Press, New York (1966)

[2] Antonsson, E., Otto, K.: Imprecision in Engineering Design. ASME Journal of Mechanical Design, 25–32 (1995)

[3] Arrow, K.J.: Social Choice and Individual Values, 2nd edn. Yale University Press, New Haven (1963)

[4] Baccelli, F., Cohen, G., Olsder, G.J.: Quadrat: Synchronization and Linearity: an Algebra for Discrete Event Systems. John-Wiley and Sons, New York (1992)

[5] Bellman, R.E., Dreyfus, S.E.: Applied Dynamic Programming. Princeton University Press, Princeton (1962)

[6] Chen, J., Otto, K.: A Tool for Imprecise Calculations in Engineering Design. In: Proc. of the Fourth IEEE International Conference on Fuzzy Systems, vol. I, pp. 389–394 (1995)

[7] Choquet, G.: Theory of capacities. Ann. Inst. Fourier (Grenoble) 5, 131–295 (1953)

[8] Cuninghame-Green, R.A.: Minimax Algebra. Lecture Notes in Economics and Math. Systems, vol. 166. Springer, Heidelberg (1979)

[9] Drakopoulos, J.A.: Probabilities, possibilities, and fuzzy sets. Fuzzy Sets and Systems 75, 1–15 (1995)

[10] Dubois, D., Prade, H.: A Class of Fuzzy Measures Based on Triangular Norms. Internat. J. Gen. System 8, 43–61 (1982)

[11] Dubois, D., Prade, H.: Fuzzy Sets and Systems: Theory and Applications. Academic Press, New-York (1980)

[12] Dubois, D., Fodor, J.C., Prade, H., Roubens, M.: Aggregation of Decomposable Measures with Applications to Utility Theory. Theory and Decision 41, 59–95 (1996)

[13] Dubois, D., Godo, L., Prade, H., Zapico, A.: Making Decision in a Qualitative Setting: from Decision under Uncertainty to Case-based Decision. In: Cohn, A.G., Schubert, L., Shapiro, S.C. (eds.) Proceedings of the Sixth International Conference on Principles of Knowledge Representation and Reasoning (KR 1998), pp. 594–605. Morgan Kaufman Publishers Inc., San Francisco (1998)

[14] Dubois, D., Pap, E., Prade, H.: Hybrid Probabilistic-Possibilistic Mixtures and Utility Functions. In: Fodor, J., de Baets, B., Perny, P. (eds.) Preferences and Decisions under Incomplete Knowledge. Studies in Fuzziness and Soft Computing, vol. 51, pp. 51–73. Springer, Heidelberg (2000)

[15] Dubois, D., Pap, E., Prade, H.: Pseudo-Additive Measures and the Independence of Events. In: Bouchon-Meuner, B., Gutierrez-Rios, J., Magdalena, L., Yager, R.R. (eds.) Technologies for Constructing Intelligent Systems 1. Studies in Fuzziness and Soft Computing, vol. 89, pp. 179–191. Springer, Heidelberg (2001)

[16] Dyckhoff, H., Pedrycz, W.: Generalized Means as Model for Compensative Connectives. Fuzzy Sets and Systems 14, 143–154 (1984)

[17] Fodor, J., Roubens, M.: Fuzzy Preference Modeling and Multi-Criteria Decision Aid. Kluwer Academic Publisher, Norwell (1994)

[18] Golan, J.S.: The Theory of Semirings with Applications in Mathematics and Theoretical Computer Sciences. Pitman Monog. and Surveys in Pure and Appl. Maths. 54 (1992)

[19] Grabisch, M.: A New Algorithm for Identifying Fuzzy Measures and its Application to Pattern Recognition. Proc. IEEE, 145–150 (1995)

[20] Grabisch, G., Marichal, J.L., Mesiar, R., Pap, E.: Aggregation Functions. Cambridge University Press, Cambridge (2009)

[21] Grabisch, M., Nguyen, H.T., Walker, E.A.: Fundamentals of Uncertainity Calculi with Application to Fuzzy Inference. Kluwer Academic Publishers, Dordrecht (1995)
[22] Ishii, K., Sugeno, M.: A Model of Human Evaluation Process Using Fuzzy Measure. Int. J. Man-Machine Studies 22, 19–38 (1985)
[23] Hazelrigg, G.A.: The Implications of Arrow's Impossibility Theorem and Approaches to Optimal Engineering Design. J. of Mechanical Design 118, 161–164 (1996)
[24] Hazelrigg, G.A.: Approaches to Improve Engineering Design. The National Academic Press, Washington (2001)
[25] Hsu, Y.-L., Lin, Y.-F., Sun, T.-L.: Engineering Design Optimization as a Fuzzy Control Process. In: Proc. of the Fourth IEEE International Conference on Fuzzy Systems, vol. IV, 2001–2008 (1995)
[26] Keeny, R.L.: Multiplicative Utility Functions. Operation Research 1, 23–25 (1974)
[27] Klement, E.P., Mesiar, R., Pap, E.: On the Relationship of Associative Compensatory Operators to Triangular Norms and Conorms. Inter. J. Uncertainty, Fuzziness and Knowledge Based Systems 4, 129–144 (1996)
[28] Klement, E.P., Mesiar, R., Pap, E.: Triangular Norms. Kluwer Academic Publisher, Dordrecht (2000)
[29] Klir, G.J., Folger, T.A.: Fuzzy Sets, Uncertainty, and Information. Prentice-Hall International Inc., Englewood Cliffs (1992)
[30] Knosala, R., Pedrycz, W.: Evaluation of Design Alternatives in Mechanical Engineering. Fuzzy Sets and Systems 47, 269–280 (1992)
[31] Kuich, W.: Semirings, Automata, Languages. Springer, Berlin (1986)
[32] Luce, R.D., Raiffa, H.: Games and Decisions. John Wiley Sons Inc., New York (1957)
[33] Luhandjula, M.K.: Compensatory operators in fuzzy linear programming with multiple objectives. Fuzzy Sets and Systems 32, 45–79 (1982)
[34] Murofushi, T., Sugeno, M.: Fuzzy t–conorm integrals with respect to fuzzy measures: generalization of Sugeno integral and Choquet integral. Fuzzy Sets and Systems 42, 57–71 (1991)
[35] de Neufville, R.: Applied System Analysis. McGraw-Hill Inc., Singapore (1990)
[36] Osycska, A.: Multi-Criterion Optimization in Engineering with Fortran Examples. Halstad Press, New York (1984)
[37] Otto, K., Antonsson, E.: Trade-Off Strategies in Engineering Design. Research in Engineering Design 3(2), 87–104 (1991)
[38] Pap, E.: Null-Additive Set Functions. Kluwer Academic Publishers, Dordrecht (1995)
[39] Pap, E.: Pseudo-Analysis as a Mathematical Base for Soft Computing. Soft Computing 1, 61–68 (1997)
[40] Pap, E.: Pseudo-Additive Measures and Their Applications. In: Pap, E. (ed.) Handbook of Measure Theory, pp. 1403–1465. Elsevier, Amsterdam (2002)
[41] Pap, E.: A Generalization of the Utility Theory using a Hybrid Idempotent-Probabilistic Measure. In: Litvinov, G.L., Maslov, V.P. (eds.) Proceedings of the Conference on Idempotent Mathematics and Mathematical Physics, Contemporary Mathematics, AMS, Providence, Rhode Island, vol. 377, pp. 261–274 (2005)
[42] Pap, E., Devedžic, G.: Multicriteria-Multistages Linguistic Evaluation and Ranking of Machine Tools. Fuzzy Sets and Systems 102, 451–461 (1999)
[43] Pap, E., Jegdic, K.: Pseudo-Analysis and Its Applications in Railway Routing. Fuzzy Sets and Systems 116, 103–118 (2000)
[44] Pap, E., Roca, M.: An Axiomatization of the Hybrid Probabilistic-Possibilistic Utility Theory. In: Proceedings of the 4th Serbian-Hungarian Joint Symposium on Intelligent Systems, Subotica (SISY 2006), pp. 229–235 (2006)

[45] Pugh, S.: Total Design. Addison-Wesley, New York (1990)

[46] Pugh, S.: Creting Innovative Products Using Total Design: The Living Legacy of Stuart Pugh. In: Clausing, D., Mass, R.A. (eds.). Addison-Wesley, Mass. (1996)

[47] Quiggin, J.: Generalized Expected Utility Theory. Kluwer Academic, Dordrecht (1993)

[48] Saaty, T.L.: Exploring the Interfaces between Hierarchies, Multiple Objectives and Fuzzy Sets. Fuzzy Sets and Systems 1, 57–68 (1978)

[49] Sage, A.P.: Methodology for Large-Scale Systems. McGraw-Hill Inc., New York (1977)

[50] Scott, M.J., Antonsson, E.K.: Aggregation Functions for Engineering Design Trade-offs. In: 9th International ASME Conference on Design Theory and Methodology, Boston, MA, pp. 389–396 (1995)

[51] Scott, M.J., Antonsson, E.K.: Aggregation Functions for Engineering Design Trade-offs. Fuzzy Sets and Systems (1997)

[52] Sugeno, M.: Theory of Fuzzy Integrals and its Applications. Ph. D. Thesis, Tokyo Institute of Technology (1974)

[53] Sugeno, M., Kwon, S.-H.: A New Approach to Time Series Modeling With Fuzzy Measures and the Choquet Integral. Proc. IEEE (1995)

[54] Taguchi, G.: Introduction to Quality Engineering. In: Asian Productivity Organization. Unipub., White Plains (1986)

[55] Terano, T., Asai, K., Sugeno, M.: Fuzzy Systems Theory and Its Applications. Academic Press Inc., London (1992)

[56] Tribus, M.: Rational Descriptions, Decisions and Designs. Pergamon Press, Elmsford (1969)

[57] von Neumann, J., Morgenstern, O.: The Theory of Games and Economic Behavior, 3rd edn. Princeton University Press, Princeton (1953)

[58] Wang, Z., Klir, G.J.: Fuzzy Measure Theory. Plenum Press, New York (1992)

[59] Whithey, D.E.: Manufacturing by Design. Harvard Business Review 66(4), 83–91 (1988)

[60] Yager, R.R.: Connectives and Quantifiers in Fuzzy Sets. Fuzzy Sets and Systems 40, 39–75 (1991)

[61] Yager, R.R., Filev, D.P.: Essentials of Fuzzy Modelling and Control. John Willey, New York (1994)

[62] Zimmermann, H.-J., Sebastian, H.J.: Intelligent System Design Support by Fuzzy-Multi-Criteria Decision Making and/or Evolutionary Algorithms. In: Proc. of the Fourth IEEE International Conference on Fuzzy Systems, vol. I, pp. 367–374 (1995)

[63] Zimmermann, H.-J., Zysno, P.: Latent Connectives in Human Decision Making. Fuzzy Sets and Systems 4, 37–51 (1980)

[64] Zimmermann, H.-J., Zysno, P.: Decisions and Evolutions by Hierarchical Aggregation of Information. Fuzzy Sets and Systems 10, 243–260 (1983)

On Possibilistic Mean Value, Variance, Covariance and Correlation of Fuzzy Numbers

Christer Carlsson and Robert Fullér

IAMSR, Åbo Akademi University, Joukahaisenkatu 3-5, FIN-20520 Åbo, Finland
christer.carlsson@abo.fi, robert.fuller@abo.fi

Abstract. In 2001 we introduced the notions of possibilistic mean value and variance of fuzzy numbers. In this paper we give a short survey of these notations and show some examples of their application from the literature.

1 Introduction

In *probability theory* the expected value of functions of random variables plays a fundamental role in defining the basic characteristic measures of probability distributions. For example, the variance, covariance and correlation of random variables can be computed as the expected value of their appropriately chosen real-valued functions. In possibility theory we can use the principle of *expected value* of functions on fuzzy sets to define variance, covariance and correlation of possibility distributions. Marginal probability distributions are determined from the joint one by the principle of 'falling integrals' and marginal possibility distributions are determined from the joint possibility distribution by the principle of 'falling shadows'. Probability distributions can be interpreted as carriers of *incomplete information* [17], and possibility distributrions can be interpreted as carriers of *imprecise information*. In 1987 Dubois and Prade [12] defined an interval-valued expectation of fuzzy numbers, viewing them as consonant random sets. They also showed that this expectation remains additive in the sense of addition of fuzzy numbers. In 2005 Carlsson, Fuller and Majlender [9] introduced a measure of *possibilistic correlation* between fuzzy numbers A and B by their joint possibility distribution C as an average measure of their interaction (introduced earlier by Fuller and Majlender in [15]) compared to their respective marginal variances. In particular, they presented the concept of possibilistic correlation in a probabilistic setting and pointed out the fundamental difference between the standard probabilistic approach and their proposed possibilistic approach to computing and interpreting the correlation coefficient in these environments. They also showed that the *possibilistic covariance* between fuzzy numbers A and B is nothing else but the

I.J. Rudas et al. (Eds.): Towards Intelligent Engineering & Information Tech., SCI 243, pp. 17–36.
springerlink.com   © Springer-Verlag Berlin Heidelberg 2009

weighted average of the probabilistic covariances between random variables with uniform joint distribution on the level sets of their joint possibility distribution C. Furthermore, the *possibilistic variance* of a fuzzy number A computes the *weighted average of the probabilistic variances of uniformly distributed random variables* on its level sets. Finally, the *possibilistic mean value* of a fuzzy number A computes the *weighted average of the probabilistic mean values of uniformly distributed random variables* on its level sets.

In this Section we will recall the possibilistic mean value and variance of fuzzy numbers, which are consistent with the extension principle and with the well-known defintions of expectation and variance in probability theory. Furthermore, following Carlsson, Fuller and Majlender [9], we will introduce a measure of *possibilistic correlation* between fuzzy numbers by their joint possibility distribution as an average measure of their interaction compared to their respective marginal variances. In particular, we will present the concept of possibilistic mean value, variance, covariance and possibilistic correlation in a probabilistic setting and point out the fundamental difference between the standard probabilistic approach and the possibilistic one.

A *fuzzy number* A is a fuzzy set $\mathbb{R}$ with a normal, fuzzy convex and continuous membership function of bounded support. The family of fuzzy numbers is denoted by $\mathcal{F}$. Fuzzy numbers can be considered as possibility distributions [13, 18]. A fuzzy set C in $\mathbb{R}^2$ is said to be a joint possibility distribution of fuzzy numbers $A, B \in \mathcal{F}$, if it satisfies the relationships $\max\{x \mid C(x, y)\} = B(y)$ and $\max\{y \mid C(x, y)\} = A(x)$ for all $x, y \in \mathbb{R}$. Furthermore, A and B are called the marginal possibility distributions of C. Let $A \in \mathcal{F}$ be fuzzy number with a γ-level set denoted by $[A]^\gamma = [a_1(\gamma), a_2(\gamma)]$, $\gamma \in [0, 1]$. A function $f: [0, 1] \to \mathbb{R}$ is said to be a weighting function if f is non-negative, monoton increasing and satisfies the following normalization condition

$$\int_0^1 f(\gamma) d\gamma = 1.$$

Different weighting functions can give different (case-dependent) importances to γ-levels sets of fuzzy numbers. It is motivated in part by the desire to give less importance to the lower levels of fuzzy sets [16] (it is why f should be monotone increasing).

The possibilistic mean (or expected value), variance, covariance and correlation were originally defined from the measure of possibilistic interactivity (as shown in [9, 15]) but for simplicity, we will present the concept of possibilistic mean value, variance, covariance and possibilistic correlation in a probabilistic setting and point out the fundamental difference between the standard probabilistic approach and the possibilistic one. Let $A \in \mathcal{F}$ be fuzzy number with $[A]^\gamma = [a_1(\gamma), a_2(\gamma)]$ and let U_γ denote a uniform probability distribution on $[A]^\gamma$, $\gamma \in [0,1]$. Recall that the probabilistic mean value of U_γ is equal to

$$M(U_\gamma) = \frac{a_1(\gamma) + a_2(\gamma)}{2},$$

and its probabilistic variance is computed by

$$\mathrm{var}(U_\gamma) = \frac{(a_2(\gamma) - a_1(\gamma))^2}{12}.$$

The f-weighted *possibilistic mean value* of $A \in \mathcal{F}$, defined in [14], is defined as

$$E_f(A) = \int_0^1 E(U_\gamma) f(\gamma) d\gamma = \int_0^1 \frac{a_1(\gamma) + a_2(\gamma)}{2} f(\gamma) d\gamma,$$

where U_γ is a uniform probability distribution on $[A]^\gamma$ for all $\gamma \in [0,1]$. If $f(\gamma) = 1$ for all $\gamma \in [0,1]$ then we get

$$E(A) = \int_0^1 E(U_\gamma) f(\gamma) d\gamma = \int_0^1 \frac{a_1(\gamma) + a_2(\gamma)}{2} d\gamma,$$

which the possibilistic mean value of A originally introduced in 2001 by Carlsson and Fullér [3].

In 1986 Goetschel and Voxman introduced a method for ranking fuzzy numbers as [16]

$$A \le B \iff \int_0^1 \gamma(a_1(\gamma) + a_2(\gamma)) d\gamma \le \int_0^1 \gamma(b_1(\gamma) + b_2(\gamma)) d\gamma$$

As was pointed out by Goetschel and Voxman this definition of ordering was motivated in part by the desire to give less importance to the lower levels of fuzzy numbers. In the terminology introduced by Carlsson and Fullér, the ordering by Goetschel and Voxman can be written as

$$A \le B \iff E(A) \le E(B).$$

We note further that from the equality

$$E(A) = \int_0^1 \gamma(a_1(\gamma) + a_2(\gamma))d\gamma = \frac{\int_0^1 \gamma \cdot \frac{a_1(\gamma) + a_2(\gamma)}{2} d\gamma}{\int_0^1 \gamma \, d\gamma},$$

it follows that $E(A)$ is nothing else but the level-weighted average of the arithmetic means of all γ-level sets, that is, the weight of the arithmetic mean of $a_1(\gamma)$ and $a_2(\gamma)$ is just γ.

Example 1.1 If $A = (a, \alpha, \beta)$ is a triangular fuzzy number with center a, left-width $\alpha > 0$ and right-width $\beta > 0$ then a γ-level of A is computed by

$$[A]^\gamma = [a - (1 - \gamma)\alpha, a + (1 - \gamma)\beta], \ \forall \gamma \in [0,1],$$

Then,

$$E(A) = \int_0^1 \gamma[a - (1 - \gamma)\alpha + a + (1 - \gamma)\beta]d\gamma = a + \frac{\beta - \alpha}{6}.$$

When $A = (a, \alpha)$ is a symmetric triangular fuzzy number we get $E(A) = a$.

The f-weighted *possibilistic variance* of $A \in \mathcal{F}$, defined in [14], can be written as

$$Var_f(A) = \int_0^1 \sigma_{U_\gamma}^2 f(\gamma)d\gamma = \int_0^1 \frac{(a_2(\gamma) - a_1(\gamma))^2}{12} f(\gamma)d\gamma.$$

If $f(\gamma) = 1$ for all $\gamma \in [0,1]$ then we get

$$Var(A) = \int_0^1 \sigma_{U_\gamma}^2 d\gamma = \int_0^1 \frac{(a_2(\gamma) - a_1(\gamma))^2}{12} d\gamma.$$

which the possibilistic variance of A originally introduced in 2001 by Carlsson and Fullér [3].

Example 1.2 If $A = (a, \alpha, \beta)$ is a triangular fuzzy number then

$$Var(A) = \frac{1}{6} \int_0^1 \gamma(a + \beta(1 - \gamma) - (a - \alpha(1 - \gamma)))^2 d\gamma = \frac{(\alpha + \beta)^2}{72}.$$

In 2004 Fullér and Majlender [15] introduced a measure of possibilistic co-variance between marginal distributions of a joint possibility distribution C as the expected value of the interactivity relation between the γ-level sets of its marginal distributions. In 2005 Carlsson, Fullér and Majlender [9] showed that the possi-bilistic covariance between fuzzy numbers A and B can be written as the weighted average of the probabilistic covariances between random variables with uniform joint distribution on the level sets of their joint possibility distribution C.

The f-weighted *measure of possibilistic covariance* between $A, B \in \mathcal{F}$, (with respect to their joint distribution C), defined by [15], can be written as

$$Cov_f(A,B) = \int_0^1 Cov(X_\gamma, Y_\gamma) f(\gamma) d\gamma,$$

where X_γ and Y_γ are random variables whose joint distribution is uniform on $[C]^\gamma$ for all $\gamma \in [0,1]$.

Now we show how the possibilistic variance can be derived from possibilistic covariance. Let $A \in \mathcal{F}$ be fuzzy number with $[A]^\gamma = [a_1(\gamma), a_2(\gamma)]$ and let U_γ denote a uniform probability distribution on $[A]^\gamma$, $\gamma \in [0,1]$. First we compute the level-wise covariances by

$$\mathrm{cov}(U_\gamma, U_\gamma) = M(U_\gamma^2) - (M(U_\gamma))^2$$

$$= \frac{1}{a_2(\gamma) - a_1(\gamma)} \int_{a_1(\gamma)}^{a_2(\gamma)} x^2 dx - \left(\frac{1}{a_2(\gamma) - a_1(\gamma)} \int_{a_1(\gamma)}^{a_2(\gamma)} x dx \right)^2$$

$$= \frac{a_1^2(\gamma) + a_1(\gamma)a_2(\gamma) + a_2^2(\gamma)}{3} - \left(\frac{a_1(\gamma) + a_2(\gamma)}{2} \right)^2$$

$$= \frac{a_1^2(\gamma) - 2a_1(\gamma)a_2(\gamma) + a_2^2(\gamma)}{12}$$

$$= \frac{(a_2(\gamma) - a_1(\gamma))^2}{12},$$

and we get

$$Var_f(A) = Cov_f(A, A) = \int_0^1 \mathrm{cov}(U_\gamma, U_\gamma) f(\gamma) d\gamma$$

$$= \int_0^1 \frac{(a_2(\gamma) - a_1(\gamma))^2}{12} f(\gamma) d\gamma.$$

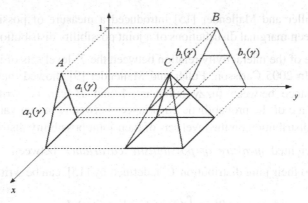

Fig. 1. The case of Cov $f(A,B)=0$

Let C be a joint possibility distribution in $\mathbb{R}^2$ with marginal possibility distributions $A, B \in \mathcal{F}$, and let $[A]^\gamma = [a_1(\gamma), a_2(\gamma)]$ and $[B]^\gamma = [b_1(\gamma), b_2(\gamma)]$, $\gamma \in [0,1]$. Let us assume that A and B are non-interactive, i.e. $C = A \times B$. This situation is depicted in Fig. 1. Then $[C]^\gamma = [A]^\gamma \times [B]^\gamma$ *(that is, $[C]^\gamma$ is rectangular subset of $\mathbb{R}^2$)* for any $\gamma \in [0,1]$, and from

$$\mathrm{cov}(X_\gamma, Y_\gamma) = 0$$

for all $\gamma \in [0,1]$ we have

$$Cov_f(A, B) = \int_0^1 \mathrm{cov}(X_\gamma, Y_\gamma) f(\gamma) d\gamma = 0$$

If A and B are non-interactive then $Cov_f(A, B) = 0$ for any weighting function f .

Example 1.3 *Now consider the case when*

$$A(x) = B(x) = (1-x) \cdot \chi_{[0,1]}(x)$$

for $x \in \mathbb{R}$, that is, $[A]^\gamma = [B]^\gamma = [0, 1-\gamma]$ for $\gamma \in [0,1]$. Suppose that their joint possibility distribution is given by

$$F(x, y) = (1 - x - y) \cdot \chi_T(x, y),$$

where

$$T = \{(x,y) \in \mathbb{R}^2 \mid x \geq 0, y \geq 0, x+y \leq 1\}.$$

This situation is depicted on Fig. 2, where we have shifted the fuzzy sets to get a better view of the situation.

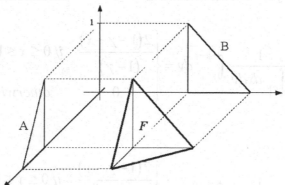

Fig. 2. Illustration of joint possibility distribution F

It is easy to check that A and B are really the marginal distributions of F. A γ-level set of F is computed by

$$[F]^\gamma = \{(x,y) \in \mathbb{R}^2 \mid x \geq 0, y \geq 0, x+y \leq 1-\gamma\}.$$

The density function of a uniform distribution on $[F]^\gamma$ can be written as

$$g(x,y) = \begin{cases} \dfrac{1}{\displaystyle\int_{[F]^\gamma} dxdy} & if\,(x,y) \in [F]^\gamma \\[2mm] 0 & otherwise \end{cases}$$

in details,

$$g(x,y) = \begin{cases} \dfrac{1}{\displaystyle\int_{0}^{1-\gamma}\int_{0}^{1-\gamma-x} dxdy} & if\,(x,y) \in [F]^\gamma \\[2mm] 0 & otherwise \end{cases}$$

that is,

$$g(x,y) = \begin{cases} \dfrac{2}{(1-\gamma)^2} & if\,(x,y) \in [F]^\gamma \\ 0 & otherwise \end{cases}$$

The marginal functions are obtained as

$$g_1(x) = \frac{1}{\int_{[F]^\gamma} dxdy} \int_0^{1-\gamma-x} dy = \begin{cases} \dfrac{2(1-\gamma-x)}{(1-\gamma)^2} & if\,0 \le x \le 1-\gamma \\ 0 & otherwise \end{cases}$$

and

$$g_2(y) = \frac{1}{\int_{[F]^\gamma} dxdy} \int_0^{1-\gamma-y} dx = \begin{cases} \dfrac{2(1-\gamma-y)}{(1-\gamma)^2} & if\,0 \le y \le 1-\gamma \\ 0 & otherwise \end{cases}$$

Furthermore, the probabilistic expected values of marginal distributions of X_γ and Y_γ are equal to $(1-\gamma)/3$ see (Fig. 3). Really,

$$M(X_\gamma) = \frac{2}{(1-\gamma)^2} \int_0^{1-\gamma} x(1-\gamma-x)dx = (1-\gamma)/3.$$

$$M(Y_\gamma) = \frac{2}{(1-\gamma)^2} \int_0^{1-\gamma} y(1-\gamma-y)dy = (1-\gamma)/3.$$

And the covariance between X_γ and Y_γ is positive on H_1 and H_4 and negative on H_2 and H_3. In this case we get (see Fig. 3 for a geometrical interpretation),

$$\mathrm{cov}(X_\gamma,Y_\gamma) = M\left(X_\gamma Y_\gamma\right) - M\left(X_\gamma\right)M\left(Y_\gamma\right) = \frac{1}{\int_{[F]^\gamma} dxdy} \int_{[F]^\gamma} xydxdy$$

$$-\frac{1}{\int_{[F]^\gamma} dxdy} \int_{[F]^\gamma} xdxdy \times \frac{1}{\int_{[F]^\gamma} dxdy} \int_{[F]^\gamma} ydxdy.$$

That is,

$$\mathrm{cov}(X_\gamma, Y_\gamma) = \frac{2}{(1-\gamma)^2} \int_0^{1-\gamma} \int_0^{1-\gamma-x} xydxdy - \frac{(1-\gamma)^2}{9}.$$

$$\mathrm{cov}(X_\gamma, Y_\gamma) = \frac{2}{(1-\gamma)^2} \int_0^{1-\gamma} x(1-\gamma-x)dx - \frac{(1-\gamma)^2}{9}.$$

$$\mathrm{cov}(X_\gamma, Y_\gamma) = \frac{(1-\gamma)^2}{12} - \frac{(1-\gamma)^2}{9} = -\frac{(1-\gamma)^2}{36}.$$

Therefore we get

$$Cov_f(A, B) = -\frac{1}{36} \int_0^1 (1-\gamma)^2 f(\gamma)d\gamma,$$

and

$$Var_f(A) = Var_f(B) = \frac{1}{12} \int_0^1 (1-\gamma)^2 f(\gamma)d\gamma.$$

for any weighting function f.

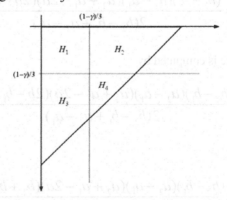

Fig. 3. Partition of $[F]^\gamma$

Example 1.4 *Let C be a joint possibility distribution (see Fig. 4) defined from* non-symmetrical marginal distributions of quasi-triangular form A *and* B *by*

$$C(x, y) = T_W(A(x), B(y)),$$

where T_W denotes the weak t-norm, that is,

$$T_W(x, y) = \begin{cases} \min\{x, y\} & \text{if } \max\{x, y\} = 1 \\ 0 & \text{otherwise} \end{cases}$$

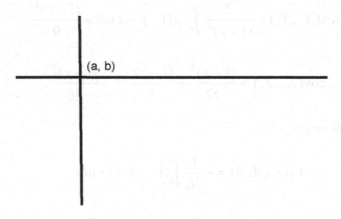

Fig. 4. A non-symmetrical γ-level set of C

Then with the notations $a_1 = a_1(\gamma), a_2 = a_2(\gamma), b_1 = b_1(\gamma), b_2 = b_2(\gamma)$ we get

$$\text{cov}(X_\gamma, Y_\gamma) = \frac{(b_2 - b_1)(a_2 - a_1)(a_2 + a_1 - 2a)(2b - b_1 - b_2)}{2(b_2 - b_1 + a_2 - a_1)}.$$

and their covariance is computed by

$$Cov(A, B) = \int_0^1 \frac{(b_2 - b_1)(a_2 - a_1)(a_2 + a_1 - 2a)(2b - b_1 - b_2)}{2(b_2 - b_1 + a_2 - a_1)} 2\gamma d\gamma$$

that is,

$$Cov(A, B) = -\int_0^1 \frac{(b_2 - b_1)(a_2 - a_1)(a_2 + a_1 - 2a)(b_2 + b_1 - 2b)}{(b_2 - b_1 + a_2 - a_1)} \gamma d\gamma$$

If the γ-level set of B is symmetrical, i.e.,

$$2b - b_1 - b_2 = 0,$$

or the γ-level set of A is symmetrical, i.e.,

$$a_2 + a_1 - 2a = 0$$

then

$$\mathrm{cov}(X_\gamma, Y_\gamma) = 0.$$

where a is the maximizing point of A and b is the maximizing point of B.

The f-weighted *possibilistic correlation* of $A, B \in \mathcal{F}$, (with respect to their joint distribution C), defined in [9], can be written as

$$\rho_f(A, B) = \frac{Cov_f(A, B)}{\sqrt{Var_f(A)}\sqrt{Var_f(B)}} = \frac{\int_0^1 \mathrm{cov}(X_\gamma Y_\gamma) f(\gamma) d\gamma}{\left(\int_0^1 \sigma_{U_\gamma}^2 f(\gamma) d\gamma\right)^{1/2} \left(\int_0^1 \sigma_{V_\gamma}^2 f(\gamma) d\gamma\right)^{1/2}}$$

where V_γ is a uniform probability distribution on $[B]^\gamma$. Thus, the possibilistic correlation represents an average degree to which X_γ and Y_γ are linearly associated as compared to the dispersions of U_γ and V_γ.

It is clear that the standard probabilistic calculation is not used here. Really, a standard probabilistic calculation might be the following

$$\frac{\int_0^1 \mathrm{cov}(X_\gamma, Y_\gamma) f(\gamma) d\gamma}{\left(\int_0^1 \sigma_{X_\gamma}^2 f(\gamma) d\gamma\right)^{1/2} \left(\int_0^1 \sigma_{Y_\gamma}^2 f(\gamma) d\gamma\right)^{1/2}}$$

That is, the standard probabilistic approach would use the marginal distributions, X_γ and Y_γ, of a uniformly distributed random variable on the level sets of $[C]^\gamma$.

Theorem 1.1 ([9]) *If $[C]^\gamma$ is convex for all $\gamma \in [0,1]$ then $-1 \le \rho_f(A, B) \le 1$ for any weighting function f.*

In the case, depicted in Fig. 5, the covariance of A and B with respect to their joint possibility distribution C is

$$Cov_f(A,B) = \frac{1}{12} \int_0^1 [a_2(\gamma) - a_1(\gamma)][b_2(\gamma) - b_1(\gamma)]f(\gamma)d\gamma,$$

and

$$\rho_f(A,B) = 1,$$

for any weighted function f.

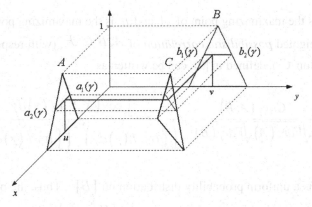

Fig. 5. The case of $\rho_f(A,B) = 1$,

In the case, depicted in Fig. 6, the covariance of A and B with respect to their joint possibility distribution D is

$$Cov_f(A,B) = -\frac{1}{12} \int_0^1 [a_2(\gamma) - a_1(\gamma)][b_2(\gamma) - b_1(\gamma)]f(\gamma)d\gamma,$$

and

$$\rho_f(A,B) = -1,$$

for any weighted function f.

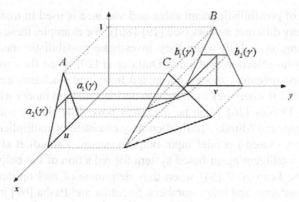

Fig. 6. The case of $\rho_f(A, B) = -1$,

Example 1.5 *Let us consider again the case when*

$$A(x) = B(x) = (1-x) \cdot \chi_{[0,1]}(x)$$

for $x \in \mathbb{R}$, that is, $[A]^\gamma = [B]^\gamma = [0, 1-\gamma]$ for $\gamma \in [0,1]$. Suppose that their joint possibility distribution is given by

$$F(x, y) = (1-x-y) \cdot \chi_T(x, y),$$

where

$$T = \{(x, y) \in \mathbb{R}^2 \mid x \ge 0, y \ge 0, x + y \le 1\}.$$

Then we get $\rho_f(A, B) = -1/3$ for any weighting function f.

2 Some Examples of Application

The possibilistic mean value, variance, covariance and correlation have been extensively used in our later works for real option valuation [4, 8], project selection [2, 5, 6, 10], capital budgeting [1] and optimal portfolio selection [7]. In [11] we developed a methodology for valuing options on R&D projects, when future cash flows are estimated by trapezoidal fuzzy numbers. In particular, we presented a fuzzy mixed integer programming model for the R&D optimal portfolio selection problem, and discussed how our methodology can be used to build decision support tools for optimal R&D project selection in a corporate environment.

The concept of possibilistic mean value and variance is used in many different areas and by many different authors (see [19]-[86]). For example, these notions are applied by Zhang, et al [19] when they investigate possibilistic mean-variance models in portfolio selection problems; Dutta et al [20] when they investigate a continuous review inventory model in a mixed fuzzy and stochastic environment; Thiagarajah et al [23] when they introduce an option valuation model with adaptive fuzzy numbers; Dubois [26] when he discusses possibility theory and statistical reasoning; Beynon and Munday [28] when they elucidate of multipliers and their moments in fuzzy closed Leontief input-output systems; Zarandi et al [44] when they present an intelligent agent-based system for reduction of the bullwhip effect in supply chains; Lazo et al [54] when they determine of real options value by Monte Carlo simulation and fuzzy numbers: Saeidifar and Pasha [67] introduced a defuzzification method to find the possibilistic moments and partial possibilistic moments of a fuzzy number.

The notion of *possibilistic correlation* has been used in [87]-[92]. Mizukoshi et al [91] discussed the fuzzy differential equations obtained from a deterministic differential equation introducing an uncertainty coefficient and fuzzy initial condition. The notions of f-weighted possibilistic mean value and variance are used in [93]-[103]. For example, Zhang and Li [99] use these notions to portfolio selections problems with quadratic utility function in a fuzzy environment; and Garcia [102] uses these notions to fuzzy real option valuation in a power station reengineering project.

References

[1] Carlsson, C., Fullér, R.: Capital Budgeting Problems with Fuzzy Cash Flows. Mathware and Soft Computing 6, 81–89 (1999)

[2] Carlsson, C., Fullér, R.: Real Option Evaluation in Fuzzy Environment. In: Proceedings of the International Symposium of Hungarian Researchers on Computational Intelligence, Budapest Polytechnic, pp. 69–77 (2000)

[3] Carlsson, C., Fullér, R.: On Possibilistic Mean Value and Variance of Fuzzy Numbers. Fuzzy Sets and Systems 122, 315–326 (2001)

[4] Carlsson, C., Fullér, R.: On Optimal Investment Timing with Fuzzy Real Options. In: Proceedings of the EUROFUSE 2001 Workshop on Preference Modelling and Applications, pp. 235–239 (2001)

[5] Carlsson, C., Fullér, R., Majlender, P.: Project Selection with Fuzzy Real Options. In: Proceedings of the Second International Symposium of Hungarian Researchers on Computational Intelligence, pp. 81–88 (2001)

[6] Carlsson, C., Fullér, R.: Project Scheduling with Fuzzy Real Options. In: Trappl, R. (ed.) Cybernetics and Systems '2002, Proceedings of the Sixteenth European Meeting on Cybernetics and Systems Research, Vienna, April 2-4, pp. 511–513. Austrian Society for Cybernetic Studies (2002)

[7] Carlsson, C., Fullér, R., Majlender, P.: A Possibilistic Approach to Selecting Portfolios with Highest Utility Score. Fuzzy Sets and Systems 131, 13–21 (2002)

[8] Carlsson, C., Fullér, R.: A Fuzzy Approach to Real Option Valuation. Fuzzy Sets and Systems 139, 297–312 (2003)

[9]   Carlsson, C., Fullér, R., Majlender, P.: On Possibilistic Correlation. Fuzzy Sets and Systems 155, 425–445 (2005)

[10] Carlsson, C., Fullér, R., Majlender, P.: A Fuzzy Real Options Model for R&D Project Evaluation. In: Liu, Y., Chen, G., Ying, M. (eds.) Proceedings of the Eleventh IFSA World Congress, Beijing, China, July 28-31, pp. 1650–1654. Tsinghua University Press and Springer, Beijing (2005)

[11] Carlsson, C., Fullér, R., Heikkil, M., Majlender, P.: A Fuzzy Approach to R&D Project Portfolio Selection. International Journal of Approximate Reasoning 44, 93–105 (2007)

[12] Dubois, D., Prade, H.: The Mean Value of a Fuzzy Number. Fuzzy Sets and Systems 24, 279–300 (1987)

[13] Dubois, D., Prade, H.: Possibility Theory: An Approach to Computerized Processing of Uncertainty. Plenum Press, New York (1988)

[14] Fullér, R., Majlender, P.: On Weighted Possibilistic Mean and Variance of Fuzzy Numbers. Fuzzy Sets and Systems 136, 363–374 (2003)

[15] Fullér, R., Majlender, P.: On Interactive Fuzzy Numbers. Fuzzy Sets and Systems 143, 355–369 (2004)

[16] Goetschel, R., Voxman, W.: Elementary Fuzzy Calculus. Fuzzy Sets and Systems 18, 31–43 (1986)

[17] Jaynes, E.T.: Probability Theory: The Logic of Science. Cambridge University Press, Cambridge (2003)

[18] Zadeh, L.A.: L. A. Zadeh, Concept of a Linguistic Variable and its Application to Approximate Reasoning. I, II, III, Information Sciences 8, 199–249, 301-357 (1975); 9, 43–80 (1975)

[19] Zhang, W.-G., Wang, Y.-L., Chen, Z.-P., Nie, Z.-K.: Possibilistic Mean-Variance Models and Efficient Frontiers for Portfolio Selection Problem. Information Sciences 177(13), 2787–2801 (2007)

[20] Dutta, P., Chakraborty, D., Roy, A.R.: Continuous Review Inventory Model in Mixed Fuzzy and Stochastic Environment. Applied Mathematics and Computation 188(1), 970–980 (2007)

[21] Thavaneswaran, A., Thiagarajah, K.: Appadoo SS Fuzzy Coefficient Volatility (FCV) Models with Applications. Mathematical and Computer Modelling 45(7-8), 777–786 (2007)

[22] Vercher, E., Bermudez, J.D., Segura, J.V.: Fuzzy Portfolio Optimization under Downside Risk Measures. Fuzzy Sets and Systems 158(7), 769–782 (2007)

[23] Thiagarajah, K., Appadoo, S.S., Thavaneswaran, A.: Option Valuation Model with Adaptive Fuzzy Numbers. Computers and Mathematics with Applications 53, 831–841 (2007)

[24] Fang, Y., Lai, K.K., Wang, S.Y.: Portfolio Rebalancing Model with Transaction Costs Based on Fuzzy Decision Theory. European Journal of Operational Research 175(2), 879–893 (2006)

[25] Yoshida, Y., Yasuda, M., Nakagami, J.I., et al.: A New Evaluation of Mean Value for Fuzzy Numbers and its Application to American Put Option under Uncertainty. Fuzzy Sets and Systems 157(19), 2614–2626 (2006)

[26] Dubois, D.: Possibility Theory and Statistical Reasoning. Computational Statistics & Data Analysis 51(1), 47–69 (2006)

[27] Stefanini, L., Sorini, L., Guerra, M.L.: Parametric Representation of Fuzzy Numbers and Application to Fuzzy Calculus. Fuzzy Sets and Systems 157(18), 2423–2455 (2006)

[28] Beynon, M.J., Munday, M.: The Elucidation of Multipliers and their Moments in Fuzzy Closed Leontief Input-Output Systems. Fuzzy Sets and Systems 157(18), 2482–2494 (2006)

[29] Hashemi, S.M., Modarres, M., Nasrabadi, E., et al.: Fully Fuzzified Linear Programming, Solution and Duality. Journal of Intelligent & Fuzzy Systems 17(3), 253–261 (2006)

[30] Facchinetti, G., Pacchiarotti, N.: Evaluations of Fuzzy Quantities. Fuzzy Sets and Systems 157(7), 892–903 (2006)

[31] Bodjanova, S.: Median Alpha-Levels of a Fuzzy Number. Fuzzy Sets and Systems 157(7), 879–891 (2006)

[32] Sheen, J.N.: Generalized Fuzzy Numbers Comparison by Geometric Moments. WSEAS Transactions on Systems 5(6), 1237–1242 (2006)

[33] Liu, X.: On the Maximum Entropy Parameterized Interval Approximation of Fuzzy Numbers. Fuzzy Sets and Systems 157, 869–878 (2006)

[34] Cheng, C.B.: Fuzzy Process Control: Construction of Control Charts with Fuzzy Numbers. Fuzzy Sets and Systems 154(2), 287–303 (2005)

[35] Ayala, G., Leon, T., Zapater, V.: Different Averages of a Fuzzy Set with an Application to Vessel Segmentation. IEEE Transactions on Fuzzy Systems 13(3), 384–393 (2005)

[36] Hong, D.H., Kim, K.T.: A Note on Weighted Possibilistic Mean. Fuzzy Sets and Systems 148(2), 333–335 (2004)

[37] Jahanshahloo, G.R., Soleimani-damaneh, M., Nasrabadi, E.: Measure of Efficiency in DEA with Fuzzy Input-Output Levels: a Methodology for Assessing. Ranking and Imposing of Weights Restrictions, Applied Mathematics and Computation 156(1), 175–187 (2004)

[38] Roventa, E., Spircu, T.: Averaging Procedures in Defuzzification Processes. Fuzzy Sets and Systems 136, 375–385 (2003)

[39] Facchinetti, G.: Ranking Functions Induced by Weighted Average of Fuzzy Numbers. Fuzzy Optimization and Decision Making 1, 313–327 (2002)

[40] Zhang, W.-G., Chen, Q., Lan, H.-L.: A Portfolio Selection Method Based on Possibility Theory. In: Cheng, S.-W., Poon, C.K. (eds.) AAIM 2006. LNCS, vol. 4041, pp. 367–374. Springer, Heidelberg (2006)

[41] Yoshida, Y.: A Defuzzification Method of Fuzzy Numbers Induced from Weighted Aggregation Operations. In: Torra, V., Narukawa, Y., Valls, A., Domingo-Ferrer, J. (eds.) MDAI 2006. LNCS (LNAI), vol. 3885, pp. 161–171. Springer, Heidelberg (2006)

[42] Yoshida, Y.: Mean Values of Fuzzy Numbers with Evaluation Measures and the Measurement of Fuzziness. In: Proceedings of the 9th Joint Conference on Information Sciences, JCIS 2006 (2006) (art. no. 298), doi:10.2991/jcis.2006.298

[43] Yoshida, Y.: Mean Values of Fuzzy Numbers and the Measurement of Fuzziness by Evaluation Measures. In: 2006 IEEE Conference on Cybernetics and Intelligent Systems, pp. 1–6 (2006) (art. no. 4017804)

[44] Zarandi, M.H.F., Pourakbar, M., Turksen, I.B.: An Intelligent Agent-based System for Reduction of Bullwhip Effect in Supply Chains. In: IEEE International Conference on Fuzzy Systems, pp. 663–670 (2006) (art. no. 1681782)

[45] Liu, H., Brown, D.J.: An Extension to Fuzzy Qualitative Trigonometry and its Application to Robot Kinematics. In: IEEE International Conference on Fuzzy Systems, pp. 1111–1118 (2006) (art. no. 1681849)

[46] Wang, X., Xu, W., Zhang, W.-G., Hu, M.: Weighted Possibilistic Variance of Fuzzy Number and its Application in Portfolio Theory. In: Wang, L., Jin, Y. (eds.) FSKD 2005. LNCS, vol. 3613, pp. 148–155. Springer, Heidelberg (2005)

[47] Dubois, D., Fargier, H., Fortin, J.: The Empirical Variance of a Set of Fuzzy Intervals. In: IEEE International Conference on Fuzzy Systems, pp. 885–890 (2005)

[48] Zhang, W.-G., Wang, Y.: Using fuzzy possibilistic mean and variance in portfolio selection model. In: Hao, Y., Liu, J., Wang, Y.-P., Cheung, Y.-m., Yin, H., Jiao, L., Ma, J., Jiao, Y.-C. (eds.) CIS 2005. LNCS (LNAI), vol. 3801, pp. 291–296. Springer, Heidelberg (2005)

[49] Zhang, W.-G., Liu, W.-A., Wang, Y.-L.: A Class of Possibilistic Portfolio Selection Models and Algorithms. In: Deng, X., Ye, Y. (eds.) WINE 2005. LNCS, vol. 3828, pp. 464–472. Springer, Heidelberg (2005)

[50] Sato, T., Takahashi, S., Huang, C., Inoue, H.: Option Pricing with Fuzzy Barrier Conditions. In: Liu, Y., Chen, G., Ying, M. (eds.) Proceedings of the Eleventh International Fuzzy systems Association World Congress, July 28-31, pp. 380–384. Tsinghua University Press and Springer, Beijing (2005)

[51] Zhang, J.-P., Li, S.-M.: Portfolio Selection with Quadratic Utility Function under Fuzzy Envirornment. In: 2005 International Conference on Machine Learning and Cybernetics, ICMLC 2005, pp. 2529–2533 (2005)

[52] Yoshida, Y.: Mean Values of Fuzzy Numbers by Evaluation Measures and its Measurement of Fuzziness. In: Proceedings - International Conference on Computational Intelligence for Modelling, Control and Automation, CIMCA 2005 and International Conference on Intelligent Agents, Web Technologies and Internet, vol. 2, pp. 163–169 (2005) (art. no. 1631462)

[53] Blankenburg, B., Klusch, M.: BSCA-F: Efficient Fuzzy Valued Stable Coalition Forming among Agents. In: Proceedings - 2005 IEEE/WIC/ACM International Conference on Intelligent Agent Technology, IAT 2005, pp. 732–738 (2005) (art. no. 1565632)

[54] Lazo, J.G.L., Vellasco, M.M.B.R., Pacheco, M.A.C.: Determination of Real Options Value by Monte Carlo Simulation and Fuzzy Numbers. In: Proceedings - HIS 2005: Fifth International Conference on Hybrid Intelligent Systems, pp. 488–493 (2005) (art. no. 1587794)

[55] Yoshida, Y.: A mean estimation of fuzzy numbers by evaluation measures. In: Negoita, M.G., Howlett, R.J., Jain, L.C. (eds.) KES 2004. LNCS, vol. 3214, pp. 1222–1229. Springer, Heidelberg (2004)

[56] Garcia, F.A.A.: Fuzzy Real Option Valuation in a Power Station Reengineering Project Soft Computing with Industrial Applications. In: Proceedings of the Sixth Biannual World Automation Congress, pp. 281–287 (2004)

[57] Collan, M.: Fuzzy Real Investment Valuation Model for Very Large Industrial Real Investments. In: Soft Computing with Industrial Applications - Proceedings of the Sixth Biannual World Automation Congress, pp. 379–384 (2004)

[58] Spircu, L., Spircuin, T.: Fuzzy Treatment of American Call Options. In: Proceedings of the Tenth International Conference on Information Processing and Management of Uncertainty in Knowledge-based Systems IPMU 2004, Perugia, Italy, July 4-9, pp. 1841–1846 (2004)

[59] Wang, X., Xu, W.-J., Zhang, W.-G.: A Class of Weighted Possibilistic Mean-Variance Portfolio Selection Problems. In: Proceedings of 2004 International Conference on Machine Learning and Cybernetics, vol. 4, pp. 2036–2040 (2004)

[60] Wang, G.-X., Zhao, C.-H.: Characterization of Discrete Fuzzy Numbers and Application in Adaptive Filter Algorithm. In: Proceedings of 2004 International Conference on Machine Learning and Cybernetics, vol. 3, pp. 1850–1854 (2004)

[61] Fang, Y., Lai, K.K., Wang, S.-Y.: A Fuzzy Approach to Portfolio Rebalancing with Transaction Costs. In: Sloot, P.M.A., Abramson, D., Bogdanov, A.V., Gorbachev, Y.E., Dongarra, J., Zomaya, A.Y. (eds.) ICCS 2003. LNCS, vol. 2658, pp. 10–19. Springer, Heidelberg (2003)

[62] Zhang, W.-G., Nie, Z.-K.: On Possibilistic Variance of Fuzzy Numbers. In: Wang, G., Liu, Q., Yao, Y., Skowron, A. (eds.) RSFDGrC 2003. LNCS, vol. 2639, pp. 398–402. Springer, Heidelberg (2003)

[63] Dai, Z., Scott, M.J., Mourelatos, Z.P.: Incorporating Epistemic Uncertainty in Robust Design. In: Proceedings of the 2003 ASME Design Engineering Technical Conferences (DETC 2003), Chicago, Illinois, USA, September 2-6 (2003) (DAC-48713.pdf)

[64] Peschland, G.M., Schweiger, H.F.: Reliability Analysis in Geotechnics with Finite Elements - Comparison of Probabilistic, Stochastic and Fuzzy Set Methods. In: Proceedings of the 3rd International Symposium on Imprecise Probabilities and Their Applications (ISIPTA 2003), Carleton Scientific Proceedings in Informatics, July 14–17, 2003, vol. 18, University of Lugano, Lugano (2003)

[65] Zhang, W.-G., Zhang, Q.-M., Nie, Z.-K.: A Class of Fuzzy Portfolio Selection Problems. In: International Conference on Machine Learning and Cybernetics, vol. 5, pp. 2654–2658 (2003)

[66] Daz-Hermida, F., Carinena, P., Bugarn, A., Barro, S.: Modelling of Task-oriented Vocabularies: An Example in Fuzzy Temporal Reasoning. In: IEEE International Conference on Fuzzy Systems, vol. 1, pp. 43–46 (2001)

[67] Saeidifar, A., Pasha, E.: The Possibilistic Moments of Fuzzy Numbers and their Applications. Journal of Computational and Applied Mathematics 223, 1028–1042 (2009)

[68] Chrysas, K.A., Papadopoulos, B.K.: On Theoretical Pricing of Options with Fuzzy Estimators. Journal of Computational and Applied Mathematics 223, 552–566 (2009)

[69] Thavaneswaran, A., Appadoo, S.S., Paseka, A.: Weighted Possibilistic Moments of Fuzzy Numbers with Applications to GARCH Modeling and Option Pricing. Mathematical and Computer Modelling 9, 352–368 (2009)

[70] Gong, Y.-B.: Method for Fuzzy Multi-Attribute Decision Making with Preference on Alternatives and Partial Weights Information. Kongzhi yu Juece/Control and Decision 23, 507–510 (2008)

[71] Beynon, M.J., Munday, M.: Considering the effects of imprecision and uncertainty in ecological footprint estimation: An approach in a fuzzy environment. Ecological Economics 67, 373–383 (2008)

[72] Liu, H., Brown, D.J., Coghill, G.M.: Fuzzy Qualitative Robot Kinematics. IEEE Transactions on Fuzzy Systems 16, 808–822 (2008)

[73] Huang, X.: Mean-Semivariance Models for Fuzzy Portfolio Selection. Journal of Computational and Applied Mathematics 217, 1–8 (2008)

[74] Appadoo, S.S., Bhatt, S.K., Bector, C.R.: Application of Possibility Theory to Investment Decisions. Fuzzy Optimization and Decision Making 7, 35–57 (2008)

[75] Lalla, M., Facchinetti, G., Mastroleo, G.: Vagueness Evaluation of the Crisp Output in a Fuzzy Inference System. Fuzzy Sets and Systems 159, 3297–3312 (2008)

[76] Cong, G.D., Zhang, J.L., Chen, T., Lai, K.K.: A Variable Precision Fuzzy Rough Group Decision-Making Model for IT Offshore Outsourcing Risk Evaluation. Journal of Global Information Management 16(2), 18–34 (2008)

[77] Allahviranloo, T., Hosseinzadeh Lotfi, F., Kiasary, M.K., Kiani, N.A., Alizadeh, L.: Solving Fully Fuzzy Linear Programming Problem by the Ranking Function. Applied Mathematical Sciences 2(1), 19–32 (2008)

[78] Huang, X.: Risk Curve and Fuzzy Portfolio Selection. Computers and Mathematics With Applications 55, 1102–1112 (2008)

[79] Soleimani-damaneh, M.: Fuzzy Upper Bounds and their Applications. Chaos, Solitons And Fractals 36(2), 217–225 (2008)

[80] Zarandi, M.H.F., Pourakbar, M., Turksen, I.B.: A Fuzzy Agent-based Model for Reduction of Bullwhip Effect in Supply Chain Systems. Expert Systems With Applications 34(3), 1680–1691 (2008)

[81] Yoshida, Y.: A Risk-Sensitive Portfolio with Mean and Variance of Fuzzy Random Variables. In: Huang, D.-S., Wunsch II, D.C., Levine, D.S., Jo, K.-H. (eds.) ICIC 2008. LNCS, vol. 5227, pp. 358–366. Springer, Heidelberg (2008)

[82] Yanbing, G., Jiguo, Z.: A Method for Fuzzy Multi-Attribute Decision Making with Preference Information in the Form of Fuzzy Complementary Judgment Matrix. In: Proceedings of 5th International Conference on Fuzzy Systems and Knowledge Discovery, FSKD 2008, vol. 3, pp. 336–340 (2008) (art. no. 4666265)

[83] Gong, Y., Zhang, J.: A new Method for Assessing the Weights of Fuzzy Opinions in Group Decision Environment. In: Proceedings of International Conference on Intelligent Computation Technology and Automation, ICICTA 2008, vol. 1, pp. 836–839 (2008) (art. no. 4659604)

[84] Hsu, H.-W., Wang, H.-F.: Closed-Loop Green Supply Chain Logistics Model with Uncertain Reverse Parameters. In: 2008 International Conference on Wireless Communications, Networking and Mobile Computing, WiCOM 2008 (2008) (art. no. 4679358)

[85] Yoshida, Y.: A Risk-Minimizing Portfolio Model with Fuzziness. In: IEEE International Conference on Fuzzy Systems (FUZZ-IEEE 2008), June 1-6, 2008, pp. 909–914 (2008)

[86] Tupac, Y.J., Lazo, J.G., Faletti, L., Pacheco, M.A., Vellasco, M.M.B.R.: Decision Support System for Economic Analysis of E&P Projects under Uncertainties. In: Society of Petroleum Engineers - Intelligent Energy Conference and Exhibition: Intelligent Energy 2008, vol. 2, pp. 1115–1124 (2008)

[87] Saeidifar, A., Pasha, E.: The Possibilistic Moments of Fuzzy Numbers and their Applications. Journal of Computational and Applied Mathematics 223, 1028–1042 (2009)

[88] Chen, C.-C., Tang, H.-C.: Degenerate Correlation and Information Energy of Interval-valued Fuzzy Numbers. International Journal of Information and Management Sciences 19, 119–130 (2008)

[89] Bede, B., Bhaskar, T.G., Lakshmikantham, V.: Perspectives of Fuzzy Initial Value Problems. Communications in Applied Analysis 11(3-4), 339–358 (2007)

[90] Hong, D.H., Kim, K.T.: A Maximal Variance Problem. Applied Mathematics Letters 20(10), 1088–1093 (2007)

[91] Mizukoshi, M.T., Barros, L.C., Chalco-Cano, Y., Roman-Flores, H., Bassanezi, R.C.: Fuzzy Differential Equations and the Extension Principle. Information Sciences 177(17), 3627–3635 (2007)

[92] Matia, F., Jimenez, A., Al-Hadithi, B.M., et al.: The Fuzzy Kalman Filter: State Estimation Using Possibilistic Techniques. Fuzzy Sets and Systems 157(16), 2145–2170 (2006)

[93] Liu, X.: On the Maximum Entropy Parameterized Interval Approximation of Fuzzy Numbers. Fuzzy Sets and Systems 157, 869–878 (2006)

[94] Ayala, G., Leon, T., Zapater, V.: Different Averages of a Fuzzy Set with an Application to Vessel Segmentation. IEEE Transactions on Fuzzy Systems 13(3), 384–393 (2005)

[95] Cheng, C.B.: Fuzzy Process Control: Construction of Control Charts with Fuzzy Numbers. Fuzzy Sets and Systems 154(2), 287–303 (2005)

[96] Bodjanova, S.: Median Value and Median Interval of a Fuzzy Number. Information Sciences 172(1-2), 73–89 (2005)

[97] Hong, D.H., Kim, K.T.: A Note on Weighted Possibilistic Mean. Fuzzy Sets and Systems 148(2), 333–335 (2004)

[98] Wang, X., Xu, W., Zhang, W.-G., Hu, M.: Weighted Possibilistic Variance of Fuzzy Number and its Application in Portfolio Theory. In: Wang, L., Jin, Y. (eds.) FSKD 2005. LNCS (LNAI), vol. 3613, pp. 148–155. Springer, Heidelberg (2005)

[99] Zhang, W.-G., Wang, Y.-L.: Portfolio Selection: Possibilistic Mean-Variance Model and Possibilistic Efficient Frontier. In: Megiddo, N., Xu, Y., Zhu, B. (eds.) AAIM 2005. LNCS, vol. 3521, pp. 203–213. Springer, Heidelberg (2005)

[100] Zhang, J.-P., Li, S.-M.: Portfolio Selection with Quadratic Utility Function under Fuzzy Enviornment. In: 2005 International Conference on Machine Learning and Cybernetics, ICMLC 2005, pp. 2529–2533 (2005)

[101] Blankenburg, B., Klusch, M.: BSCA-F: Efficient Fuzzy Valued Stable Coalition Forming Among Agents. In: Proceedings - 2005 IEEE/WIC/ACM International Conference on Intelligent Agent Technology, IAT 2005, pp. 732–738 (2005) (art. no. 1565632)

[102] Garcia, F.A.A.: Fuzzy Real Option Valuation in a Power Station Reengineering Project Soft Computing with Industrial Applications. In: Proceedings of the Sixth Biannual World Automation Congress, pp. 281–287 (2004)

[103] Wang, X., Xu, W.-J., Zhang, W.-G.: A Class of Weighted Possibilistic Mean-Variance Portfolio Selection Problems. In: Proceedings of 2004 International Conference on Machine Learning and Cybernetics, vol. 4, pp. 2036–2040 (2004)

Problems and Results in Matrix Perturbation Theory

Aurél Galántai

Budapest Tech
Bécsi út 96/B, H-1034 Budapest, Hungary
galantai.aurel@nik.bmf.hu

Abstract. The perturbation theory is important in applications and theoretical investigations as well. Here we investigate three groups of perturbation problems which are related to computational methods of importance. The first section is related to the solution of linear systems of equations and a posteriori error estimates of the computed solution. The second section gives optimal bounds for the perturbations of LU factorizations. The final section gives a sharp upper bound for the eigenvalue perturbation of general matrices, which is better than the classical result of Ostrowski. We also show two applications of this result. The first application gives a sharp perturbation bound for the zeros of polynomials. The second application is related to a result of Edelman and Murakami on the backward stability of companion matrix type polynomial solvers.

1   Introduction

Numerical computations can formally be written in the form

$$compute \ \xi \equiv f(\theta),$$

where f defines the problem, θ represents the particular data and ξ is the solution of the problem. A problem is called *ill-conditioned* if small perturbations $\delta\theta$ may cause huge changes $\delta\xi \equiv f(\theta + \delta\theta) - f(\theta)$. The solution of ill-conditioned problems requires special care and techniques if one wishes to get reliable numerical solutions on a digital computer with floating point arithmetic, which is otherwise another source of related numerical stability problems.

The nature of ill-conditioning is perhaps best understood for the solution of nonsingular linear systems of the form

$$Ax = b \ \left(A \in \mathbb{R}^{n \times n}, \ x, b \in \mathbb{R}^n \right). \tag{1}$$

I.J. Rudas et al. (Eds.): Towards Intelligent Engineering & Information Tech., SCI 243, pp. 37–53.
springerlink.com   © Springer-Verlag Berlin Heidelberg 2009

Here the exact solution is $x = A^{-1}b$. Assuming a perturbation of the linear system in the form

$$(A + \Delta A)(x + \Delta x) = b + \Delta b \tag{2}$$

we have the following well-known bound for the relative error of the computed solution

$$\frac{\|\Delta x\|}{\|x\|} \leq \frac{\kappa(A)\left(\frac{\|\Delta A\|}{\|A\|} + \frac{\|\Delta b\|}{\|b\|}\right)}{1 - \kappa(A)\frac{\|\Delta A\|}{\|A\|}} \tag{3}$$

provided that $\kappa(A)\frac{\|\Delta A\|}{\|A\|} < 1$, $b \neq 0$. Quantity $\kappa(A) = \|A\|\|A^{-1}\|$ is the condition number of matrix A and plays a key role in the numerical stability. This relative error bound implies the following "*rule of thumb*" (see, e.g. [44], [23]):

"*If the entries of A and b are accurate about s decimal places and $\kappa(A) \approx 10^t$, where $t < s$, then the entries of the computed solution are accurate to about $s - t$ decimal places.*"

Consider the following "simple" example of Nievergelt [31]:

$$888445x_1 + 887112x_2 = 1,$$
$$887112x_1 + 885781x_2 = 0.$$

The exact solution of the system is $x_1 = 885781$ and $x_2 = -887112$. The coefficients (and the solution) can be exactly represented in the Matlab system (version 7.5), which uses the double precision IEEE 754-1985 floating point standard with machine epsilon $2.2204e - 016$. The computed solution is $\hat{x}_1 = \mathbf{8.856440}223037928e + 005$ and

$\hat{x}_2 = -\mathbf{8.869748}164771678e + 005$ with the relative error

$$\frac{\|x - \hat{x}\|_2}{\|x\|_2} \approx 1.5464e - 004.$$

The result is accurate to, say, three decimal digits. Since $s \approx 16$ and $\kappa(A) \approx 3.15e + 012$, the result essentially corresponds to the rule of thumb.

Unfortunately this is not the only example. Rump [35] constructed a class of arbitrarily ill-conditioned real matrices that are exactly representable in a floating-point system. For such matrices one clearly has a problem with the precision or reliability of the approximate solution in floating-point arithmetic.

However ill-conditioning is an intrinsic problem of the matrix space that might be further worsened by the floating-point representation. Eckart, Young and Gastinel showed (see, e.g. Kahan [27], Demmel [12]) that

$$\frac{1}{\kappa(A)} = \min_{A+\Delta A \text{ is singular}} \frac{\|\Delta A\|}{\|A\|}. \tag{4}$$

Hence a matrix A is ill-conditioned if A is close to a nonsingular matrix, which is entirely a self-property of the matrix.

In practice, instead of estimating the distance of a matrix A from the space of singular matrices, a posteriori estimates are used to evaluate the reliability of the approximate solution. One frequently used technique is the estimation of the condition number $\kappa(A)$ plus the rule of thumb (see, e.g. [23], [24]).

A possible alternative way a posteriori estimation is the use of the residual error $r(\hat{x}) = A\hat{x} - b$. Auchmuty [1] proved that if $\hat{x}$ is an approximate solution of $Ax = b$, then

$$\|x - \hat{x}\|_2 = \frac{c\|r(\hat{x})\|_2^2}{\|A^T r(\hat{x})\|_2}, \tag{5}$$

where $c \geq 1$.

It was shown in Galántai [18] that the error constant c depends on A and the direction of error vector $\hat{x} - x$. Furthermore

$$1 \leq c \leq C_2(A) = \frac{1}{2}\left(\kappa(A) + \frac{1}{\kappa(A)}\right), \tag{6}$$

where the upper bound is approximately half of $\kappa(A)$, which is a clear gain over the estimate (3). The error constant c takes the upper value $C_2(A)$ only in exceptional cases (the Nievergelt example is such a case). The computational experiments on a wide class of test matrices indicated that the average value of c grows slowly with the order of A and it depends more strongly on n than the condition number of A. The following experimental estimate

$$\|x - \hat{x}\|_2 \lessgtr 0.5 \dim(A)\|r(\hat{x})\|_2^2/\|A^T r(\hat{x})\|_2 \tag{7}$$

seems to hold with a high degree of probability.

The above a posteriori error estimates are satisfactory only on the average. Demmel, Diament and Malajovich [13] showed that for such estimators there are always cases when the estimate is unreliable (the error of the estimate exceeds a given order).

2   Perturbations of Multiplicative Matrix Factorizations

Matrix $A \in \mathbb{R}^{n \times n}$ has an LU decomposition, if A can be written in the form $A = LU$, where L is a lower triangular matrix and U is an upper triangular matrix. For the unicity of the LU factorization we assume that L is unit lower triangular, i.e., its main diagonal entries are 1's. The unique LDU decomposition of A is defined by $A = LDU$, where L is unit lower triangular, D is diagonal and U is unit upper triangular. These triangular decompositions play major roles in many numerical algorithms for solving linear systems of equations or the eigenvalue problem. It is remarkable that von Neumann, Goldstine [30] and Turing [43] discovered first that the Gaussian elimination method implicitly computes an LU decomposition of the coefficient matrix.

The perturbations of triangular matrix factorizations, which satisfy certain nonlinear equations, were studied by several authors [2], [5], [8], [9], [10], [14], [16], [17], [19], [20], [36], [37], [39], [40], [41].

Let $\delta_A \in \mathbb{R}^{n \times n}$ be a perturbation such that $A + \delta_A$ also has the LU factorization. Then

$$A + \delta_A = \left(L + \delta_L \right)\left(U + \delta_U \right) \tag{8}$$

is the corresponding unique LU factorization. Since $L + \delta_L$ is unit lower triangular, the perturbation matrix δ_L is strict lower triangular. The other perturbation matrix δ_U is upper triangular.

We use the following notations. For $A = \left[a_{ij} \right]_{i,j=1}^{n}$, let $|A| = \left[|a_{ij}| \right]_{i,j=1}^{n}$,

$$diag\left(A \right) = diag\left(a_{11}, a_{22}, \ldots, a_{nn} \right),$$

$$tril\left(A, l \right) = \left[\alpha_{ij} \right]_{i,j=1}^{n}, \quad triu\left(A, l \right) = \left[\beta_{ij} \right]_{i,j=1}^{n} \quad \left(0 \le |l| < n \right)$$

with

$$\alpha_{ij} = \begin{cases} a_{ij}, & i \ge j - l \\ 0, & i < j - l \end{cases}, \quad \beta_{ij} = \begin{cases} a_{ij}, & i \le j - l \\ 0, & i > j - l \end{cases}.$$

Related special notations are

$$tril(A) = tril(A,0), \quad tril^*(A) = tril(A,-1), \quad triu(A) = triu(A,0).$$

$|A|$ is sometimes called the matricial norm. The spectral radius of A will be denoted by $\rho(A)$. For two matrices $A, B \in \mathbb{R}^{m \times n}$, the relation $A \leq B$ holds if and only if $a_{ij} \leq b_{ij}$ for all $i = 1, \ldots, m$, $j = 1, \ldots, n$.

The following results hold.

Theorem 1. *(Galántai [19]). Assume that A and $A + \delta_A$ are nonsingular and have LU factorizations $A = LU$ and $A + \delta_A = (L + \delta_L)(U + \delta_U)$, respectively. Let $B = L^{-1}\delta_A U^{-1}$. The exact perturbation terms δ_L and δ_U are then given by $\delta_L = Ltril^*(F)$ and $\delta_U = triu(G)U$, where F and G are the unique solutions of the equations $F = B - Btriu(F)$ and $G = B - tril^*(G)B$, respectively.*

Theorem 2. *(Galántai [19]). Assume that A and $A + \delta_A$ are nonsingular and have LU factorizations $A = LU$ and $A + \delta_A = (L + \delta_L)(U + \delta_U)$, respectively. Let $B = L^{-1}\delta_A U^{-1}$. Then*

$$\left|\delta_L\right| \leq |L| tril^*\left(|F|\right), \quad \left|\delta_U\right| \leq |L| triu\left(|G|\right)|U|, \tag{9}$$

where F and G are given by equations $F = B - Btriu(F)$ and $G = B - tril^*(G)B$, respectively. If $\rho(|B|) < 1$, then

$$\left|\delta_{L_1}\right| \leq |L_1| tril^*\left(F^{b,1}\right), \quad \left|\delta_U\right| \leq triu\left(G^{b,1}\right)|U|, \tag{10}$$

where $F^{b,1}$ and $G^{b,1}$ are the unique solutions of the equations

$$F = |B| + |B| triu(F), \quad G = |B| + tril^*(G)|B| \tag{11}$$

respectively.

The matrix equations $F = B - Btriu(F)$, $G = B - tril^*(G)B$, $F = |B| + |B| triu(F)$ and $G = |B| + tril^*(G)|B|$ can be solved exactly (see [19]). The upper bounds (10), which are improvements of those of Sun [40], can be computed by fixed point iterations.

Theorem 3. *(Galántai [19]). Consider equation*

$$W = C + Btriu\ (W, l),\tag{12}$$

where

$B, C, W \in \mathbb{R}^{n \times n}$ *and* $l \geq 0$. *If* $\rho(|B|) < 1$, *then for every* $W_0 \in \mathbb{R}^{n \times n}$ *the*

sequence $W_{k+1} = C + Btriu(W_k, l)$ *converges to* W *and*

$$|W_k - W| \leq \left(I - |B|\right)^{-1} |B|^k |W_1 - W_0| \quad (k = 1, 2, \ldots).\tag{13}$$

Furthermore, if $B \geq 0$ *and* $C \geq 0$, *then for* $X_0 = 0$ *and* $Y_0 = \left(I - B\right)^{-1} C$,

the iterates $X_{k+1} = C + Btriu(X_k, l)$ *and* $Y_{k+1} = C + Btriu(Y_k, l)$ *($k \geq 0$)*

are monotone and satisfy

$$0 \leq X_k \leq X_{k+1} \leq W \leq Y_{k+1} \leq Y_k \leq \left(I - B\right)^{-1} C.\tag{14}$$

If $B \geq 0$ *and* $C \leq 0$, *then for* $X_0 = \left(I - B\right)^{-1} C$ *and* $Y_0 = 0$, *the iterates*

$\{X_k\}$ *and* $\{Y_k\}$ *satisfy*

$$\left(I - B\right)^{-1} C \leq X_k \leq X_{k+1} \leq W \leq Y_{k+1} \leq Y_k \leq 0 \quad (k \geq 0).\tag{15}$$

Theorem 4. *(Galántai [19]). Consider equation* $W = C + tril(W, -l)B$, *where*

$B, C, W \in \mathbb{R}^{n \times n}$ *and* $l \geq 0$. *If* $\rho(|B|) < 1$, *then for every* $W_0 \in \mathbb{R}^{n \times n}$ *the se-*

quence $W_{k+1} = C + tril(W_k, -l)B$ *converges to* W *and*

$$|W_k - W| \leq |W_1 - W_0| |B|^k \left(I - |B|\right)^{-1} \quad (k = 1, 2, \ldots).\tag{16}$$

Furthermore, if $B \geq 0$ *and* $C \geq 0$, *then for* $X_0 = 0$ *and* $Y_0 = C\left(I - B\right)^{-1}$,

the iterates $X_{k+1} = C + tril(X_k, -l)B$ *and* $Y_{k+1} = C + tril(Y_k, -l)B$

($k \geq 0$) are monotone and satisfy

$$0 \leq X_k \leq X_{k+1} \leq W \leq Y_{k+1} \leq Y_k \leq C\left(I - B\right)^{-1}.\tag{17}$$

If $B \geq 0$ and $C \leq 0$, then for $X_0 = C(I - B)^{-1}$ and $Y_0 = 0$, the iterates $\{X_k\}$ *and* $\{Y_k\}$ *satisfy*

$$C(I - B)^{-1} \leq X_k \leq X_{k+1} \leq W \leq Y_{k+1} \leq Y_k \leq 0 \quad (k \geq 0). \tag{18}$$

In the case of $F^{b,1}$, $G^{b,1}$ and $\tilde{F}^{b,1}$ we always have monotone convergence in the partial ordering "$\leq$" for the given initial matrices.

Consider the following example for the *LU* perturbation bound computed in Matlab. Let

$$A = \begin{bmatrix} 2 & 1 & 1 & 2 \\ 1 & 2 & 1 & 1.5 \\ 1 & 1 & 2 & 1.5 \\ 1 & 1 & 1 & 2.5 \end{bmatrix}$$

and

$$\delta_A = \begin{bmatrix} 0.0107 & -0.3193 & -0.0176 & -0.0738 \\ -0.2032 & 0.3638 & -0.0637 & -0.0823 \\ -0.0002 & 0.0420 & -0.0630 & 0.0236 \\ -0.0403 & 0.0148 & -0.0351 & -0.0301 \end{bmatrix}.$$

In this case $\rho\left(\left|L^{-1}\delta_A U^{-1}\right|\right) = 0.5001$, $\|\delta_A\|_F = 0.55$,

$$\delta_L = \begin{bmatrix} 0 & 0 & 0 & 0 \\ -0.1037 & 0 & 0 & 0 \\ -0.0028 & 0.0027 & 0 & 0 \\ -0.0227 & -0.0039 & -0.003 & 0 \end{bmatrix} \quad \left(\|\delta_L\|_F = 0.1063\right),$$

and

$$\delta_U = \begin{bmatrix} 0.0107 & -0.3193 & -0.0176 & -0.0738 \\ 0 & 0.5940 & 0.0470 & 0.1544 \\ 0 & 0 & -0.0686 & 0.0126 \\ 0 & 0 & 0 & -0.0014 \end{bmatrix} \quad \left(\|\delta_U\|_F = 0.7011\right).$$

The error estimates of Sun [40] (case $i = 0$ of the iteration) are given by

$$\delta_L^{Sun} = \begin{bmatrix} 0 & 0 & 0 & 0 \\ 0.1958 & 0 & 0 & 0 \\ 0.1014 & 0.0195 & 0 & 0 \\ 0.0798 & 0.0193 & 0.0027 & 0 \end{bmatrix} \quad \left(\left\| \delta_L^{Sun} \right\|_F = 0.2361 \right),$$

and

$$\delta_U^{Sun} = \begin{bmatrix} 0.1007 & 0.6379 & 0.4064 & 0.3451 \\ 0 & 1.1909 & 0.7164 & 0.4955 \\ 0 & 0 & 0.0763 & 0.0509 \\ 0 & 0 & 0 & 0.0016 \end{bmatrix} \quad \left(\left\| \delta_U^{Sun} \right\|_F = 1.6990 \right).$$

The error estimates given by the first iterate are

$$\delta_{L_1}^{(1)} = \begin{bmatrix} 0 & 0 & 0 & 0 \\ 0.1095 & 0 & 0 & 0 \\ 0.0701 & 0.0182 & 0 & 0 \\ 0.0490 & 0.0190 & 0.0027 & 0 \end{bmatrix} \quad \left(\left\| \delta_{L_1}^{(1)} \right\|_F = 0.1414 \right),$$

$$\delta_U^{(1)} = \begin{bmatrix} 0.0107 & 0.3301 & 0.1989 & 0.1427 \\ 0 & 0.6913 & 0.4070 & 0.2812 \\ 0 & 0 & 0.0725 & 0.0482 \\ 0 & 0 & 0 & 0.0016 \end{bmatrix} \quad \left(\left\| \delta_U^{(1)} \right\|_F = 0.9482 \right).$$

The error estimate (10) (case $i \to \infty$) yields the bounds

$$\delta_L^{best} = \begin{bmatrix} 0 & 0 & 0 & 0 \\ 0.1048 & 0 & 0 & 0 \\ 0.0671 & 0.0181 & 0 & 0 \\ 0.0469 & 0.0189 & 0.0027 & 0 \end{bmatrix} \quad \left(\left\| \delta_L^{best} \right\|_F = 0.1356 \right),$$

$$\delta_U^{best} = \begin{bmatrix} 0.0107 & 0.3301 & 0.1989 & 0.1427 \\ 0 & 0.6617 & 0.3894 & 0.2692 \\ 0 & 0 & 0.0721 & 0.0481 \\ 0 & 0 & 0 & 0.0016 \end{bmatrix} \quad \left\| \delta_U^{best} \right\|_F = 0.9157.$$

For the relative errors we have

$$\left\|\delta_L^{Sun}\right\|_F / \left\|\delta_L\right\|_F = 2.2211 > \left\|\delta_L^{best}\right\|_F / \left\|\delta_L\right\|_F = 1.2753$$

and

$$\left\|\delta_U^{Sun}\right\|_F / \left\|\delta_U\right\|_F = 2.4232 > \left\|\delta_U^{best}\right\|_F / \left\|\delta_U\right\|_F = 1.3059.$$

Hence estimate (10) is indeed an improvement over Sun's. One can also observe that the first iterate gives estimates almost as good as the best estimates.

For the perturbations of other types of triangular (LDU, $Cholesky$) and full rank factorizations we refer to [17], [19], [20].

3 Eigenvalue Perturbation Bounds and Applications

The perturbation theory of matrix eigenvalues is perhaps the most complicated and diverse matter of linear algebra (see, e.g. [3], [6], [29], [38]). The eigenvalues are continuous functions of the matrix entries. Hence the eigenvalue perturbations are also continuous functions of the entries of the perturbation matrix. For practical purposes we seek for quantitative estimates of the eigenvalue perturbations. The first of such estimates is due to Ostrowski [33], who obtained it from his famous polynomial perturbation result [32].

Let $A, B \in \mathbb{C}^{n \times n}$. Denote the spectrum of A and B by $\sigma(A) = \{\lambda_1, \dots, \lambda_n\}$ and $\sigma(B) = \{\mu_1, \dots, \mu_n\}$, repectively. Let S_n be the set of all permutations of $\{1, 2, \dots, n\}$. The eigenvalue variation of A and B is then defined by

$$v(A, B) = \min_{\pi \in S_n} \left\{ \max_i \left| \mu_{\pi(i)} - \lambda_i \right| \right\}. \tag{19}$$

Ostrowski's result and its later improvement by Bhatia, Elsner and Krause [7] say that

$$v(A, A + \delta_A) \le 4 \times 2^{-1/n} \left(\|A\| + \|A + \delta_A\| \right)^{1-1/n} \|\delta_A\|^{1/n}. \tag{20}$$

This essentially means that the perturbation size of the eigenvalues is $O\left(\sqrt[n]{\varepsilon}\right)$ for perturbations of the size $\|\delta_A\| = O(\varepsilon)$. If, for example, one takes the machine epsilon $\varepsilon_{machine} = 2.2204e - 016$ and a matrix of modest size, say $n = 50$, then $\sqrt[50]{\varepsilon_{machine}} \approx 0.48$.

For matrices similar to a diagonal form, the perturbation error is $O(\varepsilon)$ by the Bauer-Fike theorem (see, e.g. [23]). For general matrices, the Ostrowski-Bhatia-Elsner-Krause perturbation bound is widely believed to be sharp. However this is not so. The following theorem is an improvement on Chu [11].

Theorem 5. *(Galántai, Hegedűs [21]). Assume that*

$$A = X\Omega X^{-1}, \quad \Omega = diag(U_1,...,U_k), \tag{21}$$

where $U_i = \Lambda_i + N_i \in \mathbb{C}^{n_i \times n_i}$ is upper triangular, Λ_i is diagonal and N_i is strict upper triangular for $i = 1,...,k$. For any $\mu \in \sigma(A+\delta_A)$ there exists $\lambda_j \in \sigma(A)$ such that

$$\left|\mu - \lambda_j\right| \le \left\|N_j\right\|_2 \max\left\{n_j\theta,(n_j\theta)^{1/n_j}\right\} \tag{22}$$

holds with $\theta = \left\|X^{-1}\delta_A X\right\|_2 / \left\|N_j\right\|_2$.

Observe that $\sigma(\Lambda_i)$'s are not necessarily disjoint. The result is global in the sense that δ_A is not restricted. The following results are readily obtained.

Let $J_k(\lambda) \in \mathbb{C}^{k \times k}$ be an upper Jordan block. For any $A \in \mathbb{C}^{n \times n}$, there exists a nonsingular matrix X such that

$$X^{-1}AX = diag\left(J_{n_1}(\lambda_1), J_{n_2}(\lambda_2),...,J_{n_k}(\lambda_k)\right) \tag{23}$$

and $\sum_{j=1}^{k} n_j = n$. The eigenvalues λ_i, $i = 1,...,k$ are not necessarily distinct.

Corollary 1. *(Chu [11]). If $A \in \mathbb{C}^{n \times n}$ has the Jordan canonical form (23), then for any $\mu \in \sigma(A+\delta_A)$ there exists $\lambda_j \in \sigma(A)$ such that*

$$\left|\mu - \lambda_j\right| \le \max\left\{n_j\theta,(n_j\theta)^{1/n_j}\right\} \tag{24}$$

holds with $\theta = \left\|X^{-1}EX\right\|_2$.

Corollary 2. *(Galántai, Hegedűs [21]). If $A \in \mathbb{C}^{n \times n}$ has the Jordan canonical form (23), then*

$$v\left(A, A + \delta_A\right) \le \left(2n - 1\right) \max\left\{m\theta, \left(m\theta\right)^{1/m}\right\}, \tag{25}$$

where $m = \max_i n_i$ and $\theta = \left\|X^{-1}\delta_A X\right\|_2$.

For $k \ge 2$ different eigenvalues, this estimate is asymptotically better than those of the Ostrowski-Elsner type. For the example $\varepsilon \approx 2.2204e - 016$, $n = 50$ and double eigenvalues, one has $\sqrt[2]{\varepsilon_{machine}} \approx 1.49e - 008$, which is better than 0.48.

The upper bound of Theorem 5 is generally sharp. This is easily verified by taking $\delta_A = X$ diag $\left(E_1, \ldots, E_k\right) X^{-1} = O(\varepsilon)$ with $E_i = \varepsilon e_{n_i} e_1^T$ for $i = 1, \ldots, k$ (Wilkinson's example).

Finally we show two applications of Theorem 5.

Assume that

$$p(z) = z^n + a_1 z^{n-1} + \cdots + a_{n-1}z + a_n \tag{26}$$

has the distinct zeros $z_1, \ldots, z_k$ with multiplicity $n_1, \ldots, n_k$. Then the Jordan form of its companion matrix

$$C = C(p) = \begin{bmatrix} -a_1 & -a_2 & & & -a_n \\ 1 & 0 & & & 0 \\ & 1 & \ddots & & \\ & & \ddots & \ddots & \\ 0 & & & 1 & 0 \end{bmatrix} \tag{27}$$

is given by

$$C = \Pi V J V^{-1} \Pi^T, \tag{28}$$

where $\Pi = \left[e_n, e_{n-1}, \ldots, e_2, e_1\right]$, $J = $ diag $\left(J_{n_1}(z_1), \ldots, J_{n_k}(z_k)\right)$ and
$$V = \left[V_1, \ldots, V_k\right] \quad (V_i \in \mathbb{C}^{n \times n_i})$$
with entries

$$(V_i)_{pq} = \begin{cases} 0, & if \ p < q \\ \\ \binom{p-1}{q-1} z_i^{p-q}, & if \ p \geq q \end{cases} \qquad (29)$$

The matrix V is called the confluent Vandermonde matrix.

Theorem 6. *(Galántai, Hegedűs [21]). Assume that* $p(z) = z^n + a_1 z^{n-1} + \cdots + a_{n-1} z + a_n$ *has the distinct roots* $z_1, \ldots, z_k$ *with multiplicity* $n_1, \ldots, n_k$.

Let $\tilde{p}(z) = z^n + \tilde{a}_1 z^{n-1} + \cdots + \tilde{a}_{n-1} z + \tilde{a}_n$ *be a perturbation of* p *with* $\tilde{a}_i = a_i + \varepsilon_i$, $|\varepsilon_i| \leq \varepsilon$, $i = 1, \ldots, n$. *For any root* $\tilde{z}_i$ *of* $\tilde{p}(z)$, *there exists a root* z_j *of* $p(z)$ *such that*

$$\left| \tilde{z}_i - z_j \right| \leq \max \left\{ n_j \theta, (n_j \theta)^{1/n_j} \right\} \qquad (30)$$

with $\theta = \left\| V^{-1} \Delta V \right\|_2$ *and* $\Delta = -e_n w^T$ ($w^T = [\varepsilon_n, \varepsilon_{n-1}, \ldots, \varepsilon_1]$). *There also exists a permutation* $\pi \in S_n$ *such that for* $i = 1, \ldots, n$,

$$\left| \tilde{z}_{\pi(i)} - z_i \right| \leq (2n-1) \max \left\{ m\theta, (m\theta)^{1/m} \right\}, \qquad (31)$$

where $m = \max_i n_i$.

Since $\|\Delta\|_2 = \|w^T\|_2 \leq \varepsilon \sqrt{n}$ and $\theta = \|V^{-1} \Delta V\|_2 \leq \varepsilon \sqrt{n} \kappa_2(V)$, we obtained an $O(\varepsilon^{1/m})$ perturbation bound for the zeros. Hence it is better than those classical results of Ostrowski [32] and Bhatia, Elsner, Krause [7]. A similar but local result was obtained by Beauzamy [4].

Today it is a standard and widely used technique to compute the polynomial roots from the eigenvalues of the companion matrix (see, e.g. Matlab). Edelman and Murakami [15] investigated the backward stability of this algorithm. They assumed that the companion matrix C of $p(z)$ is perturbed by a dense matrix E with small entries. The result of Edelman and Murakami as interpreted by Toh and Trefethen [42] is the following.

If C is a companion matrix, then any infinitesimal dense matrix perturbation $C + \delta C$ is similar to a perturbed matrix $C + \delta C'$ of companion matrix form, with $\|\delta C'\| \le \gamma \|\delta C\|$ for some γ depending on C but not on δC.

Theorems 5 and 6 imply a formally weaker result.

Theorem 7. *(Galántai, Hegedűs [21]). If C is a companion matrix and m is the maximum size of its Jordan blocks, then for any sufficiently small perturbation E, $C + E$ is similar to a perturbed matrix $C + E'$ of companion matrix form, with $\|E'\| \le \gamma \|E\|^{1/m}$ for some γ depending on C but not on E.*

Using the proof of the latter claim we can also prove the formally stronger result as well.

Theorem 8. *If C is a companion matrix, then for any sufficiently small perturbation E, $C + E$ is similar to a perturbed matrix $C + E'$ of companion matrix form, with $\|E'\| \le \gamma \|E\|$ for some γ depending on C but not on E.*

Proof. C is an unreduced upper Hessenberg matrix. Thus for small perturbations $C + E$ is similar to an unreduced upper Hessenberg matrix (see Theorem 12 of [21]) whose characteristic polynomial is identical with the minimal polynomial. A matrix is similar to the companion matrix of its characteristic polynomial if and only if the minimal and characteristic polynomials are identical (see, e.g. Horn, Johnson). Hence $C + E$ is similar to its companion matrix, which can be written in the unique form $C + E'$ where $E' = e_1 w^T$ and vector w^T contains the perturbations of the polynomial's coefficients. Let $p(z) = \det(C - zI)$ and $\tilde{p}(z) = \det(C + E - zI)$. We use the identity

$$\det(A - \lambda I) = (-1)^n \lambda^n + \sum_{k=1}^{n} (-1)^{n-k} S_k(A) \lambda^{n-k},$$

where $S_k(A)$ is the sum of all the $k \times k$ principal minors of A. We can write

$$\det(C - zI) = (-1)^n z^n + \sum_{k=1}^{n} (-1)^{n-k} S_k(C) z^{n-k} \quad \text{and}$$

$$\det(C + E - zI) = (-1)^n z^n + \sum_{k=1}^{n} (-1)^{n-k} S_k(C + E) z^{n-k}. \text{ Hence}$$

$$w^T = \left[(-1)^{n-1} \left(S_1(C + E) - S_1(C) \right), \ldots, \left(S_n(C + E) - S_n(C) \right) \right].$$

Thus we need to estimate $\left| S_k(C + E) - S_k(C) \right|$. Given $\ell \ge 2$, $\varepsilon > 0$ and complex numbers $b_1, \ldots, b_\ell$ and $a_1, \ldots, a_\ell$, then $|b_i - a_i| \le \varepsilon$ ($i = 1, \ldots, \ell$) implies that

$$\left| \prod_{i=1}^{l} b_i - \prod_{i=1}^{l} a_i \right| \le \sum_{i=0}^{l-1} \sigma_i \varepsilon^{l-i}, \tag{32}$$

where σ_i is the ith elementary symmetric function of $|a_1|, \ldots, |a_\ell|$ $(\sigma_0 := 1)$.
For $A, D \in \mathbb{C}^{\ell \times \ell}$, let $\max_{ij} |a_{ij}| = a$, and $\max_{ij} |d_{ij}| = d$. Since

$$\det(A+D) - \det(A) = \sum_{\pi \in S_l} (-1)^{\text{sgn}(\pi)} \left[\prod_{i=1}^{l} \left(a_{i\pi(i)} + d_{i\pi(i)} \right) - \prod_{i=1}^{l} a_{i\pi(i)} \right],$$

we can apply inequality (32). The ith elementary symmetric function of the
elements $|a_{1\pi(1)}|, \ldots, |a_{\ell,\pi(\ell)}|$ is bounded by $\binom{l}{i} a^l$ and

$$\left| \prod_{i=1}^{l} \left(a_{i\pi(i)} + d_{i\pi(i)} \right) - \prod_{i=1}^{l} a_{i\pi(i)} \right| \le \sum_{i=0}^{l-1} \sigma_i d^{l-i} \le \sum_{i=0}^{l-1} \binom{l}{i} a^l d^{l-i} = \alpha(a,l,d) d.$$

Hence $|\det(A+D) - \det(A)| \le \ell! \alpha(a,\ell,d) d = O(d)$, where $O(d)$
depends only on A and D. If $\max_{ij} |c_{ij}| = c$, and $\max_{ij} |e_{ij}| = \varepsilon$, then

$$|S_k(C+E) - S_k(C)| \le k! \alpha(c,k,\varepsilon) \varepsilon \le \gamma \varepsilon \quad (k=1,\ldots,n)$$

holds with a suitable constant $\gamma > 0$. Thus we proved the theorem.

This backward stability result is important from a theoretical point of view. But it
has no direct practical consequences. Consider the polynomial $p(z) = (z-1)^3$
and the related companion matrix

$$C = \begin{bmatrix} 3 & -3 & 1 \\ 1 & 0 & 0 \\ 0 & 1 & 0 \end{bmatrix}.$$

Although this matrix is exactly represented in the floating point arithmetic, the
eigensolver of the Matlab system (version 7.5) gives the approximate eigenvalues
(zeros)

$\lambda_1 = 1.000004888842545e+000 + 8.467811199808864e-006i$

$\lambda_2 = 1.000004888842545e+000 - 8.467811199808864e-006i$

$\lambda_3 = 9.999902223149081e-001$

with an absolute error of the size $9.7778e-006$. Since
$\sqrt[3]{\varepsilon_{machine}} \approx 6.0555e-006$, the obtained error corresponds to a perturbation of

the matrix (or polynomial) of the size $\varepsilon_{macheps}$ (see Theorems 5, 6). Consider now the following perturbed C:

$$C + E = \begin{bmatrix} 3 & -3 & 1+\varepsilon \\ 1 & 0 & 0 \\ 0 & 1 & 0 \end{bmatrix}$$

already in companion matrix form. The following figure shows the maximum error of Matlab's eigensolver for $\varepsilon = 10^{-i}$ ($i = 1,\ldots,16$).

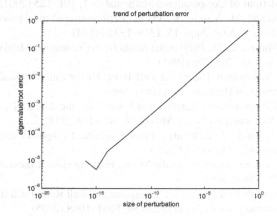

We can see that the error trend corresponds to $\varepsilon^{1/3}$ and not ε. Hence for multiple eigenvalues/roots, Theorems 5 and 6 are the ones which describe the asymptotic behavior of the perturbation effects.

For other perturbation bounds and comparisons we refer to Galántai, Hegedűs [21]. For an analysis of the Matlab eigensolver, see also Galántai, Hegedűs [22].

References

[1] Auchmuty, G.: A Posteriori Error Estimates for Linear Equations. Numerische Mathematik 61, 1–6 (1992)
[2] Barrlund, A.: Perturbation Bounds for the LDL^H and LU Decompositions. BIT 31, 358–363 (1991)
[3] Baumgartel, H.: Analytic Perturbation Theory for Matrices and Operators. Birkhauser Verlag, Basel (1985)
[4] Beauzamy, B.: How the Roots of a Polynomial Vary with its Coefficients: a Local Quantitative Result. Canadian Mathematical Bulletin 42(1), 3–12 (1999)
[5] Bhatia, R.: Matrix Factorizations and their Perturbations. Linear Algebra and its Applications 197,198, 245–276 (1994)
[6] Bhatia, R.: Perturbation Bounds for Matrix Eigenvalues. SIAM, Philadelphia (2007)
[7] Bhatia, R., Elsner, L., Krause, G.: Bounds for the Variation of the Roots of a Polynomial and the Eigenvalues of a Matrix. Linear Algebra and its Applications 142, 195–209 (1990)

[8] Chang, X.-W., Paige, C.: On the Sensitivity of the *LU* Factorization. BIT 38, 486–501 (1998)

[9] Chang, X.-W., Paige, C.: Sensitivity Analyses for Factorizations of Sparse or Structured Matrices. Linear Algebra and its Applications 284, 53–71 (1998)

[10] Chang, X.-W., Paige, C., Stewart, G.W.: New Perturbation Analyses for the Cholesky Factorization. IMA J. Numer. Anal. 16, 457–484 (1996)

[11] Chu, E.K.: Generalization of the Bauer-Fike Theorem. Numerische Mathematik 49, 685–691 (1986)

[12] Demmel, J.W.: On Condition Numbers and the Distance to the Nearest Ill-posed Problem. Numerische Mathematik 51, 251–289 (1987)

[13] Demmel, J., Diament, B., Malajovich, G.: On the Complexity of Computing Error Bounds. Foundations of Computational Mathematics 1, 101–125 (2001)

[14] Drmač, Z., Omladič, M., Veselič, K.: On the Perturbation of the Cholesky Factorization. SIAM J. Matrix. Anal. Appl. 15, 1319–1332 (1994)

[15] Edelman, A., Murakami, H.: Polynomial Roots from Companion Matrix Eigenvalues. Math. Comp. 64(210), 763–776 (1995)

[16] Galántai, A.: Perturbation Theory for Full Rank Factorizations, Quaderni DMSIA, 1999/40. University of Bergamo, Bergamo (1999)

[17] Galántai, A.: Componentwise perturbation bounds for the LU, LDU, and LDL^T decompositions. Mathematical Notes, Miskolc 1, 109–118 (2000)

[18] Galántai, A.: A Study of Auchmuty's Error Estimate. Computers and Mathematics with Applications 42, 1093–1102 (2001)

[19] Galántai, A.: Perturbations of Triangular Matrix Factorizations. Linear and Multilinear Algebra 51, 175–198 (2003)

[20] Galántai, A.: Perturbation Bounds for Triangular and Full Rank Factorizations. Computers and Mathematics with Applications 50, 1061–1068 (2005)

[21] Galántai, A., Hegedűs, C.J.: Perturbation Bounds for Polynomials. Numerische Mathematik 109, 77–100 (2008)

[22] Galántai, A., Hegedűs, C.: Hyman's Method Revisited. Journal of Computational Mathematics and Applied Mathematics 226, 246–258 (2009)

[23] Golub, G.H., Van Loan, C.F.: Matrix Computations, 2nd edn. The Johns Hopkins University Press, Baltimore (1993)

[24] Higham, N.: Accuracy and Stability of Numerical Algorithms. SIAM, Philadelphia (1996)

[25] Hogben, L.: Handbook of Linear Algebra. Chapman & Hall/CRC (2007)

[26] Horn, R., Johnson, C.: Matrix Analysis. Cambridge University Press, Cambridge (1985)

[27] Kahan, W.M.: Numerical Linear Algebra. Canadian Mathematical Bulletin 9, 757–801 (1966)

[28] Kahan, W.: Conserving Confluence Curbs Ill-Condition, technical report, AD-766 916, Computer Science, University of California, Berkeley (1972)

[29] Kato, T.: Perturbation Theory for Linear Operators. Springer, Heidelberg (1966)

[30] von Neumann, J., Goldstine, H.: Numerical Inverting of Matrices of High Order. Bull. Amer. Math. Soc. 53, 1021–1099 (1947)

[31] Nievergelt, Y.: Numerical Linear Algebra on the HP-28 or How to Lie with Super-calculators. American Mathematical Monthly, 539–544 (1991)

[32] Ostrowski, A.: Recherches sur la méthode de Gräffe et les zeros des polynômes et des series de Laurent. Acta Math. 72, 99–257 (1940)

[33] Ostrowski, A.: Über die Stetigkeit von charakteristischen Wurzeln in Abhängigkeit von den Matrizenelementen. Jahresber. deut. Mat.-Ver. 60, 40–42 (1957)

[34] Prasolov, V.V.: Problems and Theorems in Linear Algebra. American Mathematical Society, Providence (1994)

[35] Rump, S.M.: A Class of Arbitrary ill Conditioned Floating-Point Matrices. SIAM J. Matrix Anal. Appl. 12, 645–653 (1991)

[36] Stewart, G.W.: On the perturbation of LU, Cholesky and QR factorizations. SIAM J. Matrix. Anal. Appl. 14, 1141–1145 (1993)

[37] Stewart, G.W.: On the perturbation of LU and Cholesky factors. IMA J. Numer. Anal. 17, 1–6 (1997)

[38] Stewart, G., Sun, J.: Matrix Perturbation Theory. Academic Press, London (1990)

[39] Sun, J.-G.: Perturbation Bounds for the Cholesky and QR Factorizations. BIT 31, 341–352 (1991)

[40] Sun, J.-G.: Componentwise Perturbation Bounds for some Matrix Decompositions. BIT 32, 702–714 (1992)

[41] Sun, J.-G.: Rounding-Error and Perturbation Bounds for the Cholesky and LDL^T Factorizations. Linear Algebra and its Applications 173, 77–97 (1992)

[42] Toh, K., Trefethen, L.N.: Pseudozeros of Polynomials and Pseudospectra of Companion Matrices. Numerische Mathematik 68, 403–425 (1994)

[43] Turing, A.: Rounding-Off Errors in Matrix Processes. Quart. J. Mech. Appl. Math. 1, 287–308 (1948)

[44] Watkins, D.S.: Fundamentals of Matrix Computations. John Wiley & Sons, Chichester (1991)

The New Computationalism –
A Lesson from Embodied Agents

Jozef Kelemen* and Alica Kelemenová**

*Institute of Computer Science, Silesian University, Opava
Czech Republic and College of Management, Bratislava, Slovakia
kelemen@fpf.slu.cz

**Institute of Computer Science, Silesian University, Opava
Czech Republic and Department of Informatics, Catholic University, Ružomberok, Slovakia
kelemenova@fpf.slu.cz

Abstract. Computationalism is traditionally considered in the context of cognitive science as perhaps the dominant contemporary approach to understand cognition and cognitive phenomena. It consists in application of concepts and methods of theoretical computer science for understanding and (re)constructing phenomena appearing in much broader fields of science, including the natural sciences and also economics, and some other branches of social sciences. The contribution sketches this new situation, and provides an example of a theoretical model rooted in the traditional computationalism which reflects some new requirements.

1   Introduction

In the context of cognitive science, *computationalism* is traditionally considered as perhaps the dominant contemporary approach to understand cognition and cognitive phenomena. In present days computationalism plays a crucial role not only in cognitive science, but also in the field of cognitive psychology, artificial intelligence and also in an important part of advanced cognitive robotics.

According to [6] (p. 71) the central doctrine of the *traditional computationalism* considered as the basic paradigm for the study of cognition consists in the view that cognition is essentially a matter of the computations that a cognitive system performs in certain situations. In this context, computation is considered as the activity performable by Turing machine, so as computation in the Turing sense generally accepted in the field of theoretical computer science. Computationalists also maintain that *neural computations* are Turing-computable, that is, computable by Turing machines, which means that all of the *connectionism* present in cognitive science becomes a part of the traditional computationalism.

Having at hand the collection of notions and results formulated and discovered during the 50 years of the existence of *theoretical computer science* research based

I.J. Rudas et al. (Eds.): Towards Intelligent Engineering & Information Tech., SCI 243, pp. 55–66.
springerlink.com   © Springer-Verlag Berlin Heidelberg 2009

crucially on the concept of the Turing machine proposed in [18] we are trying to explain the nature of phenomena of (human) intelligence (usual esp. in artificial intelligence and in advanced robotics) at the level which provides the real base for engineering (re)production of it. The crucial hypothesis accepted almost generally as a base for such considerations is the *Church-Turing hypothesis* which says, roughly speaking, that *a function is computable, in the intuitive sense, if and only if it is Turing-computable* [2, 18].

2   A Slump in the Traditional Computationalism?

However, it is possible to treat computationalism in more general context. In this *broader meaning*, computationalism consists in a conceptual framework originated and formulated in theoretical computer science to understand some aspects of (re)construction (some fragments of) phenomena appearing in broader fields of science as those related to cognition and mind. As parts of these fields we recognize (some branches of) natural sciences, like biology, chemistry, physics, astronomy, also some subfield of economy, some branches of social sciences and arts (e.g. some sub-area of the field of new media, computer art, robotic art, etc.). As the present day state-of-the-art in science and engineering signalizes, both the traditional computationalism and the connectionism have some problems how to react to many new situation appearing in some fields of science and engineering.

Let us emphasize R. Brooks' appeal formulated during his plenary talk for the 8th *International Conference on Artificial Life* (Sydney, Australia, December 11, 2002) which focused the attention on a need of a new understanding of computing and computability, in other words to reconsider the actual form of computationalism and push our understanding of computation closer to the present-day requirements. Earlier, in *Nature* (p. 410), R. Brooks wrote:

We have become very good at modeling fluids, materials, planetary dynamics, nuclear explosions and all manner of physical systems. Put some parameters into the program, let it crank, and outcome accurate predictions of the physical character of modeled system. But we are not good at modeling living systems, at small or large scales. Something is wrong. What is wrong?

There are a number of possibilities: (1) we might just be getting a few parameters wrong; (2) we might be building models that are below some complexity threshold; (3) perhaps it is still a lack of computing power; and (4) we might be missing something fundamental and currently unimaginable in our models of biology.

The important and general lesson from the fields like artificial intelligence, advanced robotics, artificial life and cognitive science is, that *the Turing machine universality* as a mathematical concept which states that all kinds of computers are equally good devices for performing computational tasks, might be misleading in situations, when we consider *machines embedded in their real physical environments.*

The fact that an active agent is embedded in its dynamically changing environment may cause two fundamental consequences:

(1) The *input-output relation*, required when we consider processes as Turing machine computations, seems to be an unrealistic requirement, because of the often unpredictable *environment dynamics*, and because of the fact that real agents form at least in some extent *open systems* functioning in this environment.

(2) The *potentially infinite tape* of the Turing machine as a computing device cannot be required as a realistic part of any real physical system.

The matter is discussed in more details in [16] or in [8], where particular examples were presented for embodied agents in connection with some computationally relevant questions.

Moreover, going through Turing's pioneering paper of artificial intelligence [19] we recognize that Turing seems to have no difficulty in accepting the reality of some - let us use the present day trendy terminology - hypercomputational behaviors. His goal, as it is stated in [17], consists in finding some machine that can perform well the *imitational game* (then called *Turing test*). The core of the Turing test lies, roughly speaking, in the conviction that *(human) intelligence can be expressed as a computable function*. If that machine should happen to be a relatively simple one, like the Turing machine, so much better! But if not, then no reason to resign, let's go to construct a much suitable one!

In the core of the traditional computationalism there exists the believe (a) in the power of symbolic representation, and (b) in the approximation of a large number of processes as computational processes, which transform structures created from symbols into the structures of the same type.

For building a concise computational theory the above convictions seem to be realistic. So, their inclusions into different new computational frameworks are required. In the following we will concentrate on a model in which the above sketched limitations (1) and (2) will be suppressed in certain extent, and which will preserve the requirements (a) and (b), but eliminates in certain extent the requirement of the infinity from the basic computational units.

3   An Example on How to Cope with Limits

In [17] an overview of different approaches how to overcome the limitations of the traditional Turing machine is presented. As it is emphasized there (p. 140), any computation in a Turing machine depends on the controlled manipulation of internal configuration, where each configuration encodes a finite amount of information as *state*, a finite amount of information as *memory*, and a finite amount of information as a *program*. The Church-Turing hypothesis tells us that "cosmetic" changes in the architecture of the Turing machine have no relevance to computational power. So, as Stannett states in [17], in order to go behind the Turing's computability, we can consider changing the information contents of the temporal structure of computation, or the information contents of the memory, programs, or states.

We add the possibility to consider *decentralization*, and to consider the influence of the *contingent*, the *random*, and similar situations, which appear very often in architecture and functioning of real embodied systems (like real computers, robots, as well as biological systems).

In [8] we have illustrated, using an interesting result by D. Wtjen [21] on the generative power of a specific type of decentralized grammar-theoretic model of language generation, on the so called teams [5] in eco-grammar systems [4], that there exist formalized (formal grammar like) systems set up from decentralized components with higher computational power than those of Turing machines. We have also expressed our conviction that there are no principal reasons to reject the hypothesis that it is possible to construct real robots as a certain kind of implementations of these formalized systems. If we include into the functioning of such robots the activation of their functional modules according a non-recursive (in Turing sense) computation, the behavior of the agents might be non-recursive.

We suppose that this situation may appear if some of the functional parts of the robots are switched on or switch off on the base of the random behavior of the robots environments, for instance. So, we exclude the situation when a computer simulation of randomness are included into the functional architecture of robots. Rather, we suppose the randomness appearing in the environment, a randomness which follows from the ontology of robots situated in their environments.

The just mentioned *ontological randomness*, might be caused by different reasons - e.g. by imprecise work of sensors and actuators of robots, by erroneous behavior of their hardwired or software parts, so by the general cause of their embodiment, by nondeterminism of the behavior of the environment, by the lack of resources necessary for executing the required computations, so by their finite nature, and so. All these influences may be reflected in the specific behavior of the robots and we cannot reject the *hypothesis* that just these kinds of irregularities cause also the phenomenon called robot consciousness. It is also possible, that the organic, effective, and enough rigorous inclusion of this type of randomness, caused in fact by the embodiment of computing systems, and by their finiteness, as well, can contribute to our new understanding of computing machineries, and computing as well. If this possibility turns to reality, then we will have at hand the required new model for rigorous formal study and understanding of a new type of computationalism.

In the following we present a formal framework of the eco-colonies [20] as a simple example of formalized systems which respect the finite nature of their basic building components and we prove - following the idea presented in [21] that in the case of inclusion of a formalized variant of the above mentioned ontological randomness into the framework of the eco-colonies we receive a computational power beyond the classical Turing machines.

4 Eco-colonies Working in Time Varying Teams

In present section we propose the formal model to illustrate previous ideas. For this purpose we have chosen the model of cooperating grammars called eco-colonies.

Eco-colonies are collections of simple grammars (called components or agents) working on common environment. A *component* is specified by its start symbol (an object from the environment, the component can process it) and by its finite set of strings, the finite language of the grammar, which determines actions of the component. The component substitutes its start symbol by some of these strings. *The environment* in the eco-colonies are able to develop itself using its own developing rules in totally parallel way like in Lindenmayer systems. All symbols not substituted by components are changed by inner rules of the environment.

Eco-colonies were chosen for following reasons:

(1) *Behavior of an individual component*, an agent, in this model is very simple. Each component acts on the environment common for whole system and it posses with *finite behavior*.

(2) The *behavior of the environment* itself is characterized as classical $0L$ *behavior*, it means the environment has its own rules and all symbols of the environment (on the places not influenced by the agents) are changed simultaneously using these rules. No interaction between neighboring symbols is considered.

(3) The model allows to study team behavior of the agents and it makes possible to consider *function dependent changing of teams*, which can be used to model and discuss the ontological randomness mentioned above.

To present our formal model we assume that the reader is familiar with the formal language theory [15] including the L system theory [7]. The following model is based on the notation of the colony introduced in [9, 10], which was extended to the notion of the $0L$ eco-colony in [20]. While in colonies, only agents take a part in development of the environment and environment itself is passive, the environment in an eco-colony posses own developmental rules, too. Eco-colonies may be considered as special type of eco-grammar systems, where each agent can have opportunity to rewrite one of the several possible symbols. For an information on eco-grammar systems consult [4, 5].

We start with formal definition of $0L$ eco-colony.

Definition 1. An $0L$ eco-colony is a tuple $\sum = \left(V, E, \left(S_1, F_1\right), ..., \left(S_n, F_n\right), w_0\right)$, where

V is a finite non-empty alphabet,

$E = (V, P)$ is an 0L scheme (of the environment), $P \subset V \times V^*$ is a finite set of rules over V,

$\left(S_i, F_i\right)$, $1 \le i \le n$, is the ith component, where $S_i \in V$ is the start symbol of the component and $F_i \subseteq (V - \{S_i\})^*$ is a finite set of words (the action language of the component),

$w_0 \in V^*$ is the axiom.

There were various types of derivation modes (sequential, parallel, team, etc.) introduced to describe the behavior of the grammar systems including colonies [12] and eco-colonies [20]. For our purpose we choose here *the team* behavior for the $0L$ eco-colony. Team behavior was introduced generally for grammar systems [3] and used for eco-grammar systems in [5] and [21].

By a team we mean here a collection of active components. It will be determined by its size, *the number* of active components, nondeterministically chosen for a derivation step from all components. Teams varying in time can be determined by a function f defined on natural numbers, where the value $f(t)$ determines size of active components in the tth step of the computation. We speak on the *function depending* team behavior in this case. We formalize the derivation step of an $0L$ eco-colony:

Definition 2. Let $\sum = \left(V, E, \left(S_1, F_1\right), ..., \left(S_n, F_n\right), w_0\right)$ be an $0L$ eco-colony.

$A = k$ team derivation step is a binary relation $\overset{=k}{\Rightarrow}$ on $V^* \times V^*$ defined as follows: $\alpha \overset{=k}{\Rightarrow} \beta$ iff $\alpha = \gamma_0 S_{i_1} \gamma_1 S_{i_2} \gamma_2 ... \gamma_{k-1} S_{i_k} \gamma_k$, and

$\beta = \gamma_0 `f_{i_1} \gamma_1 `f_{i_2} \gamma_2` ... \gamma_{k-1} `f_{i_k} \gamma_k`$, where

$-f_{i_s} \in F_{i_s}$ for (S_{i_s}, F_{i_s}), $1 \leq s \leq k$, with $\{i_1, i_2, ..., i_k\} \subseteq \{1, 2, ..., n\}$ and

$i_s \neq i_m$ for all s, m, $1 \leq s \neq m \leq k$, and

$-\gamma_s \overset{E}{\Rightarrow} \gamma_s`$, $\gamma_s, \gamma_s` \in V^*$, $0 \leq s \leq k$, is the derivation step of 0L scheme E, i.e.

$\gamma \overset{E}{\Rightarrow} \gamma`$ iff $\gamma = a_1 ... a_s$, $\gamma` = \alpha_1 ... \alpha_s$ and $\left(a_i \to \alpha_i\right) \in P$ for $1 \leq i \leq s$.

Definition 3. A $\leq k$ team derivation step is the relation $\overset{\leq k}{\Rightarrow}$ defined by

$$\alpha \overset{\leq k}{\Rightarrow} \beta \text{ iff } \alpha \overset{=l}{\Rightarrow} \beta \text{ for some } l \leq k.$$

Definition 4. Let $\sum = \left(V, E, \left(S_1, F_1\right), ..., \left(S_n, F_n\right), w_0\right)$ be an $0L$ eco-colony and let $f : N \to \{1, ..., n\}$ be the function defined on the natural numbers.

The language generated by f teams of Σ denoted by $L(\Sigma, f)$, is determined

by $L\left(\sum, f\right) = \left\{ w \in V^* \middle| w_0 \overset{=f(1)}{\Rightarrow} w_1 \overset{=f(2)}{\Rightarrow} w_2 ... \overset{=f(r)}{\Rightarrow} w_r = w, r \geq 0 \right\}$

The language generated by $\leq f$ teams of Σ denoted $L(\Sigma, \leq f)$, is deter-mined by $L\left(\sum, f\right) = \left\{ w \in V^* \middle| w_0 \overset{=f(1)}{\Rightarrow} w_1 \overset{=f(2)}{\Rightarrow} w_2 \dots \overset{=f(r)}{\Rightarrow} w_r = w, r \geq 0 \right\}$

We illustrate the behavior of eco-colonies in the following example.

Example 1. Let $\sum_n = \left(V, E, (S_1, F_1), \dots, (S_n, F_n), w_0 \right)$ be an 0L eco-colony, where

$V = \{a, b, c\}$,

$E = \{a \to a^2, b \to b, c \to b\}$,

$(S_i, F_i) = (b, \{c\})$ for $1 \leq i \leq n$ are n identical components and

$w_0 = a^2 b^{2n}$.

Typical derivation in $\sum_n$ has form

$$w_0 = a^2 b^{2n} \overset{=f(1)}{\Rightarrow} a^4 u_1 \overset{=f(2)}{\Rightarrow} a^8 u_2 \overset{=f(3)}{\Rightarrow} \dots \overset{=f(t)}{\Rightarrow} a^{2^{t+1}} u_t \overset{=f(t+1)}{\Rightarrow} \dots$$

with $u_s \in \{b, c\}^{2n}$ and $|u_s|_c = f(s)$.

The $0L$ eco-colony $\sum_n$ generates the languages

$$L\left(\sum_n, f\right) = \left\{a^2 b^{2n}\right\} \cup \bigcup_{k \in N} \left\{a^{2^{k+1}}\right\} perm(\underbrace{b, \dots, b}_{2n - f(k)}, \underbrace{c, \dots, c}_{f(k)})$$

and

$$L\left(\sum_n, \leq f\right) = \left\{a^2 b^{2n}\right\} \cup \bigcup_{k \in N, r \leq f(k)} \left\{a^{2^{k+1}}\right\} perm(\underbrace{b, \dots, b}_{2n - r}, \underbrace{c, \dots, c}_{r})$$

where by $perm(a_1, \dots, a_n)$ we mean all the words containing each of the letters $a_1, \dots, a_n$ exactly ones concatenated in an arbitrary order, i.e.

$$perm(a_1, \dots, a_n) = \left\{ \left(a_{i_1} \dots a_{i_n}\right) \middle| \text{ for all permutations } (i_1, \dots, i_n) \text{ of } (1, \dots, n) \right\}.$$

Note that the number of occurrences of letter a in w determines uniquely the length of derivations of w in $\sum_n$. Moreover, we have

$|w_k|_c = f(k)$ for $w_k = a^{2^{k+1}} u_k \in L(\sum_n, f)$ and

$|w_k|_c \leq f(k)$ for $w_k = a^{2^{k+1}} u_k \in L(\sum_n, \leq f)$.

This property will be used in the proofs in next section.

5   Computability of Functions versus Recursivity of Environments

In the present section we will study the influence of the computability and non computability of function f to the status of recursivity of languages $L(\Sigma, f)$ and $L(\Sigma, \leq f)$, i.e. we are looking to classify $L(\Sigma, f)$ and $L(\Sigma, \leq f)$ to be recursive, recursively enumerable, or not a recursively enumerable set.

We will use notions of computable function, non computable function, recursive set, recursively enumerable set and not recursively enumerable set as in [14].

Function f is *computable* iff following conditions (i) and (ii) are satisfied for some Turing machine M

(i) if $f(x)$ is defined, then M eventually halts in the final states with $f(x)$ symbols of 1 on the tape,

(ii) if $f(x)$ is undefined, then M never reaches the final state.

A set is *recursively enumerable* iff it is the domain of a computable function $f(x)$.

A set is *recursive* iff it equals the domain of a total computable function $f(x)$. Equivalently, L is recursive iff it is decidable whether $w \in L$ for every w.

Ideas presented in this section follow analogous results of Wätjen in [21]. To prove our results we use the languages of the eco-colonies over fixed (three letters) alphabet from the example presented in previous section. This simplify the systems used in [21], where the alphabets of the systems increase linearly with the number of components.

For every eco-colony Σ and for every computable function $f : N \rightarrow \{1, ..., n\}$ are languages $L(\Sigma, f)$ and $L(\Sigma, \leq f)$ recursively enumerable.

We present a stronger result for the languages from the example above. For this purpose we will use denotation $L(\Sigma_n, f) = L_{n,f}$ and $L(\Sigma_n, \leq f) = L_{n,\leq f}$.

Theorem 1. *Let* $f : N \rightarrow \{0, ..., n\}$ *be a function.*

$L_{n,f}$ *is recursive if and only if* f *is computable.*

$L_{n,\leq f}$ *is recursive if and only if* f *is computable.*

Proof. Let $L_{n,f}$ $(L_{n,\leq f})$ be recursive. We choose an arbitrary $k \in N$ and consider the words $w_r = a^{2^{k+1}} c^r b^{2n-r}$ for all $r \in \{0, ..., n\}$. Because of the form of the words in $L_{n,f}$ $(L_{n,\leq f})$ presented in the previous section we know that at least one of such w_r are in $L_{n,f}$ $(L_{n,\leq f})$. Since $L_{n,f}$ $(L_{n,\leq f})$ is recursive we can decide whether $w_r \in L_{n,f}$ ($w_r \in L_{n,\leq f}$) for all w_r . We set $f(k) = \max\{r \mid w_r \in L_{n,f}\}$ (resp. $f(k) = \max\{r \mid w_r \in L_{n,\leq f}\}$.) Since maximum of finite set is computable f is computable as well.

To prove the other implication assume that f is computable. Let $w \in V^*$.

If $w = a^2 b^{2n}$ then $w \in L_{n,f}$ and $w \in L_{n,\leq f}$. Otherwise it is decidable whether $w \in a^{2^{k+2}} perm\left(\underbrace{b,...,b}_{2n-r},\underbrace{c,...,c}_{r}\right)$

for some $k, r \in N$ and $0 \leq r \leq n$. For k determined by w we compute $f(k)$. We have $w \in L_{n,f}$ if and only if $r = f(k)$ and $w \in L_{n,\leq f}$ if and only if $r \leq f(k)$. Thus both $L_{n,f}$ and $L_{n,\leq f}$ are recursive.

A stronger result holds for $L_{n,f}$ but not for $L_{n,\leq f}$.

Theorem 2. *Let $f : N \rightarrow \{0,...,n\}$ be a function. If f is not computable then $L_{n,f}$ is not recursively enumerable.*

Proof. Assume contrary that $L_{n,f}$ is recursively enumerable for not computable f . So there exists an effective listing of all words of $L_{n,f}$. We choose an arbitrary $k \in N$. There exists a word $w_k \in L_{n,f}$ with prefix $a^{2^{k+1}} b$ and this word is listed after a finite number of steps. We can compute $f(k) = |w_k|_c$. This gives that f is computable, a contradiction.

6   Some Concluding Remarks and Questions

In the case of the above mentioned model, and more generally, in all cases when formal models are built on the conceptual base of formal grammars and languages, the rules governing the dynamics of the behavior of agent-like entities are described in the form of rewriting rules. This kind of description defines extremely simple, the so called purely reactive, agents. Each purely reactive agent has its own sensor capacity represented by left-hand side of the corresponding rule, and its own action capacity represented by the right-hand side of rules. The ways of interactions of agents are specified by different derivation modes and rewriting regulations in grammar-like structures.

Understanding of rewriting rules as agents is a fundamental advantage at least from the methodological point of view. We know very well, that some specific multi-agent systems (formal grammars) define well-specified behaviors (formal languages) with interesting relation to different models of computation (to different types of automata) which have important relations to real engineered (computing) machines. What we do not know, is the answer to the question concerning the universality of the approach accepted for describing languages (behaviors). What kind of behaviors are we able to describe using the just sketched framework of agent (and multi-agent system) inspired grammar-like models behind the Turing-computable ones?

The second question follows from an incorporation of dynamics of the environment in which our simple agents act. In the traditional formal language theory we do not consider any inner dynamics of changes of the strings under rewriting. The only changes result from applying rewritings using rules. In the case of eco-grammar systems, however, the situation is substantially modified by providing an "independent" dynamics describing the environment changes using a specific parallel rewriting mechanism (modeled by L-systems) working independently on the activities of agents. What will happen when more complicated mechanisms of changes will be included into the models? What we know about the situation when, for instance, some finite subsequences (belonging to a language with specific Turing-computability properties) will be randomly replaced by words from another set of words (of known Turing-computability property)? Of course, there are more similar questions which can be formulated in a more or less formal ways. We provided some examples, only.

However, the most fundamental question is, according to our meaning, the following one: Is it possible to receive (define) some stabilized (well defined in the framework of Turing-computability, for instance) behavior in the hardly-predictable behavior of the environment in which the agents act? If yes, in which cases, and what are the conditions of such behavior. From the standpoint of the practice, this question is very important, because to design stabilizing multi-agent systems working in the unstable environment is the main goal of many engineering activities. What are we able to say about the possibility of such design in our theoretical framework?

The last question leads us from the speculations about the universality of our models to the question of their realism. Real (embodied or software) multi-agent systems are never perfectly reliable. To be more concrete, let us mention some interesting phenomena related with reliability of (multi-agent) systems. One among the most often appearing is the phenomenon of dysfunction of some agents which are parts of some whole system. Suppose that some of the components of a complicated multi-agent machine go down. Will the whole machine work well (in some acceptable enough way) after this reduction of its components? What kind of changes will appear in its behavior after dysfunction of some of its parts? How to preserve some appropriate level of the functionality of the machine (its resistance with respect of the "small" changes in its architecture)?

An approach to incorporate reliability into the formal models might be inspired by the incorporation of the fuzzy approaches into the traditional grammar-theoretic models. It seems to be possible to use fuzzy rewriting rules, and in the consequence of the derived strings, and to receive formal languages as fuzzy sets, in such a way.

It is possible to use fuzzy components of grammar systems, or the regions of membrane systems, and then to propagate the fuzziness toward the generated sets of behaviors, etc. It is also possible to compare the behavior of such models with the behavior (generative capacity) of the not fuzzified models. How to define the necessary notions?

In other type of systems, despite of the reliability of the agents, their involvement into the work of the whole system is important. In a society of ants, for instance, it is practically impossible to organize the work of any particular agent. Some ants work in some time period, some of them not, and, moreover, we have absolutely no predictive power to know exactly which ant will or will not act in the next time period. This problem is the problem of randomness in multi-agent systems.

How to cope with the randomness of the impact of particular components of multi-grammar models to the derivative capacity of the whole systems, this is illustrated in certain extent by the present article, in which one example of the form of team-forming function is presented and studied. On the base of the presented result, it seems to be realistic to suppose that the relation of other particular forms of team-forming functions and their computational properties considerably influence the behavior of the multi-grammar models. However, exactly what are the ways of this influence, and in what extent, in dependence on the computational properties of the team-forming functions?

Timing is another way of incorporation the dynamics of components into the behavior of models as defined in [11] for colonies, e.g. by functions defined of the length of the derivation chains. Note that the similar approaches are incorporable also into the fuzzy models, so it seems to be realistic also the ability to combine different models of reliabilities and randomness inside one theoretical model. What is the most perspective way of doing that?

Acknowledgments. JK's work on the subject was supported by the grant MSM 4781305903, and by Gratex International Corp., Bratislava. AK's work on the subject was supported by Scientific Grant Agency of Slovakia, VEGA, project no 1/0692/08.

References

[1] Brooks, R.A.: The Relationship between Matter and Life. Nature 406, 409–414 (2001)

[2] Church, A.: An Unsolvable Problem in Elementary Number Theory. The American Journal of Mathematics 58, 345–363 (1936)

[3] Csuhaj-Varjú, E., Dassow, J., Kelemen, J., Paun, G.: Grammar Systems - A Grammatical Approach to Distribution and Cooperation. Gordon and Breach, London (1994)

[4] Csuhaj-Varjú, E., Kelemen, J., Kelemenova, A., Păun, G.: Eco-Grammar Systems - a Grammatical Framework for Studying Life-like Interactions. Artificial Life 3, 1–28 (1997)

[5] Csuhaj-Varjú, E., Kelemenová, A.: Team Behaviour in Eco-Grammar Systems. Theoretical computer science 209, 213–224 (1998)

[6] Giunti, M.: Beyond Computationalism. In: Cottler, G.W. (ed.) Proc. 18th Annual Conference of the Cognitive Science Society, pp. 171–175. Lawrence Erlbaum, Mahvah (1996)

[7] Herman, G.T., Rozenberg, G.: Developmental Systems and Languages. North-Holland/American Elsevier, Amsterdam/New York (1975)

[8] Kelemen, J.: May embodiment cause hyper-computation? In: Capcarrère, M.S., Freitas, A.A., Bentley, P.J., Johnson, C.G., Timmis, J. (eds.) ECAL 2005. LNCS, vol. 3630, pp. 31–36. Springer, Heidelberg (2005)

[9] Kelemen, J., Kelemenová, A.: A Subsumption Architecture for Generative Symbol Systems. In: Trappl, R. (ed.) Cybernetics and System Reseach 1992, pp. 1529–1536. World Scientific, Singapore (1992)

[10] Kelemen, J., Kelemenová, A.: A Grammar-Theoretic Treatment of Multiagent Systems. Cybernetics and Systems 23, 621–633 (1992)

[11] Kelemenová, A.: Timing in Colonies. In: Păun, G., Salomaa, A. (eds.) Grammatical Models of Multi-Agent Systems, pp. 136–143. Gordon and Breach, London (1999)

[12] Kelemenová, A.: Bounded Life Resources in Colonies. In: XIV Tarragona Seminar on Formal Syntax and Semantics, Report 23/02 Research group on mathematical linguistics (Universitat Rovira i Virgili) p. 24 (2002)

[13] Kelemenová, A., Csuhaj-Varjú, E.: Languages of Colonies. Theoretical Computer Science 134, 119–130 (1994)

[14] Rozenberg, G., Salomaa, A.: Cornestones of Undecidability. Prentice Hall, New York (1994)

[15] Rozenberg, G., Salomaa, A. (eds.): The Handbook of Formal Languages, vol. 3. Springer, Berlin (1996)

[16] Sloman, A.: The Irrelevance of Turing Machines to AI. In: Scheutz, M. (ed.) Computationalism - New Directions, pp. 87–127. The MIT Press, Cambridge (2002)

[17] Stannett, M.: Hypercomputational Models. In: Teuscher, C. (ed.) Alan Turing - Life and Legacy of a Great Thinker, pp. 135–157. Springer, Berlin (2004)

[18] Turing, A.M.: On Computable Numbers, with an Application to the Entscheidungsproblem. Proc. London Mathematical Society 42, 230–265 (1936); corrections in 43, 544-546 (1937)

[19] Turing, A.M.: Computing Machinery and Intelligence. Mind 59, 433–460 (1950)

[20] Vavrečková, Š.: Properties of Eco-Colonies. In: Kelemenová, A., et al. (eds.) Information Systems and Formal Models 2007, pp. 235–242. Silesian University, Opava (2007)

[21] Wätjen, D.: Function Dependent Teams in Eco-Grammar Systems. Theoretical Computer Science 306, 39–53 (2003)

Measuring Voting Power: The Paradox of New Members vs. the Null Player Axiom

László Á. Kóczy

Keleti Faculty of Economics, Budapest Tech, Budapest, Tavaszmező u. 15–17.
Department of Economics, Maastricht University
koczy.laszlo@kgk.bmf.hu

Abstract. Qualified majority voting is used when decisions are made by voters of different sizes. In such situations the voters' influence on decision making is far from obvious; power measures are used for an indication of the decision making ability. Several power measures have been proposed and characterised by simple axioms to help the choice between them. Unfortunately the power measures also feature a number of so-called paradoxes of voting power. In this paper we show that the Paradox of New Members follows from the Null Player Axiom. As a corollary of this result it follows that there does not exist a power measure that satisfies the axiom, while not exhibiting the Paradox.

1 Introduction

Power measures or, more appropriately: a priory measures of voting power give an indication of a voter's ability to change decisions. Voting bodies are everywhere from faculty councils to the UN Security Council, from national parliaments to shareholders' meetings. The most discussed case is, however, the EU Council of Ministers that uses qualified majority voting, rather complicated procedures to determine whether a subset of EU member countries is able to pass a decision or not. A key element of these procedures is the voting weight that is determined by EU treaties for all countries. How should these weights be determined is a topic of on-going discussions partly because the implications of a particular set of weights is not totally clear.

To illustrate the difficulties consider one of the simplest possible voting bodies (such as a shareholders' meeting) with only three decision makers respectively having 49, 49, and 2 votes (such as shares). A decision can be passed with plain majority, that is, if a coalition supporting the motion has at least 51 votes in total. It is easy to verify that despite the dramatic differences in weight (size) the three

I.J. Rudas et al. (Eds.): Towards Intelligent Engineering & Information Tech., SCI 243, pp. 67–78.
springerlink.com   © Springer-Verlag Berlin Heidelberg 2009

voters have equal influence on decision making as any pair of voters has majority, plus of course the grand coalition including all voters has majority, too. The problem is general: when voters have weights, these weights are difficult to translate into shares of voting power.

To complicate the arguments consider now a very similar problem, where three parties have the above number of representatives in a legislative body, such as a national parliament. When the MPs are allowed to vote freely and their preferences are largely independent of the parties, the probability that the voter that turns a losing coalition into a winning one belongs to a given party is proportional to the shares of seats the party has [20].

In order to measure voting power several approaches have been proposed. The first measured voting power by directly applying the Shapley value to simple co-operative games [31]. The most common alternative is the Banzhaf measure and index[1] [1, 5, 26] which, although predates the Shapley-Shubik index, the first two discoveries have been largely forgotten and were only connected to the mainstream literature later. Since then several alternatives have been introduced: the Johnston-index [19], the Deegan-Packel index [6], the Public Good Index [18] or the partition value [24] so the question naturally arises: which one of these should be used?

The key to answering this question is to study the properties of the various indices. As we have seen in the above example the institutional design might imply one or another index. In general, however it is rather difficult to make a direct choice. On the other hand if the various indices and measures are characterised by a handful of simple, elementary properties or axioms making a choice between these axioms is often easier. In the best case a set of axioms fully characterise an index, that is, there is a unique index that satisfies the given axioms. While in theory the idea is simple and appealing, many of the axioms used in these characterisations are rather technical and show little relevance to practical problems [22, pp37-38]. It seems other properties, that perhaps do not help towards a full characterisation, but ones that have clearer practical implications can be equally useful to make a choice [12, 13]. Unfortunately as researchers have produced positive results also some rather unattractive properties have been discovered. The so-called paradoxes of voting power [4] are three rather intuitive properties that are nevertheless not satisfied by the best-known power indices, the Shapley-Shubik and the Banzhaf index. Felsenthal and Machover [11] argue that "paradox" is perhaps a word too strong to describe these properties and argue that these are merely 'apparently strange pieces of behaviour' [12, p. 221], but then the question arises whether we should be guided by our intuition and consider the paradoxes a problem or whether we should be comforted by the theoretical underpinnings of these indices and revise our intuition.

In an earlier paper [20] we have shown that there is a rather intuitive index, the proportional index where none of the paradoxes arise. On the other hand there is

---

[1] It is common to refer to power measures normalised to 1 as *power indices*.

ample empirical evidence that suggests that in practical matters this proportional index is used as a rule of thumb to assign power[2]. We believe that the paradoxes and a number of other results stem from this proportional index, but then it is somewhat difficult to decide what "natural properties" should serve as the basis of evaluation of power indices and which ones are prejudices rather than true requirements.

In this paper we show that the problem is general: an index that satisfies the widely accepted axioms will necessarily exhibit some of the paradoxes. In particular we show that the paradox of new members is in conflict with the null player axiom.

The outline of this paper is then as follows. First we introduce voting games and some of the well-known properties. Then we present the Brams's paradoxes and an axiomatisation of the Shapley-Shubik and Banzhaf indices, including the null player axiom. Next we prove our main result. We end the paper with some conclusions.

2 Voting Games

Since Shapley and Shubik [31] it is common to study voting situations as cooperative games. A cooperative game is given by a pair (N, v) consisting of a set of n players and a real valued function, the so-called *characteristic function*[3] v defined over the set of *coalitions*: a coalition is a subset of the player set. Thus $v : 2^N \longrightarrow \mathbb{R}$. It is common to make a number of assumptions about the characteristic function. We assume that the empty set has no value, therefore $v(\emptyset) = 0$ and that the function is superadditive or

$$v(S \cup T) \geq v(S) + v(T) \quad \text{if} \quad S \cap T = \emptyset. \tag{1}$$

For such games the total value of the players is maximised by the grand coalition N and hence the purpose of the game is to find an equilibrium allocation of this payoff among the players. While strategies are not explicit in cooperative games (they correspond to forming coalitions), we still deal with the same intelligent, payoff-maximising agents as in noncoopertive games, and in fact the cooperative concepts used here have been implemented as equilibria of certain noncooperative games [17, 27].

In the following we shall be interested in *simple games*: A game is *simple* if the value of the coalition is either 0 or 1, that is, if $v : 2^N \longrightarrow \{0, 1\}$. As we shall see these values can be seen as wins and losses, and correspondingly we can talk about winning and losing coalitions. We denote the set of winning coalitions by $\mathscr{W}$, thus

$$\mathscr{W} = \{S \mid S \subseteq N, v(S) = 1\}.$$

---

[2] See [8, 16, 14] and references therein

[3] The name comes from the early literature of game theory. Von Neumann and Morgenstern in their seminal work [23] assumed that when in an n player game a subset S of players forms a coalition, this coalition must play against a natural opponent, the complementer coalition $N \setminus S$. The value the coalition S can obtain in this game is its characteristic value.

In a winning coalition we will be interested in critical players, that is, players whose presence is essential for the success of the coalition. Formally the player i is critical in coalition S if $S \in \mathscr{W}$, but $S \setminus \{i\} \notin \mathscr{W}$. A player that is never critical is called *null*. Among winning coalitions, *minimal winning coalitions* containing only critical players deserve special attention. The set of such coalitions is denoted by $\mathscr{M}$, where

$$\mathscr{M} = \{S \, | \, S \in \mathscr{W}, \forall i \in S : \ S \setminus \{i\} \notin \mathscr{W} \}.$$

We assume that the grand coalition is always winning, that is, $v(N) = 1$ and that the addition of new members to a coalition does not make it loosing, formally if $v(T) = 1$ and $T \subseteq S$ then $v(S) = 1$. While this assumption is a standard one, in some cases the addition of a new member to a winning coalition may also result an *infeasible* (winning) coalition. In such cases the coalition, though has the power to make decisions, cannot. We will use infeasible coalitions to formulate one of the paradoxes below, but this idea is used in the literature of games over convex geometries [2, 3] and in the models of strategic power indices, where players have the ability to block coalitions [21]. The set of feasible coalitions is denoted by $\mathscr{F}$. Unless otherwise stated we assume that $\mathscr{F} = 2^N$.

A *weighted voting game* $G = (N, (w_i)_{i \in N}, q)$ consists of a collection N of n voters having $w_1, w_2, \ldots, w_n > 0$ votes such that $w = \sum_{i=1}^{n} w_i$, and a quota q, $w \geq q > w/2$, or the number of votes *required* to pass a bill. For more on weighted voting games see [32]. It is clear that there is a unique mapping from weighted voting games to voting games, and therefore the first is a subset of the latter. Given $(N, (w_i)_{i \in N}, q)$ we define the corresponding voting game (N, v) by

$$v(S) = \begin{cases} 1 & \text{if } \sum_{i \in S} w_i \geq q \\ 0 & \text{otherwise.} \end{cases} \tag{2}$$

2.1 Power Indices

Now that the games have been defined we can move on to defining the ways to establish the power of the individual players. In this respect there are two approaches depending on the goal of the study. On the one hand we can look at the players' share of power. In this case we calculate power *indices*, that is, normalised power measures. This is the approach we take here. When the focus is on engineering the voting situation, determining the voting weights and quotas or, more generally, the set of winning coalitions, the likelihood that any coalition is winning is important too. For it is good to know if one can change the decision in a certain percentage of cases, but such cases rarely occur the power is certainly more limited. Put it differently, power is often money – it is sufficient to think of lobbying to see this. If so, it is one thing the get a good slice of the cake and another to have a large cake, that is, much lobbying.

Since our focus is not on the institutional design, we present power indices.

A *power index* is a function k that assigns to each weighted voting game a non-negative vector in $\mathbb{R}^N_+$.

The *Shapley-Shubik index* [31] is an application of the Shapley value [30] to measure voting power, motivated by a story where parties throw their support at a motion in some order until a winning coalition is reached. The last, *pivotal* party gets all the credit; the Shapley-Shubik index is then the proportion of orderings where it is pivotal

$$\Phi_i = \frac{\# \text{ times } i \text{ is pivotal}}{n!}.$$

There is also an explicit formula to express the Shapley-Shubik index:

$$\Phi_i = \sum_{S \ni i} \frac{(s-1)!(n-s)!}{n!} (v(S) - v(S \setminus \{i\})) \tag{3}$$

The *Banzhaf measure* [26, 1] is the probability that a party is *critical* for a coalition, that is, the probability that it can turn winning coalitions into losing ones.

$$\psi_i = \frac{\# \text{ times } i \text{ is critical}}{2^{n-1}}.$$

Or explicitly:

$$\psi_i = \frac{1}{2^{n-1}} \sum_{S \ni i} (v(S) - v(S \setminus \{i\})) \tag{4}$$

The *Banzhaf index* β [5] is the normalised Banzhaf measure, where the total power is scaled to 1 –already in the spirit of the Shapley-Shubik index.

The two indices can give substantially different implications, despite the fact that the main difference is in the probabilities they attach to the formation of particular coalitions or to the fact that a given player is critical for some coalition. While in the Banzhaf index all such instances of criticality happen with equal probability, for the Shapley-Shubik index the probability depends on the size of the coalition (when a player is critical in medium sized coalitions, this is taken with a smaller weight into account).

There are a few variants of the (normalised) Banzhaf index. In the *Johnston index* γ [19] the credit a critical player gets is inversely proportional to the number of critical players in the coalition. In effect, coalitions of different sizes have the same contribution to the distribution of power or the probability that a given coalition is the one making the decision is the same for all coalitions. The *Deegan-Packel index* ρ [6] is a further modification that only considers minimal winning coalitions, motivated by the idea that only minimal winning coalitions should form so that the benefits from winning should be least divided [29]. Finally the Holler-Packel or *Public Good Index h* [18] modifies the Deegan-Packel index: here the benefit of forming a winning coalition is given to each and every player in the coalition. With the normalisation in simple games the index is nothing but a normalised Banzhaf index, where only minimal coalitions are taken into account. Finally the *partition index*

[24] is motivated by decision making with multiple alternatives that then results in a partition of the voters that consists of possibly more than two coalitions. The probability that a coalition forms is then the probability that a partition containing this coalition forms. The partition index clearly favours smaller winning coalitions.

The *proportional index* α is the trivial power index given by $\alpha_i = \frac{w_i}{w}$. This measure is popularly known in political science as Gamson's Law: 'Any participant will expect others to demand from a coalition a share of the payoff proportional to the amount of resources which they contribute' [15].

2.2 Axioms

In the following we present the full characterisations of Dubey [9] and Dubey and Shapley [10] for the Shapley-Shubik and the Banzhaf index respectively.

Before we move to the different axioms we need to introduce some additional terminology. The permutation π of the players is a bijective mapping of the player set. The permutation of a game πv is given by $(\pi v)(S) = v(\pi(S))$.

Definition 1 (Anonymity Axiom). For all simple games v any permutation π of N, and any $i \in N$,

$$k_i(\pi v) = k_{\pi(i)}(v). \tag{5}$$

Definition 2 (Null Player Axiom). For any simple game v and any $i \in N$, if i is a null player in game v then

$$k_i(\pi v) = 0. \tag{6}$$

For two simple games v and w over the player set N let

$$(u \vee w)(S) = \max\{v(S), w(S)\} \quad \text{and} \quad (u \wedge w)(S) = \min\{v(S), w(S)\}.$$

Of these two conditions the latter is perhaps the more interesting one as it gives a formula for a combination game, for instance a game where a coalition must be winning in both chambers of the parliament or when there are multiple criteria to determine the winning coalitions as in the case of the EU Council of Ministers, for instance.

Definition 3 (Transfer Axiom). For any simple games v, w such that $v \vee w$ is a simple superadditive game, too the transfer axiom states that

$$k_i(v) + k_j(w) = k(v \wedge w) + k(v \vee w) \tag{7}$$

Finally there are two axioms that express a notion of efficiency for both the Banzhaf and the Shapley-Shubik case.

Definition 4 (Shapley Total Power Axiom). For any simple game v

$$\sum_{i \in N} \Phi_i(v) = 1 \tag{8}$$

Definition 5 (Banzhaf Total Power Axiom). For any simple game v

$$\sum_{i\in N}\psi_i(v) = \frac{1}{2^{n-1}}\sum_{i\in N}\sum_{S\ni i}(v(S)-v(S\setminus\{i\})) \tag{9}$$

With these definitions the Shapley-Shubik index and Banzhaf measure can be characterised as follows:

Theorem 1 (Dubey [9]). *If a power index k satisfies Anonimity, Null Player, Transfer and Shapley Total Power Axioms, then $k = \Phi$.*

That is the Shapley-Shubik index is the unique power index satisfying the above axioms.

Theorem 2 (Dubey and Shapley [10]). *If a power measure k satisfies Anonimity, Null Player, Transfer and Banzhaf Total Power Axioms, then $k = \psi$.*

The Banzhaf measure is the unique power measure satisfying the above axioms.

While there are many other axiomatisations of these (and other indices), the null player axiom is one of the central properties. In fact, the proportional index is mostly criticised for not satisfying this property (for players with nontrivial weights). These motivate our interest in the Null Player Axiom.

2.3 Paradoxes

Brams [4] lists three natural properties that any power index should satisfy (the list is extended by Felsenthal and Machover [12]), but the best-know indices satisfy none of these. These disappointing negative results are known as *paradoxes*. In the following we list them in their positive form as "properties."

Given a game (N,v) by the merger of players i and j we mean a modified game (N_{ij},v_{ij}) with one less players $N_{ij} = N\setminus i,j\cup\{ij\}$ and winning coalitions

$$\mathscr{W}_{ij} = \left\{S \in 2^{N_{ij}} \mid S \in \mathscr{W}, \text{ or } ij \in S \text{ and } S\setminus\{ij\}\cup\{i,j\} \in \mathscr{W}\right\}$$

When the game is defined as a weighted voting game the combined player ij has a weight $w_{ij} = w_i + w_j$.

Definition 6 (Property of (large) size). Let (N,v) be a voting game and k a power index. Define (N_{ij},v_{ij}) by the merger of players i and j. The game satisfies the *property of large size* if $k_i(v) + k_j(v) \le k_{ij}(v')$.

Definition 7 (Property of new members). Now define (N^+,v^+) as an extension of (N,v) by parties $n+1,\ldots,m$ such that $\mathscr{W}\setminus\mathscr{W}^+ = \emptyset$. The *property or new members* is satisfied if $k_i(N,v) \ge k_i(N^+,v^+)$, that is, the introduction of new members should not increase a party's power.

Starting from a voting game (N, v) consider the game (N, v^{ij}), which only differs in the fact that players i and j refuse to cooperate, thus

$$\mathscr{W}^{ij} = \{S \in \mathscr{W} \,|\, \{i, j\} \not\subseteq S\}$$

Definition 8 (Property of quarrelling members [29]). If two parties refuse to vote together, this should not increase their total power. Formally $k_h(N, v) \geq k_h(N, v^{ij})$ for $h \in i, j$.

Note that the game defined here is not a proper voting game, since necessarily $N \notin \mathscr{W}^{ij}$, so the grand coalition is not a feasible winning coalition.

2.4   Axioms and Paradoxes

The properties above turn into paradoxes when we find that none of the well-known power indices satisfy all these. In fact the Shapley-Shubik and Banzhaf indices fail all three [4]. This result is well-known, and we leave it to the reader to find examples of games where the paradoxes appear. Unfortunately the paradoxes do not only appear in made-up examples, but van Deemen and Rusinowska found numerous real life instances in Dutch politics [7].

In fact, we show that the Null Player Axiom implies the Paradox of New Members and therefore any index based on the aforementioned axiom will be unreliable in circumstances where the player set is likely to expand. Unfortunately, yet again, such examples are common, in fact, the most common application of power measures is the EU Council of Ministers that is expected to expand further in the coming years, moreover the recent surge of interest in power indices is largely due to the "problems" caused by the extensions.

Theorem 3. *The Null Player Axiom implies the Paradox of New Members.*

Proof. Consider a power index k that satisfies the Null Player Axiom. For such an index all players that do not contribute to any of the winning coalitions receive a value of 0. Now consider an extension of the game and we show that for all games there is an extension such that a null player becomes non-null.

On the other hand the Paradox of New Members implies that there exist games that fail the Property of New Members. We therefore restrict our attention to weighted voting games of the form $(N, (w_i)_{i \in N}, q)$. Consider an arbitrary game with a null player that we denote by i. Instead of minimal winning coalitions consider maximal losing coalitions, in the following sense

$$L(i) \in \arg\max_{\substack{S \ni i \\ S \notin \mathscr{W}}} \sum_{j \in S} w_j, \tag{10}$$

That is, losing coalitions that lose with the smallest margin. Then consider the extension $(N \cup \{k\}, (w_i)_{i \in N \cup \{k\}}, q')$ where w_k and q' are determined as

$$w_k = (q' - q) + \left(q - \sum_{j \in L(i)} w_j \right) \tag{11}$$

Which can be reorganised as follows:

$$q' = w_k + \sum_{j \in L(i)} w_j. \tag{12}$$

For w_k sufficiently large (or, rather, not too small) the game is a proper simple game. On the other hand observe that the coalition $L(i) \cup \{k\}$ is a minimal winning coalition in the extended game, moreover, per definition, $i \in L(i) \cup \{k\}$. In a minimal winning coalition all players are critical, so i is critical in $L(i) \cup \{k\}$ and a player that is critical in any coalition is not null. Therefore i is not null in the extended game $(N \cup \{k\}, (w_i)_{i \in N \cup \{k\}}, q')$ and therefore its power increases as a result of a new member. Therefore the index exhibits the Paradox of New Members.

In the more general case considering winning coalitions, but no voting weights the argument is not so easy to present formally and would rely on some additional assumptions, but it is clear that if a new member is very weak, the coalition $\{i, k\}$ is losing, while if it is very powerful k can be a dictator. Assuming a "smooth" change in the size of k (and the corresponding update of $\mathscr{W}$) at one point the coalition $\{i, k\}$ becomes winning, while k, in itself is not yet, and then the player i is critical.

To illustrate that such extensions are not completely artificial, consider the EU Council of Ministers. While in the early days of –what is now called– the EU most decisions were made unanimously, already in the 6-member Community the rules of qualified majority voting to make decisions were laid down. The weights were specified as follows: 4 votes for each of the large members (Germany, France, Italy), 2 for the medium-sized members (Belgium, Netherlands) and 1 for the smallest country, Luxembourg. The Shapley-Shubik and Banzhaf indices are presented in Table 1. Observe that Luxembourg is a null player.

Table 1. The 1958 Council of Ministers and two possible expansions.

Country	votes	Original Φ_i	β_i	votes	Expansion I Φ'_i	β'_i	votes	Expansion II Φ''_i	β''_i
Germany	4	23.3%	23.8%	4	22.4%	21.6%	4	21.9%	22.0%
France	4	23.3%	23.8%	4	22.4%	21.6%	4	21.9%	22.0%
Italy	4	23.3%	23.8%	4	22.4%	21.6%	4	21.9%	22.0%
Netherlands	2	15.0%	14.3%	2	10.7%	11.8%	2	13.6%	12.2%
Belgium	2	15.0%	14.3%	2	10.7%	11.8%	2	13.6%	12.2%
Luxembourg	1	0	0	1	5.7%	5.9%	1	3.6%	4.9%
new member				1	5.7%	5.9%	1	3.6%	4.9%
Total	17	100.0%	100.0%	18	100.0%	100.0%	18	100.0%	100.0%
Quota	12			12			13		

Now consider an extension by a small new member, who receives the same number of votes as Luxembourg.[4] We look at two scenarios depending on the quota in the new union. Observe that in either of these scenarios Luxembourg is not any more a null player, its power has increased. While this is a hypothetical extension, since Denmark, Ireland and the UK have joined the EU none of the players are null.

3   Conclusion

In democracies decisions are commonly made by (qualified majority) voting, be those decisions in a national parliament, at a shareholder meeting or university senate. When the voters are of different sizes voting weights can be used to express the differences and the voting rules can then be expressed as qualified majority voting. As soon as the voting situation is asymmetric it is not so straightforward to see the actual influence of a single voter in the voting process not to mention complex voting situations such as in the UN Security Council, where the permanent members have veto rights [25, Chapter XII], or the International Monetary Fund where some countries vote directly some via voting coalitions that themselves share power in varying ways [28]. The topic has entered even the popular press when the extension of the EU made an update to voting in the Council of Ministers necessary. What is then the power of the individual member states? The Council uses currently a very complicated voting mechanism with three criteria, so it is hardly surprising that one needs powerful tools to evaluate this and similar voting situations.

The theory of a priory measures of voting power has developed much since Shapley and Shubik, and it does have a number of answers to such and similar questions. Unfortunately more than one answer has been given and a selection is far from obvious. The commonly used indices have been axiomatised by a number of elementary properties hoping that based on the attractiveness of simple properties a choice can be made. On the other hand a number of shortcomings of current theories have surfaced and been summarised as "paradoxes". The word paradox may be too strong, the behaviour these properties describe is certainly odd. In this paper we have shown that the Paradox of New Members is a direct consequence of the Null Player Axiom therefore of one insists on the latter, the first is not at all paradoxical.

Acknowledgements. The author thanks the funding of the OTKA (Hungarian Fund for Scientific Research) for the project "The Strong the Weak and the Cunning: Power and Strategy in Voting Games" (NF-72610) and of the European Commission under a Marie Curie Reintegration Grant (PERG-GA-2008-230879).

---

[4] This was a realistic scenario in the late 60s: Soon after the formation of the EC, Denmark, Ireland, Norway and the UK have applied. The application of the UK was vetoed by France, Norway withdrew due to a referendum, while the remaining two due to the veto on the UK. Had any of the smaller countries joined, the result would have been close to one of the above alternatives.

References

1. Banzhaf III, J.F.: Weighted voting doesn't work: A mathematical analysis. Rutgers Law Review 19(2), 317–343 (1965)
2. Bilbao, J.M., Edelman, P.H.: The Shapley value on convex geometries. Discrete Applied Mathematics 103, 33–40 (2000)
3. Bilbao, J.M., Jiménez, A., López, J.J.: The Banzhaf power index on convex geometries. Mathematical Social Sciences 36, 157–173 (1998)
4. Brams, S.J.: Game theory and politics, 2nd edn. Dover, Mineola (1975/2003)
5. Coleman, J.S.: Control of collectives and the power of a collectivity to act. In: Lieberman, B. (ed.) Social Choice, pp. 192–225. Gordon and Breach, New York (1971)
6. Deegan, J., Packel, E.W.: A new index of power for simple n-person games. International Journal of Game Theory 7(2), 113–123 (1978)
7. van Deemen, A., Rusinowska, A.: Paradoxes of voting power in Dutch politics. Public Choice 115(1-2), 109–137 (2003)
8. Diermeier, D., Merlo, A.: An empirical investigation of coalitional bargaining procedures. Journal of Public Economics 88(3-4), 783–797 (2004)
9. Dubey, P.: On the uniqueness of the Shapley value. International Journal of Game Theory 4(3), 131–139 (1975),
   http://www.springerlink.com/content/k5up7485268n3353/
10. Dubey, P., Shapley, L.S.: Mathematical properties of the Banzhaf index. Mathematics of Operations Research 4(2), 99–131 (1979)
11. Felsenthal, D.S., Machover, M.: Postulates and paradoxes of relative voting power – A critical re-appraisal. Theory and Decision 38(2), 195–229 (1995)
12. Felsenthal, D.S., Machover, M.: The Measurement of Voting Power: Theory and Practice, Problems and Paradoxes. Edward Elgar, Cheltenham (1998)
13. Felsenthal, D.S., Machover, M.: A priori voting power: What is it all about? Political Studies Review 2(1), 1–23 (2004)
14. Fréchette, G.R., Kagel, J.H., Morelli, M.: Gamson's Law versus non-cooperative bargaining theory. Games and Economic Behavior 51(2), 365–390 (2005)
15. Gamson, W.A.: A theory of coalition formation. American Sociological Review 26(3), 373–382 (1961)
16. Gelman, A., Katz, J.N., Bafumi, J.: Standard voting power indexes do not work: An empirical analysis. British Journal of Political Science 34, 657–674 (2004)
17. Gul, F.: Bargaining foundations of Shapley value. Econometrica 57(1), 81–95 (1989)
18. Holler, M.J., Packel, E.W.: Power, luck and the right index. Journal of Economics (Zeitschrift für Nationalökonomie) 43(1), 21–29 (1983)
19. Johnston, R.J.: On the measurement of power: Some reactions to Laver. Environment and Planning A 10(8), 907–914 (1978)
20. Kóczy, L.Á.: Proportional power is free from paradoxes. Working Paper 0806, Budapest Tech, Keleti Faculty of Economics, Budapest (2008)
21. Kóczy, L.Á.: Strategic power indices: Quarrelling in coalitions. Working paper 0803, Budapest Tech, Keleti Faculty of Economics, Budapest (2008)
22. Laruelle, A., Valenciano, F.: A critical reappraisal of some voting paradoxes. Public Choice 125(1), 14–41 (2005)
23. von Neumann, J., Morgenstern, O.: Theory of Games and Economic Behavior. Princeton University Press, Princeton (1944)
24. Neyman, A., Tauman, Y.: The partition value. Mathematics of Operations Research 4(3), 236–264 (1979)

25. Owen, G.: Game Theory, 3rd edn. Academic Press, San Diego (1995)
26. Penrose, L.S.: The elementary statistics of majority voting. Journal of the Royal Statistical Society 109(1), 53–57 (1946)
27. Pérez-Castrillo, J.D., Wettstein, D.: Bidding for the surplus: A non-cooperative approach to the Shapley value. Journal of Economic Theory 96(1/2), 1–21 (2001)
28. Reynaud, J., Lange, F., Gatarek, L., Thimann, C.: Proximity in coalition building. Working Paper Series 0808, Budapest Tech, Keleti Faculty of Economics (2007), http://ideas.repec.org/p/pkk/wpaper/0808.html
29. Riker, W.H.: The Theory of Political Coalitions. Yale University Press, New Haven (1962)
30. Shapley, L.S.: A value for n-person games. In: Kuhn, H.W., Tucker, A.W. (eds.) Contributions to the Theory of Games II. Annals of Mathematics Studies, vol. 28, pp. 307–317. Princeton University Press, Princeton (1953)
31. Shapley, L.S., Shubik, M.: A method for evaluating the distribution of power in a committee system. American Political Science Review 48(3), 787–792 (1954)
32. Straffin Jr., P.D.: Power and stability in politics. In: Aumann, R.J., Hart, S. (eds.) Handbook of Game Theory with Economic Applications, ch. 32, vol. II, pp. 1127–1151. Elsevier, Amsterdam (1994)

On Generation of Digital Fuzzy Parametric Conjunctions

Ildar Z. Batyrshin[1], Imre J. Rudas[2], and Alexandra Panova[3]

[1] Mexican Petroleum Institute
   batyr1@gmail.com
[2] Budapest Tech, Hungary
   rudas@bmf.hu
[3] Mexican Polytechnic Institute, Computer Research Center
   panova2@list.ru

Abstract. A new method of generation of digital fuzzy parametric conjunctions by means of basic t-norms is proposed. Fuzzy conjunctions also referred to as conjunctors and semicopulas. Digital fuzzy conjunctions are defined on the set of integer membership values $L=\{0,1,2,\dots,2^m-1\}$, where m is a number of bits used in presentation of membership values and $I = 2^m-1$ denotes a maximal membership value corresponding to 1 in traditional set of true values $L= [0,1]$. The proposed method referred to as the monotone sum of fuzzy conjunctions generalizes the method of ordinal sum of t-norms and gives possibility to construct a wide class of digital fuzzy parametric conjunctions that have effective digital hardware implementation. The classes of simplest commutative digital fuzzy parametric conjunctions obtained by this method are described. These classes of operations can constitute a part of a library of basic blocks for generation of digital fuzzy systems.

Keywords: Digital fuzzy conjunction, conjunctor, semicopula, finite scale, t-norm, commutativity.

1   Introduction

In construction of digital fuzzy systems it is convenient instead of traditional set on membership values $L=[0,1]$ to use the set of true values $L= \{0,1,2,\dots,I\}$, where $I= 2^m-1$ and m is a number of bits used in presentation of membership values. Such replacement of the set of true values on the one hand simplifies digital hardware implementation of fuzzy logic operations. On the other hand digital representation of true values preserves most of traditional definitions and properties of fuzzy concepts.

For hardware implementation of digital systems it is important to have the basic blocks that can be used as "bricks" in construction of digital system [1]. As such blocks in construction of digital fuzzy conjunctions in [2] they are used basic t-norms and t-conorms together with the set of simple generators. These basic functions have simple and effective digital hardware implementation because they

I.J. Rudas et al. (Eds.): Towards Intelligent Engineering & Information Tech., SCI 243, pp. 79–89.
springerlink.com   © Springer-Verlag Berlin Heidelberg 2009

use only simple operations such as comparison, minimum, maximum, bounded sum and bounded difference.

In this paper we introduce a new method of generation of digital fuzzy parametric conjunctions referred to as a monotone sum of fuzzy conjunctions that generates fuzzy conjunctions by means of only basic t-norms (drastic, Lukasiewicz and minimum t-norms) and one parameter p that can take values in the set of true values L. This method generalizes the method of generation of t-norms by means of ordinal sum of t-norms [3-6]. We describe all classes of commutative digital fuzzy parametric conjunctions that can be obtained as monotone sum of basic t-norms with one parameter. These classes of operations can constitute a part of a library of simple basic blocks for generation of digital fuzzy systems.

The paper has the following structure. Section 2 contains basic definitions of digital fuzzy conjunctions. Section 3 proposes a new method of generation of digital fuzzy parametric conjunctions referred to as the monotone sum of basic t-norms. Section 4 describes classes of commutative digital fuzzy parametric conjunctions that can be obtained as monotone sum of basic t-norms with parameter p. Section 5 describes the classes of commutative digital fuzzy parametric conjunctions obtained by monotone sum of basic t-norms when the value of parameter p is used together with the value I - p in definition of partition of domain $L \times L$ on sections. In Conclusion we discuss obtained results and future directions of research.

2   Basic Definitions

Suppose m bits are used in digital representation of membership values $L = \{0, 1, 2, \ldots, 2^m - 1\}$ with maximal value I = $2^m - 1$. This value will represent the full membership corresponding to the value 1 in traditional set of membership values [0,1]. For example, for m = 4 we have I = 15. Most of definitions of fuzzy operations have straightforward extension on digital case by replacing the set of membership values [0,1] by $L = \{0, 1, 2, \ldots, 2^m - 1\}$ and maximal membership value 1 by I.

Fuzzy conjunction operation is a function $T : L \times L \rightarrow L$ satisfying on L conditions:

$$T(x,I) = x, \qquad T(I,y) = y, \qquad \text{(boundary conditions)}$$

$$T(x,y) \leq T(u,v), \qquad \text{if } x \leq u,\ y \leq v. \qquad \text{(monotonicity)}$$

Fuzzy conjunctions have been studied in [7, 8]. Such functions are also referred to as conjunctors [6] or semicopulas [9]. Note that from the properties of fuzzy conjunctions it follows:

$$T(x,0) = 0, \quad T(0,y) = 0.$$

T-norms are fuzzy conjunctions satisfying on L commutativity and associativity conditions [3]:

$$T(x,y)=T(y,x),$$ (commutativity)

$$T(x,T(y,z)) = T(T(x,y),z).$$ (associativity)

Associativity property of fuzzy conjunctions usually does not used in applied fuzzy systems. For this reason we do not require associativity from digital fuzzy conjunctions. Note also that most of known associative parametric t-norms are sufficiently complicated for digital hardware implementation and we consider here the methods of generation of simple digital fuzzy parametric conjunctions. Commutativity of fuzzy conjunctions can be useful in some applications of fuzzy systems so we will describe further the classes of such operations obtained by proposed methods.

Below are basic t-norms [3] that will be used further in generation of digital fuzzy parametric conjunctions:

$$T_M(x,y) = min\{x,y\},$$ (minimum)

$$T_L(x,y)= max\{x+y-I, 0\},$$ (Lukasiewicz t-norm)

$$T_D(x,y) = \begin{cases} x, & if\ y = I \\ y, & if\ x = I \\ 0, & if\ x, y < I \end{cases}.$$ (drastic product)

Lukasiewicz t-norm is also known as a bounded product [10]. The definition of these t-norms uses only simple operations like comparison, minimum, maximum, bounded sum and bounded difference, for this reason these t-norms have efficient digital hardware implementation [11].

It can be shown that any fuzzy conjunction T satisfies the following inequalities:

$$T_D(x,y) \leq T(x,y) \leq T_M(x,y).$$ (1)

We will say that $T_1 \leq T_2$ if $T_1(x,y) \leq T_2(x,y)$ for all x,y from L. For example, we have:

$$T_D \leq T_L \leq T_M.$$ (2)

An example of parametric t-norm having efficient digital hardware implementation is the Mayor-Torrens t-norm [3] depending on parameter $p \in L$:

$$T(x,y) = \begin{cases} max(x+y-p,0), & if\ p > 0, \quad x \leq p, y \leq p \\ min(x,y), & otherwise \end{cases}.$$ (3)

Another example of parametric fuzzy conjunction having efficient hardware implementation gives the following fuzzy conjunction introduced in [8] and depending on two parameters $p, q \in L$:

$$T(x, y) = \begin{cases} \min(x, y), & if \quad p \le x \; or \; q \le y \\ 0, & otherwise \end{cases}. \qquad (4)$$

This conjunction will be a t-norm when $p = q$. Both parametric t-norm and fuzzy conjunction considered above are obtained by using suitable generator function in some way. The methods of generation of digital fuzzy parametric conjunctions by means of basic t-norms, t-conorms and simple generators are considered in [2]. In the following section we propose new method of generation of digital fuzzy parametric conjunctions without generator functions by means of basic t-norms and one parameter p.

3   Monotone Sum of Digital Fuzzy Conjunctions

A set X containing one number or a sequence of consecutive numbers from L is called an interval in L. Let $J = \{1,...,n\}$, $1 < n \le I+1$, be a set of indexes and $(X_j)_{j \in J}$, is a partition of L on pairwise disjoint intervals such that from $i < j$ it follows $x < y$ for all $x \in X_i$ and $y \in X_j$. Denote $D_{ij} = X_i \times X_j$. Suppose Q is some set of indexes and $(T_q, \le)_{q \in Q}$ is a partially ordered set of fuzzy conjunctions. Assign to each D_{ij} some $T_{ij} = T_q$ from this set such that

$$T_{ij}(x, y) \le T_{st}(u, v) \quad if \quad i \le s, j \le t, \text{ and } x \le u, y \le v, \quad x, y \in D_{ij}, u, v \in D_{st}. \quad (5)$$

Define a function T on $L \times L$ by $T(x, y) = T_{ij}(x, y)$ if $(x, y) \in D_{ij}$, $i, j \in J$. Then T is called the monotone sum of $(D_{ij}, T_{ij})_{i,j \in J}$ or monotone sum of fuzzy conjunctions T_{ij}, $i, j \in J$.

Theorem 1. A monotone sum of fuzzy conjunctions is a fuzzy conjunction.

Proof. Since all fuzzy conjunctions used in construction of T satisfy boundary conditions $T(x, 1) = x$, $T(1, y) = 1$ then a resulting function T also will satisfy these conditions. Monotonicity of T follows from definition (5).

Note that the monotone sum of fuzzy conjunctions can be considered as a generalization of the ordinal sum of t-norms [3-6]. The ordinal sum can be obtained from monotone sum by suitable selection of diagonal sections $(D_{ii}, T_{ii})_{i \in J}$ when T_M is used in other sections.

Two conjunctions T_{ij} and T_{st} satisfying condition (5) in the definition of monotone sum may be not related to each other by ordering relation $\le$ on all domain $L \times L$. If in the definition above we will replace condition (5) by the more simple condition:

$$T_{ij} \le T_{st} \quad if \quad i \le s \quad \text{and} \quad j \le t \qquad (6)$$

then due to monotonicity of fuzzy conjunctions from (6) it will follow (5). The function T defined by (6) will be referred to as simple monotone sum of $(D_{ij}, T_{ij})_{i,j \in J}$.

Corollary 2. A simple monotone sum of fuzzy conjunctions is a fuzzy conjunction.

Based on Theorem 1 we will consider two methods of construction of monotone sum of fuzzy conjunctions using one parameter $p \in L$. Fig. 1 illustrates the methods of partition of L on intervals X_j and partition of $L \times L$ on corresponding sections $D_{ij} = X_i \times X_j$. In the first method we define the partition of L as follows (see Fig. 1, on the left): $X_1 = [0,p]$, $X_2 = [p+1,I]$, $p \in \{0,1,\ldots, 2^m-2\}$. In the second method partitions are defined as follows (see Fig. 1, on the right): $X_1 = [0,p]$, $X_2 = [p+1,I-p]$, $X_3 = [I-p+1,I]$, $p \in \{0, 1,\ldots, 2^{m-1}-1\}$. In the second method we have $p < I-p$ and for $m= 4$, $I= 15$, p can take values $0,1,\ldots,7$. The first method will be referred to as (p)-monotone sum and the second method as $(p,I-p)$-monotone sum.

4 (p)-Monotone Sum of Digital Fuzzy Conjunctions

(p)-monotone sum of fuzzy conjunctions can be defined as follows. Select a set of fuzzy conjunctions $\{T_{11}, T_{21}, T_{12}, T_{22}\}$ ordered as follows: $T_{11} \leq T_{12} \leq T_{22}, T_{11} \leq T_{21} \leq T_{22}$. Define fuzzy conjunction T as follows (see also Fig. 1, on the left):

$$T(x,y) = \begin{cases} T_{11}(x,y), & if \quad x \leq p, \quad y \leq p \\ T_{21}(x,y), & if \quad x > p, \quad y \leq p \\ T_{12}(x,y), & if \quad x \leq p, \quad y > p \\ T_{22}(x,y), & if \quad x > p, \quad y > p \end{cases} \tag{7}$$

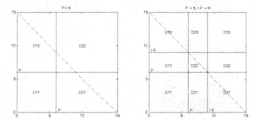

Fig. 1. Partition of $L \times L$ on sections $D_{ij} = X_i \times X_j$ defined by parameter values: 1) p (on the left); 2) p and $I-p$ (on the right)

Consider all nontrivial fuzzy commutative conjunctions that can be obtained by this method by means of basic t-norms T_D, T_L, T_M. Remember that these t-norms are ordered as follows: $T_D \leq T_L \leq T_M$. To obtain commutative conjunction it should be $T_{21} = T_{12}$. Fig. 2 depicts all commutative digital fuzzy parametric conjunctions obtained by considered method. On the left there are shown assignments of basic t-norms to sections D_{ij}. On the right there are shown the shapes of obtained digital fuzzy parametric conjunctions for parameter value $p= 9$. Here we use four bits for digital representation of conjunctions: $m= 4$, and $I= 15$. Each of obtained fuzzy conjunctions coded by the name depending on the list of t-norms used in its definition: $(T_{11}, T_{21}, T_{12}, T_{22})$. For example, conjunction T_{DLLM} is defined by the sequence of t-norms (T_D, T_L, T_L, T_M). It can be seen that some of obtained fuzzy conjunctions will be t-norms, for example conjunctions T_{DDDM} and T_{DMMM} are t-norms (see [3] and (4) above). An example of non-commutative digital fuzzy conjunction is depicted in Fig. 3.

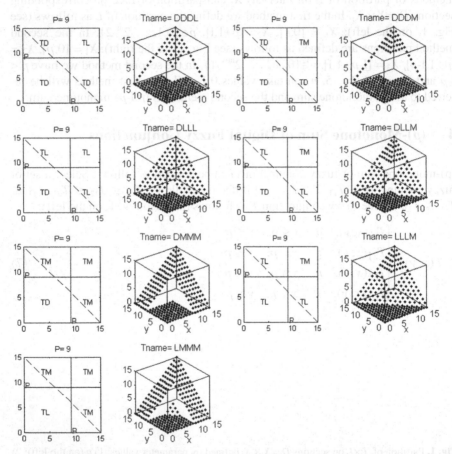

Fig. 2. Simplest commutative digital fuzzy parametric conjunctions obtained by (p)-monotone sum of basic t-norms: TD - drastic, TL - Lukasiewicz and TM – minimum t-norms

Note that both commutative and non-commutative digital fuzzy parametric conjunctions obtained from basic t-norms by means of (7) have very simple digital hardware implementation. We need to have blocks implementing basic t-norms and we need to have selector that will compare values of x and y with the value of parameter p and select corresponding t-norm given in the sequence $(T_{11}, T_{21}, T_{12}, T_{22})$.

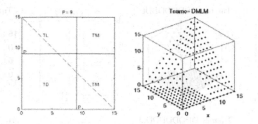

Fig. 3. Example of non-commutative digital fuzzy parametric conjunction

5   (p,I-p)-Monotone Sum of Digital Fuzzy Conjunctions

$(p,\mathrm{I}\text{-}p)$-monotone sum of fuzzy conjunctions can be defined as follows (see Fig. 1 on the right). Select a set of fuzzy conjunctions $\{T_{11}, T_{21}, T_{31}, T_{12}, T_{22}, T_{32}, T_{13}, T_{23}$

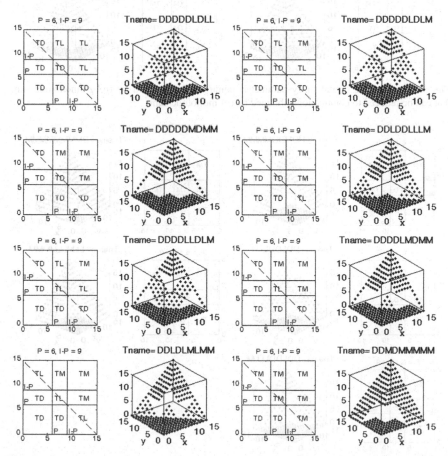

Fig. 4. Non-trivial commutative digital fuzzy parametric conjunctions obtained by $(p,\mathrm{I}\text{-}p)$-monotone sum

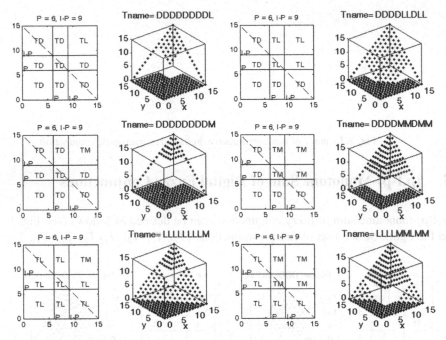

Fig. 5. Commutative digital fuzzy parametric conjunctions obtained by $(p,\text{I-}p)$-monotone sum that can be reduced to conjunctions obtained by (p)-monotone sum

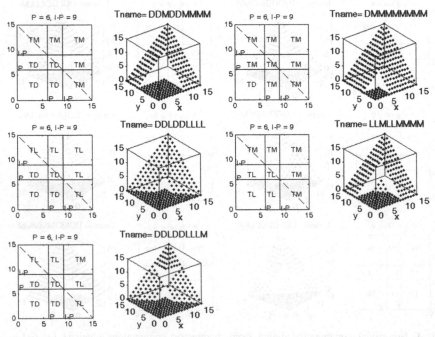

Fig. 6. Commutative digital fuzzy parametric conjunctions obtained by $(p,\text{I-}p)$-monotone sum that can be reduced to conjunctions obtained by (p)-monotone sum

$T_{33}\}$ ordered as follows: $T_{ij} \leq T_{st}$ if $i \leq s$ or $j \leq t$. Define fuzzy conjunction T as follows:

$$
T(x,y) = \begin{cases}
T_{11}(x,y), & \text{if} & x \leq p, & y \leq p \\
T_{21}(x,y), & \text{if} & p < x \leq I-p, & y \leq p \\
T_{31}(x,y), & \text{if} & x > I-p, & y \leq p \\
T_{12}(x,y), & \text{if} & x \leq p, & p < y \leq I-p \\
T_{22}(x,y), & \text{if} & p < x \leq I-p, & p < y \leq I-p \\
T_{32}(x,y), & \text{if} & x > I-p, & p < y \leq I-p \\
T_{13}(x,y), & \text{if} & x \leq p, & y > p \\
T_{23}(x,y), & \text{if} & p < x \leq I-p, & y > p \\
T_{33}(x,y), & \text{if} & x > p, & y > p
\end{cases}
$$

Consider all nontrivial fuzzy commutative conjunctions that can be obtained by this method by means of basic t-norms T_D, T_L, T_M. To obtain commutative conjunction it should be $T_{21} = T_{12}$, $T_{31} = T_{13}$ and $T_{32} = T_{23}$. Fig. 4 depicts all non-trivial commutative digital fuzzy parametric conjunctions obtained by this method. Note that because the maximal value of p in this method is less than the value of I-p, the values of both T_D and T_L equal 0 in sections D_{11}, D_{21} and D_{12}. For this reason any two conjunctions coinciding in all other sections and containing in sections D_{11}, D_{21} and D_{12} only t-norms T_D and T_L will be equal.

Other commutative conjunctions that can be obtained by $(p,I$-$p)$-monotone sum of basic t-norms can be reduced to some of conjunctions obtained by (p)-monotone sum of basic t-norms. These conjunctions are presented in Figs. 5, 6. For example, conjunctions $DDDDDDDDL$ and $DDDDLLDLL$ obtained by $(p,I$-$p)$-monotone sum and shown on the top of Fig. 5 are partial cases of conjunction $DDDL$ obtained by (p)-monotone sum and shown on the top of Fig. 2. Remember that in the first two conjunctions parameter p is changing only till the mean of the scale L but in the last case this parameter can take any value from L less than I.

Conclusions

Digital fuzzy conjunctions considered in this work are related with conjunctors [6] and semicopulas [9] defined on the set $L= [0,1]$, with t-norms defined on finite ordinal scales [12,13] and with T-seminorms on partially ordered sets [14]. It gives possibility to extend many results obtained for these operations on digital fuzzy conjunctions and vice versa. For example, the method of generation of fuzzy conjunctions by monotone sums introduced in this work can be directly applied to conjunctors and semicopulas. As it was mentioned above the method of generation of fuzzy conjunctions by monotone sum of fuzzy conjunctions generalizes the method of ordinal sum used for generation of t-norms [3, 5]. It would be interesting to study more carefully relationships between the method of monotone sum used here for generation of fuzzy conjunctions and existing methods of generation of t-norms, e.g. ordinal sums of t-norms.

Acknowledgments. The research work was partially supported by IMP project D.00507, Bilateral CONACYT-NKTH project No. I0110/127/08 Mod. Ord. 38/08, by ICyTDF funding, award No. PICCT08-22, and by matching funding by IPN, award No. SIP/DF/2007/143.

References

[1] Tocci, R.J., Widmer, N.S., Moss, G.L.: Digital Systems: Principles and Applications, 9th edn. Prentice-Hall, Englewood Cliffs (2003)

[2] Rudas, I.J., Batyrshin, I.Z., Hernández Zavala, A., Camacho Nieto, O., Villa Vargas, L.: Digital Fuzzy Parametric Conjunctions for Hardware Implementation of Fuzzy Systems. In: Rudas, I.J. (ed.) Towards Intelligent Engineering and Information Technology. SCI, vol. 243. Springer, Heidelberg (2009)

[3] Klement, E.P., Mesiar, R., Pap, E.: Triangular Norms. Kluwer Academic Publishers, Dordrecht (2000)

[4] Birkhoff, G.: Lattice Theory. American Mathematical Society, Providence (1973)

[5] Saminger, S., Klement, E.P., Mesiar, R.: A Note on Ordinal Sums of t-Norms on Bounded Lattices. EUSFLAT – LFA, pp. 385–388 (2005)

[6] Saminger, S., De Meyer, H.: On the Dominance Relation between Ordinal Sums of Conjunctors. Kybernetika 42, 337–350 (2006)

[7] Batyrshin, I., Kaynak, O., Rudas, I.: Generalized Conjunction and Disjunction Operations for Fuzzy Control. In: Proc. 6th European Congress on Intelligent Techniques & Soft Computing, EUFIT 1998, Aachen, Germany, vol. 1, pp. 52–57 (1998)

[8] Batyrshin, I., Kaynak, O.: Parametric Classes of Generalized Conjunction and Disjunction Operations for Fuzzy Modeling. IEEE Transactions on Fuzzy Systems 7, 586–596 (1999)

[9] Durante, F., Quesada–Molina, J.J., Sempi, C.: Semicopulas: Characterizations and Applicability. Kybernetika 42, 287–302 (2006)

[10] Jang, J.-S.R., Sun, C.T., Mizutani, E.: Neuro-Fuzzy and Soft Computing. A Computational Approach to Learning and Machine Intelligence. Prentice-Hall International, Englewood Cliffs (1997)

[11] Batyrshin, I., Hernández Zavala, A., Camacho Nieto, O., Villa Vargas, L.: Generalized Fuzzy Operations for Digital Hardware Implementation. In: Gelbukh, A., Kuri Morales, Á.F. (eds.) MICAI 2007. LNCS (LNAI), vol. 4827, pp. 9–18. Springer, Heidelberg (2007)

[12] De Baets, B., Fodor, J., Ruiz-Aguilera, D., Torrens, J.: Idempotent Uninorms on Fi-
      nite Ordinal Scales. Internatioal Journal of Uncertainty, Fuzzyness and Knowledge-
      based Systems 17, 1–14 (2009)
[13] Batyrshin, I.Z., Batyrshin, I.I.: Negations, Strict Monotonic t-Norms and t-Conorms
      for Finite Ordinal Scales. In: 10th Int. Conf. Fuzzy Theory Technology. 8th Joint
      Conf. on Inform. Sciences, JCIS. Salt Lake City, USA, pp. 50–53 (2005)
[14] De Cooman, G., Kerre, E.E.: Order Norms on Bounded Partially Ordered Sets. J.
      Fuzzy Math. 2, 281–310 (1994)

Part II
Control

Quantum Control Systems

From the von Neumann Architecture to Quantum Computing

László Nádai and József Bokor

Computer and Automation Research Institute, Hungarian Academy of Sciences
Kende u. 13-17, H-1111 Budapest, Hungary
{nadai,bokor}@sztaki.hu

Abstract. In this paper we consider the (state) reachability and controllability problems of special two-level quantum systems, the so-called quantum bits via externally applied electro-magnetic field. The system is described by special a bilinear right-invariant model whose state varies on the Lie group of 2×2 special unitary matrices. We show that if two or more independent controls are used, then every state can be achieved in arbitrary time using finite energy. The mathematical construction is motivated by the demand of manipulating (or logically operating on) the state of quantum bits, and the results provide some insight into the feasibility of realizing given operations in quantum computers.

1 Introduction

John von Neumann (born December 28, 1903 in Budapest, Austria-Hungary; died February 8, 1957 in Washington D.C., United States) was a Hungarian-born mathematician and polymath who made contributions to many mathematics-related fields as one of history's outstanding mathematicians. Most notably, von Neumann was a pioneer of the application of operator theory to quantum mechanics, and the commonly known von Neumann architecture is the de facto standard of nowadays computers.

In the paper we first outline von Neumann's contributions to both fields, then we show that the ever-increasing need for computer speed inevitably leads to the formulation of a new scientific and technological field, the so-called quantum information technology. However, operating on information physically associated to quantum mechanical phenomena is fundamentally different to "classical" computer technology, and several theoretical and pragmatical questions have to be answered. Out of them we aim at the (state) reachability and controllability problems of

I.J. Rudas et al. (Eds.): Towards Intelligent Engineering & Information Tech., SCI 243, pp. 93–102.
springerlink.com   © Springer-Verlag Berlin Heidelberg 2009

quantum bits using the ideas of another famous Hungarian Rudolph E. Kalman. We will show that if two or more independent controls (externally applied electro-magnetic fields) are used, then every state of the quantum bit can be achieved in arbitrary time using finite energy.

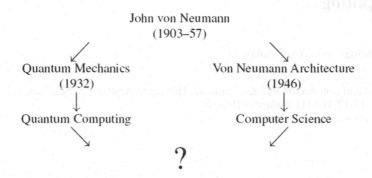

Fig. 1. The "heritage" of John von Neumann – quantum information technology

2   Quantum Mechanics

At the *International Congress of Mathematicians of 1900*, David Hilbert presented his famous list of twenty-three problems considered central for the development of the mathematics of the new century. The sixth of these was the axiomatization of physical theories. Among the new physical theories of the century the only one which had yet to receive such a treatment by the end of the 1930's was quantum mechanics. QM found itself in a condition of foundational crisis similar to that of set theory at the beginning of the century, facing problems of both philosophical and technical natures. On the one hand, its apparent non-determinism had not been reduced to an explanation of a deterministic form. On the other, there still existed two independent but equivalent heuristic formulations, the so-called *matrix mechanical formulation* due to Werner Heisenberg and the *wave mechanical formulation* due to Erwin Schrödinger, but there was not yet a single, unified satisfactory theoretical formulation.

After having completed the axiomatization of set theory, von Neumann began to confront the axiomatization of QM. He immediately realized, in 1926, that a quantum system could be considered as a point in a so-called *Hilbert-space*, analogous to the $6N$ dimension (N is the number of particles, 3 general coordinate and 3 canonical momentum for each) phase space of classical mechanics but with infinitely many dimensions (corresponding to the infinitely many possible states of the system) instead: the traditional physical quantities (e.g. position and momentum) could therefore be represented as particular linear operators operating in these spaces. The physics of quantum mechanics was thereby reduced to the mathematics of the linear Hermitian operators on Hilbert spaces. For example, the famous *indeterminacy principle of Heisenberg*, according to which the determination of the position of a particle prevents the determination of its momentum and vice versa, is translated into the non-commutativity of the two corresponding operators.

This new mathematical formulation included as special cases the formulations of both Heisenberg and Schrdinger, and culminated in the 1932 classic *The Mathematical Foundations of Quantum Mechanics* [Grundlagen]. However, physicists generally ended up preferring another approach to that of von Neumann (which was considered elegant and satisfactory by mathematicians). This approach was formulated in 1930 by Paul Dirac and was based upon a strange type of function (the so-called *Dirac delta function*) which was harshly criticized by von Neumann.

In any case, von Neumann's abstract treatment permitted him also to confront the foundational issue of determinism vs. non-determinism and in the book he demonstrated a theorem according to which quantum mechanics could not possibly be derived by statistical approximation from a deterministic theory of the type used in classical mechanics. This demonstration contained a conceptual error, but it helped to inaugurate a line of research which, through the work of John Stuart Bell in 1964 on *Bell's Theorem* [3] and the experiments of Alain Aspect in 1982 [2], demonstrated that quantum physics requires a notion of reality substantially different from that of classical physics.

3   Computer Science

The earliest computing machines had fixed programs. Some very simple computers still use this design, either for simplicity or training purposes. To change the program of such a machine, one have to re-wire, re-structure, or even re-design the machine. Indeed, the earliest computers were not so much "programmed" as they were "designed". "Reprogramming", when it was possible at all, was a very manual process, starting with flow charts and paper notes, followed by detailed engineering designs, and then the often-arduous process of implementing the physical changes.

The idea of the stored-program computer changed all that. In 1945 John von Neumann published a now-famous paper, the *First Draft of a Report on the EDVAC* [4], describing a computer architecture in which data and program memory are mapped into the same address space. The *von Neumann architecture* became the de facto standard and can be contrasted with the so-called Harvard architecture, which has separate program and data memories on a separate bus. By creating an instruction set architecture and detailing the computation as a series of instructions (the program), the machine becomes much more flexible. By treating those instructions in the same way as data, a stored-program machine can easily change the program, and can do so under program control. *The majority of home computers, microcomputers, minicomputers and mainframe computers use the single-memory (a.k.a. Von Neumann) computer architecture.*

However, the ever-increasing need for computer speed drives the design of integrated circuits into the direction of using smaller and smaller units of physical state-space to represent units of information. This technological process in the near future will lead to reach atomic (or sub-atomic) dimensions, therefore there is a need for a new paradigm in computer design: the so-called *quantum information technology*.

Von Neumann worked together with other immigrant Hungarian scientists, who also participated in the development of computer science.

- János Kemény (1926-92) as the Rector of Dartmouth Collage obligated students of arts and laws to use personal computers, and for helping them formulated the *BASIC* language. Kemény is also known for the time-shared computer network that was honored by the first Robinson Prize of IBM.
- Von Neumann was also collaborating with Leó Szilárd who introduced the term *bit* as the elementary unit of information (yes/no).
- This list would not be complete without Andy Grove (Gróf András), who was nominated as "The Man of the Year" in 1997 by the journal *Time*. He was the CEO of INTEL, and almost yearly doubled the speed of its microprocessors.

The ever-increasing need for computer speed implies using *smaller and smaller units of physical state-space* to represent units of information. This technological process in the near future will lead to reach *atomic (or sub-atomic) dimensions*. There is a need for a new paradigm: *quantum information technology*.

4   Quantum Information Technology

A *quantum bit*, or *qubit* (sometimes qbit) is a unit of quantum information. That information is described by a state vector in a two-level quantum mechanical system which is formally equivalent to a two-dimensional vector space over the complex numbers.

Benjamin Schumacher discovered a way of interpreting quantum states as information. He came up with a way of compressing the information in a state, and storing the information on a smaller number of states. This is now known as *Schumacher compression*. In the acknowledgments of his paper [Schumacher], Schumacher states that the term qubit was invented in jest, during his conversations with Bill Wootters.

A bit is the base of computer information. Regardless of its physical representation, it is always read as either a '0' or a '1'. An analogy to this is a light switch – the down position can represent '0' (normally equated to *off*) and the up position can represent '1' (normally equated to *on*).

A qubit has some similarities to a classical bit, but is overall very different. Like a bit, a qubit can have only two possible values – normally a '0' or a '1'. The difference is that whereas a bit must be either '0' *or* '1', a qubit can be '0', '1', or a *superposition* of both.

4.1 Physical Representation

Any two-level system can be used as a qubit. Multilevel systems can be used as well, if they possess two states can be effectively decoupled from the rest (e.g., ground state and first excited state of a nonlinear oscillator). There are various proposals. Several physical implementations which approximate two-level systems to various degrees were successfully realized. Similarly to a classical bit where the state of a transistor in a processor, the magnetization of a surface in a hard disk and the presence of current in a cable can all be used to represent bits in the same computer, an eventual *quantum computer* is likely to use various combinations of qubits in its design. *Table tab:qubit* contains an incomplete list of possible physical implementation of qubits.

Table 1. Possible physical implementation of qubits

Physical support	Name	Information support	'0'	'1'
Single photon (Fock states)	Polarization encoding	Polarization of light	Horizontal	Vertical
	Photon number	Photon number	Vacuum	Single photon state
	Time-bin encoding	Time of arrival	Early	Late
Coherent state of light	Squeezed light	Quadrature	Amplitude-squeezed state	Phase-squeezed state
Electrons	Electronic spin	Spin	Up	Down
	Electron number	Charge	No electron	One electron
Nucleus	Nuclear spin addr. through NMR	Spin	Up	Down
Optical lattices	Atomic spin	Spin	Up	Down
Josephson junction	Superconducting charge qubit	Charge	Uncharged superconducting island (Q=0)	Charged super-conducting island (Q=2e)
	Superconducting flux qubit	Current	Clockwise current	Counterclockwise current
Singly-charged quantum dot pair	Electron localization	Charge	Electron on left dot	Electron on right dot

4.2 Mathematical Representation

Each physical system is associated with a (topologically) *separable complex Hilbert-space* H with inner product $\langle \psi \mid \phi \rangle$. Physical observables are represented by densely-defined *self-adjoint operators* on H. The expected value (in the sense of probability theory) of the observable A for the system in state represented by the unit vector $\mid \psi \rangle \in H$ is $\langle \psi \mid A \mid \psi \rangle$.

The states a qubit may be measured in are known as basis states (or vectors). As is the tradition with any sort of quantum states, *Dirac* (or bra-ket) *notation* is used to represent them. This means that the two computational basis states are conventionally written as $|0\rangle$ and $|1\rangle$.

The state at any time t is given by:

$$\psi(t) = X(t)\psi(0),\tag{1}$$

where X is the so-called *evolution operator* (matrix), solution of the Schrdinger-equation

$$i\hbar\dot{X} = HX.\tag{2}$$

with initial condition equal to the identity operator.

A pure qubit state is a linear superposition of those two states. This means that the qubit can be represented as a *linear combination* of $|0\rangle$ and $|1\rangle$:

$$|\psi\rangle = \alpha\,|0\rangle + \beta\,|1\rangle$$

where α and β are *probability amplitudes* and can in general both be complex numbers.

When we measure this qubit in the standard basis, the probability of outcome $|0\rangle$ is $|\alpha|^2$ and the probability that the outcome is $|1\rangle$ is $|\beta|^2$. Because the absolute squares of the amplitudes equate to probabilities, it follows that α and β must be constrained by the equation

$$|\alpha|^2 + |\beta|^2 = 1,$$

simply because this ensures you must measure *either* one state *or* the other.

In implementations of quantum computers, the operation given by the evolutionary operator X represents a (logic) operation to be performed on a quantum bit, i.e. the *reachability* question can be seen as the feasibility of logic operations on a quantum bit. In practical terms: can all operations be achieved on a quantum bit by opportunely shaping an input electro/magnetic field? (Using finite energy.)

5   Reachability of Quantum States

Consider the Schrdinger equation for the evolution operator (2) in the common situation where the Hamiltonian operator H can be written as $H = H_0 + \sum_{i=1}^{m} H_i u_i$,

$$i\hbar\dot{X} = \left(H_0 + \sum_{i=1}^{m} H_i u_i(t) \right) X.\tag{3}$$

The operators H_i ($i = 0, \ldots, m$) are Hermitian operators on a finite dimensional vector space and the overall phase of the solution of (3) does not have physical meaning. Considering the above, Eq. (3) can be transformed into a differential system [6] of the form

$$\dot{X}(t) = AX(t) + \sum_{i=1}^{m} B_i X(t) u_i(t), \tag{4}$$

where A, B_i are elements of the Lie algebra of 2×2 skew-Hermitian matrices with zero trace, which is denoted by su(2).

Definition 1. Lie group is a group G that is also a differentiable manifold such that for any $a, b \in G$ the multiplication $(a, b) \mapsto ab$ and the inverse $a \mapsto a'$ are smooth maps.

Proposition 1. *All compact finite dimensional Lie groups can be represented as matrix Lie groups.*

The solution of (4) with initial condition equal to identity varies in the Lie group associated to su(2), namely in the Lie group of 2×2 unitary matrices with determinant 1. This group is called the group of *special unitary matrices* and is denoted by SU(2).

Definition 2. The set of reachable states $R(T)$ consists of all the possible values for $X(T)$ (solution of (4) at time T with initial condition equal to identity) obtained varying the controls $u_1, \ldots, u_m$ in the set of all the piecewise continuous functions defined in $[0, T]$.

Theorem 1 [7]. *Consider system (4) with $3 \geq m \geq 2$ and assume that $B_1, \ldots, B_m$ are linearly independent. Then, for any time $T > 0$ and for any desired final state X_f there exist a set of piecewise continuous control functions $u_1, \ldots, u_m$ driving the state of the system X to $X(T) = X_f$ at time T. This means that in this case $R(T) = SU(2)$ for every $T > 0$.*

We present a general approach to derive reachability/controllability results like was given in Theorem 1.

6   Open-Loop Unconstrained Controllability

First we cite the Kalman controllability results for LTI systems. The fundamental matrix for zero initial time is

$$\Phi(t) = e^{At} = \sum_{i=1}^{n} \psi_i(t) A^{i-1},$$

and the reachability subspace is

$$\mathcal{R} = \sum_{k=0}^{n-1} \operatorname{Im} A^k B.$$

Proposition 2. *It is possible to generate linearly independent functions* $\psi_i, i = 1, \ldots, n$ *if the Kalman-rank condition*

$$rank\left[B, AB, \ldots, A^{n-1}B \right] = n$$

is satisfied.

We show a general method for systems over Lie groups using Lie algebraic approach. Write the fundamental matrix (locally) as exponential function of the "coordinates of second kind" associated with the equation

$$\dot{x} = \sum_{i=0}^{N} \rho_i(t) A_i x.$$

Using the Wei--Norman equation:

$$\dot{g}(t) = \left(\sum_{i=1}^{K} e^{\Gamma_1 g_1} \cdots e^{\Gamma_{i-1} g_{i-1}} E_{ii} \right)^{-1} \rho(t), \quad g(0) = 0,$$

where $\{\hat{A}_1, \ldots, \hat{A}_K\}$ is a basis of the Lie-algebra $\mathcal{L}(A_1, \ldots, A_N)$,

$$[\hat{A}_i, \hat{A}_j] = \sum_{l=1}^{K} \Gamma_{i,j}^{l} \hat{A}_l, \quad \Gamma_i = [\Gamma_{i,j}^{l}]_{j,l=1}^{K}.$$

Proposition 3 (Generalized Kalman-rank condition). *For systems* $A(\rho), B(\rho)$
the points attainable from the origin are those from the subspace spanned by the vectors

$$\mathcal{R}_{(\mathcal{A},\mathcal{B})} := \operatorname{span} \left\{ \prod_{j=1}^{K} A_{l_j}^{i_j} B_k \right\}$$

where $K \geq 0, l_j, k \in \{0, \cdots, N\}, i_j \in \{0, \cdots, n-1\}$, i.e., $\mathcal{R} \subset \mathcal{R}_{(\mathcal{A},\mathcal{B})}$.

Denote by $\mathcal{L}(A_0, \ldots, A_N)$ the finitely generated Lie-algebra containing the matrices $A_0, \ldots, A_N$, and let $\hat{A}_1, \ldots, \hat{A}_K$ be a basis of this algebra, then the points attainable from the origin are in the subspace

$$\mathcal{R}_{(A,B)} = \sum_{l=0}^{N} \sum_{n_1=0}^{n-1} \cdots \sum_{n_K=0}^{n-1} \operatorname{Im}(\hat{A}_1^{n_1} \cdots \hat{A}_K^{n_K} B_l).$$

The question is that under what condition is $\mathcal{R} = \mathcal{R}_{A,B}$?

The fundamental matrix can be written in exponential form:

$$\Phi(t) = \sum_{n_1=0}^{n-1} \cdots \sum_{n_K=0}^{n-1} \hat{A}_1^{n_1} \ldots \hat{A}_K^{n_K} \psi_{n_1,\cdots,n_K}(t).$$

$$\Phi(t) = \sum_{\mathbf{j} \in \mathbf{J}} \hat{A}_{\mathbf{j}} \varphi_{\mathbf{j}}(t), \quad \hat{A}_{\mathbf{j}} := \hat{A}_1^{j_1} \ldots \hat{A}_K^{j_K}.$$

The subspace $\mathcal{R}_{A,B}$ is the image space of the matrix $R_{A,B} := [\hat{A}_{\mathbf{j}} B]_{\mathbf{j} \in \mathbf{J}}$. The controllability Grammian is given as

$$W(\sigma, \tau) = R_{A,B} \left(\int_{\sigma}^{\tau} [\varphi_{\mathbf{j}}(s)]_{\mathbf{j} \in \mathbf{J}} [\varphi_{\mathbf{j}}(s)]_{\mathbf{j} \in \mathbf{J}}^* ds \right) R_{A,B}^*.$$

Theorem 2 [8]. *The quantum system is controllable, iff*
(i) The generalized Kalman-rank condition is satisfied:

$$\operatorname{rank} R_{\mathcal{A},\mathcal{B}} = \operatorname{rank} [\hat{A}_{\mathbf{j}} B]_{\mathbf{j} \in \mathbf{J}}$$

(ii) The set of functions $\{\varphi_j(\sigma) | \mathbf{j} \in \mathbf{J}\}$ contains n linearly independent functions.

7 Switching System's Controllability

Hybrid models characterize systems governed by continuous differential and difference equations and discrete variables. Such systems are described by several operating regimes (modes) and the transition from one mode to another is governed by the evolution of internal or external variables or events.

Depending on the nature of the events there are two big classes of hybrid systems that are considered in the control literature: *switching systems* and *impulsive systems*.

A switching system is composed of a family of different (smooth) *dynamic modes* such that the switching pattern gives continuous, piecewise smooth trajectories. Moreover, it is assumed that one and only one mode is active at each time instant.

In a broader sense every time-varying system with measurable variations in time can be cast as a switching system, therefore it is usually assumed that the number of switching modes is finite and for practical reasons the possible switching functions (sequences) are restricted to be piecewise constant, i.e. only a finite number of transition is allowed on a finite interval. Moreover, sometimes the frequency of the transitions is also bounded – dwell time.

Formally, these systems can be described as:

$$\dot{x}(t) = f_{\sigma(t)}(x(t), u(t)),$$

$$y(t) = h_{\sigma(t)}(x(t), u(t)), \quad x(\tau^+) = \iota(x(\tau^-), u(\tau), \tau),$$

where $x \in \mathbb{R}^n$ is the state variable, $u \in \Omega \subset \mathbb{R}^m$ is the input variable and $y \in \mathbb{R}^p$ is the output variable.

The $\sigma : \mathbb{R}^+ \to S$ is a measurable switching function mapping the positive real line into $S = \{1, \cdots, s\}$. The impulsive effect can be described by the relation $(\tau, x(\tau^-)) \in \mathcal{I} \times \mathcal{A}$ with $\mathcal{I}$ a set of time instances and $\mathcal{A} \in \mathbb{R}^n$ a certain region of the state space.

Consider a bimodal system

$$\dot{X} = A_{\sigma(t)} X, \quad X(0) = I, \quad \sigma(t): \mathbb{R}^+ \mapsto \{1, 2\}$$

and $A_1, A_2 \in \mathcal{U}(n)$ that is SU(n).

A set of gates is called *universal* if – by switching $\{A_1, A_2\}$ – it is possible to generate all (special) unitary evolutions.

Proposition 4 [8]. *Since A_1, A_2 generate the whole Lie-algebra $u(n)$ or $su(n)$, therefore almost every couple of skew-Hermitian matrices generate $u(n)$, i.e. almost every quantum gate is universal.*

8 Outlook

We certainly know that the above results are only the first steps into the direction of physically realizing quantum computers. There are several theoretical problems to be solved, among them we outline state-estimation, state-observation, and *Kalman filtering*, that are all necessary to relax the inconsistency between the fundamental indeterministic (or statistical) character of quantum mechanics, and the natural need against computers to execute exact calculations.

Besides the above, there are practical considerations that should be taken into account in order to build functionally adequate quantum computers: the control of multi-level quantum systems, minimal-time control, minimal-energy control.

References

[1] von Neumann, J.: Mathematische Grundlagen der Quantenmechanik. Mathematical Foundations of Quantum Mechanics. Springer, Berlin (1932)
[2] Aspect, A., Dalibard, J., Roger, G.: Experimental Test of Bell's inequalities Using Time-Varying Analyzers. Phys. Rev. Lett. 49(25), 1804 (1982)
[3] Bell, J.S.: On the Einstein-Podolsky-Rosen Paradox. Physics 1(3), 195 (1964)
[4] First Draft of a Report on the EDVAC by John von Neumann. Contract No. W-670-ORD-4926, Between the United States Army Ordnance Department and the University of Pennsylvania Moore School of Electrical Engineering, University of Pennsylvania, June 30 (1945)
[5] Schumacher, B.: Quantum Coding. Phys. Rev. A 51, 2738–2747 (1995)
[6] Ramakrishna, V., Salapaka, M.V., Dahleh, M., Rabitz, H., Peirce, A.: Controllability of Molecular Systems. Phys. Rev. A 51(2), 960–966 (1995)
[7] D'Alessandro, D., Dahleh, M.: Optimal Control of Two-Level Quantum Systems. In: Proc. American Control Conference, Chicago, IL (June 2000)
[8] Bokor, J., Szabó, Z.: Controllability of Linear Time Varying Systems: the Affine Linear Parameter Varying case, Technical Report, MTA SZTAKI, Revision 3 (February 2005)

Model-Based Control Design of Integrated Vehicle Systems

Péter Gáspár

Systems and Control Laboratory, Computer and Automation Research Institute, Hungarian
Academy of Sciences, Hungary
gaspar@sztaki.hu

Abstract. This paper presents a model-based control design for an integrated vehicle system in which several active components are operated in co-operation. In
control-oriented modeling vehicle dynamics is augmented with the performance
specifications of the controlled system and the uncertainties of the model. In control
design performance specifications must be formalized in such a way that the performance demands are guaranteed, conflicts between performances are achieved
and a priority between different actuators is created and various fault information is
taken into consideration. As an illustration an integrated control, which includes an
active steering, active anti-roll bars and an active brake system, is proposed for
tracking the path of the vehicle and guaranteeing road holding and roll stability. In
the paper the operations of the integrated vehicle systems are presented in simulated
vehicle maneuvers.

Keywords: Vehicle dynamics and control, integrated vehicle control, performances, uncertainties, robust control.

1 Introduction

Conventionally, the control systems of vehicle functions to be controlled are designed separately. These control systems contain hardware components such as
sensors, actuators, communication links, power electronics, switches and mi-
cro-processors. One of the problems of independent design is that the performance
demands, which are met by independent controllers, are sometimes in conflict with
each other in terms of the full vehicle. The braking action affects the longitudinal
dynamics of the vehicle, the velocity and the pitch angle. However, due to the
geometry of the vehicle, the braking action causes changes in both the yaw and roll

I.J. Rudas et al. (Eds.): Towards Intelligent Engineering & Information Tech., SCI 243, pp. 103–119.
springerlink.com   © Springer-Verlag Berlin Heidelberg 2009

dynamics, see Figure 1a. Similarly, the steering angle also modifies the yaw angle of the vehicle. Since the center of gravity is high the consequence of the steering maneuver is that the roll angle and the pitch angle of the sprung mass will also change, see Figure 1b. Moreover, the second problem in the independent control design is that control hardware can be grouped into discrete subsets with sensor information and control demands operating in parallel processes and these solutions can lead to unnecessary hardware redundancy. The integrated control methodologies were presented in [15, 20, 24, 33]. The purpose of the integrated vehicle control is to combine and supervise all controllable subsystems affecting vehicle dynamic responses. In more details it means that multiple-objective performance from available actuators must be improved, sensors must be used in several control tasks, the number of independent control systems must be reduced, at the same time the flexibility of control systems must be enhanced.

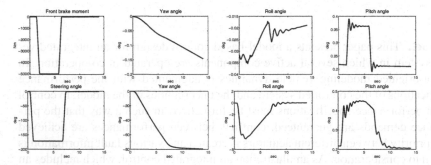

Fig. 1. The effects of braking and steering on vehicle dynamics (brake moment/steering angle, yaw angle, roll angle and pitch angle

An integrated control system is designed in such a way that the effects of a control system on other vehicle functions are taken into consideration in the design process by selecting the various performance specifications. In line with the requirements of the vehicle industry several performance specifications are in the focus of the research, e.g. improving road holding, enhancing passenger comfort and improving roll and pitch stability, proposing fault-tolerant solutions, see [14]. Another solution to the integrated control is to design a high-level controller which is able to supervise the effects of individual control components on vehicle dynamics. In this case there is a logical relationship between the supervisor and the individually-designed local controllers, thus the control design leads to hybrid and switching methods with a large number of theoretical difficulties. Significant open problems are the construction of the hybrid control architectures, the analysis of the influence of communication mechanisms, distributed computational algorithms or the dynamic task management, see [6].

Several researchers have focused on the integration of control systems. A combined use of brakes and rear-steer to augment the driver's front-steer input in controlling the yaw dynamics is proposed by [20, 23]. A four-wheel steer and four-wheel drive ($4WS/4WD$) controller via feed-forward and feedback compensators is proposed by [19]. A possible integration of the brake, steering and suspension system is presented by [33]. An integrated control by using steering and suspension systems is created by [22]. A global chassis control involving an active suspension and ABS to improve road holding and passenger comfort is proposed by [25, 37]. A road-friendly solution of the chassis control must also be provided in order to reduce dynamic loads on roads, see e.g. [8]. The adaptive cruise control is also a traffic-friendly integrated vehicle system, in which the vehicle follows a preceding vehicle along the highway at a safe distance and a desired speed. The model-based control design requires a vehicle model in which the powertrain system (motor, gear-box and retarder) and the brake are integrated, e.g. [2, 26, 31]. An important feature of integrated vehicle control is that it is able to adapt to changes during operation and to different fault operations. Fault scenarios are the consequence of a sensor failure, the lock failure in the actuator or the loss in effectiveness. There are also numerous papers dealing with the design of reconfigurable controls, which include FDI filters, reconfigurable controllers and reconfiguration mechanisms, see e.g. [10].

Model-based control for integrated vehicle systems is formalized in the following steps. First a vehicle model which contains longitudinal, lateral and vertical dynamics is formalized. The vehicle model contains nonlinear components, i.e. the dynamics of the dampers and springs and relatively fast nonlinear actuator dynamics, see e.g. [1]. The difficulty in the integrated control is that it usually yields high-complexity vehicle systems. Second the model is augmented with performance specifications of the controlled system. They contain formalisms of the possible conflicts and constrains of the system, e.g. actuator saturation. The integrated vehicle control is not a simple control in which several inputs must be designed simultaneously. Here the priority between actuators must also be taken into consideration. Then the model uncertainty that is covered by unmodelled dynamics and parametric uncertainties is also built into the model. Finally, in the closed-loop interconnection structure weighting functions with parameter-dependent gains are designed in order to guarantee all of the performances and the trade-off between performances, see e.g. [3, 4]. Since the model is of high-complexity, the control design procedure often leads to numerical difficulties.

The structure of the paper is as follows. In Section 2 the motivation example is presented. In Section 3 the control-oriented vehicle model with the vertical, yaw and roll dynamics is constructed. In Section 4 the performance specifications and the uncertainty structures are formalized in a closed-loop interconnection structure. In Section 5 the integrated control mechanism is presented through simulation examples. Finally, Section 6 contains some concluding remarks.

2 Motivation Example

Consider the following example, which is a motivation example of the integrated vehicle control. The objective of the control design is to minimize the tracking error and prevent rollovers. The chassis control integrates the active steering and the active brake. A possible solution to the tracking problem is proposed by using active steering and in this operation the objective is to follow a pre-defined path, i.e. minimize the tracking error between the predefined and the actual paths of the vehicle. In normal cruising the brake is not activated. When a rollover is imminent and this emergency persists the brake system is activated to reduce the rollover risk. When the brake is used, however, the real path significantly deviates from the desired path due to the brake moment, which affects the yaw motion. This deviation must be compensated by the active steering system. To perform tracking and rollover prevention at the same time poses a difficult problem since these tasks are in contradiction with each other. A possible solution to reduce the interaction between performance specifications is to apply a third control input. Thus, active anti-roll bars, which focus only on the rollover problem, are also applied in the integrated vehicle system. The active anti-roll bars generate a stabilizing moment to balance an overturning moment and they operate all the time to improve roll stability and road holding.

In the following the main steps of a control design of an integrated vehicle system with steering, brake and anti-roll bars are presented. When the vehicle is cruising it only performs the tracking. In an emergency when a rollover is imminent, it also performs the prevention of rollovers. The combined yaw-roll model, which is the basis of the control design, is nonlinear with respect to the forward velocity of the vehicle and the adhesion coefficient. The model is augmented with the signals defined by the performance specifications and the uncertainty structure defined by the difference between the plant and its model. The active brake switches on an emergency and it switches off when the emergency is over. In order to compensate for deviation of the yaw motion caused by braking a slight modification of the tracking command for the steering subsystem is needed to avoid under- or oversteering.

The example is also motivated by the adaptive cruise control. The predefined distance between the two vehicles is presented in Figure 2. During the simulation the leading vehicle modifies its velocity, it performs acceleration, travelling at a constant speed and deceleration. Moreover, the a-priori defined distance between the two vehicles must be maintained. According to the specification the distance must be constant in the first part of the simulation and then it must be increased. The functions of the powertrain system are the driving force, the engine revolution, the gearshift of the transmission. The control design of this system is a difficult multi-objective optimization problem, see e.g. [11]. Since the distance between the vehicles must be increased according to the task, the brake must also be activated.

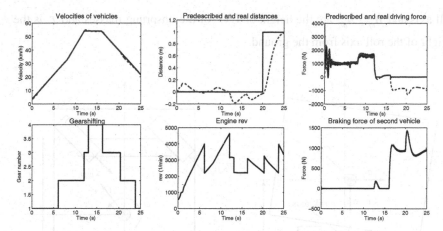

Fig. 2. Time responses of the adaptive cruise control (velocity, distance, throttle force, gearshift, engine revolution, brake force)

3 Control-Oriented Modeling of Vehicle Dynamics

In vehicle modeling the motion differential equations of the combined yaw and roll dynamics of the single-unit vehicle are formalized. Figure 3 illustrates the combined yaw and roll dynamics of the vehicle, which is modelled by a three-body system with a sprung mass m_s and two unsprung masses at the front m_{uf} and the rear m_{ur} including the wheels and axles. The vehicle can translate longitudinally and laterally and it can also yaw. The sprung and the unsprung masses can rotate around a horizontal axis. The suspension springs, dampers and active anti-roll bars generate moments between the sprung and unsprung masses in response to roll motions. The tires produce lateral forces that vary linearly with the side-slip angles.

The signals involved are the lateral acceleration a_y, the side slip angle of the vehicle body β, the heading angle of the sprung mass ψ, the yaw rate $\dot{\psi}$, the roll angle ϕ, the roll rate $\dot{\phi}$, the roll angle of the unsprung mass at the front axle ϕ_{tf} and at the rear axle ϕ_{tr}. δ_f is the front wheel steering angle. v is the forward velocity. The roll motion of the sprung mass is damped by suspensions with damping coefficients b_{sf}, b_{sr} and stiffness coefficients k_{sf}, k_{sr}. The tire stiffnesses are denoted by k_{tf} and k_{tr}. I_{xx}, I_{xz}, I_{zz} are the roll moments of the inertia of the sprung mass, the yaw-roll product, and the yaw moment of inertia, respectively. h is the height of the center of gravity (CG) of the sprung mass from roll axis and h_{uf}, h_{ur} are the height of CG of the unsprung masses and r is the height of the roll axis from the ground.

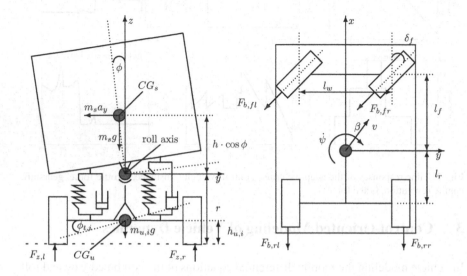

Fig. 3. The full car model with yaw and roll dynamics

The control inputs are the front-wheel steering angle δ_f the difference between the brake forces of the vehicle ΔF_b and the roll moments at the front and at the rear generated by the active anti-roll bars u_f, and u_r. The control input provided by the brake system generates a yaw moment, which affects the lateral tire forces directly. The motion differential equations are the following. The first equation is a lateral force balance, the second equation is a yaw moment balance for the entire vehicle, the third equation is the balance of roll moments on the sprung mass, and the last two equations are the balances of roll motions of the front and rear unsprung masses.

$$mv(\dot{\beta}+\dot{\psi})-m_s h\ddot{\phi} = F_{yf} + F_{yr}, \tag{1a}$$

$$-I_{xz}\ddot{\phi}+I_{zz}\ddot{\psi} = F_{yf}l_f - F_{yr}l_r + l_w\Delta F_b, \tag{1b}$$

$$(I_{xx}+m_s h^2)\ddot{\phi}-I_{xz}\ddot{\psi} = m_s gh\phi + m_s vh(\dot{\beta}+\dot{\psi}) - F_{zf} + u_f - F_{zr} + u_r, \tag{1c}$$

$$-rF_{yf} = m_{uf}v(r - h_{uf})(\dot{\beta} + \dot{\psi}) + m_{uf}gh_{uf}\phi_{t,f} - k_{tf}\phi_{tf} + F_{zf} + u_f, \quad (1d)$$

$$-rF_{yr} = m_{ur}v(r - h_{ur})(\dot{\beta} + \dot{\psi}) - m_{ur}gh_{ur}\phi_{tr} - k_{t,r}\phi_{tr} + F_{zr} + u_r. \quad (1e)$$

with the suspension forces $F_{zf} = k_f(\phi - \phi_{tf}) + b_f(\dot{\phi} - \dot{\phi}_{tf})$ and $F_{zr} = k_r(\phi - \phi_{tr}) + b_r(\dot{\phi} - \dot{\phi}_{t,r})$. The lateral tire forces F_{yf} and F_{yr} in the direction of the wheel ground contact are approximated linearly to the tire side-slip angles α_f and α_r, respectively: $F_{yf} = \mu C_f \alpha_f$ and $F_{yr} = \mu C_r \alpha_r$, where C_f and C_r are tire side slip constants and μ is the adhesion coefficient. The tire side slip angles for the front and rear wheels in the lateral and in the longitudinal directions are approximated as follows: $\alpha_f = -\beta + \delta_f - l_f \cdot \dot{\psi}/v$ and $\alpha_r = -\beta + l_r \cdot \dot{\psi}/v$.

These equations are expressed in a state space representation with the following state vector: $x = \begin{bmatrix} \beta & \dot{\psi} & \phi & \dot{\phi} & \phi_{t,f} & \phi_{t,r} \end{bmatrix}^T$. The vector of the control inputs is $u = \begin{bmatrix} \delta_f & \Delta F_b & u_f & u_r \end{bmatrix}$. The disturbance w is caused by the changing of the adhesion coefficient, side wind or road irregularities. Then the state equation arises in the following form: $E(v)\dot{x} = A_0(v,\mu)x + B_{10}w + B_{20}u$, where $E(v)$, which contains masses and inertias, is an invertible matrix.

The nonlinear effects of the forward velocity and those of the adhesion coefficient are taken into consideration in the vehicle dynamics. Forward velocity is approximately equivalent to the velocity in the longitudinal direction while the side-slip angle is small. It is assumed that the forward velocity is available, i.e. it is estimated on-line by using the on-board sensors, see e.g [29]. The adhesion coefficient depends on several factors, e.g. the type of road surface, the maneuver of the vehicle. The changes in the adhesion coefficient occurring at only one of the tires are ignored, their influence is taken into consideration in terms of the entire vehicle. Since this coefficient is time-varying, an adaptive observer-based grey-box identification method has been proposed for its estimation, see [17].

If forward velocity and the adhesion coefficient are selected as scheduling variables, the differential equations of the vehicle model are transformed into a Linear Parameter Varying (LPV) model:

$$\dot{x} = A(\rho)x + B_1(\rho)w + B_2(\rho)u \quad (2)$$

where $\qquad$ $\rho = \begin{bmatrix} \rho_1 & \rho_2 & \rho_3 & \rho_4 \end{bmatrix}$ $\qquad$ with

$\rho_1 = \mu, \rho_2 = \mu/v, \rho_3 = \mu/v^2, \rho_4 = 1/v$ $\qquad$ and

$A(\rho) = A_0 + \rho_1 A_1 + \rho_2 A_2 + \rho_3 A_3 + \rho_4 A_4,$ $\qquad$ $B_1(\rho) = B_{10} + \rho_1 B_{11},$

$B_2(\rho) = B_{20} + \rho_1 B_{21}.$

The LPV methods assume that the nonlinearities of vehicle systems can be hidden by scheduling signals, which are assumed to be measured or achieved. The LPV modeling approaches allow us to take into consideration the highly nonlinear effects in the state space description in such a way that the model structure is nonlinear in the parameters, but linear in the states. Furthermore this state space representation of the LPV model is valid in the entire operating region of interest The design is based on an H_∞ control synthesis extended to LPV systems that use parameter dependent Lyapunov functions. The LPV method in the early applications is based on a single Lyapunov function (SLF) approach, in which the variation of the scheduling variables can be arbitrarily fast, see [27, 12]. In later applications parameter dependent Lyapunov functions (PDLF) are applied. The motivation reason is that ignoring the bandwidth of the actuators or the signals leads to an infinite rate bound on the scheduling variables in an impractical way. If the rate bound is assumed a less conservative result for the class of systems is yielded, see [4, 5, 7, 36]. The advantage of LPV methods is that the controller meets robust stability and nominal performance demands in the entire operational interval, since the controller is able to adapt to the current operational conditions.

4   Control Design Based on an Interconnection Structure

The purpose of the integrated vehicle control is to track a predefined path of the vehicle, reduce the rollover risk and guarantee performances even in fault scenarios. Two variables must be monitored and added to the scheduling vector in order to improve the safety of the vehicle: a variable is needed to reduce the rollover risk and the harmful effects of the abrupt braking; another variable is also required to take a detected failure of an active component into consideration.

Roll stability is achieved by limiting the lateral load transfers on both axles, i.e. at the front and rear axles ΔF_{zi}, $i \in (f, r)$, to below the levels for wheel lift-off. The lateral load transfer is calculated: $\Delta F_{zi} = k_t \phi_{ti}$, where ϕ_{ti} is the roll angle of the unsprung mass. This requirement leads to the definition of the normalized lateral load transfer, which is the ratio of the lateral load transfer and the mass: $R_i = \Delta F_{zi}/(m_i g)$. If the R_i takes on the value ± 1 then the inner wheels in the bend lift off. The limit cornering condition occurs when the load on the inside wheels has dropped to zero and all the load has been transferred onto the outside wheels. This event does not necessary result in the rolling over of the vehicle. However, the aim of the control design is to prevent rollover in all cases and thus the lift-off of the wheels must also be prevented. Thus, the normalised load transfer is critical when the vehicle is stable but the tendency of the dynamics is unfavourable in terms of a rollover. The maximal value of the normalized lateral load transfers $\rho_R = \max\{R_f, R_r\}$ is selected as a scheduling variable. The aim of the control design is to reduce the normalized lateral load transfer ρ_R if it exceeds a predefined critical value.

In the calculation of the normalized lateral load transfer the roll angles of the unsprung masses both at the front and rear axles must be applied. The method requires the measurement of the roll angle of the unsprung masses, see [28]. However, this is not a cheap measurement to carry out. It would therefore be more useful if the roll rate or the roll acceleration could be measured using a rate gyro and fed back into the controller. A method is proposed for the estimation of the roll angles of the unsprung masses based on an observer design, see [16].

Since the fault-tolerant control requires fault information in order to guarantee performances it modifies its operation. It is assumed that the fault information is provided by a fault detection and isolation (FDI) filter in the following way: $\rho_f = f_{act} / f_{max}$, where f_{act} is an estimation of the failure (output of the FDI filter) and f_{max} is an estimation of the maximum value of the potential failure (fatal error). The estimated value f_{act} means the measure of the performance degradation of an active suspension component. The value of a possible fault is normalized into the interval $[0, 1]$. The value 0 corresponds to the non-faulty operation and the value 1 means that the active anti-roll bars are not able to work. This fault information is used in the reconfiguration mechanism. Significant research results have been published for the general FDI problem and several methods have been proposed, e.g. the parity space approach, the multiple model method, detection filter design using a geometric approach, or the dynamic inversion based detection, see [21, 13]. Most of the design approaches refer to linear, time-invariant (LTI) systems, but references to some nonlinear cases are also found in the literature, see [30, 9]. An H_∞ approach to design a fault detection and isolation gain-scheduled filter for LPV systems was presented by [7, 32].

The scheduling vector is augmented with two variables: $\rho = \begin{bmatrix} \rho_0 & \rho_R & \rho_f \end{bmatrix}^T$. It is noted that the monitored scheduling variables are not in the original scheduling vector, which is defined in order to transform the nonlinear vehicle model into an LPV form. The variables, which are used only for monitoring purposes, can be calculated by using additional components such as an FDI filter.

Consider the closed-loop system which includes the feedback structure of the model, the compensator and elements associated with the uncertainty models and performance objectives, see Figure 4. The command signal is a pre-defined yaw rate signal $\dot{\psi}_{cmd}$. The control inputs are the front wheel steering angle and the difference between the brake forces and the roll moments of the active anti-roll bars. The performance signals are the tracking error $e_{\dot{\psi}}$, the lateral acceleration a_y

and   the   control   inputs: $z = \begin{bmatrix} e_{\dot{\psi}} & a_y & u^T \end{bmatrix}$,   where

$u = \begin{bmatrix} \delta_f & \Delta F_b & u_f & u_r \end{bmatrix}^T$. The measured outputs are the lateral acceleration of the sprung mass a_y, the yaw rate $\dot{\psi}$ and the roll rate $\dot{\phi}$. The noises $n = \begin{bmatrix} n_a & n_{\dot{\psi}} \end{bmatrix}$ are from the measurements. In order to solve the yaw-rate tracking problem, the command signal must be fed forward to the controller. Hence, the measured output vector contains the yaw-rate command signal $\dot{\psi}_{cmd}$.

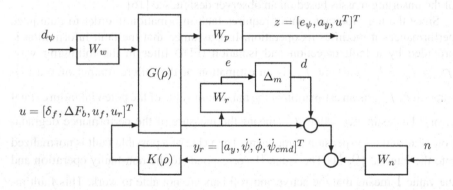

Fig. 4. The closed-loop structure for the design of an active steering

The mechanism of the integrated control requires a priority between the actuators, which can be guaranteed by using a weighting strategy. The weighting function for the tracking error is selected in such a way that in steady state the tracking error must be below an acceptable limit. A parameter dependent gain ϕ_{ay} is applied in the weighting function of the lateral acceleration, since the brake is activated only in an emergency, i.e. the normalized lateral load transfer ρ_R has reached its critical value. Thus, the normalized lateral load transfer is applied as a scheduling variable. The weighting functions of the tracking error and the lateral acceleration are selected as:

$$W_{p,e\dot{\psi}} = 0.05 \frac{0.05s+1}{10s+1} \tag{3a}$$

$$W_{p,ay} = \phi_{ay}(\rho_R, \rho_f)\frac{0.005s+1}{0.083s+1}. \tag{3b}$$

When the vehicle is in normal cruising the weighting function $W_{p,ay}$ should be small and when the lateral acceleration increases and the normalized lateral load transfer has reached its critical value the weighting function $W_{p,ay}$ should be large to avoid the rollover. Thus, a parameter-dependent gain $\phi_{ay}(\rho_R, \rho_f)$ in equation (3b), which reflects the relative importance of the normalized lateral load transfer is applied. A large gain $\phi_{ay}(\rho_R, \rho_f)$ corresponds to a design that reduces rollover risk. When the vehicle is in normal cruising $\phi_{ay}(\rho_R, \rho_f)$ is small and the minimization of lateral acceleration is not needed. In order to take into consideration a nonlinear function of the controller with respect to the operating domain a parameter dependent weighting function must be used. The parameter dependent gain $\phi_{ay}(\rho_R, \rho_f)$ is as follows:

$$\phi_{ay}(\rho_R, \rho_f) = \begin{cases} 0 & \text{if } |\rho_R| < R_1, \rho_f = 0, \text{or} |\rho_R| < R_3, \rho_f = 1 \\ \frac{1}{R_2-R_1}(|\rho_R|-R_1) & \text{if } R_1 \le |\rho_R| \le R_2, \rho_f = 0 \\ \frac{1}{R_4-R_3}(|\rho_R|-R_3) & \text{if } R_3 \le |\rho_R| \le R_4, \rho_f = 1 \\ 1 & |\rho_R| > R_2, \rho_f = 0, \text{or} |\rho_R| > R_4, \rho_f = 1 \end{cases}$$

Here, the normalized lateral load transfer is applied to monitor the rolling over of the vehicle. R_1 defines the critical status when the vehicle is close to rolling over, i.e. the active anti-roll bars are not capable of generating more stabilizing moment. The closer R_1 is to 1 the later the brake will be activated. Parameter R_2 shows how fast the brake should focus on minimizing the lateral acceleration. The selection of R_1 and R_2 is critical. If the brake is activated at a large R_1 the probability of rollover increases. If the value R_1 was small, the brake would be activated very frequently. Moreover, in the event of a fault the range of the operation of the brake system must be extended. In the fault case the design parameter R_3 (and R_4) are chosen from R_1 (and R_2) to be scheduled on fault information ρ_f: $R_3 = R_1 - 0.1\rho_f$ (and $R_4 = R_2 - 0.1\rho_f$) where ρ_f is the normalized value of the fault information.

In this paper the control inputs, i.e. control forces and moments are designed. The designed control inputs are required ones, which must be created by the electro-hydraulic actuators. E.g. the required force must be tracked by an actuator level controller. Note that for the sake of simplicity the dynamics of the actuators is ignored. An example of the solution of tracking problem by using an actuator is presented in [18].

In order to describe the control objective, the parameter dependent augmented plant $P(\varrho)$ must be built up using the closed-loop interconnection structure. The augmented plant $P(\varrho)$ includes the yaw-roll dynamics of the vehicle $G(\varrho)$ and the weighting functions. Using the controller K the closed-loop system $M(\varrho)$ is given by a lower Linear Fractional Transformation (LFT) structure: $M(\varrho) = F_\ell(P(\varrho), K(\varrho))$. The control goal is to minimize the induced L_2 norm of a LPV system $M(\varrho)$, with zero initial conditions, which is given by

$$\inf_{K} \sup_{\varrho \in F_P} \sup_{\|w\|_2 \neq 0, w \in L_2} \frac{\|z\|_2}{\|w\|_2}$$

The quadratic LPV performance problem is to choose the parameter-varying controller in such a way that the resultant closed-loop system is quadratically stable and the induced L_2 norm from w to z is less than γ. The existence of a controller that solves the quadratic LPV γ-performance problem can be expressed as the feasibility of a set of Linear Matrix Inequalities (LMIs), which can be solved numerically. The state space representation of the LPV control $K(\varrho)$ can be constructed, see [4, 5, 35, 36].

The constraints set by the LMIs are infinite dimensional, as is the solution space. The infinite-dimensionality of the constraints is relieved by a finite, sufficiently fine grid. To specify the grid of the performance weights for the LPV design the scheduling variables are defined through lookup-tables. The grid is determined by v, μ, ρ_R and ρ_f as follows: $v = [20, \ldots, 120]$, $\mu = [0.1, \ldots, 1.1]$, $\rho_R = [0, R_1, R_2, 1]$ and $\rho_f = [0, 1]$. The gridding reflects the qualitative changes of the performance weights, i.e. the scheduling variables. Weighting functions for both the performance and uncertainty specifications are defined in all of the grid points. The stability and performance are guaranteed by the LPV design process, see [4, 5, 35].

The realization of an LPV controller poses a problem, which must be handled. The control design is performed in continuous-time, in which it is assumed that the scheduling variable ϱ is known in continuous-time. ϱ is measured only at sampling times. Instead of having a fixed dependence of system matrices on ϱ, the matrices are only known at discrete ϱ values. The suitable sampling time must be selected according to the physical system; however, the real sampling time is modified by the implementation possibilities. The theoretical investigation of the quantization effects has not been completed. Thus the determination of the parameters during the intervals between sampling times is a difficult theoretical problem. A simple procedure applied in practice uses a zero-order hold method between sampling times, see [34. A better solution of the approximation is based on polynomial or rational functions through curve fitting.

If parameter-dependent Lyapunov functions are used, the controller designed depends explicitly on $\dot{\varrho}$. Thus, in order to construct a parameter-dependent controller, both ϱ and $\dot{\varrho}$ must be measured or available. When $\dot{\varrho}$ is not measured in practice, a suitable extrapolation algorithm must be used to achieve an estimation of the parameter $\dot{\varrho}$. The disadvantage of this approach is that the sources of the scheduling variables are not independent. Balas et al. proposed a possible method to perform a ϱ-dependent change of variables to remove $\dot{\varrho}$ dependence, see [4].

5 Simulation Examples

In this section, the operation of the integrated control is illustrated in a double-lane-changing manoeuver, which is defined by the signal yaw-rate. The size of the path deviation is chosen to test a real obstacle avoidance in an emergency on a road. The maneuver has a 2 m path deviation over 100 m. The velocity of the vehicle is 90 $kmph$. The steering angle is a ramp signal which reaches the maximum value (3.5 deg) in 0.5 sec and filtered at 4 rad/s to represent the finite bandwidth of the driver. The steering angle input is generated in such a way that the vehicle with no roll control comes close to a rollover during the maneuver and its normalized load transfers are above the value ± 1.

Figure 5 shows the time responses in cases when active anti-roll bars (dash-dot), an active braking mechanism (dash) and an integrated controller (solid) are used. The controller must guarantee both the tracking performance of yaw rate command and rollover prevention. The tracking error of all the control cases is acceptable in the yaw rate channel. Using only active anti-roll bars the control does not have a direct effect on the change in acceleration. However, they generate stabilizing lateral displacement moments, which balance the destabilizing overturning moment caused by lateral acceleration. The control torque is approximately 60 kNm at both the front and the rear axles. In the case of an active brake, the lateral acceleration is the same as when the normalized load transfers do not reach the critical value, which is determined by $R_1 (= 0.75)$ but once the critical value has been exceeded the control algorithm is activated and the active brake system reduces the lateral acceleration. The time when the brake control is activated can be seen in the brake force figure, which shows that the rear-left wheel is braked to avoid the rollover of the vehicle. The brake forces are 100 kN at the left and 20 kN at the right at the rear axle.

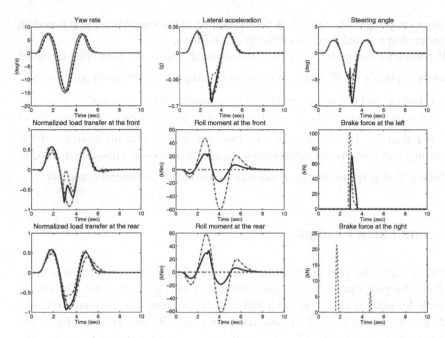

Fig. 5. Time responses of the steering maneuver (yaw rate, lateral acceleration, steering angle, normalized lateral load transfers both at the front and the rear, roll moments at the front and the rear and the brake force at the left and the right on the rear side)

In the case of the integrated control, only the active anti-roll bars are working and generating a stabilizing lateral displacement moment when the normalized load transfer does not reach the critical value and the brake system is not activated. The integrated control can be considered as simple active anti-roll bars when the normalized lateral load transfer is less than R_1. Hence, the brake force required to prevent the rollover of the vehicle is less than when using only the brake system. It can also be observed that the critical value of the lateral acceleration is the second peak from the rollover point of view. The engineering interpretation of this phenomenon is that the vehicle generates larger lateral acceleration when the vehicle starts returning into the lane because the driver must set a double steering angle with -180 phase shifting to steer back the vehicle into its original position. The control torque is approximately 30 kNm at the front and 20 kNm at the rear and the brake force is 60 kN on the left at the rear axle. The advantage of the integrated control is that it meets performance demands with smaller control energy, i.e. the control force and control moments required are smaller than without an integration.

In the second example a float failure in the active anti-roll bars is assumed to have occurred. In this case the relative displacement of the hydraulic actuator changes the suspension travel instantaneously. This means that the active anti-roll bars cannot generate lateral displacement moment to balance the overturning moment and the piston of hydraulic actuator can move freely in the cylinder. This situation may arise when the power supply is cut off and sufficient oil pressure is not provided. Figure 6 shows the time responses in cases when there is a float failure at the front (dash), float failure at the rear (dash-dot), and fault-free operation (solid).

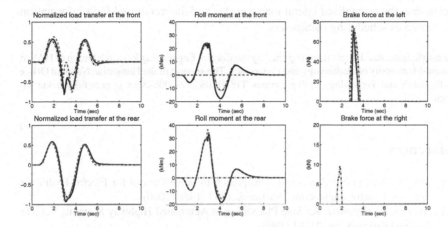

Fig. 6. Time responses of the steering maneuver in fault operation (normalized lateral load transfers both at the front and the rear, roll moments at the front and the rear and the brake force at the left and the right on the rear side)

The active anti-roll bars are not able to generate control torque over 2.5 sec, so the reconfigured controller structure is identical to the controller in which only the brake system is used to prevent the rollover of the vehicle. When there is a float failure at the front of the anti-roll bars the brake generates slightly larger force than in a fault-free case since it is not critical in this maneuver. However, when there is a failure at the rear of the anti-roll bars the brake is activated on the right (at 1.8 sec) and on both sides (at 2.5 sec). The reason for this is that in the fault case the critical value of $R_1(=0.65)$ is smaller than in the fault-free case.

Conclusions

In the paper, a model-based control design for an integrated vehicle system has been presented. As an illustration an integrated control, which includes an active steering, active anti-roll bars and an active brake system, is proposed for tracking the path of the vehicle and guaranteeing road holding and roll stability. In cruising mode, the controller minimises the tracking error and when the normalised load transfer has reached its critical value, the brake control is also activated in order to prevent the rollover, while the active anti-roll bars are operating continuously.

The modelling and the control design are based on the LPV method. In the control-oriented modeling the vehicle dynamics is augmented with the performance specifications of controlled system and uncertainties of the model. In the control design performance specifications must be formalized in such a way that the performance demands are guaranteed, conflicts between performances are handled, priority between different actuators is created and the various fault information is taken into consideration. In the LPV model, the forward velocity, the adhesion coefficient, the normalised lateral load transfer and the monitored fault information are chosen as scheduling parameters.

Acknowledgments. This work is supported by the Control Engineering Research Group of HAS at Budapest University of Technology and Economics. The support of the Hungarian National Office for Research and Technology through grants TECH_08_2/2-2008-0088 is gratefully acknowledged.

References

[1] Alleyne, A., Liu, R.: A Simplified Approach to Force Control for Electro-Hydraulic Systems. Control Engineering Practice 8, 1347–1356 (2000)

[2] Alvarez, L., Horowitz, R.: Safe Platooning in Automated Highway Systems. Vehicle System Dynamics, pp. 23–84 (1999)

[3] Balas, G., Doyle, J.: Robustness and Performance Tradeoffs in Control Design for Flexible Structures. IEEE Transactions on Control Systems Technology 2(4), 352–361 (1994)

[4] Balas, G., Fialho, I., Lee, L., Nalbantoglu, V., Packard, A., Tan, W., Wolodkin, G., Wu, F.: Theory and Application of Linear Parameter Varying Control Techniques. In: Proc. of the American Control Conference (1997)

[5] Becker, G., Packard, A.: Robust Performance of Linear Parametrically Varying Systems Using Parametrically-Dependent Linear Feedback. System Control Letters 23, 205–215 (1994)

[6] Bemporad, A.: Modeling, Control and Reachability Analysis of Discrete-Time Hybrid Systems. PhD Thesis, Department of Information Engineering, University of Siena (2003)

[7] Bokor, J., Balas, G.: Linear Parameter Varying Systems: A Geometric Theory and Applications. In: 16th IFAC World Congress, Prague (2005)

[8] Cebon, D.: Interaction between Heavy Vehicles and Roads. SAE-SP 951 (1993)

[9] Chen, J., Patton, R.: Robust Model-based Fault Diagnosis for Dynamics Systems. Kluwer Academic Publishers, Dordrecht (1999)

[10] Fischer, D., Isermann, R.: Mechatronic Semi-Active and Active Vehicle Suspensions. Control Engineering Practice (2004)

[11] Freriksson, J., Egardt, B.: Nonlinear Control Applied to Gearshifting in Automated Manual Transmissions. In: Proc. Conference on Decision and Control, Sydney (2000)

[12] Gahinet, P., Apkarian, P.: A linear Matrix Inequality Approach to H_∞ Control. International Journal of Robust and Nonlinear Control 4, 421–448 (1994)

[13] Gertler, J.: Fault Detection and Isolation Using Parity Relations. Control Engineering Practice 5(5), 653–661 (1997)

[14] Gillespie, T.: Fundamentals of Vehicle Dynamics. Society of Automotive Engineers Inc. (1992)

[15] Gordon, T., Howell, M., Brandao, F.: Integrated Control Methodologies for Road Vehicles. Vehicle System Dynamics 40, 157–190 (2003)

[16] Gáspár, P., Szabó, Z., Bokor, J.: Brake Control Combined with Prediction to Prevent the Rollover of Heavy Vehicles. In: Proc. of the IFAC World Congress, Praha (2005)

[17] Gáspár, P., Szabó, Z., Bokor, J.: Continuous-Time Parameter Identification Using Adaptive Observers for LPV Models with Vehicle Dynamics Applications. In: Robust Control Conference, Toulouse (2006)

[18] Gáspár, P., Szabó, Z., Bokor, J.: The Design of a Two-Level Controller for Suspension Systems. In: IFAC World Congress, Seoul, Korea (2008)

[19] Hirano, Y., Harada, H., Ono, E., Takanami, K.: Development of an Integrated System of 4WS and 4WD by H_∞ Control. SAE Journal, 79–86 (1993)

[20] Kiencke, U.: Integrated Vehicle Control Systems. In: Proc. of the Intelligent Components for Autonomous and Semi-Autonomous Vehicle, Tolouse, pp. 1–5 (1995)

[21] Massoumnia, M.A.: A Geometric Approach to the Synthesis of Failure Detection Filters. IEEE Transactions on Automatic Control AC-31(9), 839–846 (1986)

[22] Mastinu, G., Babbel, E., Lugner, P., Margolis, D.: Integrated Controls of Lateral Vehicle Dynamics. Vehicle System Dynamics 23, 358–377 (1994)

[23] Nagai, M., Hirano, Y., Yamanaka, S.: Intergated Robust Control of Active Rear Wheel Steering and Direct Yaw Moment Control. Vehicle System Dynamics 28, 416–421 (1998)

[24] Palkovics, L., Fries, A.: Intelligent Electronic Systems in Commercial Vehicles for Enhanced Traffic Safety. Vehicle System Dynamics 35, 227–289 (2001)

[25] Poussot-Vassal, C., Sename, O., Dugard, L., Gáspár, P., Szabó, Z., Bokor, J.: Attitude and Handling Improvements Trough Gain-scheduled Suspensions and Brakes Control. In: IFAC World Congress, Seoul (2008)

[26] Rajamani, R.: Vehicle Dynamics and Control. Springer, Heidelberg (2005)

[27] Rough, W., Shamma, J.: Research on Gain Scheduling. Automatica 36, 1401–1425 (2000)

[28] Sampson, D., Cebon, D.: Active Roll Control of Single Unit Heavy Road Vehicles. Vehicle System Dynamics 40, 229–270 (2003)

[29] Song, C., Uchanski, M., Hedrick, J.: Vehicle Speed Estimation Using Accelerometer and Wheel Speed Measurements. In: Proc. of the SAE Automotive Transportation Technology, Paris (2002)

[30] Stoustrup, J., Niemannn, H.: Fault Detection for Nonlinear Systems - a Standard Problem Approach. In: Proc. of the Conference on Decision and Control, Tampa, pp. 96–101 (1998)

[31] Swaroop, D.: String Stability of Interconnected Systems: An Application to Platooning in Automated Highway Systems. Research Report of PATH: Paper UCB-ITS-PRR-97-14 (1997)

[32] Szabó, Z., Bokor, J., Balas, G.: Inversion of LPV Systems and its Application to Fault Detection. In: Proc. of the Safeprocess-2003, Washington, USA (2003)

[33] Trachtler, A.: Integrated Vehicle Dynamics Control Using Active Brake, Steering and Suspension Systems. International Journal of Vehicle Design 36, 1–12 (2004)

[34] Tóth, R., Heuberger, P., den Hof, P.V.: Crucial Aspects of Zero-Order Hold LPV State-Space System Discretization. In: IFAC World Congress, Seoul, Korea (2008)

[35] Wu, F.: Control of Linear Parameter Varying Systems. PhD Thesis, Mechanical Engineering, University of California at Berkeley (1995)

[36] Wu, F., Yang, X., Packard, A., Becker, G.: Induced L_2 Norm Controller for LPV Systems with Bounded Parameter Variation Rates. International Journal of Robust and Nonlinear Control 6, 983–988 (1996)

[37] Zin, A., Sename, O., Gáspár, P., Dugard, L., Bokor, J.: An LPV/Hinf Active Suspension Control for Global Chassis Technology, Design and Performance Analysis. Vehicle System Dynamics, 889–912 (2008)

Design and Applications of Cerebellar Model Articulation Controller

Chih-Min Lin

Department of Electrical Engineering, Yuan Ze University
135, Far-East Rd., Chung-Li, Tao-Yuan, 320, Taiwan, Republic of China
cml@saturn.yzu.edu.tw

Abstract. This study presents a control system design based on cerebellar-model-articulation-controller (CMAC) for a class of multiple-input multiple-output (MIMO) uncertain nonlinear systems. The proposed control system merges a CMAC and sliding mode control, so the input space dimension of CMAC can be simplified. The control system consists of a CMAC-based principal controller (CMPC) and a robust controller. CMPC containing a CMAC uncertainty observer is used as the principal controller and the robust controller is designed to dispel the effect of approximation error. The gradient descent method is used to on-line tune the parameters of CMAC and the Lyapunov function is applied to guarantee the stability of the system. An experimental result of linear ultrasonic motor motion control and a simulation study of biped robot fault tolerance control show that favorable control performance can be achieved by using the proposed control system.

Keywords: cerebellar model articulation controller (CMAC), gradient descent method, Lyapunov stability theorem, uncertain nonlinear systems, linear ultrasonic motor, biped robot.

1 Introduction

Sliding mode control (SMC) is an effective robust approach to the problem of maintaining stability and consistent performance of a controlled system with imprecise modeling [1, 2]. Another important advantage of SMC is that by integrating several state variables into a sliding surface, the number of input variables of SMC system can be reduced.

Cerebellar model articulation controller (CMAC) has been proposed for the identification and control of complex dynamical systems, due to its fast learning property and good generalization capability [3-7]. CMAC is a non-fully connected perceptron-like associative memory network with overlapping receptive fields.

I.J. Rudas et al. (Eds.): Towards Intelligent Engineering & Information Tech., SCI 243, pp. 121–135.
springerlink.com   © Springer-Verlag Berlin Heidelberg 2009

This network has already been shown to be able to approximate a nonlinear function over a domain of interest to any desired accuracy [7]. Several literatures have demonstrated the applications of CMAC on control problems [8-13]. However, in the above CMAC literatures, some approaches are too complicated and some lacks of on-line real time adaptation ability.

This study proposes a CMAC-based control system for a class of MIMO uncertain nonlinear systems. This control system combines the advantages of CMAC and SMC. The developed system is comprised of a CMAC-based principal controller and a robust controller. CMAC is used as an uncertainty observer of the principal controller, and a gradient-descent learning method is applied to adjust the parameters of CMAC. In the design of robust controller, a Lyapunov function is applied to guarantee the system's stability. Finally, an experiment of linear ultrasonic motor motion control and a simulation study of biped robot fault tolerance control are demonstrated to illustrate the effectiveness of the proposed control method.

This study is organized as follows. Problem formulation is described in Section 2. Section 3 expresses the design of CMAC-based control system. Simulation and experimental results are provided to validate the effectiveness of the proposed control system in Section 4. Finally, Section 5 concludes the study.

2 Problem Formulation

Consider a class of multi-input multi-output (MIMO) nonlinear dynamic system described in the following form

$$\begin{cases} x^{(n)} = f_0(\underline{x}) + G_0(\underline{x})u + L(\underline{x}) \\ y = x \end{cases} \tag{1}$$

where $x \in \mathfrak{R}^m$ is the state, $u \in \mathfrak{R}^m$ is the control input, $y \in \mathfrak{R}^m$ is the system output. Define $\underline{x} = [x^T \quad \dot{x}^T \quad \cdots \quad x^{(n-1)^T}]^T \in \mathfrak{R}^{nm}$ as the system state vector, and it is assumed to be available for measurement. In addition, $f_0(\underline{x}) \in \mathfrak{R}^m$ and $G_0(\underline{x}) \in \mathfrak{R}^{m \times m}$ are system nominal nonlinear vector-valued and matrix-valued functions, respectively, and assumed they are bounded and available. Meanwhile, assume the nonlinear system of (1) is controllable and $G_0^{-1}(\underline{x})$ exists for all $\underline{x}$. $L(\underline{x}) \in \mathfrak{R}^m$ denotes the unknown uncertainty, which is assumed to be bounded. If there exist mismodelings between practical systems and the nominal functions, they can be absorbed into the uncertainty.

The control purpose is to design a control system such that the system output can track a desired trajectory signal $y_d \in R^m$. Define the tracking error as

$$e = y_d - y \tag{2}$$

and the system tracking error vector is defined as

$$\underline{e} \underline{\triangleq} \left[e^T, \dot{e}^T, \cdots, e^{(n-1)T} \right]^T \in \Re^{nm} \tag{3}$$

Define an integrated sliding function as

$$\underline{s} \underline{\triangleq} e^{n-1} + K_1 e^{n-2} + \cdots + K_n \int_0^t e(\tau) d\tau \tag{4}$$

where $K_i \in \Re^{m \times m}, i = 1, 2, \ldots, n$ are positive constant matrices and define $K = [K_1^T, \cdots, K_n^T]^T \in \Re^{nm \times m}$.

If the nominal functions $f_0(\underline{x})$, $G_0(\underline{x})$ and the uncertainty $L(\underline{x})$ are exactly known, then an ideal controller can be designed as

$$u^* = G_0^{-1}(\underline{x}) [y_d^{(n)} - f_0(\underline{x}) - L(\underline{x}) + K^T \underline{e}]. \tag{5}$$

Substituting the ideal controller in (5) into (1), gives the error dynamic equation

$$\dot{s} = e^{(n)} + K^T \underline{e} = 0. \tag{6}$$

In (6), if K is chosen to let all the roots of the polynomial $P(\lambda) = I\lambda^n + K_1 \lambda^{n-1} + \cdots + K_n$ correspond to the coefficients of Hurwitz polynomial, that is a polynomial whose roots lie strictly in the open left half of the complex plane, then it implies the tracking error will converge to zero when time tends to infinity. However, the uncertainty $L(x)$ is generally unknown for practical applications, so u^* in (5) is unavailable. Thus, a CMAC-based control system will be proposed to achieve trajectory tracking control. The proposed control system is shown in Fig. 1, which is comprised of a CMAC-based principal controller (CMPC) u_{cmpc} and a robust controller u_r as follows:

$$u = u_{cmpc} + u_r. \tag{7}$$

where u_{cmpc} is the principal controller used to approximate the ideal control in (5) and the robust controller u_r is utilized to compensate for the approximation error between the ideal controller and u_{cmpc}.

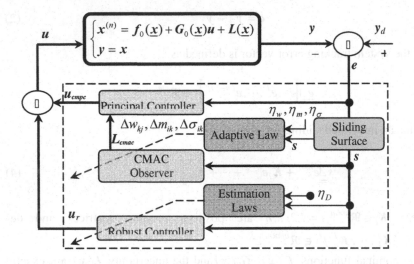

Fig. 1. Block diagram of CMAC-based adaptive control system

3   Adaptive CMAC-Based Control System Design

This section proposes the CMAC-based control system design. In this design, a CMAC is introduced and is used to estimate the system's uncertainty $L(\underline{x})$.

3.1 The CMAC Model

The scheme of CMAC architecture is shown in Fig. 2, consisting of two consequent mappings and one output computation. The CMAC model and its functional mapping are introduced as follows:

$$\text{Mapping } I : I \rightarrow A \tag{8}$$

$$\text{Mapping } A : A \rightarrow R \tag{9}$$

$$\text{Mapping } R : R \rightarrow W \tag{10}$$

$$\text{Output Computation } O : O(I) = W^T \Gamma(I) \tag{11}$$

where I is a continuous n-dimensional input space, A denotes an association memory, R is a receptive-field space with n_k adjustable receptive-field functions where n_k is the number of blocks, W is a weight memory with $m \times n_k$ adjustable weights, and O is a m-dimensional output space.

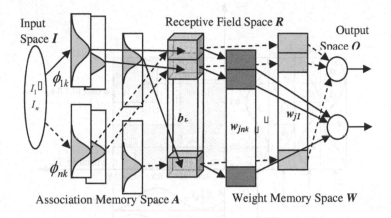

Fig. 2. Architecture of a CMAC

1) The first mapping I relates the input state variable $I = [I_1 \quad I_2 \quad \cdots \quad I_n]^T$ to an association memory A. The Gaussian function is chosen as the receptive-field basis function for each block and is given as follows:

$$\phi_{ik}(I_i) = exp\left[-\frac{(I_i - m_{ik})^2}{\sigma_{ik}^2}\right], \quad i = 1,2,\cdots,n, \quad k = 1,2,\cdots,n_k \qquad (12)$$

where m_{ik} is the mean and σ_{ik} is the variance of the k -th block basis function corresponding to the i -th input variable.

2) The second mapping A relates each location of A corresponding to a recep-tive-field R. The mechanism for this mapping is shown in Fig. 3 for a two-dimensional (2-D) input with 5 elements and 4 elements is accumulated as a block. Areas formed by multi-input regions are called hypercubes. Each activated element in each layer becomes a firing element, thus, the weight of each block can be obtained. The content of hypercube can be expressed as $b_k(I)$, which is the general basis function of the kth hypercube, that is,

$$b_k(I) \equiv \prod_{i=1}^{n} \phi_{ik}(I_i). \qquad (13)$$

In a two-dimensional case shown in Fig. 3, the output of CMAC is the sum of the value in receptive-fields Bb, Dd, Ff and Gg, where the input state is (0.7,0.8).

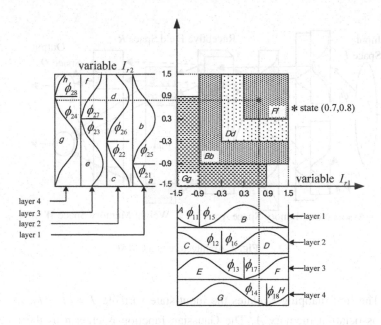

Fig. 3. A two-dimensional CMAC

3) The third mapping R relates each location of R to a particular value in the weight memory W. In CMAC, the content of weighted hypercube can be expressed as $v_{jk}(I) \equiv w_{jk}b_k(I)$, where w_{jk} is the weight of the kth hypercube for the jth output.

4) The CMAC output O is the algebraic sum of the hypercube contents with activated weights. The jth output of CMAC can be expressed as

$$O_j = \begin{bmatrix} w_{j1} & w_{j2} & \cdots & w_{jn_k} \end{bmatrix} \begin{bmatrix} b_1(I) \\ b_2(I) \\ \vdots \\ b_{n_k}(I) \end{bmatrix}$$

$$= \sum_{k=1}^{n_k} w_{jk}b_k(I) = \sum_{k=1}^{n_k} w_{jk} \prod_{i=1}^{n} \phi_{ik}(I_i), \quad j = 1,2,...,m \tag{14}$$

The output of CMAC can be expressed in a vector notation as

$$O(I) = \begin{bmatrix} w_{11} & w_{12} & \cdots & w_{1n_k} \\ w_{21} & w_{22} & \cdots & w_{2n_k} \\ \vdots & \vdots & \ddots & \vdots \\ w_{m1} & w_{m2} & \cdots & w_{mn_k} \end{bmatrix} \begin{bmatrix} b_1(I) \\ b_2(I) \\ \vdots \\ b_{n_k}(I) \end{bmatrix} \tag{15}$$

3.2 Adaptive CMAC Control System Design

In (5), the uncertainty $L(\underline{x})$ is always unknown, so u^* can not be implemented. A CMAC approximator will be used to estimate the uncertainty $L(\underline{x})$. By the universal approximation theorem, there exists a CMAC $L_{cmac}(\underline{x})$ to approximate $L(\underline{x})$ [14]

$$L(\underline{x}) = L_{cmac}(I, w_{jk}, m_{ik}, \sigma_{ik}) + \varepsilon \qquad (16)$$

where ε denotes the approximation error. In the sliding mode control, the sliding condition is derived as $s^T \dot{s} < 0$ so the stability and convergence of $s \to 0$ as $t \to \infty$ can be guaranteed [1, 2]. The adjustable parameters of CMAC observer are tuned by using the gradient decent method, which aims to minimize $s^T \dot{s}$ for achieving fast convergence of s. Therefore, $s^T \dot{s}$ is selected as the error function. Taking the derivative of s and using (1), yields

$$\dot{s} = e^{(n)} + K^T \underline{e} = -f_0(\underline{x}) - G_0(\underline{x})u + y_d^{(n)} - L(\underline{x}) + K^T \underline{e} \qquad (17)$$

Substituting (7) into (17) and multiplying both sides by s^T, yields

$$s^T \dot{s} = -s^T f_0(\underline{x}) - s^T G_0(\underline{x})[u_{cmpc} + u_r] + s^T(y_d^{(n)} - L(\underline{x}) + K^T \underline{e}) \qquad (18)$$

where

$$u_{cmpc} = G_0^{-1}(\underline{x})[y_d^{(n)} - f_0(\underline{x}) - L_{cmac}(\underline{x}) + K^T \underline{e}] \qquad (19)$$

With this representation of the CMAC-based control system, it becomes straightforward to apply the back-propagation idea to adjust the parameters. The hypercube weight w_{jk} and the mean m_{ik} and variance σ_{ik} of Gaussian function are updated by the following equations:

$$w_{jk}(t+1) = w_{jk}(t) + \Delta w_{jk} \qquad (20)$$

$$m_{ik}(t+1) = m_{ik}(t) + \Delta m_{ik} \qquad (21)$$

$$\sigma_{ik}(t+1) = \sigma_{ik}(t) + \Delta \sigma_{ik} \qquad (22)$$

The training algorithms in (20), (21) and (22) perform error back-propagation by using chain rule, that is,

$$\Delta w_{jk} = -\eta_w \frac{\partial s^T \dot{s}}{\partial w_{jk}} = -\eta_w \frac{\partial s^T \dot{s}}{\partial u_{cmpc}} \frac{\partial u_{cmpc}}{\partial w_{jk}} = -\eta_w s^T \cdot \left(\prod_{i=1}^{n} \phi_{ik}(I_i) \right) \quad (23)$$

$$\Delta m_{ik} = -\eta_m \frac{\partial s^T \dot{s}}{\partial m_{ik}}$$
$$= -\eta_m \frac{\partial s^T \dot{s}}{\partial u_{cmpc}} \frac{\partial u_{cmpc}}{\partial \phi_{ik}} \frac{\partial \phi_{ik}}{\partial m_{ik}} = -\eta_m s^T \cdot w_{jk} \cdot \left(\prod_{i=1}^{n} \phi_{ik}(I_i) \right) \cdot \frac{2(x_j - m_{ik})}{\sigma_{ik}^2} \quad (24)$$

$$\Delta \sigma_{ik} = -\eta_\sigma \frac{\partial s^T \dot{s}}{\partial \sigma_{ik}}$$
$$= -\eta_\sigma \frac{\partial s^T \dot{s}}{\partial u_{cmpc}} \frac{\partial u_{cmpc}}{\partial \phi_{ik}} \frac{\partial \phi_{ik}}{\partial \sigma_{ik}} = -\eta_\sigma s^T \cdot w_{jk} \cdot \left(\prod_{i=1}^{n} \phi_{ik}(I_i) \right) \cdot \left[\frac{2(x_i - m_{ik})^2}{\sigma_{ik}^3} \right] \quad (25)$$

where η_w, η_m and η_σ are positive learning-rates for the weight, mean and variance, respectively.

3.3 Robust Controller Design

The most useful property of CMAC is its ability to approximate linear or nonlinear mapping through learning. In (16), the approximation error ε is assumed to be bounded by $0 \le \|\varepsilon\| \le D$ where D is a positive constant and $\|\cdot\|$ denotes an induced-norm. The error bound is assumed to be a constant during the observation; however, it is difficult to measure it in practical applications. Therefore, a bound estimation is developed to estimate this error bound. Define the estimation error of the bound

$$\tilde{D} = D - \hat{D} \quad (26)$$

where $\hat{D}$ is the estimated value of D. The robust controller is designed to compensate for the effect of the approximation error and is chosen as

$$u_r = -G_0^{-1}(\underline{x})\hat{D} sgn(s) \quad (27)$$

By substituting (7) into (1), yields

$$x^{(n)} = f_0(\underline{x}) + G_0(\underline{x})\left(u_{cmpc} + u_r\right) + L(\underline{x}) \tag{28}$$

After some straightforward manipulations, the error equation governing the system can be obtained through (5), (7), (16) and (1) as follows

$$e^{(n)} + K^T \underline{e} = G_0(\underline{x})u_r + \varepsilon = \dot{s} \tag{29}$$

Define a Lyapunov function as

$$V(s,\widetilde{D}) = \frac{s^T s}{2} + \frac{\widetilde{D}^2}{2\eta_D} \tag{30}$$

where the positive constant η_D is a learning rate. Differentiating (30) with respect to time and using (7), (27), (18) and (19), yields

$$\dot{V}(s,\widetilde{D}) = s^T \dot{s} + \frac{\widetilde{D}\dot{\widetilde{D}}}{\eta_D} = s^T\left(\varepsilon - \hat{D}\,sgn(s)\right) + \frac{\widetilde{D}\dot{\widetilde{D}}}{\eta_D} = \left(s^T\varepsilon - \hat{D}\|s\|\right) + \frac{\widetilde{D}\dot{\widetilde{D}}}{\eta_D} \tag{31}$$

If the adaptive law of the error bound is chosen as

$$\dot{\widetilde{D}} = -\dot{\hat{D}} = -\eta_D \|s\| \tag{32}$$

then (31) can be rewritten as

$$\dot{V}(s,\widetilde{D}) = s^T\varepsilon - \hat{D}\|s\| - \left(D - \hat{D}\right)\|s\| = \left(s^T\varepsilon - D\|s\|\right) \leq \left(\|\varepsilon\|\|s\| - D\|s\|\right)$$

$$= -\left(D - \|\varepsilon\|\right)\|s\| \leq 0 \tag{33}$$

Since $\dot{V}(s,\widetilde{D})$ is negative semi-definite that is $\dot{V}(s(t),\widetilde{D}(t))$ $\leq \dot{V}(s(0),\widetilde{D}(0))$, it implies s and $\widetilde{D}$ are bounded. Let function $\Omega \equiv \left(D - \|\varepsilon\|\right)s \leq \left(D - \|\varepsilon\|\right)\|s\| \leq -\dot{V}(s,\widetilde{D})$, and integrate $\Omega(t)$ with respect to time, then it is obtained that

$$\int_0^t \Omega(\tau)d\tau \leq \dot{V}(s(0),\widetilde{D}(0)) - \dot{V}(s(t),\widetilde{D}(t)) \tag{34}$$

Because $\dot{V}(s(0), \tilde{D}(0))$ is bounded, and $\dot{V}(s(t), \tilde{D}(t))$ is non-increasing and bounded, the following result can be obtained:

$$\lim_{t \to \infty} \int_0^t \Omega(\tau) d\tau < \infty \tag{35}$$

Also, $\dot{\Omega}(t)$ is bounded, so by Barbalat's Lemma [1], $\lim_{t \to \infty} \Omega = 0$. That is, $s \to 0$ as $t \to \infty$. Hence, the CMAC-based control system is asymptotically stable.

4   Simulation and Experimental Results

An experiment of a linear ultrasonic motor system and a simulation of a biped robot are examined to illustrate the effectiveness of the proposed design method.

4.1 Linear Ultrasonic Motor System

The proposed CMAC-based control system is applied to control a linear ultrasonic motor (LUSM). The nonlinear dynamic equation of LUSM is given by [15]

$$\ddot{x}(t) = f(x) + G(x)u(t) + L(t) \tag{36}$$

where $x(t) = [x(t), \dot{x}(t)]^T$ represent the position and velocity of the moving table, respectively; $G(x)$ is the gain of the LC resonant inverter; $f(x)$ denotes a nonlinear dynamic function related to the components of stress, strain and electric field; $u(t)$ is the input force, and $L(t)$ is the normalized lump force of the uncertain nonlinearities such as friction, ripple force and external disturbance. Since the dynamic characteristic of LUSM is difficult to obtain, the dynamic functions $f(x)$, $G(x)$ and $L(t)$ are assumed to be unknown. The proposed CMAC-based control system is applied to control the system by letting $f_0(x) = 1$ and $G_0(x) = 1$.

The control objective is to control the moving table to move ± 2 cm periodically for a sinusoidal and a periodic step command. Moreover, a second-order transfer function is chosen as the reference model for a periodic step command:

$$\frac{w_n^2}{S^2 + 2\xi w_n S + w_n^2} = \frac{36}{S^2 + 12S + 36} \tag{37}$$

where S is the Laplace operator; ξ and w_n are the damping ratio and undamped natural frequency, respectively.

The control parameters are selected as $\eta_w = \eta_m = \eta_\sigma = 0.1$, $\textbf{\textit{K}}_1 = \textbf{\textit{I}}$, $\textbf{\textit{K}}_2 = 4\textbf{\textit{I}}$ and $\eta_D = 0.1$. The initial conditions of the system are given as $\textbf{\textit{x}}(0) = [0 \quad 0]^T$, and the initial values of system parameters are given as; the inputs of CMAC are s_1 and s_2. The experimental results of the CMAC-based control system due to a sinusoidal and periodic step commands are shown in Fig. 4. The tracking response of sinusoidal command is shown in Fig. 4(a); the associated control effort is shown in Fig. 4(b). Moreover, 1 Kg iron is put as load, the responses are given in Figs. 4(c), (d). The tracking response and control effort for periodic step commands are shown in Figs. 4 (e)-(h). The experimental results indicate that the high-accuracy motion tracking responses can be achieved by using the proposed CMAC-based control system, even for load variation.

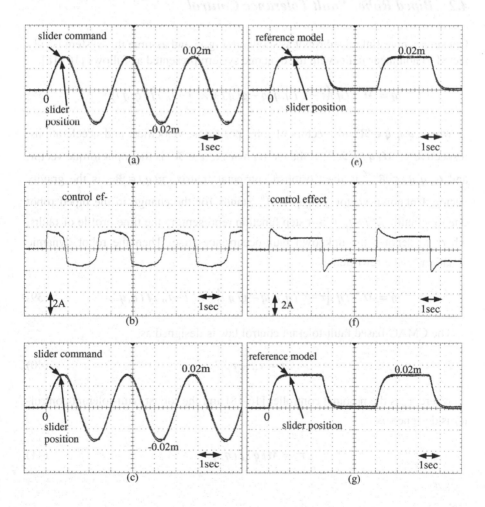

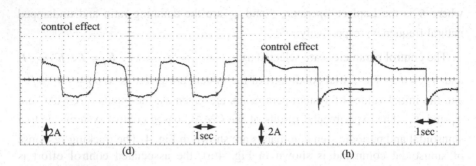

Fig. 4. Experimental results of CMAC-based control for LUSM due to a sinusoidal and a periodic step command

4.2 Biped Robot Fault Tolerance Control

Consider a nine-link biped robot as shown in Fig. 5 and assume this system is subjected to nonlinear faults with the dynamic system presented as follows [16- 18]:

$$\ddot{q} = M^{-1}(q)\big(\tau - C(q,\dot{q})\dot{q} - g(q) + \lambda(t-t_0)\bar{f}(q,\dot{q})\big) \tag{38}$$

where $q,\dot{q},\ddot{q} \in \Re^6$ are vectors of joint positions, velocities, and accelerations, respectively, $M(q) \in \Re^{6\times6}$ is the inertia matrix, $\tau \in \Re^6$ is the input torque vector, and $C(q,\dot{q}) \in \Re^{6\times6}$ is the Coriolis/Centripetal matrix, $g(q) \in \Re^6$ is the gravity vector. Unknown vector $\bar{f}(q,\dot{q}) \in \Re^6$ stands for the change in the biped robot due to a fault. $\lambda(t-t_0)$ is a step function representing the time profile of faults, where t_0 denotes the unknown fault-occurrence time. Then the robot dynamic equation can be rewritten as

$$\ddot{q} = M^{-1}(q)\big[\tau - C(q,\dot{q})\dot{q} - g(q)\big] + \lambda(t-t_0)f(q,\dot{q}) \tag{39}$$

The CMAC-based fault-tolerant control law is designed as

$$\tau = \tau_0 - \tau_r \tag{40}$$

where τ_0 is the nominal controller [16-18] and the robust fault-tolerant controller is designed as

$$\tau_r = M(q)\hat{f}(q,\dot{q}) \tag{41}$$

In which $\hat{f}(q,\dot{q})$ is the output of CMAC that is used to estimate the nonlinear fault $f(q,\dot{q})$. A fault with the nonlinear change in link 1 and link 2 occurs at the 2.5^{th} sec with the following failure function:

$$f_t(q,\dot{q}) = \begin{bmatrix} 75q_1^2 + 100q_1^2\dot{q}_2^2 + 7q_2 + 17 \\ 100q_1q_2 + 25 \\ 0 \\ 0 \\ 0 \\ 0 \end{bmatrix} \qquad (42)$$

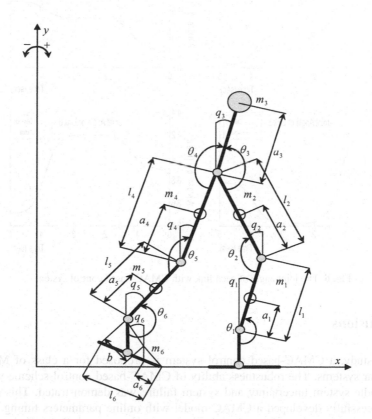

Fig. 5. A nine-link biped robot

By equipping with CMAC, the simulation results of biped robot control system are shown in Fig. 6. The simulation results show that CMAC can provide fast and accurate estimation of fault; thus, the control system can effectively achieve the fault accommodation.

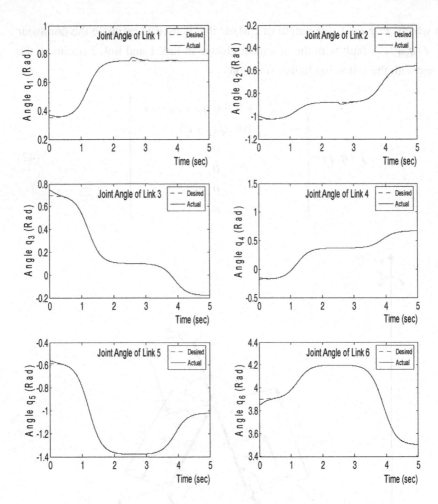

Fig. 6. The joint angle of each link with CMAC-based control system

Conclusions

In this study, a CMAC-based control system is proposed for a class of MIMO
nonlinear systems. The robustness ability of CMAC-based control scheme which
can handle system uncertainty and system failure was demonstrated. This study
has successfully developed a CMAC model with online parameters tuning algorithm. System stability of the uncertain system is also guaranteed by using the developed control system. The effectiveness of the proposed control system is illustrated by controlling a linear ultrasonic motor and a biped robot. The simulation
and experimental results confirm that the proposed CMAC-based system can
achieve favorable tracking performance for these systems.

References

[1] Slotine, J.J.E., Li, W.P.: Applied Nonlinear Control. Prentice-Hall, NJ (1991)

[2] Hung, J.Y., Gao, W., Hsu, J.C.: Variable Structure Control: A Survey. IEEE Trans. Ind. Electron. 40(1), 2–22 (1993)

[3] Albus, J.S.: A New Approach to Manipulator Control: the Cerebellar Model Articulation Controller (CMAC). J. Dyn. Syst., Meas., Contr. 97(3), 220–227 (1975)

[4] Lane, S.H., Handelman, D.A., Gelfand, J.J.: Theory and Development of Higher-Order CMAC Neural Network. IEEE Control Syst. Mag. 12(2), 23–30 (1992)

[5] Shiraishi, H., Ipri, S.L., Cho, D.D.: CMAC Neural Network Controller for Fuel-Injection Systems. IEEE Trans. Contr. Syst. Technol. 3(1), 32–38 (1995)

[6] Jagannathan, S., Commuri, S., Lewis, F.L.: Feedback Linearization Using CMAC Neural Networks. Automatica 34(3), 547–557 (1998)

[7] Chiang, C.T., Lin, C.S.: CMAC with General Basis Functions. J. Neural Netw. 9(7), 1199–1211 (1996)

[8] Kim, Y.H., Lewis, F.L.: Optimal Design of CMAC Neural-Network Controller for Robot Manipulators. IEEE Trans. Syst., Man, Cybern. C 30(1), 22–31 (2000)

[9] Su, S.F., Tao, T., Hung, T.H.: Credit-assigned CMAC and its Application to Online Learning Robust Controllers. IEEE Trans. Syst., Man and Cybern. B 33(2), 202–213 (2003)

[10] Lin, C.M., Peng, Y.F.: Adaptive CMAC-based Supervisory Control for Uncertain Nonlinear Systems. IEEE Trans. Syst., Man, and Cybern. B 34(2), 1248–1260 (2004)

[11] Commuri, S., Jagannathan, S., Lewis, F.L.: CMAC Neural Network Control of Robot Manipulators. Journal of Robotics Syst. 14(6), 465–482 (1997)

[12] Wu, T.F., Tsai, P.S., Chang, F.R., Wang, L.S.: Adaptive Fuzzy CMAC Control for a Class of Nonlinear System with Smooth Compensation. IEE Proc. Cont. Theory Appl. 153(6), 647–657 (2006)

[13] Lin, C.M., Peng, Y.F.: Missile Guidance Law Design Using Adaptive Cerebellar Model Articulation Controller. IEEE Trans. Neural Netw. 16(3), 636–644 (2005)

[14] Wang, L.X.: Adaptive Fuzzy Systems and Control: Design and Stability Analysis. Prentice-Hall, Englewood Cliffs (1994)

[15] Hagood, N.W., Mcfarland, A.J.: Modeling of a Piezoelectric Rotary Ultrasonic Motor. IEEE Trans. Ultrason., Ferroelect., Freq. Contr. 42, 210–224 (1995)

[16] Vemuri, A.T., Polycarpou, M.M.: Neural-Network-based Robust Fault Diagnosis in Robotic Systems. IEEE Trans. Neural Netw. 8(6), 1410–1420 (1997)

[17] Liu, Z., Li, C.: Fuzzy Neural Networks Quadratic Stabilization Output Feedback Control for Biped Robots via H∞ Approach. IEEE Trans. Syst., Man, Cybern. B 33(1), 67–84 (2003)

[18] Lin, C.M., Chen, L.Y., Chen, C.H.: RCMAC Hybrid Control for MIMO Uncertain Nonlinear Systems Using Sliding-Mode Technology. IEEE Trans. Neural Netw. 18(3), 708–720 (2007)

Model-Based Design Issues in Fuzzy Logic Control

Stefan Preitl[1], Radu-Emil Precup[1], Marius-Lucian Tomescu[2],
Mircea-Bogdan Rădac[1], Emil M. Petriu[3], and Claudia-Adina Dragoş[1]

[1] Department of Automation and Applied Informatics,
  "Politehnica" University of Timisoara
  Bd. V. Parvan 2, 300223 Timisoara, Romania
  {stefan.preitl,radu.precup,mircea.radac,
  claudia.dragos}@aut.upt.ro
[2] Computer Science Faculty, "Aurel Vlaicu" University of Arad
  Complex Universitar M, Str. Elena Dragoi 2, 310330 Arad, Romania
  tom_uav@yahoo.com
[3] School of Information Technology and Engineering, University of Ottawa
  800 King Edward, Ottawa, Ontario, Canada, K1N 6N5
  petriu@site.uottawa.ca

Abstract. Model-based design issues of fuzzy logic control systems for Single Input-Single Output (SISO) nonlinear time-varying plants are discussed. The emphasis is given to the stable design of fuzzy logic controllers (FLCs). The accepted FLCs belong to the classes of type-II fuzzy systems and type-III fuzzy systems according to Sugeno's classification. Two original theorems that ensure the uniformly stability and the uniformly asymptotically stability of fuzzy logic control systems are given. The stability analyses are done in the sense of Lyapunov and the approaches are expressed in terms of sufficient inequality-type stability conditions. The effectiveness of the theoretical results is proved by their application in the stable design of Takagi-Sugeno FLCs for two SISO nonlinear time-varying plants, the Lorenz chaotic system and a laboratory Anti-lock Braking System. Digital simulation and real-time experimental results are included.

Keywords: fuzzy logic control, model-based design, stability.

1 Introduction

Use is made in a lot of applications of fuzzy logic control for direct feedback control or on the low level in hierarchical control system structures. However fuzzy

I.J. Rudas et al. (Eds.): Towards Intelligent Engineering & Information Tech., SCI 243, pp. 137–152.
springerlink.com   © Springer-Verlag Berlin Heidelberg 2009

logic control can be used at the supervisory level exemplified by the popular adaptive control system structures. Nowadays fuzzy logic control is no longer only used to directly express the knowledge on the controller plant or in other words to do model-free fuzzy logic control. A fuzzy logic controller (FLC) can be calculated from a fuzzy model obtained in terms of system identification techniques and thus it can be regarded as doing model-based fuzzy logic control. The most often used controller configurations are

- Mamdani FLCs, referred to also as linguistic FLCs, with either fuzzy consequents that represent the class of type-I fuzzy systems according to the classifications given in (Sugeno 1999) and suggested by (Kóczy 1996) or singleton consequents belonging to the class of type-II fuzzy systems. Those FLCs are usually used as direct closed-loop controllers.
- Takagi-Sugeno (TS) FLCs, referred to also as the class of type-III fuzzy systems (Sugeno 1999) especially when affine consequents are employed. They are typically used as supervisory controllers.

The model-based design of fuzzy logic control system is applied for Mamdani and TS FLCs as well. Much research on the model-based design of fuzzy logic control systems making use of TS fuzzy models has been carried out in the recent years (Sun and Wang 2006), (Johanyák et al. 2006), (Blažič and Škrjanc 2007), (Oblak et al. 2007), (Pang and Lur 2008), (Vaščák 2008), (Tanaka et al. 2009), (Yuan and Wang 2009). Their advantages come from the fact that the TS fuzzy models can express highly nonlinear functional relations with a relatively small number of rules. The FLCs considered in the paper belong to the classes of type-II and type-III fuzzy systems.

The model-based design of FLCs is difficult in applications that cope with control problems related to complex plants including the linear time-varying (LTV) and nonlinear time-varying (NTV) ones. LTV and NTV systems are widely used in practice because real-world applications make use of time-varying as a result of the parametric modifications. As particular NTV systems the LTV ones may also be a result of linearizing nonlinear systems in the vicinity of sets of operating points or trajectories. Several techniques are employed in the analysis and design of control systems meant for LTV systems dealing mainly with the eigenstructure assignment (Van der Kloet and Neerhoff 2002), (Lee and Choi 2004). In turn not so much attention has been focused on NTV systems. Several methods have been proposed recently to deal with the FLC design employing the stability analysis when the fuzzy logic control systems are viewed as convenient classes of nonlinear systems in the Linear Matrix Inequality (LMI) framework (Tanaka et al. 1998), (Tanaka et al. 2007), (Yoneyama 2007), (Lam and Leung, 2008), (Lendek et al. 2008), (Wei et al. 2009), (Yang and Zhang 2009).

One of the current trends in model-based design of fuzzy logic control systems is to derive less conservative conditions in order to prove their stability and guarantee their performance indices quadratic by means of Lyapunov functions, piecewise quadratic and non-quadratic ones (Sala et al. 2005), (Wang and Sun 2005), (Michels et al. 2006), (Kruszewski et al. 2008), (Lam and Seneviratne 2009), (Kim and Park 2009). The fuzzy partitions are the combinations of the products of rather simple arguments expressed as membership functions, and in real-world applications one particular case concerns fuzzy modeling of nonlinear systems under the form of Tensor Product fuzzy systems (Baranyi 2004), (Arino and Sala 2007), (Petres et al. 2007), (Precup et al. 2008b), (Baranyi et al. 2009).

The paper suggests two theorems that ensure the uniformly stability and the uniformly asymptotically stability of fuzzy logic control systems. They are part of the above mentioned trend in terms of offering sufficient inequality-type stability conditions characterized by low conservativeness. Another advantage of the theorems concerns the avoidance of solving the LMIs. Although the LMIs are computationally solvable they require numerical algorithms embedded in well acknowledged software tools. The two theorems are based on Lyapunov's theorem for time-varying system starting with the formulation presented in (Khalil 2002). They are applied in the model-based design of fuzzy logic control systems which is illustrated by two case studies.

The paper discusses the following topics. Section 2 deals with the definition of the accepted class of fuzzy logic control systems. Next the stability conditions for fuzzy logic control systems based on two stability analysis theorems are derived in Section 3. Section 4 offers two case studies to validate the theoretical approaches by digital simulation and real-time experimental results. They deal with the fuzzy logic control of two Single Input-Single Output (SISO) NTV plants, the Lorenz chaotic system and a laboratory Anti-lock Braking System (ABS). Section 5 is focused on the conclusions.

2 Fuzzy Logic Control Systems

The structure of the fuzzy logic control system, accepted as stabilized control system for the SISO NTV plant, is presented in Fig. 1, where: r – the reference input, $\mathbf{x} = [x_1 \quad x_2 \quad \dots \quad x_n]^T \in D$ – the state vector, $n \in IN^*$, $D \subset IR^n$ – the universe of discourse, y – the controlled output, u – the control signal, t – the independent time variable, t_0 – the initial time moment, $t \geq t_0$, and the superscript T stands for matrix transposition. The dynamics of the actuator and measuring elements of the state variables $x_i, i = \overline{1, n}$, are supposed to be included in the controlled plant.

The structure presented in Fig. 1 can be viewed as a nonlinear state-feedback control system. Other variables can be considered as inputs of the FLC instead of the state variables. They are referred to as scheduling variables and may include among them the control error, its derivative, etc. However those variables should be separated by different dynamics.

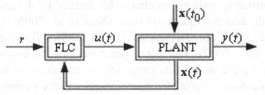

Fig. 1. Fuzzy logic control system structure

The disturbance inputs are absent in Fig. 1 because the main control aim is the regulation. The tracking can be ensured by adding integral components to the FLC which ensures additional dynamics in the controller structure and design.

The controlled plant is modeled by the SISO NTV system with the following state-space equations:

$$\dot{x} = f(x,t) + b(x,t)u,$$
$$x(t_0) = x_0,$$

(1)

where $\dot{x} = [\dot{x}_1 \quad \dot{x}_2 \quad ... \quad \dot{x}_n]^T$ is the derivative of x with respect to t, the domain D contains the equilibrium point in origin $x = 0$, and $f, b : D \times [0,\infty) \to IR^n$ are piecewise continuous functions in t and locally Lipschitz in x on $D \times [0,\infty)$. The two functions f and b describe the nonlinear and time-varying dynamics of the plant, and they are expressed in

$$f(x,t) = [f_1(x,t) \quad f_2(x,t) \quad ... \quad f_n(x,t)]^T,$$
$$b(x,t) = [b_1(x,t) \quad b_2(x,t) \quad ... \quad b_n(x,t)]^T.$$

(2)

The FLC consists of r fuzzy logic control rules. The ith rule in the rule base of the FLC is

Rule i: IF x_1 IS $\widetilde{X}_{i,1}$ AND x_2 IS $\widetilde{X}_{i,2}$ AND ... AND x_n IS $\widetilde{X}_{i,n}$

THEN $u = u_i(x)$, $i = \overline{1,r}$, $r \in IN$, $r \geq 2$,

(3)

where $\widetilde{X}_{i,1}$, $\widetilde{X}_{i,2}$, ..., $\widetilde{X}_{i,n}$ are the fuzzy sets that describe the linguistics terms (LTs) of the input variables, $u = u_i(x)$ is the control signal in the consequent of the ith rule, and the function AND, standing for the conjunction operator, is a t-norm. $u_i(x)$ can be either a constant when the FLC is considered to belong to the class of type-II fuzzy systems according to (Kóczy 1996) and (Sugeno 1999), or a function of the state vector as shown in (3) when the FLC is considered to belong to the class of type-II fuzzy systems. Therefore the FLC accepted here paper belong to the classes of type-II and type-III fuzzy systems. A Ruspini partition

(Ruspini 1996), (Thiele 1997) of the input space D is accepted for the FLCs belonging to the class of type-II fuzzy systems.

As mentioned previously the scheduling variables can be employed generally as premise variables in the FLC instead of the state variables. That is also the situation of highly nonlinear plants (Johansen et al. 2000), (Horváth and Rudas 2004), where the scheduling variables make the difference between the local models of the controlled plant.

Each rule in (3) generates the firing strength $\alpha_i(\mathbf{x}) \in [0,1]$:

$$\alpha_i(\mathbf{x}) = \min(\mu_{\tilde{X}_{i,1}}(x_1), \mu_{\tilde{X}_{i,2}}(x_2), ..., \mu_{\tilde{X}_{i,n}}(x_n)), \ i = \overline{1, r}, \tag{4}$$

where $\mu_{\tilde{X}_{i,j}}, j = \overline{1, n}$, are the continuous membership functions of the LTs. The following assumption is made:

$$\forall \mathbf{x} \in D \ \exists i = \overline{1, r} \text{ such that } \alpha_i(\mathbf{x}) \neq 0, \tag{5}$$

The defuzzification in the FLC is done according to:
- the center of gravity defuzzification method for the FLCs belonging to the class of type-II fuzzy systems,
- the weighted-sum defuzzification method for the FLCs belonging to the class of type-III fuzzy systems.

The output of the FLC is the control signal $u(\mathbf{x})$ applied to the controlled plant viz. the actuator is the following function of $\alpha_i(\mathbf{x})$ and $u_i(\mathbf{x})$:

$$u(\mathbf{x}) = \frac{\sum_{i=1}^{r} \alpha_i(\mathbf{x}) u_i(\mathbf{x})}{\sum_{i=1}^{r} \alpha_i(\mathbf{x})}. \tag{6}$$

Definition 1: An active region of the ith fuzzy logic control rule is defined as the set $D_i^A = \{\mathbf{x} \in D \mid \alpha_i(\mathbf{x}) \neq 0\}$.

3 Stability Conditions for Fuzzy Logic Control Systems

The two stability analysis theorems to be presented here are based on Lyapunov's theorem dedicated to the stability analysis of time-varying systems expressed for non-autonomous systems in the attractive formulation due to (Khalil 2002). The theorems offer sufficient stability conditions for the fuzzy logic control systems with the structure presented in Section 2.

Theorem 1: Let $\mathbf{x} = \mathbf{0} \in D \subset IR^n$ be an equilibrium point of the system (1) and the fuzzy logic control system defined in Section 2. If there exists a continuously differentiable function $V : D \times [0, \infty) \to IR$ such that the conditions (7) and (8) are fulfilled:

$$W^1(\mathbf{x}) \leq V(\mathbf{x}, t) \leq W^2(\mathbf{x}) \ \forall t \geq 0, \tag{7}$$

$$\dot{V}_i(\mathbf{x}, t) = \frac{\partial V}{\partial t} + \frac{\partial V}{\partial \mathbf{x}}[\mathbf{f}(\mathbf{x}, t) + \mathbf{b}(\mathbf{x}, t)u_i(\mathbf{x})] \leq -W_i^3(\mathbf{x}) \ \forall \mathbf{x} \in D_i^A \ \forall t \geq 0, i = \overline{1, r}, \tag{8}$$

where $W^1(\mathbf{x})$, $W^2(\mathbf{x})$ and $W_i^3(\mathbf{x}), i = \overline{1, r}$, are continuous positive definite functions on D, then $\mathbf{x} = \mathbf{0}$ is uniformly asymptotically stable. In addition, if $W^1(\mathbf{x})$ is radially unbounded, then the equilibrium point $\mathbf{x} = \mathbf{0}$ is globally uniformly asymptotically stable.

Proof: The derivative with respect to time of the time-varying scalar function with continuous first-order partial derivatives $V : D \times [0, \infty) \to IR$, calculated along the trajectories of the system (1), is

$$\dot{V}(\mathbf{x}, t) = \frac{\partial V}{\partial t} + \frac{\partial V}{\partial \mathbf{x}} \dot{\mathbf{x}} = \frac{\partial V}{\partial t} + \frac{\partial V}{\partial \mathbf{x}}[\mathbf{f}(\mathbf{x}, t) + \mathbf{b}(\mathbf{x}, t)u(\mathbf{x})] =$$

$$= \frac{\partial V}{\partial t} + \sum_{j=1}^{n} \frac{\partial V}{\partial x_j}[f_j(\mathbf{x}, t) + b_j(\mathbf{x}, t)u(\mathbf{x})] = \frac{\partial V}{\partial t} + F(\mathbf{x}, t) + B(\mathbf{x}, t)u(\mathbf{x}), \tag{8}$$

where

$$F(\mathbf{x}, t) = \sum_{j=1}^{n} \frac{\partial V}{\partial x_j} f_j(\mathbf{x}, t) \in IR, \ B(\mathbf{x}, t) = \sum_{j=1}^{n} \frac{\partial V}{\partial x_j} b_j(\mathbf{x}, t) \in IR \cdot \tag{9}$$

The time derivative of $V(\mathbf{x}, t)$ for $u = u_i(\mathbf{x})$ is given by

$$\dot{V}_i(\mathbf{x}, t) = \frac{\partial V}{\partial t} + F(\mathbf{x}, t) + B(\mathbf{x}, t)u_i(\mathbf{x}), \ i = \overline{1, r} \cdot \tag{10}$$

Next the hypotheses of Theorem 1 result in

$$\frac{\partial V}{\partial t} + F(\mathbf{x}, t) + B(\mathbf{x}, t)u_i(\mathbf{x}) \leq -W_i^3(\mathbf{x}) \ \forall \mathbf{x} \in D_i^A \ \forall t \geq 0, i = \overline{1, r} \cdot \tag{11}$$

Multiplying (11) by $\alpha_i(\mathbf{x})$ and calculating the sum, the result will be

$$\frac{\partial V}{\partial t}\sum_{i=1}^{r}\alpha_i(\mathbf{x})+F(\mathbf{x},t)\sum_{i=1}^{r}\alpha_i(\mathbf{x})+B(\mathbf{x},t)\sum_{i=1}^{r}\alpha_i(\mathbf{x})u_i(\mathbf{x})\leq-\sum_{i=1}^{r}W_i^3(\mathbf{x})\alpha_i(\mathbf{x}), \quad (12)$$

and the division of (12) by $\sum_{i=1}^{r}\alpha_i(\mathbf{x})>0$ leads to

$$\frac{\partial V}{\partial t}+F(\mathbf{x},t)+B(\mathbf{x},t)\frac{\sum_{i=1}^{r}\alpha_i(\mathbf{x})u_i(\mathbf{x})}{\sum_{i=1}^{r}\alpha_i(\mathbf{x})}\leq-\frac{\sum_{i=1}^{r}W_i^3(\mathbf{x})\alpha_i(\mathbf{x})}{\sum_{i=1}^{r}\alpha_i(\mathbf{x})}. \quad (13)$$

Therefore the following relationship is obtained from (8) and (13):

$$\dot{V}(\mathbf{x},t)\leq-W^3(\mathbf{x})\;\forall\mathbf{x}\in D,\;W^3(\mathbf{x})=\frac{\sum_{i=1}^{r}W_i^3(\mathbf{x})\alpha_i(\mathbf{x})}{\sum_{i=1}^{r}\alpha_i(\mathbf{x})}. \quad (14)$$

Since $\alpha_i(\mathbf{x})$ and $W_i^3(\mathbf{x})$ are continuous functions on D, $W_i^3(\mathbf{x})>0\;\forall\mathbf{x}\neq\mathbf{0}$ and $W_i^3(\mathbf{0})=0$, $i=\overline{1,r}$, if they are defined in $\mathbf{x}=\mathbf{0}$, it results that $W^3(\mathbf{x})$ is a continuous positive definite function on D. Summing up, the conditions (7) and (14) satisfy Lyapunov's theorem for time-varying systems expressed in (Khalil 2002). Therefore, the equilibrium point at the origin is uniformly asymptotically stable. The proof is complete.

Theorem 1 ensures the sufficient conditions (7) and (8) (and eventually the condition concerning the radially unbounded $W^1(\mathbf{x})$) for the globally (uniformly) asymptotically stability of the accepted class of fuzzy logic control systems defined in Section 2. The following theorem ensures two sufficient conditions for the uniformly stability of the accepted class of fuzzy control systems.

Theorem 2: Let $\mathbf{x}=\mathbf{0}\in D\subset IR^n$ be an equilibrium point of the system (1) and the fuzzy logic control system defined in Section 2. If there exists a continuously differentiable function $V:D\times[0,\infty)\to IR$ such that the conditions (15) and (16) are fulfilled:

$$W^1(\mathbf{x})\leq V(\mathbf{x},t)\leq W^2(\mathbf{x})\;\forall t\geq0, \quad (15)$$

$$\dot{V}_i(\mathbf{x},t)=\frac{\partial V}{\partial t}+\frac{\partial V}{\partial \mathbf{x}}[\mathbf{f}(\mathbf{x},t)+\mathbf{b}(\mathbf{x},t)u_i(\mathbf{x})]\leq0\;\forall\mathbf{x}\in D_i^A\;\forall t\geq0, i=\overline{1,r}, \quad (16)$$

where $W^1(\mathbf{x})$ and $W^2(\mathbf{x})$ are continuous positive definite functions on D, then $\mathbf{x} = \mathbf{0}$ is uniformly stable.

Proof: The relations (8) to (10) in the proof of Theorem 1 are applied again in the hypotheses of Theorem 2, but (11) is replaced by

$$\frac{\partial V}{\partial t} + F(\mathbf{x},t) + B(\mathbf{x},t)u_i(\mathbf{x}) \le 0 \; \forall \mathbf{x} \in D_i^A \; \forall t \ge 0, i = \overline{1,r} \cdot \tag{17}$$

Next (17) is multiplied by $\alpha_i(\mathbf{x})$, the sum is calculated, the result is divided by $\sum_{i=1}^{r} \alpha_i(\mathbf{x}) > 0$ and the following result, similar to (14), can be expressed:

$$\dot{V}(\mathbf{x},t) \le 0 \; \forall \mathbf{x} \in D \cdot \tag{18}$$

The conditions (15) and (18) satisfy Lyapunov's theorem for time-varying systems expressed in (Khalil 2002), so the equilibrium point at the origin is uniformly stable. Therefore the proof is complete.

Since Theorem 1 is sufficient for Theorem 2 and thus stronger, the following Section will be concentrated on two applications of Theorem 1.

4   Case Studies

This Section is dedicated to the validation of the theoretical results presented in the previous Section. The stable design of fuzzy logic control systems will be illustrated in two case studies in terms of offering low-cost FLCs that belongs to the class of type-III fuzzy systems to stabilize the Lorenz chaotic system i.e. Takagi-Sugeno FLCs.

4.1 Case Study 1

The case study 1 concerns the model-based design of a fuzzy control system to stabilize the Lorenz chaotic system. The Lorenz equations (Lorenz 1963), (Lorenz 1993) are transformed into the following state-space equations of an SISO NTV controlled plant:

$$\dot{\mathbf{x}} = \begin{pmatrix} \sigma(x_2 - x_1) \\ x_1(\rho(t) - x_3) - x_2 \\ x_1 x_2 - \beta x_3 \end{pmatrix} + \begin{pmatrix} 1 \\ 0 \\ 0 \end{pmatrix} u, \mathbf{x}(t_0) = \mathbf{x}_0, \tag{19}$$

where $\mathbf{x} = [x_1 \ \ x_2 \ \ x_3]^T$, the three parameters $\sigma, \rho, \beta > 0$ are called the Prandtl number, the Rayleigh number, and a physical proportion, respectively, and they affect the system's behavior. The plant defined in (19) exhibits chaotic behaviors being extremely sensitive to the initial conditions associated to its associated Cauchy problem. The classical values used to demonstrate the chaotic behavior and applied here are $\sigma = 10$, $\beta = 8/3$ and ρ is variable with respect to time, $0 \le \rho(t) \le 100 \ \forall \ t \ge t_0 = 0$.

Use is made of the inputs x_1 and x_2 in the design of the FLC. The fuzzification is done according to Fig. 2 which illustrates the membership functions corresponding to the LTs of the linguistic variables x_1 and x_2, with the parameters $S_i = 10, T_i = 40, i = \overline{1,2}$.

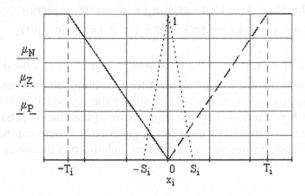

Fig. 2. Membership functions of x_1 and x_2

The complete rule base of the Takagi-Sugeno FLC is $(r = 9)$

Rule 1 : IF x_1 IS P AND x_2 IS P THEN $u = u_1(x_1, x_2)$,

Rule 2 : IF x_1 IS N AND x_2 IS N THEN $u = u_2(x_1, x_2)$,

Rule 3 : IF x_1 IS P AND x_2 IS N THEN $u = u_3(x_1, x_2)$,

Rule 4 : IF x_1 IS N AND x_2 IS P THEN $u = u_4(x_1, x_2)$,   (20)

Rule 5 : IF x_1 IS P AND x_2 IS Z THEN $u = u_5(x_1, x_2)$,

Rule 6 : IF x_1 IS N AND x_2 IS P THEN $u = u_6(x_1, x_2)$,

Rule 7 : IF x_1 IS Z AND x_2 IS P THEN $u = u_7(x_1, x_2)$,

Rule 8 : IF x_1 IS Z AND x_2 IS N THEN $u = u_8(x_1, x_2)$,

Rule 9 : IF x_1 IS Z AND x_2 IS Z THEN $u = u_9(x_1, x_2)$.

Theorem 1 has been applied to calculate the expressions of the consequents $u_i(x_1, x_2)$, $i = \overline{1,9}$. The domain D has been set to $D = [-100,100] \times [-100,100]$ and the following function $V(\mathbf{x}, t)$, which is a continuously differentiable positive function on D, has been set to fulfill the condition

$$W^1(\mathbf{x}) = (x_1^2 + x_2^2 + x_3^2) \le V(\mathbf{x}, t) = (x_1^2 + x_2^2 + x_3^2)(1 + e^{-t}) \le \atop \le 2(x_1^2 + x_2^2 + x_3^2) = W^2(\mathbf{x}), \tag{21}$$

and the functions $W^3(\mathbf{x}) = W_i^3(\mathbf{x}) = (\sigma x_1^2 + x_2^2 + \beta x_3^2), i = \overline{1,9}$, have been set for all rules.

The following control signals in the consequents of the rules satisfy (8):

$$u_1(x_1, x_2) = -100[\sigma + \rho(t)], u_2(x_1, x_2) = 100[\sigma + \rho(t)], u_3(x_1, x_2) = 0, \atop u_4(x_1, x_2) = 0, u_5(x_1, x_2) = -10[\sigma + \rho(t)], u_6(x_1, x_2) = 10[\sigma + \rho(t)], u_7(x_1, x_2) = \atop = -x_2[\sigma + \rho(t)], u_8(x_1, x_2) = -x_2[\sigma + \rho(t)], u_9(x_1, x_2) = -x_2[\sigma + \rho(t)]. \tag{22}$$

Considering $\rho(t) = t$, $0 \le t \le 100$, $r = 0$ and $\mathbf{x}_0 = [1 \quad -1 \quad 1]^T$, the digital simulation results are presented in Figs. 3 and 4 for the controlled plant and the fuzzy logic control system, respectively. Fig. 4 shows the performance improvement with respect to Fig. 3 reflected by the suppression of the chaotic behavior. However the performance indices of the fuzzy logic control system can be improved further by imposing performance specifications and inserting dynamics in the FLC structure.

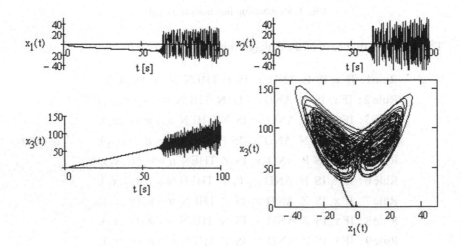

Fig. 3. Behavior of Lorenz chaotic system

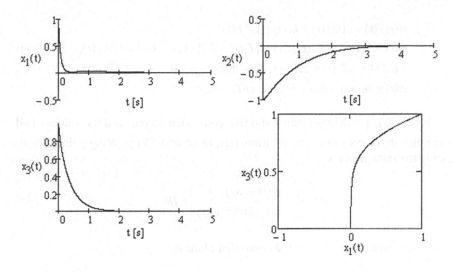

Fig. 4. Behavior of Lorenz chaotic system stabilized by type-III fuzzy logic control system

4.2 Case Study 2

The case study 2 concerns the model-based design of a fuzzy control system for the longitudinal slip control of the INTECO ABS laboratory equipment (Fig. 5) implemented in the Intelligent Control Systems Laboratory with the "Politehnica" University of Timisoara (PUT). The controlled plant is strongly nonlinear (Rădac et al. 2008) and its model can be simplified in terms of the SISO LTV expression

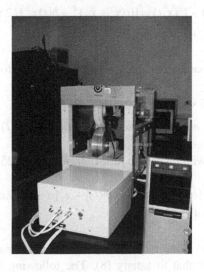

Fig. 5. Experimental setup

$$m(t)\ddot{y}(t) + c(t)\dot{y}(t) + k(t)y(t) = u(t),$$
$$m(t) = T_3(t)T_A / k_P(t), c(t) = [T_3(t) + T_A]/ k_P(t), k(t) = 1/ k_P(t), \qquad (23)$$
$$-6.234 \text{ s} \le T_3(t) \le 5.139 \text{ s}, T_A = 0.05 \text{ s}, -3.446 \le k_P(t) \le 3.405,$$
$$m(t) \ne 0, c(t)/m(t) > 0, k(t)/m(t) > 0.$$

where $y = \lambda$ is the longitudinal slip (the controlled output) and the values of all parameters have been obtained by linearization around 54 operating points. Introducing the state vector $\mathbf{x}$:

$$\mathbf{x}(t) = \begin{bmatrix} x_1(t) = y(t) - r \\ x_2(t) = \dot{y}(t) = \dot{x}_1(t) \end{bmatrix} \in IR^2, \qquad (24)$$

the state-space equations of the controlled plant are

$$\dot{\mathbf{x}}(t) = \mathbf{f}(\mathbf{x},t) + \mathbf{b}(\mathbf{x},t)u(t),$$
$$\mathbf{f}(\mathbf{x},t) = \begin{bmatrix} x_2(t) \\ -[k(t)x_1(t) + c(t)x_2(t) + k(t)r]/m(t) \end{bmatrix}, \mathbf{b}(\mathbf{x},t) = \begin{bmatrix} 0 \\ 1/m(t) \end{bmatrix}. \qquad (25)$$

The complete rule base of the Takagi-Sugeno FLC is presented in (20) and the membership function shapes are illustrated in Fig. 2, with the parameters $S_1 = T_1 = 0.3$, $S_2 = T_2 = 1.5$. The domain D has been set to $D = [-1,1] \times [-1,1]$, and the continuously differentiable positive function on D, $V(\mathbf{x},t)$, has been set to fulfill the condition

$$W^1(\mathbf{x}) = V(\mathbf{x},t) = [(\alpha x_1 + x_2)^2 + b(t)x_1^2]/2 \le \qquad (26)$$
$$\le (\alpha x_1 + x_2)^2 + b(t)x_1^2 = W^2(\mathbf{x}),$$

where the auxiliary parameters α and $b(t)$ are defined as follows:

$$\alpha = \text{const}, 0 < \alpha < \min\left(c(t)/m(t), \sqrt{k(t)/m(t)}\right), \qquad (27)$$
$$b(t) = k(t)/m(t) - \alpha^2 + \alpha c(t)/m(t), \dot{b}(t) \le 2\alpha k(t)/m(t).$$

The derivative of V with respect to time making use of (25) to (27) is

$$V(\mathbf{x},t) = x_1^2[\dot{b}(t)/2 - \alpha k(t)/m(t)] + x_2^2[\alpha - c(t)/m(t)] + \qquad (28)$$
$$+ (x_2 + \alpha x_1)[u - k(t)r]m.$$

The relationship (28) is useful in the design of the consequents of the fuzzy logic control rules such that to satisfy (8). The following functions and control signals in the consequents of the rules have been designed:

$$W_i^3(\mathbf{x},t) = W_i^3(x_1,x_2,t) = x_1^2[\dot{b}(t)/2 - \alpha k(t)/m(t)] +$$
$$+ x_2^2[\alpha - c(t)/m(t)] \; \forall \mathbf{x} \in D_i^A, i = \overline{1,9}, u_1(x_1,x_2) = k(t)r - 1, \quad (29)$$
$$u_2(x_1,x_2) = k(t)r + 1, u_i(x_1,x_2) = k(t)r - \alpha x_1 - x_2, i = \overline{3,8}, u_9(x_1,x_2) = k(t)r.$$

The scenario used in the real-time experiments is characterized by the constant reference input $r = 0.3$, without disturbance inputs, constant $k(t)$ and zero initial conditions. The real-time experimental results are presented in Fig. 6 and prove the good performance indices of the fuzzy logic control system.

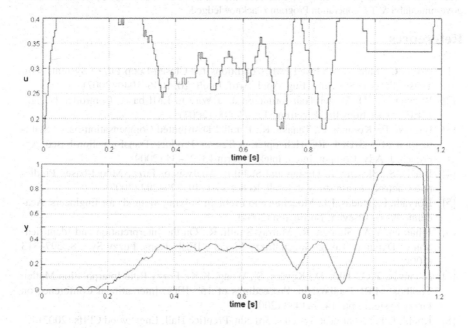

Fig. 6. Real-time experimental results: control signal (u) and controlled output (y) versus time

Conclusions

The paper has proposed two stability analysis theorems employed in the model-based design of a class of fuzzy logic control systems that cope with SISO NTV systems. FLCs belonging to the classes of type-II and type-III of fuzzy systems have been considered.

The first shortcoming of the approaches is the absence of the computer-aided design of the FLCs. However the two case studies, that validate the theoretical approaches, highlight the possibility to transform the two theorems into design algorithms (Precup et al. 2008a).

The second shortcoming is the need to compare the new approaches with the less conservative LMI-based ones. That is a future research direction although the strength of the stability conditions has been outlined in the previous Section. Another research direction is the extension of the stability analysis theorems to solving regulation and tracking control problems based on strong and clearly formulated performance specifications including robustness issues.

Acknowledgments. The paper was supported by the CNMP and CNCSIS of Romania. The fourth and sixth authors are doctoral students with the PUT and SOP HRD bursars co-financed by the European Social Fund through the project ID 6998. The support stemming from the cooperation between Budapest Tech Polytechnical Institution and PUT in the framework of the Intergovernmental S & T Cooperation Program is acknowledged.

References

[1] Arino, C., Sala, A.: Relaxed LMI Conditions for Closed-Loop Fuzzy Systems with Tensor-Product Structure. Eng. Appl. Artif. Intell. 20, 1036–1046 (2007)

[2] Baranyi, P.: TP Model Transformation as a Way to LMI-based Controller Design. IEEE Transactions Ind. Electron 51, 387–400 (2004)

[3] Baranyi, P., Korondi, P., Tanaka, K.: Parallel Distributed Compensation-based Stabilization of a 3-DOF RC Helicopter: a Tensor Product Transformation-based Approach. J. Adv. Comput. Intell. Intell. Inform 13, 25–34 (2009)

[4] Blažič, S., Škrjanc, I.: Design and Stability Analysis of Fuzzy Model-based Predictive Control - a case study. J. Intell. Robot Syst. 49, 279–292 (2007)

[5] Horváth, L., Rudas, I.J.: Modeling and Problem Solving Methods for Engineers. Academic Press, Elsevier, Burlington (2004)

[6] Johansen, T.A., Shorten, R., Murray-Smith, R.: On the Interpretation and Identification of Dynamic Takagi-Sugenofuzzy Models. IEEE Trans. Fuzzy Syst. 8, 297–313 (2000)

[7] Johanyák, Z.C., Tikk, D., Kovács, S., Wong, K.K.: Fuzzy Rule Interpolation Matlab Toolbox - FRI Toolbox. In: Proceedings of 15th IEEE International Conference on Fuzzy Systems, pp. 1427–1433 (2006)

[8] Khalil, H.K.: Nonlinear Systems, 3rd edn. Prentice Hall, Englewood Cliffs (2002)

[9] Kim, S.H., Park, P.G.: Observer-based Relaxed H_∞ Control for Fuzzy Systems Using a Multiple Lyapunov Function. IEEE Trans. Fuzzy Syst. 17, 477–484 (2009)

[10] Kóczy, L.T.: Fuzzy If-Then Rule Models and Their Transformation into one Another. IEEE Trans. Syst. Man Cybern. A 26, 621–637 (1996)

[11] Kruszewski, A., Wang, R., Guerra, T.M.: Nonquadratic Stabilization Conditions for a Class of Uncertain Nonlinear Discrete Time TS Fuzzy Models: a New Approach. IEEE Trans. Autom. Control 53, 606–611 (2008)

[12] Lam, H.K., Leung, F.H.F.: Stability Analysis of Discrete-Time Fuzzy-Model-based Control Systems with Time Delay: Time-Delay Independent Approach. Fuzzy Sets Syst 159, 990–1000 (2008)

[13] Lam, H.K., Seneviratne, L.D.: Tracking Control of Sampled-Data Fuzzy-Model-based Control Systems. IET Control Theory Appl. 3, 56–67 (2009)

[14] Lee, H.C., Choi, J.W.: Linear Time-Varying Eigenstructure Assignment with Flight Control Application. IEEE Trans. Aerosp. Eletron Eng. 40, 145–157 (2004)

[15] Lendek, Z., Babuška, R., De Schutter, B.: Stability Analysis and Observer Design for Decentralized TS Fuzzy Systems. In: Proceedings of the IEEE International Conference on Fuzzy Systems, pp. 631–636 (2008)

[16] Lorenz, E.N.: Deterministic Nonperiodic Flow. Journal Atmos. Sci. 20, 130–141 (1963)
[17] Lorenz, E.N.: The Essence of Chaos. University of Washington Press, Seattle (1993)
[18] Michels, K., Klawonn, F., Kruse, R., Nürnberger, A.: Fuzzy Control: Fundamentals, Stability and Design of Fuzzy Controllers. Springer, Heidelberg (2006)
[19] Oblak, S., Škrjanc, I., Blažič, S.: Fault Detection for Nonlinear Systems with Uncertain Parameters Based on the Interval Fuzzy Model. Eng. Appl. Artif. Intell. 20, 503–510 (2007)
[20] Pang, C.T., Lur, Y.Y.: On the Stability of Takagi-Sugeno Fuzzy Systems with Time-Varying Uncertainties. IEEE Trans. Fuzzy Syst. 16, 162–170 (2008)
[21] Petres, Z., Baranyi, P., Korondi, P., Hashimoto, H.: Trajectory Tracking by TP Model Transformation: Case Study of a Benchmark Problem. IEEE Trans. Ind. Electron 54, 1654–1663 (2007)
[22] Precup, R.E., Preitl, S., Rudas, I.J., et al.: Design and Experiments for a Class of Fuzzy Logic Controlled Servo Systems. IEEE/ASME Trans. Mechatron. 13, 22–35 (2008a)
[23] Precup, R.E., Preitl, S., Ursache, I.B., et al.: On the Combination of Tensor Product and Fuzzy Models. In: Proceedings of 2008 IEEE International Conference on Automation, Quality and Testing, Robotics, vol. 2, pp. 48–53 (2008b)
[24] Rădac, M.B., Precup, R.E., Preitl, S., et al.: Linear and Fuzzy Control Solutions for a Laboratory Anti-Lock Braking System. In: Proceedings of 6th International Symposium on Intelligent Systems and Informatics paper index, vol. 49, p. 6 (2008)
[25] Ruspini, E.H.: A new Approach to Clustering. Inf. Control 15, 22–32 (1969)
[26] Sala, A., Guerra, T.M., Babuška, R.: Perspectives of Fuzzy Systems and Control. Fuzzy Sets Syst. 156, 432–444 (2005)
[27] Sugeno, M.: On Stability of Fuzzy Systems Expressed by Fuzzy Rules with Singleton Consequents. IEEE Trans. Fuzzy Syst. 7, 201–224 (1999)
[28] Sun, C.H., Wang, W.J.: An Improved Stability Criterion for T-S Fuzzy Discrete Systems via Vertex Expression. IEEE Trans. Syst. Man Cybern. B 36, 672–678 (2006)
[29] Škrjanc, I., Blažič, S., Agamennoni, O.E.: Interval Fuzzy Modeling Applied to Wiener Models with Uncertainties. IEEE Trans. Syst. Man Cybern. B 35, 1092–1095 (2005)
[30] Tanaka, K., Ohtake, H., Wang, H.O.: A Descriptor System Approach to Fuzzy Control System Design via Fuzzy Lyapunov Functions. IEEE Trans. Fuzzy Syst. 15, 333–341 (2007)
[31] Tanaka, K., Ikeda, T., Wang, H.O.: Fuzzy Regulators and Fuzzy Observers: Relaxed Stability Conditions and LMI-based Design. IEEE Trans. Fuzzy Syst. 6, 250–265 (1998)
[32] Tanaka, K., Yamaguchi, K., Ohtake, H., Wang, H.O.: Sensor Reduction for Backing-up Control of a Vehicle with Triple Trailers. IEEE Trans. Ind. Electron 56, 497–509 (2009)
[33] Thiele, H.: A Characterization of Arbitrary Ruspini Partitions by Fuzzy Similarity Relations. In: Proceedings of Sixth IEEE International Conference on Fuzzy Systems, vol. 1, pp. 131–134 (1997)
[34] Van der Kloet, P., Neerhoff, F.L.: Dynamic Eigenvalues for Scalar Linear Time-Varying Systems. In: Proceedings of 15th International Symposium on Mathematical Theory of Networks and Systems paper index, vol. 14423, p. 8 (2002)
[35] Vaščák, J.: Fuzzy Cognitive Maps in Path Planning. Acta Tech. Jaurinensis Ser. Intell. Comput. 1, 467–479 (2008)

[36] Wang, W.H., Sun, C.H.: Relaxed Stability and Stabilization Conditions for a TS
      Fuzzy Discrete System. Fuzzy Sets Syst. 156, 208–225 (2005)
[37] Wei, G., Feng, G., Wang, Z.: Robust H_∞ Control for Discrete-Time Fuzzy Systems
      with Infinite-distributed Delays. IEEE Trans. Fuzzy Syst. 17, 224–232 (2009)
[38] Yang, C., Zhang, Q.: Multiobjective Control for T-S Fuzzy Singularly Perturbed Sys-
      tems. IEEE Trans. Fuzzy Syst. 17, 104–115 (2009)
[39] Yoneyama, J.: Robust Stability and Stabilization for Uncertain Takagi-Sugeno Fuzzy
      Time-Delay Systems. Fuzzy Sets Syst. 158, 115–134 (2007)
[40] Yuan, X., Wang, Y.: A Novel Electronic-Throttle-Valve Controller-based on Ap-
      proximate Model Method. IEEE Trans. Ind. Electron. 56, 883–890 (2009)

Situational Control, Modeling and Diagnostics of Large Scale Systems

Ladislav Madarász*, Rudolf Andoga*, Ladislav Fozo*, and Tobiáš Lazar **

* Dep. of Cybernetics and A.I., Technical University of Košice, Letná 9, Košice, Slovakia
ladislav.madarasz@tuke.sk,rudolf.andoga@tuke.sk,ladislav.fozo@tuke.sk
** Dep. of Avionics, Technical University of Košice, Rampová 7, Košice, Slovakia
tobias.lazar@tuke.sk

Abstract. A large scale system in general is a high dimensional high parametric system with complex dynamics. For efficient and optimal function of such systems, it is necessary to propose and implement newest knowledge from the areas of cybernetics and artificial intelligence. Present control systems are often limited to control a complex system only at some given conditions. However in real-world applications these systems find themselves in very different working conditions, what influences parameters of their operation and characteristics of behavior and may lead to errors and critical states. The article deals with overview of methods of situational control and is aimed on implementation of these methods in the area of turbojet propulsion.

1 Introduction

A complex system in general is a high dimensional high parametric system with complex dynamics. For effective and optimal function of such systems it is necessary to propose and implement newest knowledge from the areas of cybernetics and artificial intelligence. Present control systems are often limited to control a complex system only at some given conditions. However in real-world applications these systems find themselves in very different working conditions, what influences parameters of their operation and characteristics of behavior and may lead to errors and critical states.

It is needed to handle all these conditions and situations in such way, the system would work economically and effectively, thus optimally. It is possible to secure this factor of optimal operation of a controlled system in all eventual states of the environment and its inner states, represented by its inner parameters with use of situational control methodology. The terms, situational, situational control intuitively tell, that they represent control of a chosen complex system in its different situational states. In an ideal case this will represent the control of a complex

I.J. Rudas et al. (Eds.): Towards Intelligent Engineering & Information Tech., SCI 243, pp. 153–164.
springerlink.com   © Springer-Verlag Berlin Heidelberg 2009

system in its all operational states. However such an ideal case represents existence of infinite number of operational states but we posses only a limited number of control strategies (algorithms). So we are coming to the main idea of situational control methodology. Due to limitation of number of control algorithms, it is necessary to limit the number of operational situations, so that control strategies would cover all operational situations. By declaration of this fact we are getting towards situational classification, which demands proposal of situational classes and algorithms to control a system which finds itself within states defined by these situational classes. Situational class represents then a set of similar operational states of a complex system. Usually one control strategy covers one situational class, but a case can occur where more situational classes are covered by one suitable robust control strategy.

From situational control methodology point of view, proposal of situational classes, proposal of algorithms classifying the actual state of a system into situational classes and proposal of the control strategies are the key points in this area. The proposal of situational classes is usually realized by an expert from the given problems area. The area of classifiers of actual states of a system, offers a broad field of use of the most modern knowledge from the field of artificial intelligence, as are neuro, neuro-fuzzy, or genetic algorithms systems usable in this field. By proposal of individual control strategies it is again possible to use approaches of artificial intelligence, but also the traditional approaches of control of complex systems.

Applications, where methodology of situational control may be applied cover a vast area of complex systems, as are for example electricity networks, electric energy production, control of robot technological complexes, control of jet engines, etc. This article will be aimed at the use of the methods of situational control in the field of control of the aircraft turbojet engines.

2 Methodology of Situational Control

Situational control of large scale systems as one of the alternatives of such systems control was invented and further developed in Russia (Ju. I. Klykov, J. M. Klimnik, A. I. Sokoľnikov, D. A. Pospelov; G. Osipov; A. N. Averkin; O. Citkin; A. A. Zenkin, L. S. Zagadskaja, V. F. Erlich, V. F. Gorjachuk, V. S. Lozovskij, V. F. Choroschevskij). In western countries the developers of situational control methodics were the following scientists J. Zaborszky, K. W. Whang, K. V. Prasad, D. R. Stinson, F. B. Vernadat, D. Howland, S. Beer, A. P. Sage. Problems of situational control of complex systems were also solved in conditions of former Czech-slovak federation by scientists as J. Beneš, J. Spal, L. Madarász and others. [1].

Situational control was designed for the control of complex systems, where the traditional cybernetic models weren't sufficient. [2]. The model proposed in [2] wasn't sufficient to control systems characterized by features like unique dynamics, incompleteness and indeterminacy of description, ambiguity and presence of a free will.

More general approach to situational control is the following structural scheme of formatter control of complexes shown in Figure 1 [3]. Formatter control of a complex means not only the control of its parameters, but also the control of the form of the complex system.

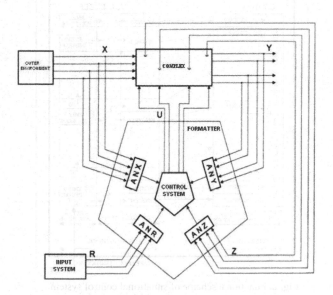

Fig. 1. Formatter control of a complex system

The central element of this system is the control component, which is represented by a structure of a control component (Fig. 1) using methods of situational control. This means that it is a system, which makes situational classification and chooses the appropriate control strategy upon the basis of incoming signals from analyzers ANX, ANY, ANR and ANZ. The following figure shows the functional scheme of situational control of a complex system.

The control process is composed of two phases, decision and the control phase, where every of them is divided into classification and the action phase. Processed situation is analyzed in the selection part of the decision phase. According to analysis results the situation is then assigned to one of the "N" standard situations, which are designed to process according emergency situations. Every standard situation has a certain file of algorithms which are saved in memory to its disposal. During the action period of the decision-making phase, the most suitable file of algorithms is being activated to process the given situation.

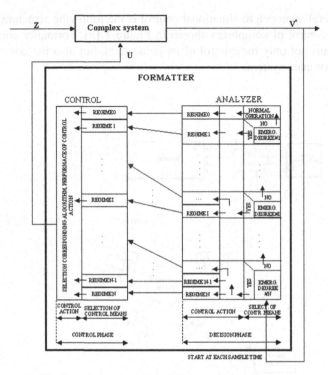

Fig. 2. Functional scheme of situational control system

During the selection interval of the control phase, these algorithms adapt themselves for solution of according situation (parameterization and other adaptation connected with activation). Realization of control activity occurs in the action period of the control phase.

For the design of a system respecting the requirements of control in anytime, the following algorithm is proposed [4]:

a)   description of the structure and function of the controlled complex system,
b)   global goal designation,
c)   classification of erroneous operational states and their causes,
d)   classification and description regimes functions of the control, that are assigned to individual erroneous states,
e)   algorithmization of individual regimes of control,
f)   implementation.

By design of regimes of control and also by the design of a classification mechanism, it is today necessary to consider use of robust intelligent methods for these tasks. In the past for situational classification methods of multi-criteria decision making [1], expert system [4], or catastrophe and chaos theory have been used [4]. Situational control in its explicit definition has been successfully used in the following applications [4]:

- control of electricity networks,
- control of robot – technological complexes,
- ablation of myocard structures,
- situational control of automated technological workstations for mechanical working.

Nowadays however, as classification systems neural networks that are able to approximate any continuous function with the ability to learn. By proposal and algorithmization it is necessary to choose a very susceptive approach to selection of an optimal modern method to control the chosen complex. The following chapters will be aimed on implementation of the mentioned approaches in the area of turbojet engines control.

3 Modern Control Systems of Turbojet Engines

The main global aim of control of turbojet engines is similar to other systems and that is increasing their safety and effectiveness by reduction of costs. This demands application of new technologies, materials, new conceptions of solutions [5] and also development in systems of control and regulation of aircraft turbojet engines and processes ongoing in them.

Demands for control and regulation systems result mainly from specific properties of the object of control – a turbojet engine. Among the basic functions of control systems of turbojet engine belong the following ones – manual control, regulation of its parameters and their limitation. Manual control and therefore choice of regime of the engine is realized by a throttle lever according to a flight situation or expected maneuver. By regulation of a turbojet engine we understand such a kind of control where the chosen parameters of the engine are maintained on certain set levels, thus keeping its regime.

In the past, the classical control systems of turbojet engines were implemented mainly by hydro-mechanical elements, which however suffered from deficiencies characteristic for such systems. Among such deficiencies were, high mass of such systems, inaccuracies due to mechanical looses and low count of regulated parameters. However development of electronic systems and elements is ongoing, which will allow to increase precision of regulation of parameters of turbojet engines and their count to secure more complex and precise regulation of turbojets.

Use of electronics and digital technologies in control systems of turbojet engines has brought [5]:

- lowering of mass of control system,
- higher complexity of control – The count of regulated parameters used to be 3 to 7 by hydro-mechanical systems, however the digital systems operate with 12 to 16 parameters,

- increasing of static precision of regulation of different parameters (for example, precision of rotations from ±0.5% to ±0.1%, precision of regulation of temperature from ±12K to ±5K,
- increase in reliability, service life and economics of operation of the driving unit of an aircraft,
- easier backup, technology of use and repairs, possibility of use of automatic diagnostics.

By design of solution of a control system for a turbojet engine, it is necessary to build an appropriate mathematical model of the engine. The ideal approach to design of electronic systems is a modular one, from hardware or software point of view. This implies use of qualitative processing units that are resistant to noises of environment and also realization of bus systems with low delays is very important in this approach. Further improvement in quality of control can be achieved by implementation of progressive algorithms of control, diagnostics and planning in electronic systems. These algorithms have to be able to asses the state of the controlled system (turbojet engine in our case), then parameterize action elements and they have to be able to control the engine under erroneous conditions represented in outer environment or as errors in subsystems of the engine itself. Prediction of such states represents an area to incorporate predictive control system. Methods of situational control bound with elements of artificial intelligence supply many robust tools for solution of afore mentioned problems and sub-problems.

From the point of view of use of electrical and electronic systems in controls the turbojet control systems can be roughly hierarchically divided into following sets: [6]:

1 electronic limiters,
2 Partial Authority Flight Control Augmentation (PAFCA)
3 „High Integration Digital Electronic Control" (HIDEC); „Digital Engine Control" - (DEC); „Full Authority Digital Engine Control" – (FADEC)).

The division of control systems into these three levels is not absolutely distinct, as systems on higher level as for example HIDEC system can utilize control mechanisms as electronic limiters. For example FADEC systems are often realized as single or double loop control systems with utilization of PI control algorithms or electronic limiters with estimation filters.

4 Implementation of Situational Control Methodology in Turbojet Engines

4.1 Characteristics of a Small Turbojet Engine MPM 20

It wouldn't be economically favorable to test new control methodologies with expensive and also very complex big turbojet engines, taking also in regard the safety of such testing. Therefore a special class of turbojet engines designated as small turbojet engines (usually used to start normal sized engines) can be used as an ideal test-bed for differently aimed experiments in this area. Our research is headed towards three basic aims [6].

1   Digital measurement of turbojet engines, which means digital real-time measurement of different state and diagnostic parameters of such engines.

2   Design and implementation of new dynamic models and control algorithms of turbo-jet engines, especially the situational control algorithms incorporating methods of artificial intelligence.

3   The aim resulting from the previous points is to explore possibilities of use of alternative fuels in turbojet engines.

The experimental engine MPM 20 has been derived from the TS – 20 engine, what is a turbo-starter turbo-shaft engine previously used for starting engines AL-7F. The engine has been rebuilt to a state, where it represents a single stream engine with radial compressor and single one stage non-cooled turbine and outlet jet. The basic scheme of the engine is shown in Figure 3

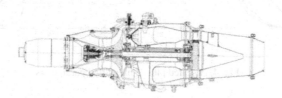

Fig. 3. Structural scheme of small turbojet engine MPM 20

The engine is located in "Laboratory of intelligent control systems of aircraft engines" on a specially built mount with digital data acquisition system designed by authors. At present 15 parameters of the engine are measured, among these are the following ones:

- air temperature at the outlet from the diffuser of the radial compressor - t_{2C} [^{0}C];
- gas temperatures in front of the gas turbine - t_{3C} [^{0}C];
- gas temperature beyond the gas turbine - t_{4C} [^{0}C];
- static pressure of air beyond the compressor p_2 [Ata];
- static pressure of gases in front of the gas turbine p_3 [Ata];
- static pressure of gases beyond the gas turbine p_4 [Ata];
- fuel flow Q_{pal} [l/min];
- thrust **Th** [kg];
- rotations of the turbine/compressor, n_1 [rpm]

The following figure shows a scaled plot of courses of measured parameters during one operational run of the engine from startup to shutdown.

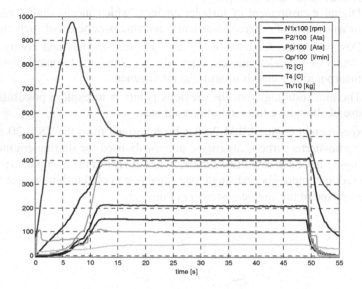

Fig. 4. A scaled plot of measured parameters

The plot also shows the basic parameters of the engine with temperatures reaching as high as 980 C, speed of the engine is at about 40,000 rpm and thrust of the engine is in this case around 40 kg of force.

4.2 Design of Situational Control System for MPM 20 Engine

Methods of artificial intelligence may offer new quality into control systems. However they can bring such benefits only after a careful model based analysis of a system where they should be applied with regards to simplicity and error free operation of such control system. Because on the lowest level of control we deal mostly with data and raw numbers, the approaches of sub-symbolic AI are appropriate to be used in design of intelligent FADEC control systems. However, on higher level of integration some symbolic concepts could also be used. From the area of symbolic AI three basic approaches can be successfully used:

- neural networks,
- fuzzy inference systems,
- genetic algorithms.

All three approaches offer a vast number of methods and their combinations in hybrid architectures. Feed-forward topologies can be successfully employed mainly in modeling of non linearity of jet engine and as decision elements in control circuits, for example as a gating neural network in general jet engine situational control formatter scheme designed by authors as shown in Figure 5.

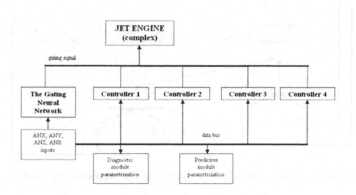

Fig. 5. General situational control system scheme to be used in turbojet engines

By design of control system for the MPM 20 engine, we consider two basic approaches. In broader scope it will be the situational control methodology and control by a single parameter and in more focused scope it will be anytime control and control by two parameters. The proposed structure of situational control system uses paradigm and schemes described in [6]. The whole conception of the situational control system is decomposition of operational states into time spaced situational frames (classes) and every situational frame has one corresponding control algorithm (or controller) assigned to it. Anytime control techniques are more focused because their aim is to control the system in specific critical states or situational frames. By those critical states we mean the deficiency in data for the controller due to system's overload or failure of certain components in control circuit.

In development and design of FADEC (Full Authority Digital Engine Control) compliant control system, situational control methVodology approach has been used. It is similar in to the one described in previous chapter, what means we use a gating neural network as a classifier of situational frames and system of controllers to handle those situational frames. We use concepts of traditional situational control and formatter control of complex systems [1, 2, 7]. The system has been described in [6]. The resulting physical architecture including analyzers of input (X), state (Z), output (Y) and desired (R) parameters is shown in Figure 6.

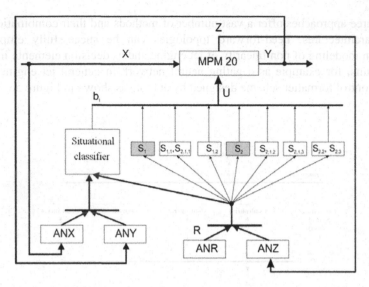

Fig. 6. Implementation of situational control system with decomposition into situational frames

Blocks designated as $S_{i,j,k}$ represent controllers for different situational frames, which result from the situational decomposition of operation of the engine in three levels:

– startup of the engine – situational frame S_1
– steady operational state – situational frame S_2
– shut down – situational frame S_3

Every situational frame is decomposed into sub – frames while only certain ones are shown in Figure 6. The whole decomposition is done with use of expert knowledge in the area and data clustering algorithms. Full situational decomposition of states is shown in Figure 7.

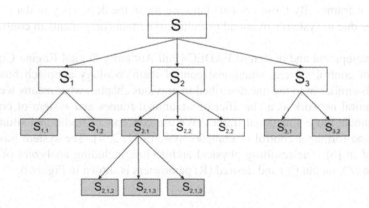

Fig. 7. Situational frames of operation of MPM 20 engine

The blocks in Figure 7 have the following meaning:

S_1 – startup of the engine:

$S_{1,1}$ – excess temperatures, $S_{1,2}$ – insufficient pressure P_{2c}

S_2 – steady state of operation:

$S_{2,1}$ – atypical state:

$S_{2,1,1}$: low compression, $S_{2,1,2}$ – low fuel flow, $S_{2,1,3}$ – unstable speed,

$S_{2,2}$ – acceleration, $S_{2,3}$ – deceleration

S_3 – shutdown

$S_{3,1}$ – stall of the engine

$S_{3,2}$ – error by run-down

The grey blocks in Figure 7 represent atypical situational frames and every frame has defined control strategy while the switching element is done by feed-forward neural network with time delays on input, trained by scaled conjugate gradient algorithm [8]. The description of all these elements is out of the scope of this paper and the whole algorithm is undergoing testing in laboratory conditions.

Within the frame of anytime control methods, we will deal with proposal of a simplified dynamic model of the small turbojet engine. The model will incorporate two input parameters compared to the more complex situational model, which is dependant on the fuel supply parameter only. The model will result from measurements of change to fuel supply and different cross-sections of exhaust nozzles. Simulation of temporary failure in input data (sampling errors) of the system (or designed dynamic model). Design and implementation of anytime control algorithm for the constructed mathematical model and critical states. This will include design of multi-parametric system of automatic control, in our case with two inputs and multiple outputs [9].

Research and observation of flexibility and quality of regulation of the designed system according to measured data and other possible critical states of action units (for example in case of blockage of outlet nozzle or total failure in data measuring current cross-section, etc.).

Conclusions

The object of a small turbojet engine MPM 20 gives us an ideal test bed for research of methods in the areas of non-linear dynamic systems modeling and design of advanced control algorithms. Further research will be done in the area of situational modeling that will be headed towards broadening of input parameters of the situational model of the engine and further refinement of situational classes designation. In this area we will be aimed at use of automatic algorithms to find boundaries between situational frames within multivariate space of parameters contrary to their setting by an expert. Anytime control algorithms represent other area of our interest with great possibilities of application of intelligent algorithms that will deal with critical states of operation of the engine and will be further embedded in the whole system of situational control of the engine. Design of such algorithms demands also further refinement of proposed models. All research in the areas of situational modeling, situational control and anytime algorithms should bring new quality of control and modeling in the area of turbojet engines and we expect this knowledge to be also expanded to other areas of technical systems.

Acknowledgments. The work is supported by the project VEGA no. 1/0394/08 – Algorithms of situational control and modeling of large scale systems.

References

[1] Spal, J., Madarász, L.: Problems of Classification in Diagnostics and Control of Complex Systems. In: 9th International World Congress, IFAC, Budapest, Colloquia 14.1, 11.1, July 2-6, 1984, vol. X, pp. 249–254 (1984)

[2] Pospelov, D.A.: Situacionnoje upravlenije. Teoria i praks. Nauka, Moskva, p. 284 (1986) (in Russian)

[3] Beneš, J.: Teorie systémů (řízení komplexů), p. 200. Academia, nakladatelství ČSAV (1974) (in Czech)

[4] Madarász, L.: Metodika situačného riadenia a jej aplikácie, p. 212. Elfa Košice (1996) (in Slovak) ISBN 80 – 88786-66-5

[5] Lazar, T., et al.: Tendencie vývoja a modelovania avionických systémov, Ministerstvo Obrany SR, p. 160 (2000) ISBN 80-8842-26-3 (in Slovak)

[6] Andoga, R., Madarász, L., Főző, L.: Situational Modeling and Control of a Small Turbojet Engine MPM 20. In: IEEE International Conference on Computational Cybernetics, Tallinn, Estonia, August 20-22, 2006, pp. 81–85 (2006) ISBN 1-4244-0071-6

[7] Madarász, L.: Inteligentné technológie a ich aplikácie v zložitých systémoch, Vydavateľstvo Elfa, s.r.o., TU Košice, p. 349 (2004) (in Slovak) ISBN 80 – 89066 – 75 - 5

[8] Moller, M.F.: A Scaled Conjugate Gradient Algorithm for Fast Supervised Learning. Neural Networks 6, 525–533 (1993)

[9] Andoga, R., Főző, L., Madarász, L.: Digital Electronic Control of a Small Turbojet Engine MPM 20. Acta Polytechnica Hungarica 4(4), 83–95 (2007)

Hybrid Approach in Power Electronics and Motion Control

Karel Jezernik

Faculty of Electrical Engineering and Computer Science, University of Maribor
Smetanova ul. 17, 2000 Maribor, Slovenia

Abstract. In this paper a unified discrete event model is given for power electronic circuits based on hybrid system theory. Based on this model, FPGA switching control strategy for a three phase inverter is developed. The functionality of a three phase inverter is discussed from a discrete-event point of view. Event driven dynamics of a three phase inverter originates from inherently switching operation of a three phase transistor bridge. It is further emphasized by accompanying logical management functions e.g. protection and steering. Recently developed hybrid based approach for modeling of discrete event systems is applied for modeling, simulations and implementation of a speed/current control, protection and steering functionality of VSI fed induction machine. A DSP / FPGA based digital control platform for inverter system built in the laboratory is presented and discussed. The reference tracking performance of speed and rotor flux is demonstrated in terms of transient characteristics by experimental results.

1   Introduction

This paper introduces recently developed hybrid based approach for modeling of discrete event systems in the field of power electronics. Power electronic circuits are hybrid dynamic systems. Because of the ON and OFF switching of power electronic devices (e.g. MOSFET, IGBT, diode, …), the operation of power electronic circuits can be described by a set of discrete sets with associated continuous dynamics [1], [2].

Special attention is paid to the new current control principle where traditional scheme consisting of discrete-time current controller and pulse-width modulator is replaced with new discrete-event current controller [3]. The key idea, used for the event-driven current control approach, is to evaluate the transistor switching pattern directly from the phase current errors [4]. The idea originates from the hysteresis current control principle. In this paper, the initial idea is further developed with introduction of switching pattern sequences for the switching frequency

I.J. Rudas et al. (Eds.): Towards Intelligent Engineering & Information Tech., SCI 243, pp. 165–177.
springerlink.com   © Springer-Verlag Berlin Heidelberg 2009

reduction. Additionally, systematic design methodology is introduced for the design of multivariable sequential hysteresis control principle. Same formalism is also used for design of functionality providing the management and the protection of the drive.

The paper starts with brief discussion of hybrid modeling for the discrete event-driven systems (DES) [5]. The case study starts with the three phase inverter current controller design and the design of inverter steering and protection functions. Evolved models are checked with simulations using MATLAB / Simulink and experimentally confirmed. During the experiment, mapping of the obtained models into the FPGA executable code is presented. The findings and the comments of the presented approach are discussed in the conclusion [6].

2 Hybrid Systems Focus

2.1 Machine Dynamics

The dynamics of IM consist of mechanical motion (1), dynamics of stator electromagnetic system (2) and the dynamics of the rotor electromagnetic system (3) with electromagnetic torque developed by machine (5):

$$\frac{d\omega_r}{dt} = \frac{1}{J}(T_e - T_L), \tag{1}$$

$$\frac{d\mathbf{i}_s^s}{dt} = \frac{1}{\sigma L_s}\left(\mathbf{u}_s^s - R_s \mathbf{i}_s^s - \frac{L_m}{L_r}\frac{d\mathbf{\Psi}_r^s}{dt}\right), \tag{2}$$

$$\frac{d\mathbf{\Psi}_r^s}{dt} = \frac{R_r}{L_r}L_m \mathbf{i}_s^s - \left(\frac{R_r}{L_r} - jp\omega_r\right)\mathbf{\Psi}_r^s, \tag{3}$$

$$\mathbf{i}_s^s = \frac{1}{\sigma L_s}\left(\mathbf{\Psi}_s^s - \frac{L_m}{L_r}\mathbf{\Psi}_r^s\right), \quad \mathbf{\Psi}_s^s = \frac{L_m}{L_r}\mathbf{\Psi}_r^s + \sigma L_s \mathbf{i}_s^s \tag{4}$$

$$T_e = \frac{2}{3}p\frac{L_m}{L_r}\left|\mathbf{\Psi}_r^s \times \mathbf{i}_s^s\right|, \tag{5}$$

where ω_r is mechanical rotor angular speed, the two dimensional complex space vectors $\boldsymbol{\Psi}_s^s = \left[\Psi_{sa}^s, \Psi_{sb}^s\right]^T$, $\boldsymbol{\Psi}_r^s = \left[\Psi_{ra}^s, \Psi_{rb}^s\right]^T$, $\mathbf{u}_s^s = \left[u_{sa}^s, u_{sb}^s\right]^T$, $\mathbf{i}_s^s = \left[i_{sa}^s, i_{sb}^s\right]^T$ are stator and rotor flux, stator voltage and current, respectively, T_e is electromagnetic motor torque, T_L is load torque, J is inertia of the rotor and p is the number of pole pairs of the machine. The meaning of the subscript and superscript are as follows: s is stator, r is rotor. The subscript denotes the location of variable and superscript denotes the frame of references.

One of the most important issue in implementing control of induction machine, either direct torque control (DTC) or field oriented control (FOC) strategies is to obtain real-time instantaneous flux magnitude and its position with sufficient accuracy for the entire speed range [7], [8], [9]. The difficulty in flux estimation lies with the non-linear induction machine dynamics, which is characterized by speed dependent and time varying parameters [10].

2.2 Control System Design

Here are given a unified model of power electronic circuits, which can be described as a network of electrical components selected from the following three groups: ideal voltage or current sources, linear elements (e.g. resistors, capacitors, inductors, transformers), and nonlinear switching elements (e.g. IGBT, power MOSFET, power diode). In the following analysis, the behavior of a switch is idealized as having two discrete states: open or close. In every discrete state, power electronic frame system behaves corresponding continuous dynamic [11].

The basic principle considered in power electronic circuits is the switching control. In switching control, one build a bank of alternative candidate voltages depends on configuration of switching elements in power electronic circuits. By VSI the switching is orchestrated by a specially designed decision logic, that uses the measurements of continuous state variables, currents, to asses the performance of the candidate voltage input vector V_i currently in use, and also the potential performance of alternative voltage input vectors V_{i-1}, V_{i+1}. Fig. 1 shows the basic architecture employed by switching control. In this figure $\mathbf{u}_s(V_i)$ represents the discontinuous control input voltage vector, $\mathbf{e}_s$ an exogenous disturbance and $\mathbf{i}_s$ the measured output. The dashed box is a conceptual representation of a switching controller. The top element in Fig. 1 is the decision logic that controls the switches, or more precisely, that generates the switching control input vector $S(S_1, S_2, S_3)$. The decision logic is called the supervisor and its purpose is to monitor the signals that can be measured ($\mathbf{u}_s(V_i)$ and current control error $\Delta \mathbf{i}_s$) and decide, at each instant of time, which candidate voltage vector V_i should be put in the feedback loop with the process. In the supervisory control, the supervisor combines continuous dynamic with discrete logic and is therefore hybrid system, i.e. discontinuous input V_i and feedback continuous current i_s output. The modeling of such systems has received considerable attention in the control and computer science literature in the last few years. Power electronics and motion control system is rather a new approach with lack of applications in hybrid based control approach.

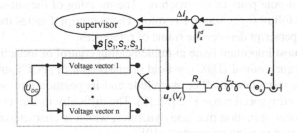

Fig. 1. Switching control supervisory

In power electronic systems based on FPGA the output voltage vector will be implemented in the form of the look up table (direct torque control approach) [12].

3   System Analysis and Control

3.1 System Analysis

The basic circuit of a voltage source inverter (VSI) feeding a Y-connected three phase load is presented in Fig. 1, where the load has been modeled by phase resistance, inductance and induced voltages. The voltage equation of the Y-connected three phase load is:

$$u_{si}(V_i) = R_s\, i_{si} + L_s \frac{d\, i_{si}}{dt} + e_{si}; \quad i = 1,2,3, \tag{6}$$

and the phase currents satisfy the linear condition

$$i_1 + i_2 + i_3 = 0. \tag{7}$$

The considered control problem is the tracking of a three phase current reference signal. After defining $\Delta \mathbf{i}_s = \mathbf{i}_s^d - \mathbf{i}_s$, (6) rewritten in error form becomes

$$L_s \frac{d}{dt}\Delta\mathbf{i}_s + R_s \, \Delta\mathbf{i}_s = \mathbf{u}_s(\mathbf{V}_i) - \mathbf{e}_s, \tag{8}$$

collects all the disturbances (exogenous and endogenous) action on the system.

The basic principle of the inverter voltage control is to manipulate the output voltage vectors such that the desired current of power electronic circuits is produced. This is achieved by choosing an inverter switch combination S_i that drives the stator current vector by directly applying the appropriate voltages $\mathbf{u}_s(\mathbf{V}_i)$ to the AC machine windings (6). The switch positions of the three-phase inverter are described using the logical variables V_i, dependent if switch S_i is ON or OFF. Each variable corresponds to one phase of the inverter (Fig. 2). Three-phase inverter can produce 2^3 voltage vector combinations; two of them are zero vectors and 6 active vectors, Fig. 3.

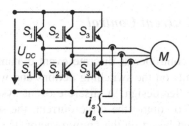

Fig. 2. Basic circuit of voltage source inverter

The energy flow between the input and output side of the three phase inverter is controlled by switching matrix [11]. By introducing the binary variables S_i that are "1" if particular switch S_i is On and "0" if switch S_i is OFF (i=1,2,3,...,6) the behavior of the switching matrix can be described by the three dimensional vector $\mathbf{u}_s = U_{DC} \, \mathbf{L} \, \mathbf{S}_i$, where matrix L and vector $S(S_1, S_2, S_3)$ are defined as

$$\mathbf{L} = \begin{bmatrix} 1 & 0 & 0 & -1 & 0 & 0 \\ 0 & 1 & 0 & 0 & -1 & 0 \\ 0 & 0 & 1 & 0 & 0 & -1 \end{bmatrix}.$$

$$\mathbf{S}^T = \begin{bmatrix} S_1 & S_2 & S_3 & \bar{S}_1 & \bar{S}_2 & \bar{S}_3 \end{bmatrix} \tag{9}$$

Relation (9) is true for the switching matrix L depicted in Fig. 2. It essentially shows that this particular switching matrix is able to generate three independent control actions denoted as the components S_1, S_2 and S_3 of the control vector $\mathbf{u}_s(\mathbf{V})_i = U_{DC}\begin{bmatrix} S_1, S_2, S_3 \end{bmatrix}^T$. The components, i.e. switch position of inverter are generated by look up table of FPGA controller (Table 1 and Fig. 6).

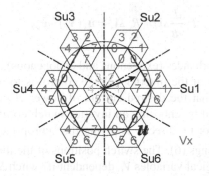

Fig. 3. Stator voltage $\boldsymbol{u}_s$ sector allocation spaces

3.2 Discrete-Event Current Control

Considering a hysteresis controller as discrete-event dynamical system allows to forms in much more details on the switching actions and will enable a better understanding of the controller design. A discrete-event system reacts only if an event is recognized [4]. To control the drive current, the sector of drive voltage $\boldsymbol{u}_s$ is recognized first, and based on the known sector, the output voltage vector (the transistor switching pattern) for the drive current control is selected respecting the current control error related with Lyapunov stability condition. Considering space vector representation of the drive stator voltage $\boldsymbol{u}_s$, the voltage is represented as vector rotating around the origin. Six active switching vectors of the three phase transistor inverter result in six active output voltage vectors denoted $\mathbf{V}_1...\mathbf{V}_6$; $\mathbf{V}_0$ and $\mathbf{V}_7$ are two zero voltage vectors. According to the signs of the phase voltages u_{s1}, u_{s2} and u_{s3}, the phase plane is divided into six sectors denoted Su1 ... Su6, Fig. 3.

Regarding the situation, the stator voltage space vector $\boldsymbol{u}_s$ is in sector 1. In this sector voltage vectors $\mathbf{V}_0$, $\mathbf{V}_1$, $\mathbf{V}_2$, $\mathbf{V}_6$ and $\mathbf{V}_7$ are selected for the current control. $\mathbf{V}_0$, $\mathbf{V}_7$ are two zero vectors, while $\mathbf{V}_1$, $\mathbf{V}_2$, $\mathbf{V}_6$ are three nearest adjacent live output voltage vectors to this sector. With the use of the discrete event system theory, five output voltage vectors $\mathbf{V}_0$, $\mathbf{V}_1$, $\mathbf{V}_2$, $\mathbf{V}_6$ and $\mathbf{V}_7$ are recognized as discrete states of the system. Events represent allowed transition among the discrete states i.e. allowed switching. The structure of the proposed strategy is represented by Petri Net graph on, Fig. 4. The Petri net formalism is propriety for research on possible deadlock or livelock of discrete event system under study [5]. Switching among the available output voltage vectors in each sector are determined by the conditions originate from the derivative of the Lyapunov function [13].

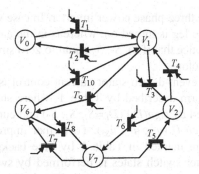

Fig. 4. PN-graph of the switching sequence in Sector 1

For the Lyapunov function candidate

$$V = (1/2)\Delta \mathbf{i_s}^T \Delta \mathbf{i_s} = (1/2)\left(\mathbf{i}_s^d - \mathbf{i}_s\right)^T \left(\mathbf{i}_s^d - \mathbf{i}_s\right), \tag{10}$$

the stability requirement will be fulfilled if control can be selected such, that the derivative of the Lyapunov function candidate is negative $\dot{V} = \Delta \mathbf{i_s}^T \Delta \dot{\mathbf{i}}_s \leq 0$. Derivatives of current control error (6) may be expressed with voltage equation

$$(d/dt)\left(\mathbf{i}_s^d - \mathbf{i}_s\right) - (d/dt)\mathbf{i}_s^d - (1/L_s)\left(\mathbf{u}_s - R_s\mathbf{i}_s - \mathbf{e}_s\right), \tag{11}$$

where $\mathbf{i}_s^d$, $\mathbf{i}_s$ are desired and actual stator current of motor, $\mathbf{u}_s$ is voltage control input, $R_s\mathbf{i}_s$ is resistive voltage drop and $\mathbf{e}_s$ is EMF of the motor.

The conditions for the sequential switching of the power inverter are selected as:

$$S_1 = (1/2)\left(1 - \text{sign}(A)\right), \ S_2 = (1/2)\left(1 - \text{sign}(B)\right),$$
$$S_3 = (1/2)\left(1 - \text{sign}(C)\right) \tag{12}$$

where

$$A = \left(i_{sa}^d - i_{sa}\right)$$
$$B = -(1/2)\left(i_{sa}^d - i_{sa}\right) - (\sqrt{3}/2)\left(i_{sb}^d - i_{sb}\right), \tag{13}$$
$$C = -(1/2)\left(i_{sa}^d - i_{sa}\right) + (\sqrt{3}/2)\left(i_{sb}^d - i_{sb}\right)$$

which is evolved from the Lyapunov function derivative. When U_{DC} has enough magnitude that $\dot{V} \leq 0$, than $V \to 0$ and $\mathbf{i}_s \to \mathbf{i}_s^d$. S_1, S_2 and S_3 represent the switching state of the three-phase power inverter. In case when their value is 1, the upper transistor in the leg is turned on, whereas in the case the value is 0, the lower transistor is on. Notice that if S_1, S_2, S_3 equal to zero simultaneously, no current is delivered to the motor.

The proposed logical event-driven stator current control is similar to DTC and can be realized in the form described by Table 1, where states of stator current control error are presented by $sign(D_i)$ ($D_i = S_1$, S_2, S_3) and currently active voltage sector is presented by $sign\ U$ (u_{S1}, u_{S2}, u_{S3}). To further improve the presentation, active voltage vectors are marked in Table 1 by blue background. Because the transition between inverter switch states is performed by switching only one inverter leg, inverter switching frequency stator current chattering (and consequently torque chattering of AC motor) are reduced.

Table 1. Look-up Table

sign U / sign Di	Su1 100	Su2 110	Su3 010	Su4 011	Su5 001	Su6 101
Sdi0 000	V7	V0	V7	V0	V7	V0
Sdi1 100	V1	V1	V7	V0	V7	V1
Sdi2 110	V2	V2	V2	V0	V7	V0
Sdi3 010	V7	V3	V3	V3	V7	V0
Sdi4 011	V7	V0	V4	V4	V4	V0
Sdi5 001	V7	V0	V7	V5	V5	V5
Sdi6 101	V6	V0	V7	V0	V6	V6
Sdi7 111	V7	V0	V7	V0	V7	V0

The reference current can now be calculated simply as:

$$\mathbf{i}_s^d = \frac{\left|\hat{\mathbf{\Psi}}_r\right|}{L_m} + j\frac{3}{2p}\frac{L_r}{L_m}\frac{T_e^d}{\left|\hat{\mathbf{\Psi}}_r\right|}. \tag{14}$$

A common used dq-ab transformation in conventional current control, the components of the rotor flux are directly employed as:

$$e^{j\Theta} = \cos\Theta + j\sin\Theta = \frac{\hat{\mathbf{\Psi}}_{ra}}{\left|\hat{\mathbf{\Psi}}_r\right|} + j\frac{\hat{\mathbf{\Psi}}_{rb}}{\left|\hat{\mathbf{\Psi}}_r\right|} = \frac{1}{\left|\hat{\mathbf{\Psi}}_r\right|}\hat{\mathbf{\Psi}}_r. \tag{15}$$

The advantage of proposed transformation is that the sin and cos function are replacement with multiplication in used FPGA algorithm.

4 Rotor Flux Observer

This rotor flux observer, is based on the stator equation (16), where the derivative of the estimated stator flux is calculated from measured stator voltage and current. The observer equation (16) represents the first order vectorial differential equation. The stator voltage $\mathbf{u}_s$ and current $\mathbf{i}_s$ serve as control input to the estimated stator flux $\hat{\mathbf{\Psi}}_s$. The measured value of the stator voltage is used instead of the commonly used reference voltage, in order to avoid voltage error influence due to power-stage non-linear behavior:

$$\frac{d\hat{\mathbf{\Psi}}_s}{dt} = \hat{\mathbf{u}}_s - \hat{R}_s \mathbf{i}_s, \qquad (16)$$

The most significant limitation of the rotor flux observer based on voltage model of IM in (16) is that it does not work at zero speed. At zero and low speed, the amplitude of the back emf is to small to accurately and reliably determine the rotor flux angle necessary for both field orientation and velocity estimation. At low speed the rotor estimation given by voltage model deteriorates owing to the effect of an inaccurate value of the stator resistance R_S. An error in the stator resistance causes a deviation of the rotor flux space vector slightly. As a consequent, the rotor and the stator flux exhibits undesired oscillations, what as consequences produces torque fluctuations and acoustic noise.

In order to maintain good performance at low speed, improved rotor flux estimator based on the stator flux model is used:

$$\frac{d\hat{\mathbf{\Psi}}_s}{dt} = \mathbf{u}_s - \frac{R_s}{\sigma L_s} \hat{\mathbf{\Psi}}_s + \frac{R_s}{\sigma L_s} \frac{L_m}{L_r} \mathbf{\Psi}_r^d \qquad (17)$$

Operation of the rotor flux estimator is based on the low pass filtering action of the stator winding $\sigma L_s / R_s$ and the rotor flux command, where in the left hand side of estimator scheme (Fig. 5) the estimated rotor flux will be replaced with its reference value $\mathbf{\Psi}_r^d$. Than the rotor flux estimator can be expressed by the equation in s-domain:

$$\hat{\mathbf{\Psi}}_r = \frac{\tau}{1+s\tau} \mathbf{u}_s + \frac{1}{1+s\tau} \mathbf{\Psi}_r^d; \quad \tau = \frac{\sigma L_s}{R_s}. \qquad (18)$$

The space vector of the rotor flux in the stator reference frame $\mathbf{\Psi}_r^d$ has the same magnitude of the reference of the rotor flux vector in a stator reference frame and can be defined as:

$$\Theta_r = \mathrm{atan} \frac{\hat{\mathbf{\Psi}}_{rb}}{\hat{\mathbf{\Psi}}_{ra}}. \qquad (19)$$

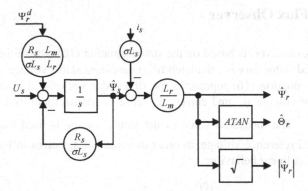

Fig. 5. Block diagram of the rotor flux observer

The voltage model (17) does not involve rotor speed ω_r. Therefore, in this work other formula for synchronous speed calculation is proposed:

$$\hat{\omega}_s = \frac{\hat{\Psi}_{sa} u_{sb} - \hat{\Psi}_{sb} u_{sa}}{\left|\hat{\Psi}_s\right|^2}. \tag{20}$$

Since the slip speed can be estimated by the torque reference T^d and the rotor flux reference Ψ_r^d :

$$\hat{\omega}_{sl} = \frac{2}{3} \frac{R_r}{p} \frac{T^d}{\left|\hat{\Psi}_r^d\right|^2}. \tag{21}$$

The rotor speed $\hat{\omega}_R$ is given by:

$$\hat{\omega}_r = \hat{\omega}_s - \hat{\omega}_{sl}. \tag{22}$$

Accurate knowledge of stator resistance is important for speed sensorless drives operating at a wide speed control range including zero speed in steady state and transients.

5   Implementation

The proposed approach is based on fast parallel processing and suitable for a Field Programmable Gate Array (FPGA) implementation [12]. In such implementation it would be possible to reproduce near ideal switching mode process. However, with FPGA implementation, designer has the difficult task to characterize and scribe the hardware architecture corresponding to the chosen control algorithm. FPGA designers must follow an efficient design methodology in order to benefit from the advantages of the FPGAs and their powerful CAD tools. From software point of view, HDL modeling system is based on using the variables that request logic values, too.

FPGA implementation of Table 1 (Fig. 3) is presented in Fig. 6, where *sign D_1*, *sign D_2* and *sign D_3* present S_1, S_2 and S_3, respectively. Voltage sector states are presented by $sign(u_{S1})$, $sign(u_{S2})$, $sign(u_{S3})$. Inverter leg switching outputs are denoted as follows: TOP1 and BOT1 present states of top and bottom transistor in inverter leg $S_1, \overline{S}_1$, TOP2 and BOT2 are presenting states of transistors in inverter leg $S_2, \overline{S}_2$, whereas TOP3 and BOT3 present states of transistors in inverter leg $S_3, \overline{S}_3$.

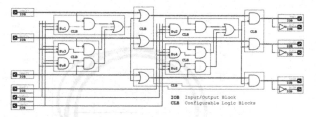

Fig. 6. FPGA schematic circuits

6   Results

The discrete event current control algorithm was implemented on the in house developed DSP/FPGA board [14]. The DSP/FPGA board contains Texas Instruments TMS 320C32 digital signal processor and Xilinx Spartan family field programmable gate array. DSP serves for A/D conversion and generating of the reference current. Replacing usual sequential calculation of algorithms on the DSP by parallel executable FPGA hardware increases the calculation speed. A/D conversion is the most critical operation regarding time and takes 5 μs. According to the fact, that A/D conversion takes most of the calculation time, switching frequencies up to 200 kHz are theoretically possible.

Fig. 7 shows reference motor speed ω_{ref} and transient response of estimated motor speed ω_{est}. Desired speed in experiment was varied following the ramp signal and additional load was introduced at the time of 0.5 s.

Satisfactory tracking of speed, even in the case of step transient at the time of 0.6 s, is shown in Fig. 7. Current, presented in Fig. 8 shows sinusoidal signals, contaminated with satisfactory current ripple cost by switching. Likewise, torque ripple is within desired limits.

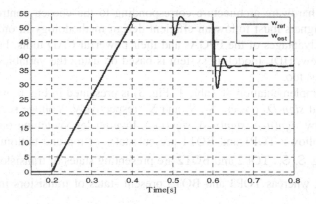

Fig. 7. Reference and estimated motor speed

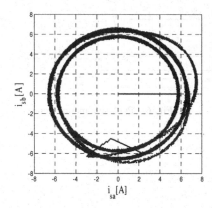

Fig. 8. Phase plane of IM current in a-b reference frame

Conclusion

The idea of hybrid based control of event driven systems is used for the design of event-driven current controller and auxiliary steering and protection functions for a three phase inverter. Individual control functions were designed using the proposed approach and integrated into the overall functionality description of the inverter. Overall functionality of the inverter and the performance of the proposed event driven current control were checked by simulations and experimentally confirmed. Special attention is paid to the mapping of the proposed design approach into the schematic form for the FPGA implementation. The simulation and experiment confirmed potentials of presented approach: traditional coding efforts are significantly reduced on one hand, and the control algorithm can be off-line verified on the other hand. Mathematically formal background of the proposed approach and its correspondence to the conventional control system theory opens further possibilities for the design, simulation, and formal analysis of the discrete event systems. The proposed approach offers a promising technique for design of complex and timely critical algorithms. Future researches are oriented toward the optimization of the introduced current control strategies on one hand, and toward the formalization of analysis and control design methods.

References

[1] Zhou, K., Wang, D.: Relationship between Space-Vector Modulation and Three-Phase Carrier-Based PWM: A Comprehensive Analysis. IEEE Transactions on Industrial Electronics 49(1), 410–420 (2002)

[2] Ramadge, P.J.G., Wonham, W.M.: The Control of Discrete Event Systems. Proceedings of the IEEE 77(1), 81–99 (1989)

[3] Tilli, A., Tonielli, A.: Sequential Design of Hysteresis Current Controller for Three-Phase Inverter. IEEE Trans. on Industrial Electronics 45(5), 771–781 (1998)

[4] Polič, A., Rodič, M., Jezernik, K.: Matrix-based Event-driven Approach for Current Control Design of VSI. In: Proc. of IEEE ISIE 2005, Dubrovnik, Croatia (2005)

[5] Holloway, L.E., Krogh, B.H., Giua, A.: A Survey of Petri Net Methods for Controlled Discrete Event Systems. Journal of Discrete Event Dynamical Systems: Theory and Applications 7(2), 151–190 (1997)

[6] Cirstea, M.N.: Modeling Environment for Power Electronic Systems Integrated Development and Controller Prototyping. In: Proc. of IEEE ISIE 2005, Dubrovnik, Croatia (2005)

[7] Holtz, J.: Estimation of the Fundamental Current in Low Switching Frequency High-Dynamic Medium Voltage Drives. In: Proc. of 42nd IAS Annual Meeting, pp. 993–1000 (2007)

[8] Buja, G.S., Kazmierkowski, M.P.: Direct Torque Control of PWM Inverter-fed AC motors – a survey. IEEE Trans. Industrial Electronics 51(4), 744–757 (2004)

[9] Lascu, C., Boldea, I., Blaabjerg, F.: Direct Torque Control of Sensorless Induction Motor Drives: A Sliding_Mode Approach. IEEE Trans. Industry Applications 40(2), 582–590 (2004)

[10] Mitronikas, E.D., Safacas, A.N.: An Improved Sensorless Vector-control Method for an Induction Motor Drive. IEEE Trans. Industrial Electronics 52(6), 1660–1668 (2005)

[11] Sabanovic, A., Jezernik, K., Sabanovic, N.: Sliding Modes Applications in Power Electronics and Electrical Drives. In: Xinghuo, Y.U., Jian-Xin, X.U. (eds.), 1st edn. LNCIS, pp. 223–251. Springer, Berlin (2002)

[12] Monmasson, E., Cirstea, M.N.: FPGA Design Methodology for Industrial Control Systems – A Review. IEEE Tran. Ind. Electronics 54(4), 1824–1842 (2007)

[13] Utkin, V.I.: Sliding Modes in Control and Optimization. Springer, Berlin (1992)

[14] Hercog, D., Čurkovič, M., Edelbaher, G., Urlep, E.: Programming of the DSP2 Board with the Matlab/Simulink. In: Proc. of IEEE ICIT 2003, Maribor, Slovenia, pp. 709–713 (2003)

Iterative Feedback Tuning in Linear and Fuzzy Control Systems

Radu-Emil Precup[1], Mircea-Bogdan Rădac[1], Stefan Preitl[1], Emil M. Petriu[2], and Claudia-Adina Dragoş[1]

[1] Department of Automation and Applied Informatics, "Politehnica" University of Timisoara
Bd. V. Parvan 2, 300223 Timisoara, Romania
{radu.precup,mircea.radac,stefan.preitl,claudia.dragos}@aut.upt.ro
[2] School of Information Technology and Engineering, University of Ottawa
800 King Edward, Ottawa, Ontario, Canada, K1N 6N5
petriu@site.uottawa.ca

Abstract. Aspects concerning the design of linear and fuzzy control systems based on the Iterative Feedback Tuning (IFT) approach are discussed. Two types of controller parametric conditions are derived to guarantee the robust stability of the control systems. The conditions are included in the steps of the IFT algorithms of linear control systems. Next an IFT-based design of a class of Takagi-Sugeno PI-fuzzy controllers (PI-FCs) is given. The design method maps the parameters of the linear PI controllers onto the parameters of the Takagi-Sugeno PI-FCs. The application of IFT in linear and fuzzy control systems is exemplified in a case study dealing with the angular position control of a DC servo system with back-lash laboratory equipment. The performance enhancement ensured by IFT and fuzzy control is illustrated by real-time experimental results.

Keywords: Iterative Feedback Tuning, fuzzy control, robust stability.

1 Introduction

The design of control systems making use of measurement data proves to be successful in many industrial applications where the mathematical models of the controlled plants are not available. Moreover that is a convenient way to avoid the time-consuming development of such models. The Iterative Feedback Tuning (IFT) is a gradient-based approach based on the input-output data measured from the closed-loop system during its operation (Hjalmarsson et al. 1994), (Gevers 1997), (Hjalmarsson et al. 1998), (Hjalmarsson 2002), (Karimi et al. 2004).

I.J. Rudas et al. (Eds.): Towards Intelligent Engineering & Information Tech., SCI 243, pp. 179–192.
springerlink.com © Springer-Verlag Berlin Heidelberg 2009

The performance specifications of the control systems can be expressed by either control system performance indices (overshoot, settling time, phase margin, maximum sensitivity functions in the frequency domain) or in terms of objective functions in the framework of appropriate optimization problems. The optimization problems can be solved by iterative gradient-based minimization implemented as IFT algorithms. An important feature of IFT is that it makes use of closed-loop experimental data to calculate the estimates of the gradients and eventually Hessians of the objective functions. Several experiments are done per iteration and the updated controller (tuning) parameters are calculated based on the input-output data. Consequently the IFT approach belongs to the direct data-based offline-adaptive controller design methods.

The Virtual Reference Feedback Tuning (VRFT) has been proposed to avoid the iterative calculation of the controller parameters (Campi et al. 2000), (Lecchini et al. 2002), (Campi et al. 2006), (Sala 2007). Thus it is viewed as one-shot/direct data-based controller design method which is complementary to the IFT.

The IFT approach can be implemented in several controller structures. The one-degree-of-freedom (1-DOF) PI controllers and PI-fuzzy controllers (PI-FCs) are considered in this paper.

Fuzzy control exhibits successful results when very good steady-state and dynamic control system performance indices are required. Although it is a relatively easily understandable nonlinear strategy its systematic design is needed (Kacprzyk 1997), (Angelov 2002), (Blažič et al. 2003), (Sala et al. 2005), (Kovačić and Bogdan 2006), (Blažič and Škrjanc 2007), (Du and Zhang 2009), (Lam and Seneviratne 2009). The application of IFT to the design of fuzzy controllers enables the systematic controller tuning and it results in the performance enhancement of fuzzy control systems (Precup et al. 2007), (Precup et al. 2008).

The second experiment specific to IFT algorithms makes use of the control error from the first experiment as reference input which is not usual for the control system operation. Therefore the first goal of this paper is to suggest a modified structure for the second experiment in IFT.

The first step in IFT algorithms sets the initial values of the controller parameters. The other steps involve the step size which determines the next controller parameters. Since all steps are important, under these conditions although the IFT is considered as a model-free design approach (Hjalmarsson et al. 1994), (Hjalmarsson et al. 1998), approximate models of the controlled plant should be known. So the second goal of this paper is to propose the use of controller parametric conditions in IFT algorithms in order to guarantee the robust stability of the closed-loop system. The models of the controlled plant are accepted to be known with approximation as belonging to classes of linear models and the parametric variations of the controlled plant occur. The robust stability conditions are obtained from Kharitonov's theorem (Kharitonov 1979) for third-order and fourth-order systems (Anderson et al. 1987), (Precup and Preitl 2006) and transformed to discrete-time conditions accepting the quasi-continuous digital control.

The third goal is an original design method of Takagi-Sugeno PI-FCs. The method is based on mapping the results from the linear controller design onto the fuzzy controller design in terms of the modal equivalence principle (Galichet and Foulloy, 1995) which can be considered also as a simplified parallel distributed compensation (Baranyi et al. 2009).

The organization of this paper is as follows: Section 2 discusses several aspects concerning the IFT algorithms including the signal processing, implementation and robust stability conditions, Section 3 presents the design method of the Takagi-Sugeno PI-FCs, the case study presented in Section 4 considers the angular position control of a DC servo system with backlash laboratory equipment by the IFT-based design of 1-DOF linear and fuzzy control systems. Experimental results are included to illustrate the application of the theoretical results as low-cost control solutions. Section 5 highlights the conclusions.

2 Iterative Feedback Tuning Algorithms in Linear Control Systems

The linear control system structure employed in IFT is presented in Fig. 1, where: r – the reference input, d – the disturbance input, y – the controlled output, $e = r - y$ – the control error, u – the control signal, ρ – the parameters vector consisting of the controller (tuning) parameters as its components, $C(\rho)$ – the transfer function of the stable linear controller that can be replaced by a fuzzy controller to improve the control system performance indices, F – the transfer function of the reference model that prescribes the desired behavior to be exhibited by the closed-loop system, P – the transfer function of the controlled plant, y_d – the desired output (of the reference model), δy – the model tracking error:

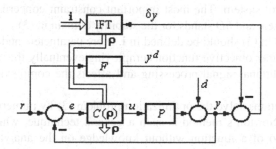

Fig. 1. Linear control system structure with Iterative Feedback Tuning

$$\delta y = y - y_d, \tag{1}$$

and IFT – the IFT algorithm which accepts the input vector $\mathbf{i}$ with important parameters, driving the control system behavior.

Two remarks are important in relation with Fig. 1. First, the operational variable in the transfer functions F, C and P has been omitted for the sake of simplicity. However that variable will be mentioned in the sequel in the well accepted notation s for continuous-time systems and z for discrete-time systems to improve the clarity of the presentation when needed. That is also the reason why the argument ρ will be for inserted or dropped out. Second, use is made of the blocks F and IFT just in the IFT experiments and not in the normal control system operation. Therefore the control system structure presented in Fig. 1 is not a model reference adaptive control one.

The controller parameterization is such that the transfer function $C(\rho)$ is differentiable with respect to ρ. The first task of the controller design is to ensure an initially stabilized control system, hence the initial controller tuning is significant.

A common accepted expression of the objective function $J(\rho)$ is

$$J(\rho) = (0.5 / N) \sum_{k=1}^{N} [\delta y(k, \rho)]^2 , \qquad (2)$$

where: N – the number of samples setting the length of each experiment. A typical objective related to $J(\rho)$ is to find a parameters vector ρ^* for which it is minimized and the error δy tends to zero. That objective is formulated analytically as the following optimization problem:

$$\rho^* = \arg \min_{\rho \in SD} J(\rho) , \qquad (3)$$

where several constraints can be imposed regarding the controlled plant or the closed-loop control system. The most important constraint concerns the stability of the control system and SD stands for the stability domain in (3).

All parameters in (3) should be defined in **i**. More parameters and variables can be accepted for other objective functions employing eventually the control signal. That leads to additional signal processing and affects the complexity of the IFT algorithms.

The IFT algorithms solve the optimization problem (3) by numerical optimization techniques. Newton's method is such a general technique, which iteratively approaches a zero of a function without knowledge on the analytical function. That method is applied here because it can be treated independently of the difficulties inherent to the model-based techniques. So it evaluates repeatedly a new solution based on a point of the function and its approximate derivative. The mathematical formulation is the following update law that calculates the next parameters vector ρ^{i+1}:

$$\rho^{i+1} = \rho^i - \gamma^i (\mathbf{R}^i)^{-1} est[\frac{\partial J}{\partial \rho}(\rho^i)], \ \mathbf{R}^i > 0, \ \det \mathbf{R}^i \neq 0 , \qquad (4)$$

where: i – the index of current iteration, $est[\frac{\partial J}{\partial \rho}(\rho^i)]$ – the estimate of the gradient vector, γ^i – the step size, and ρ^0 – the initial guess of the controller parameters. The matrix $\mathbf{R}^i$ can be the Hessian or the identity matrix to simplify the computing and reduce the complexity of the IFT algorithms.

Differentiating (2), the gradient becomes (5):

$$\frac{\partial J}{\partial \rho}(\rho^i) = (1/N)\sum_{k=1}^{N}[\delta y(k,\rho^i)\frac{\partial \delta y}{\partial \rho}(k,\rho^i)]. \qquad (5)$$

To calculate the general expressions of the gradient of the output error it is necessary to make use of the information obtained from the closed-loop system. The sensitivity function S and the complementary sensitivity function T are expressed as follows with this regard:

$$S(\rho) = 1/[1+C(\rho)P], \; T(\rho) = 1 - S(\rho) = C(\rho)P/[1+C(\rho)P], \qquad (6)$$

and Fig. 1 suggests the following relation:

$$\delta y(\rho) = T(\rho)r + S(\rho)d - Fr. \qquad (7)$$

The differentiation of (7) with respect to ρ making use of (6) and Fig. 1 yields the gradient of the model tracking error:

$$\frac{\partial \delta y}{\partial \rho}(\rho) = \frac{1}{C(\rho)} \cdot \frac{\partial C}{\partial \rho}(\rho)T(\rho)[r - y(\rho)]. \qquad (8)$$

To obtain the estimate of the gradients of the model tracking error use is made of two experiments per iteration for the PI controllers considered here. In the first experiment, referred to as the normal experiment, use is made of Fig. 1, the reference input r_1 is applied to the control system and the controlled output y_1 is measured. In the second experiment, called also the gradient experiment, the control error of the first experiment $e_1 = r_1 - y_1$ is applied as the reference input r_2 (Hjalmarsson et al. 1998) and the controlled output y_2 is measured. That processing does not match the normal control system operation which can exhibit unusual behaviors. Therefore another gradient experiment is suggested. It is characterized by the application of another reference input r_2, proportional to r_1, to the control system and the augmentation of the control signal with e_1 as illustrated in Fig. 2.

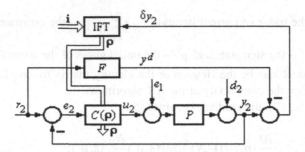

Fig. 2. Control system structure in gradient experiment

Accepting the lower subscript pointing out the index of the current experiment (1 or 2), the reference input and controlled output in the two experiments are

$$r_1 = r, \ y_1(\rho) = T(\rho)r + S(\rho)d_1,$$
$$r_2 = K r, \ y_2(\rho) = T(\rho)r_2 + \{P/[1+C(\rho)P]\}[r_1 - y_1(\rho)] + S(\rho)d_2, \tag{9}$$

where the gain $K > 0$ has been introduced to illustrate the proportional reference inputs in the two experiments with Fig. 1 for the first one. Next the first equation in (9) is multiplied by K and subtracted from the second equation in (9), the result is multiplied by $\dfrac{\partial C}{\partial \rho}(\rho)$, use is made of (8) and the estimate of the gradient of δy is

$$est[\frac{\partial \delta y}{\partial \rho}(k, \rho^i)] = \frac{\partial C}{\partial \rho}(q^{-1}, \rho^i)\{P/[1+C(\rho)P] + S(\rho^i)[d_2(k) - K d_1(k)]\}. \tag{10}$$

The second term in the right-hand side of (10) depends on the disturbance inputs, it affects the gradient be leading to shifted estimates with negative effects on the convergence of the IFT algorithms, and it should be alleviated. The alleviation of that term can be done (for example according to the minimum variance control principle) by the proper initial tuning of the controller parameters because $S(\rho^i)$ plays the role of filter. Assuming the full alleviation of that term the estimate of the gradient of δy can be expressed in terms of (11):

$$\frac{\partial \delta y}{\partial \rho}(k, \rho^i) = \frac{\partial C}{\partial \rho}(q^{-1}, \rho^i)\{P/[1+C(\rho)P]\}. \tag{11}$$

This approach is similar to that proposed in (Hildebrand et al. 2004), (Hildebrand et al. 2005) which is characterized by an additional prefilter designed as solution to optimization problems. That filter is not introduced here to simplify the signal processing, hence $K=1$.

The initial guess of the controller parameters is stressed again from the point of view of disturbance filtering and the update law (4). Accepting that the controlled plant is known approximately under the form of the transfer function P with variable parameters within certain limits, the robust stability analysis of the linear control system will be discussed as follows in relation with (4). Kharitonov's theorem (Kharitonov 1979) states the strictly Hurwitz property of a family of polynomials with coefficients varying within given intervals. The theorem guarantees that the strictly Hurwitz property of the entire family is equivalent to the strictly Hurwitz property of four specially expressed vertex polynomials called Kharitonov polynomials. However the number of polynomials can be reduced for polynomials of degree less than six (Anderson et al. 1987). Particular cases of third- and fourth-order systems will be treated in the sequel because they are widely used in practice. It is assumed as necessary condition that all coefficients of the characteristic polynomials are strictly positive.

Considering the third-order characteristic polynomial $\mu_1(s)$:

$$\mu_1(s) = s^3 + a_2 s^2 + a_1 s + a_0,$$

$$a_0 \in [a_{0\min}, a_{0\max}], a_1 \in [a_{1\min}, a_{1\max}], a_2 \in [a_{2\min}, a_{2\max}],$$

(12)

the stability of this interval polynomial can be verified by checking the stability of just one of the Kharitonov polynomials (Anderson et al. 1987), $\mu_2(s)$:

$$\mu_2(s) = s^3 + a_{2\min} s^2 + a_{1\min} s + a_{0\max},$$

(13)

thus the robust stability conditions corresponding to $\mu_1(s)$ are

$$a_{0\max} > 0, \ a_{1\min} > 0, \ a_{2\min} > 0, \ a_{1\min} a_{2\min} - a_{0\max} > 0.$$

(14)

The stability of the fourth-order interval polynomial $\mu_3(s)$:

$$\mu_3(s) = s^4 + a_3 s^3 + a_2 s^2 + a_1 s + a_0, a_0 \in [a_{0\min}, a_{0\max}],$$

$$a_1 \in [a_{1\min}, a_{1\max}], a_2 \in [a_{2\min}, a_{2\max}], a_3 \in [a_{3\min}, a_{3\max}],$$

(15)

is equivalent to the stability of only two of the Kharitonov polynomials (Anderson et al. 1987), $\mu_4(s)$ and $\mu_5(s)$:

$$\mu_4(s) = s^4 + a_{3\min} s^3 + a_{2\min} s^2 + a_{1\max} s + a_{0\max},$$

$$\mu_5(s) = s^4 + a_{3\max} s^3 + a_{2\min} s^2 + a_{1\min} s + a_{0\max},$$

(16)

so the robust stability conditions of $\mu_3(s)$ are (Precup and Preitl 2006)

$$a_{0\max} > 0, a_{1\min} > 0, a_{1\max} > 0, a_{2\min} > 0, a_{3\min} > 0, a_{3\max} > 0,$$

$$a_{1\max}(a_{2\min} a_{3\min} - a_{1\max}) > a_{0\max} a_{3\min}^2, a_{1\min}(a_{2\min} a_{3\max} - a_{1\min}) > a_{0\max} a_{3\max}^2.$$

(17)

Making use of the limits of the variable parameters of the controlled plant the conditions (14) and (17) can be transformed into parametric conditions of the continuous-time linear controllers. However accepting the conditions of quasi-continuous digital control, associated by the adequate setting of the sampling period T_s (Franklin et al. 1998), the bilinear transform can be applied because it maps the left half of the complex s-plane to the interior of the unit circle in the z-plane. Concluding that is a way to obtain convenient parametric conditions of the discrete-time linear controllers to be fulfilled when setting the step size to calculate the updated controller parameters in terms of (4).

The IFT algorithm consists of the following steps to be proceeded per iteration.

- Step 0. The initial parameters vector ρ^0 is set accounting for the parametric conditions of the discrete-time controller resulted from the robust stability conditions.
- Step 1. The two experiments are done making use of the control system structures presented in Fig. 1 and Fig. 2 and the outputs y_1 and y_2 are measured.
- Step 2. The output of the reference model is generated, y_d, and the output error δy is calculated.
- Step 3. (5), (10) and the previous step are applied to calculate the estimate of the gradient of J.
- Step 4. The step size γ^i is set and use is made of the parametric conditions of the discrete-time controller resulted from the robust stability conditions to calculate the next parameters vector ρ^{i+1} by the update law (4).

It must be highlighted that the matrix $\mathbf{R}^i$ can be either imposed prior to the steps 0 to 4 or calculated during each iteration. Other control system and controller structures require additional experiments.

3   Iterative Feedback Tuning of Fuzzy Control Systems with Takagi-Sugeno PI-Fuzzy Controllers

The Takagi-Sugeno PI-FC is a discrete-time controller constructed around the two inputs-single output fuzzy controller (TISO-FC) according to the controller structure presented in Fig. 3, where $\Delta e(k) = e(k) - e(k-1)$ is the increment of control error and $\Delta u(k) = u(k) - u(k-1)$ is the increment of control signal. The scaling factors corresponding to the input and output variables of the TISO-FC are considered to be inserted in the controlled plant thus the associated nonlinearities are transferred to the TISO-FC.

Fig. 3. Takagi-Sugeno PI-fuzzy controller structure

The Takagi-Sugeno PI-FC represents a generic element for a new class of low-cost fuzzy controllers with the following features:

- the fuzzification is assisted by the input membership functions with the shapes and parameters presented in Fig. 4,

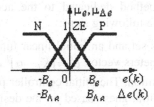

Fig. 4. Input membership functions

- the weighted average method is employed for defuzzification,
- the inference engine makes use of the MAX and MIN operators in relation with the following rule base consisting of the rules $R_i, i = \overline{1,9}$:

R_1 : IF $e(k)$ IS N AND $\Delta e(k)$ IS N THEN $\Delta u(k) = \eta K_P[\Delta e(k) + \alpha e(k)]$,

R_2 : IF $e(k)$ IS N AND $\Delta e(k)$ IS ZE THEN $\Delta u(k) = K_P[\Delta e(k) + \alpha e(k)]$,

R_3 : IF $e(k)$ IS N AND $\Delta e(k)$ IS P THEN $\Delta u(k) = K_P[\Delta e(k) + \alpha e(k)]$,

R_4 : IF $e(k)$ IS ZE AND $\Delta e(k)$ IS N THEN $\Delta u(k) = K_P[\Delta e(k) + \alpha e(k)]$,   (18)

R_5 : IF $e(k)$ IS ZE AND $\Delta e(k)$ IS ZE THEN $\Delta u(k) = K_P[\Delta e(k) + \alpha e(k)]$,

R_6 : IF $e(k)$ IS ZE AND $\Delta e(k)$ IS P THEN $\Delta u(k) = K_P[\Delta e(k) + \alpha e(k)]$,

R_7 : IF $e(k)$ IS P AND $\Delta e(k)$ IS N THEN $\Delta u(k) = K_P[\Delta e(k) + \alpha e(k)]$,

R_8 : IF $e(k)$ IS P AND $\Delta e(k)$ IS ZE THEN $\Delta u(k) = K_P[\Delta e(k) + \alpha e(k)]$,

R_9 : IF $e(k)$ IS P AND $\Delta e(k)$ IS P THEN $\Delta u(k) = \eta K_P[\Delta e(k) + \alpha e(k)]$,

where the parameters K_P and α can be obtained either directly by the direct discrete-time controller design resulting in an incremental discrete-time linear PI controller or by the continuous-time design of the linear PI controller with the transfer function

$$C(s) = k_c(1 + sT_i)/s = k_C[1 + 1/(sT_i)],$$ (19)

where k_C, $k_C = T_i k_c$, is the controller gain and T_i is the integral time constant.

Next the discretization is done according to the requirements of quasi-continuous digital control that include the setting of T_s and accounting for the presence of the zero-order hold. Tustin's method is applied here as bilinear transform to express the parameters K_P and α:

$$K_P = k_C[1 - T_s/(2T_i)], \alpha = 2T_s/(2T_i - T_s).$$ (20)

The number of rules in the complete rule base (18) can be reduced further for the sake of low-cost computing. Interpolation techniques can be applied with this regard (Johanyák and Kovács 2007). The additional parameter η with typical values within $0 < \eta < 1$ has been introduced in (18) to alleviate the overshoot of the control system when $e(k)$ and $\Delta e(k)$ have the same sign.

The IFT-based design method dedicated to the accepted class of Takagi-Sugeno PI-FCs consists of the following steps:

- Step 0. The value of T_s is set and an initial linear tuning method is applied to calculate the initial parameters vector $\rho^0 = [K_P \quad \alpha]^T$, where the superscript T stands for matrix transposition. The initial controller parameters, K_P and α, can be obtained also by an initial guess based on the designer's experience. However use is made of the linear or linearized models of the controlled plant and the parametric conditions of the discrete-time controller resulted from the robust stability conditions.
- Step 1. The initial data of the IFT algorithm and the reference model structure and parameters are set.
- Step 2. The IFT algorithm presented in Section 2 is applied resulting in the optimal controller parameters vector ρ^*.
- Step 3. The values of the parameters B_e and η are set according to the performance specifications imposed to the control system and the designer's experience. The stability analysis or the controller robustness analysis of the fuzzy control system can be taken into account.
- Step 4. The modal equivalence principle is applied (Galichet and Foulloy 1995) to map the linear controller parameters onto the Takagi-Sugeno PI-FC ones:

$$B_{\Delta e} = \alpha B_e.$$ (21)

The application of the design method leads to Takagi-Sugeno PI-FCs which exhibit as bumpless interpolators between two linear digital PI controllers which can be separately designed. The PI-FCs and the linear controllers are equivalent in certain conditions. The operation of the PI-FCs out of those conditions can be beneficial for the behavior of the fuzzy control system by ensuring the performance enhancement in terms of different behaviors.

4   Case Study: Experimental Results

The case study concerns the angular position control of the experimental setup built around the INTECO DC servo system with backlash laboratory equipment (Fig. 5) implemented in the Intelligent Control Systems Laboratory with the "Politehnica" University of Timisoara (PUT). The structure of the experimental setup, Fig. 6, is backed up by: rated amplitude of 24 V, rated current of 3.1 A, rated torque of 15 N cm and rated speed of 3000 rpm. The inertial load weighs 2.030 kg.

Fig. 5. Laboratory equipment

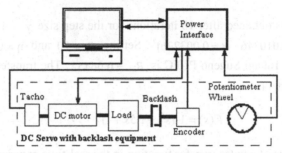

Fig. 6. Structure of experimental setup

The simplified transfer function of the controlled plant and its parameters are

$$P(s) = k_P / [s(1 + sT_\Sigma)], k_P = 139.88 \in [k_{P\min}, k_{P\max}],$$
$$T_\Sigma = 0.9198\,\text{s} \in [T_{\Sigma\min}, T_{\Sigma\max}], k_{P\min} = 125, k_{P\max} = 160, \qquad (22)$$
$$T_{\Sigma\min} = 0.75\,\text{s}, T_{\Sigma\max} = 1.1\,\text{s}.$$

The nominal values of the coefficients in the closed-loop characteristic polynomial (12) are

$$a_0 = (k_C / T_i)(k_P / T_\Sigma), a_1 = k_C(k_P / T_\Sigma), a_2 = 1/T_\Sigma, \qquad (23)$$

and the robust stability conditions (14) lead to the following parametric conditions that ensure the robust stability of the continuous-time control system with linear PI controller:

$$k_C > 0, T_i > (k_{P\max} / k_{P\min})(T_{\Sigma\max}^2 / T_{\Sigma\min}) = 2.0651\,\text{s}. \qquad (24)$$

The continuous-time PI controller has been obtained by the frequency domain design which yields the controller tuning parameters $k_C = 0.01036$ and $T_i = 3.1043\,\text{s}$ which fulfill (24). Next making use of (20) obtained by Tustin's

method the conditions (24) lead to the parametric conditions that ensure the robust
stability of the discrete-time control system with linear PI controller:

$$K_P > k_C \{1 - T_s / [2(k_{P\max} / k_{P\min})(T_{\Sigma\max}^2 / T_{\Sigma\min}) - T_s]\},$$
$$0 < \alpha < 2T_s / [2(k_{P\max} / k_{P\min})(T_{\Sigma\max}^2 / T_{\Sigma\min}) - T_s]. \tag{25}$$

Discretizing with $T_s = 0.01\,\text{s}$ the initial digital PI controller parameters are
$\rho^0 = [0.01035 \quad 0.0029]^T$ which fulfill the following particular form of (25):

$$K_P > 0.0103, 0 < \alpha < 0.0049. \tag{26}$$

The parameters obtained after 10 iterations for the step size $\gamma^i = 10^{-6}$, $i = \overline{0,9}$,
are $\rho^{10} = [K_P = 0.010346 \quad \alpha = 0.003226]^T$. Setting $B_e = 20$ and $\eta = 0.5$ the other
parameter of the Takagi-Sugeno PI-FC is $B_{\Delta e} = 0.06452$. The transfer function of
the reference model is

$$F(s) = 1/(s^2 + 1.5s + 1), \tag{27}$$

and use is made of a pulse transfer function in the real-time experiments.

The scenario used in the real-time experiments is characterized by the constant
reference input $r = 40\,\text{rad}$ and the band-limited white noise of variance 0.01 applied on the disturbance input d. The behavior of the reference model and control
system with linear PI controller before the application of IFT is illustrated in Fig.
7a, and the behavior of the control system with Takagi-Sugeno PI-FC after the application of IFT is presented in Fig. 7b. The improvement of the control system
performance indices is ensured by IFT and fuzzy control.

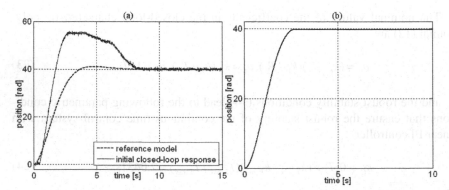

Fig. 7. a Reference model output and controlled output (position) versus time for linear control
system before IFT; **b** controlled output versus time for fuzzy control system after IFT

Conclusions

The paper has discussed several aspects concerning the IFT-based design of linear and fuzzy control systems. Several useful remarks on the signal processing and implementation of the IFT algorithms have been highlighted.

New parametric conditions resulted from the robust stability analysis have been introduced in the steps of the IFT algorithms. Real-time experiments have validated the original IFT algorithms and design methods.

The IFT-based design of fuzzy control systems has been focused on Takagi-Sugeno PI-fuzzy controllers with emphasis on low-cost implementations. Use can be made of the dynamic fuzzy models (Valente de Oliveira and Gomide 2001), (Horváth and Rudas 2004), (Vaščák 2007), (Chang et al. 2008), (Yao and Huang 2008), (Zhong et al. 2008), (Mansouri et al. 2009). So an alternative is to apply IFT directly to fuzzy control systems to avoid the mapping of the linear controller parameters onto the fuzzy controller ones. That represents one future research topic. However the differentiability with respect to the parameters of the fuzzy controllers should be fulfilled which defines additional constraints. The convergence analysis of the IFT algorithms should be done in all applications.

Acknowledgments. The paper was supported by the CNMP and CNCSIS of Romania. The second and fifth authors are doctoral students with the PUT and SOP HRD bursars co-financed by the European Social Fund through the project ID 6998. The support stemming from the cooperation between Budapest Tech Polytechnical Institution and PUT in the framework of the Intergovernmental S & T Cooperation Program is acknowledged.

References

[1] Anderson, B.D.O., Jury, E.I., Mansour, M.: On Robust Hurwitz Polynomials. IEEE Trans. Autom. Control 32, 909–913 (1987)
[2] Angelov, P.P.: Evolving Rule-based Models: A Tool for Design of Flexible Adaptive Systems. Springer, Heidelberg (2002)
[3] Baranyi, P., Korondi, P., Tanaka, K.: Parallel-distributed Compensation-based Stabilization of a 3-DOF RC Helicopter: a Tensor Product Transformation-based Approach. J. Adv. Comput. Intell. Intell. Inform. 13, 25–34 (2009)
[4] Blažič, S., Škrjanc, I.: Design and Stability Analysis of Fuzzy Model-based Predictive Control - a case study. J. Intell. Robot Syst. 49, 279–292 (2007)
[5] Blažič, S., Škrjanc, I., Matko, D.: Globally Stable Direct Fuzzy Model Reference Adaptive Control. Fuzzy Sets Syst. 139, 3–33 (2003)
[6] Campi, M.C., Lecchini, A., Savaresi, S.M.: Virtual Reference Feedback Tuning (VRFT): a New Direct Approach to the Design of Feedback Controllers. In: Proceedings of 39th Conference on Decision and Control, pp. 623–628 (2000)
[7] Campi, M.C., Lecchini, A., Savaresi, S.M.: Direct Nonlinear Control Design: the Virtual Reference Feedback Tuning (VRFT) Approach. IEEE Trans. Autom. Control 51, 14–27 (2006)
[8] Chang, Y.Z., Tsai, Z.R., Hwang, J.D., Lee, J.: Robust Fuzzy Control and Evolutionary Fuzzy Identification of Singularly Perturbed Nonlinear Systems with Parameter Uncertainty. Electr. Eng. 90, 379–393 (2008)
[9] Du, H., Zhang, N.: Fuzzy Control for Nonlinear Uncertain Electrohydraulic Active Suspensions with Input Constraint. IEEE Trans. Fuzzy Syst. 17, 343–356 (2009)
[10] Franklin, G.F., Powell, J.D., Workman, M.L.: Digital Control of Dynamic Systems. Addison-Wesley, Menlo Park (1998)

[11] Gevers, M.: Iterative Feedback Tuning: Theory and Applications in Chemical Process Control. J. A. 38, 16–25 (1997)
[12] Galichet, S., Foulloy, L.: Fuzzy Controllers: Synthesis and Equivalences. IEEE Trans. Fuzzy Syst. 3, 140–148 (1995)
[13] Hildebrand, R., Lecchini, A., Solari, G., Gevers, M.: Prefiltering in Iterative Feedback Tuning: Optimization of the Prefilter for Accuracy. IEEE Trans. Autom. Control 49, 1801–1805 (2004)
[14] Hildebrand, R., Lecchini, A., Solari, G., Gevers, M.: Optimal Prefiltering in Iterative Feedback Tuning. IEEE Trans. Autom. Control 509, 1196–1200 (2005)
[15] Hjalmarsson, H.: Iterative Feedback Tuning - an overview. Int. J. Adapt Control Signal Process 16, 373–395 (2002)
[16] Hjalmarsson, H., Gevers, M., Gunnarsson, S., et al.: Iterative Feedback Tuning: theory and applications. IEEE Control Syst. Mag 18, 26–41 (1998)
[17] Hjalmarsson, H., Gunnarsson, S., Gevers, M.: A Convergent Iterative Restricted Complexity Control Design Scheme. In: Proceedings of 33rd IEEE Conference on Decision and Control, pp. 1735–1740 (1994)
[18] Horváth, L., Rudas, I.J.: Modeling and Problem Solving Methods for Engineers. Academic Press, Elsevier, Burlington (2004)
[19] Kacprzyk, J.: Multistage Fuzzy Control: A Model-based Approach to Control and Decision-Making. Wiley, Chichester (1997)
[20] Karimi, A., Mišković, L., Bonvin, D.: Iterative Correlation-based Controller Tuning. Int. J. Adapt Control Signal Process 18, 645–664 (2004)
[21] Kharitonov, V.L.: Asymptotic Stability of an Equilibrium Position of a Family of Systems of Linear Differential Equations. Differ. Equ. 14, 1483–1485 (1979)
[22] Kovačić, Z., Bogdan, S.: Fuzzy Controller Design: Theory and Applications. CRC Press, Boca Raton (2006)
[23] Johanyák, Z.C., Kovács, S.: Sparse Fuzzy System Generation by Rule Base Extension. In: Proceedings of 11th International Conference on Intelligent Engineering Systems, pp. 99–104 (2007)
[24] Lam, H.K., Seneviratne, L.D.: Tracking Control of Sampled-Data Fuzzy-Model-based Control Systems. IET Control Theory Appl. 3, 56–67 (2009)
[25] Lecchini, A., Campi, M.C., Savaresi, S.M.: Virtual Reference Feedback Tuning for Two Degree of Freedom Controllers. Int. J. Adapt Control Signal Process 16, 355–371 (2002)
[26] Mansouri, B., Manamanni, N., Guelton, K., et al.: Output Feedback LMI Tracking Control Conditions with H_{∞} Criterion for Uncertain and Disturbed T-S Models. Inf. Sci. 179, 446–457 (2009)
[27] Precup, R.E., Preitl, S.: PI and PID Controllers Tuning for Integral-Type Servo Systems to Ensure Robust Stability and Controller Robustness. Electr. Eng. 88, 149–156 (2006)
[28] Precup, R.E., Preitl, S., Rudas, I.J., et al.: Design and Experiments for a Class of Fuzzy Controlled Servo Systems. IEEE/ASME Trans Mechatron 13, 22–35 (2008)
[29] Precup, R.E., Preitl, Z., Preitl, S.: Iterative Feedback Tuning Approach to Development of PI-Fuzzy Controllers. In: Proceedings of 2007 IEEE International Conference on Fuzzy Systems, pp. 199–204 (2007)
[30] Sala, A.: Integrating Virtual Reference Feedback Tuning into a Unified Closed-Loop Identification Framework. Automatica 43, 178–183 (2007)
[31] Sala, A., Guerra, T.M., Babuška, R.: Perspectives of Fuzzy Systems and Control. Fuzzy Sets Syst. 156, 432–444 (2005)
[32] Valente de Oliveira, J., Gomide, F.A.C.: Formal Methods for Fuzzy Modeling and Control. Fuzzy Sets Syst. 121, 1–2 (2001)
[33] Vaščák, J.: Navigation of Mobile Robots Using Potential Fields and Computational Intelligence Means. Acta Polytech. Hung. 4, 63–74 (2007)
[34] Yao, L., Huang, P.Z.: Learning of Hybrid Fuzzy Controller for the Optical Data Storage Device. IEEE/ASME Trans. Mechatron. 13, 3–13 (2008)
[35] Zhong, Q., Bao, J., Yu, Y., Liao, X.: Impulsive Control for T-S fuzzy Model-based Chaotic Systems. Math. Comput. Simul. 79, 409–415 (2008)

Part III
Robotics

Mobile Mini Robots for Space Applications

Peter Kopacek

Intelligent Handling and Robotics, Vienna University of Technology
Favoritenstr. 9-11/318, A-1040 Vienna, Austria
kopacek@ihrt.tuwien.ac.at

Abstract. The basic idea of this concept is the generation of emission-free solar energy by means of solar cells from outer space and the transmission of energy to the earth using microwave or laser beam. Instead of the conventional rigid structure a concept of a large membrane or a mesh structure which holds its corners by satellites was proposed. This paper deals with the development of low cost mobile mini robots able to move and place the solar cells as well as transmitter on a mesh structure in outer space in non gravity with huge temperature variations.

Keywords: mobile mini robots, space application, cost oriented, case study.

1   Introduction

The concept of solar power from the space (SPS) was proposed in 1968. The basic idea of this concept is the generation of emission-free solar energy by means of solar cells from outer space and the transmission of energy to the earth using microwave or laser beam. Because of high launch cost the structure – consisting of solar cells as well as microwave transmitters – should be light weight. Instead of the conventional rigid structures a new concept (Furoshiki Concept) of a large membrane or a mesh structure was proposed. Next step to be realized is the transport of solar panels and microwave transmitters on this mesh structure.

The main purpose of this project was the development of mobile mini robots that place solar cells and transmitters on the net structure to build a solar power plant based on the Furoshiki net concept. A sounding rocket launches four satellites (one mother satellite and three daughter satellites), robots, net, solar panel and the microwave transmitters in the orbit. Approximately 60 seconds after launch the rocket reaches an altitude of 60 km. The mother satellite and three daughter-satellites build the Furoshiki net. Robots transport solar cells and microwave transmitters on the net structure (Fig. 1.). In the frame work of the project a feasibility study was done to verify the performance of the Furoshiki net as well as the crawling robots.

I.J. Rudas et al. (Eds.): Towards Intelligent Engineering & Information Tech., SCI 243, pp. 195–201.
springerlink.com   © Springer-Verlag Berlin Heidelberg 2009

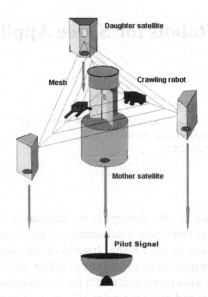

Fig. 1. Conceptual view of solar power plan by means of Furoshiki Satellite (Kaya et al. 2004)

The requirements on the robot are the limited maximum size (10 x 10 x 5 cm), a simple mechanical construction, miniaturized electronics, robustness, "low cost", and independence of the mesh's dimension (from 3 x 3 cm to 5 x 5 cm). The weight of the robot plays an important role. Even the launching cost per kilogram is very high. Another point to be considered is that in case the robot is too heavy, the satellite can not produce enough net tension. For a free movement the moving and holding mechanism of the robot should be well designed. Other difficulties are the vibration and shock during launching of the rocket. The robot should pass the vibration and shock tests up to 40 g. Last but not least the working environment of the robot is in outer space – 200 km over the earth. The high/low temperature, the radiation as well as the vacuum and others should be considered in the design phase.

It is definitely a new, innovative idea and therefore until now there are no such approaches known from the literature.

2   Roby Space-Sandwich

The first prototype Roby Space-Sandwich was built consisting of two parts – an upper part and lower part. The concept is based on using free spaces between the wires of the mesh for reference points of the movements and not the wires itself. The whole electronic and locomotion components are placed in the upper part.

Two parts are connected by magnetic force. The electronic part is built up in open architecture. The motion unit controls the motors by a desired trajectory. This desired trajectory (as well as other demand behavior like acceleration and etc.) has to be transferred to the motion unit. For the design of the robot, the considerations above mean that it consists of a motion unit and a connection via radio to its superior control unit. Remaining to the described system the radio module should be connected to a microcontroller, which selects and processes the incoming information. Afterwards a bus provides the processed information to the motion unit. The electronic part consists of a single board for power electronic, communication and a microcontroller. The task of the microcontroller is to control both DC motors and analyze the radio data. This board is universally useable and in circuit programmable by the serial port. Furthermore it contains a high-speed synchronous serial interface, which gives the possibility to connect several microcontroller boards for different tasks. The electronic part consists of the following components:

- XC167 microcontroller from Infineon with internal RAM (8 kByte) and Flash (128 kByte)
- Voltage supply by switching regulators with high efficiency
- High speed dual full bridge motor driver
- Infrared transmission module
- Bi-directional radio module for the frequencies 433, 869 or 914 MHz
- Status indication by six bright LED's in different colors
- Serial synchronous interface for communication with further modules, e.g. X Scale board.

For power supply a Lithium Ion (Li Io) rechargeable battery (1200 mAh, 10.8 V) is used.

For locomotion two PWM controlled DC motors are used. Specification of PWM controlled DC Motors:

- Mini motor Type: 1524 06 SR
- Output power: 1.70 W
- Speed up to: 10000 rpm
- Stall torque: 6.68 Nm
- Encoder resolution: 512 pr

The robot crawled on the net at 40 degrees below zero (Fig. 2.) and in micro gravity environment. In January 2005 the robot was tested in micro-gravity environment during parabolic flights in Japan.

First of all it was not easy to find out the selection of the appropriate magnetic force to connect both parts. In case the force is too high the motors have not enough power to crawl. In case the force is too low, there is a risk to divide the robot in two parts. Also the tension of the net structure plays an important role. In case the net has not enough tension there is a danger that the robot twines the net.

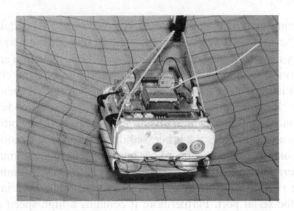

Fig. 2. Roby Space-Sandwich during temperature tests at the 40 degrees below zero

3   Roby Space Insect

In order to avoid several risks as described a new robot was built. The major changes are done in chassis and locomotion. To avoid the separation the robot consists of one part. The height of the robot is the half of the Roby Space-Sandwich. The same electronic parts are used. A specially constructed device is used for holding the net and forward movement. Six pin type grippers on each side of the robot (one on right-hand side, another on left-hand side) are responsible for the holding and the movement of the robot on the net. Always at least two devices on each side hold the net. Before the holding devices release the net following devices should catch the net and move forwards. Roby Space-Insect passed the micro gravity tests in parabolic flights. During tests the robot moved a distance of 2 m within 10 seconds (Fig. 3.).

Fig. 3. Roby Space-Insect attending micro gravity tests

4   Roby Space - Junior

During the first parabolic flight campaign Roby Space-Insect succeeded well. Nevertheless after several meetings, two major problems of Roby Space-Insect were filtered out:

- Getting out of the rocket box:
  Due to the long gripper legs it is very difficult to move out of the small box. The possibility that the robot gets stuck inside the box or on parts of the rocket frame is very high.
- Low tension of the mesh:
  Because of the huge mesh dimension the relatively small daughter satellites are not able to generate a high tension for the mesh. We may recognize that the motion of the robot influences the shape of the mesh which increases the danger of getting stuck during the gripper motion on the top of Roby Space-Insect.

This two problems led to a third prototype called Roby Space-Junior (Fig. 4.).

The robot consists of two parts – the upper part has two active driven belts, the lower one two passive driven belts. Magnetic forces push them together. The special surface between the parts prevents the lower part from moving away.

The advantage of this construction is the very low friction between mesh and robot during operation. There are no high sliding forces like in the first prototype RobySpace-Sandwich due to the passive driven belts of the lower part. The electronic system and software are the same as of Roby Space-Insect. Roby Space – Junior uses the Bluetooth communication.

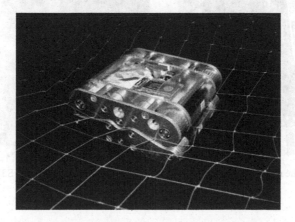

Fig. 4. Robot fixed on the mesh ready for action

5   Results of Research and Development

The robots passed following tests:

*   Microgravity tests during parabolic flights in January and March 2005 in Japan
*   Vibration and shock tests in May 2005 at the ESA Mechanical Systems Laboratory, The Netherlands
*   Mechanical verification tests, June 2005 in Japan

After the tests and technical updates of the systems a sounding rocket S310 with two Roby Space-Junior were launched on January 22, 2006 at the Uchinoura Space Center, Japan. One of two robots worked well. It crawled on the net with small resistance – the robot had a constant high voltage level and moved with constant velocity for more than 30 seconds. For the other robot we have to wait for the results from the telemetry data (Kopacek et. al., 2004).

Three cameras in the satellite delivered the video signals. The video which sent the satellite showed one of robots which moved on the net. According to European Space Agency (ESA) and Japan Aerospace Exploration Agency (JAXA) the experiment completed successfully.

Fig. 5. Set up the rocket nosecone of S310   **Fig. 6.** Sounding rocket S310

Summary

We proved that "low cost" minirobots developed by a small research team, in a short time are able to operate in outer space under zero gravity conditions. After the successful finish of the project, probably some SME's will get contracts for manufacturing such robots (Kopacek, 2009).

All the prototypes described and discussed in this paper are derived from our low cost soccer robots – in average 75% e.g the whole controller board. For a cost oriented solution and to fulfil the WEEE directive of the European Community most of the parts are from other recycled electronic devices e.g the magnets are from used hard discs of PC`s, the screws were from disassembled mobile phones. This successful project is a contribution to cost oriented automation.

Acknowledgments. This project was supported by the "European Space agency – Advanced Concept team" under contract ESTEC/Contract No. 18178/04/NL/MV- Furoshiki Net Mobility Concept and by the Austrian Space Applications Programme (ASAP) under contract "Roby Space" an autonomous mobile mini robot.

References

[1] Kaya, N., Iwashita, M., Nakasuka, S., Summerer, L., Mankins, J.: Rocket Experiment on Construction of Huge Transmitting Antenna for the SPS using Furoshiki Satellite System with Robots. In: Proceedings of 4th International Conference on Solar Power from Space SPS 2004 together with 5th International Conference on Wireless Power Transmission WPT 5, Granada, Spain, June 30 - July 2, 2004, pp. 231–236 (2004)

[2] Kopacek, P., Han, M., Putz, B., Schierer, E., Würzl, M.: A Concept of a High-Tech Mobile Minirobot for Outer Space. In: ESA SP-567 Abstracts of the International Conference on Solar Power from Space SPS 2004 together with the International Conference on Wireless Power Transmission WPT5, Granada, Spain, 06-30-2004 - 07-02-2004, pp. 132–133 (2004)

[3] Kopacek, P.: Cost-oriented Mobile Mini Robot for Space Application. In: Proceedings of " Informatica", La Habana, Cuba (February 2009)

Motion Control for Formation of Mobile Robots in Environment with Obstacles

Krzysztof Kozlowski and Wojciech Kowalczyk

Control and Systems Engineering, Poznan University of Technology
Piotrowo 3A, 60-965 Poznan, Poland
krzysztof.kozlowski@put.poznan.pl, wojciech.kowalczyk@put.poznan.pl

Abstract. Paper presents control method for the formation of nonholonomic mobile robots. Robots track desired trajectories in the environment with static convex-shaped obstacles. Algorithm includes collision-avoidance between robots and obstacles. Artificial potential functions (APF) surround objects in the environment. Virtual field orientation (VFO) method is used to track desired trajectories. Stability analysis for the system is presented. Simulation results and hardware experiments that illustrate effectiveness of presented algorithm are included in the paper.

1 Introduction

Great development in mechanics, electronics and computer sciences opens new perspectives in cooperative robotics. Many researchers focuse their attention on this subject. It is known since decades that some tasks can be executed more efficiently using multiple simple, low-cost mobile robots instead of one bigger robot, however, complexity of multi-robot systems and special demands on communication caused that they were rarely used.

On the other hand control design for multiple robots is much more complex due to distributed characteristic of the system and its complexity. There are many approaches which can be conventionally partitioned into three groups: virtual structure approach [2], [9] behavioral approach [6], [15] and leader follower scheme [3], [10] and [12]. Each of them is more suitable for particular application and less for the others.

Milti-robot system have wide range of applications: service robots, transportation, exploration, mapping and rescue systems, surveillance, security and many others.

I.J. Rudas et al. (Eds.): Towards Intelligent Engineering & Information Tech., SCI 243, pp. 203–219.
springerlink.com   © Springer-Verlag Berlin Heidelberg 2009

In this paper control method for trajectory tracking by formation of mobile robots is presented. Similar approach was described in [1], however, in that article control algorithm for dynamic model of robots is shown and control algorithm includes adaptive module, but method was not tested using hardware. Author also assumes that there are no obstacles in the environment.

In Section 2 control algorithm is described. In Subsection 2.1 kinematic model of the system is introduced, in Subsection 2.2 collision avoidance using APF is described. In Subsection 2.3 control method is shown in details. In Section 3 problem of local minima and equilibrium points is discussed. In Section 4 stability and convergence of the system is analysed. In Subsections 4.1 and 4.2 stability of position is investigated for case of collision avoidance with a single obstacle and between N robots in environment with M obstacles respectively. In Subsections 4.3 and 4.4 convergence of the orientation of robot is investigated. In Section 5 simulation and experimental results are shown. In Subsection 5.1 results of numerical simulations are presented and in Subsection 5.2 results of experimental work are described. Section 6 contains concluding remarks.

2 Control Algorithm

2.1 Model of the System

Formation is a group of N mobile robots. Kinematic model of the ith robot (Fig. 1) is given by the equation:

$$
\begin{bmatrix} \dot{x}_i \\ \dot{y}_i \\ \dot{\theta}_i \end{bmatrix} = \begin{bmatrix} \cos\theta_i & 0 \\ \sin\theta_i & 0 \\ 0 & 1 \end{bmatrix} \begin{bmatrix} u_{vi} \\ u_{\omega i} \end{bmatrix},
\tag{1}
$$

where x_i, y_i nad θ_i are position and orientation coordinates of ith robot, respectively ($i = 1,\ldots,N$), $\mathbf{u}_i = \begin{bmatrix} u_{vi} & u_{\omega i} \end{bmatrix}^T$ is control vector that includes u_{vi} - linear velocity control and $u_{\omega i}$ - angular velocity control.

2.2 Collision Avoidance

Each object is surrounded by artificial potential field that cause repelling force on robots that approach obstacle boundary to close. In this paper authors assume that all robots and obstacles can be modeled as a circle-shaped objects.

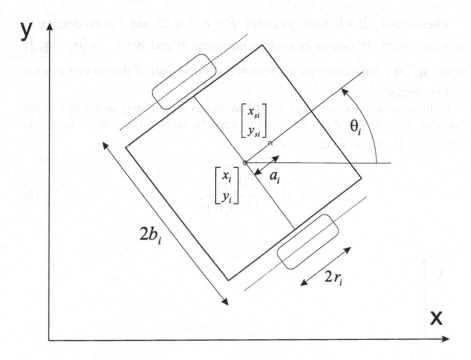

Fig. 1. Model of the differentially driven mobile robot, r_i – wheel radius, 2 b_i – distance between wheels, a_i – distance between wheel axis and center of mass, $\begin{bmatrix} x_i y_i \end{bmatrix}^T$ - robot position, θ_i - robot orientation, $\begin{bmatrix} x_{si} y_{s1} \end{bmatrix}^T$ - position of center of mass of the robot

Lets define: $\Delta = \left\{ \begin{bmatrix} x & y \end{bmatrix} : (x,y) \in \mathbb{R}^2, \left\| \begin{bmatrix} x & y \end{bmatrix}^T - \begin{bmatrix} x_a & y_a \end{bmatrix}^T \right\| \leq r \right\}$ - set of coordinates for collision area, $\Gamma = \left\{ \begin{bmatrix} x & y \end{bmatrix} : (x,y) \notin \Delta, r < \left\| \begin{bmatrix} x & y \end{bmatrix}^T - \begin{bmatrix} x_a & y_a \end{bmatrix}^T \right\| < R \right\}$ - set of coordinates for repel area and $D = \Delta \cup \Gamma$ - set that includes both areas, where r is the least radius of circle that covers obstacle or robot, R is the radius of area where repel force caused by APF acts. In case of two robots repel area will be also named interaction area.

Artificial potential function (APF) is given by the following equation:

$$B_a(l) = \begin{cases} 0 & dla \quad l < r \\ e^{\frac{l-r}{l-R}} & dla \quad r \leq l < R \\ 0 & dla \quad l \geq R \end{cases} \tag{2}$$

where $r > 0$, $R > 0$ fulfill inequality $R > r$ (Fig. 2), and l is the distance to colliding object. In case of two robots numbered n and m : $l_{nm} = \| \mathbf{q}_n - \mathbf{q}_m \|$, where $\mathbf{q}_n$, $\mathbf{q}_m$ are robots position vectors. In further part of this section indexes will be omitted.

Collision avoidance task requires APF to converge to infinity as distance to the boundary of colliding object decreases to zero. To fulfill this condition Eq. (2) is mapped to $\langle 0, \infty \rangle$ using transformation:

$$V_a(l) = \frac{B_a(l)}{1 - B_a(l)}. \qquad (3)$$

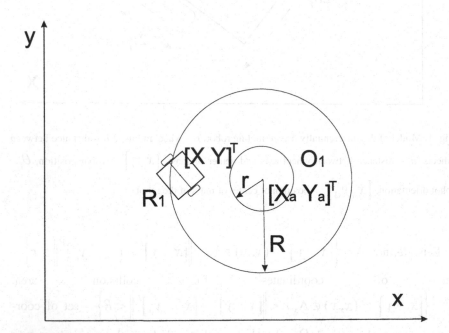

Fig. 2. Robot R_1 is in the repel area of O_1 obstacle; r – radius of the collision area, R – radius of APF's range connected with the obstacle

Differentiating 3 with respect to time once and twice one obtains respectively:

$$\frac{\partial V_a(l)}{\partial l} = \frac{(r - R)}{(l - R)^2 \left(e^{\frac{l-r}{l-R}} - 1 \right)^2} e^{\frac{l-r}{l-R}} \qquad (4)$$

Graph of resulting APF is shown in Fig. 3.

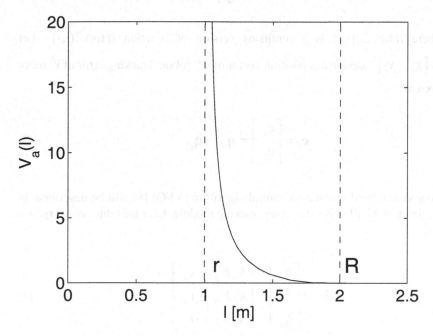

Fig. 3. Scalled APF: $V_a\left(l\right)\in\left\langle 0,\infty\right\rangle$

and

$$\frac{\partial V_a^2(l)}{\partial l^2}=\frac{\left(R-r\right)\left(r+re^{\frac{l-r}{l-R}}-2l+2le^{\frac{l-r}{l-R}}+R-3\,\mathrm{Re}^{\frac{l-r}{l-R}}\right)}{\left(l-R\right)^4\left(e^{\frac{l-r}{l-R}}-1\right)^3}e^{\frac{l-r}{l-R}}.$$ (5)

As shown in [5] numerical complexity of these functions is similar to other known in the literature.

2.3 Control

Reference position of ith robot is given by vector: $\mathbf{q}_{di}=\begin{bmatrix}x_{di}&y_{di}\end{bmatrix}^T$. Reference orientation is computed as a function of vector of reference linear velocities:

$$\theta_{di} = a \tan 2c(\dot{y}_{di}, \dot{x}_{di}), \tag{6}$$

where $a \tan 2c(\cdot, \cdot)$ is a continous version of function $a \tan 2(\cdot, \cdot)$. Let $\mathbf{q}_i = \begin{bmatrix} x_i & y_i \end{bmatrix}^T$ designates position vector of $i^{\,th}$ robot. Tracking error of ith robot is given as:

$$\mathbf{e}_i = \begin{bmatrix} e_{xi} \\ e_{yi} \end{bmatrix} = \mathbf{q}_{di} - \mathbf{q}_i. \tag{7}$$

Now virtual field orientation control algorithm (VFO) [8] will be described. In presented method VFO is a trajectory tracking module. Let's introduce convergence vector:

$$\mathbf{h}_i = \begin{bmatrix} h_{xi} \\ h_{yi} \\ h_{\theta i} \end{bmatrix} = \begin{bmatrix} k_p E_{xi} + \dot{x}_{di} \\ k_p E_{yi} + \dot{y}_{di} \\ k_\theta e_{ai} + \dot{\theta}_{ai} \end{bmatrix}, \tag{8}$$

where k_p, $k_\theta > 0$ are position and orientation control gains, respectively. Modified tracking error is computed as a difference between tracking error and sum of gradients of APF's associated with collisive objects:

$$\mathbf{E}_i = \begin{bmatrix} E_{xi} \\ E_{yi} \end{bmatrix} = \mathbf{e}_i - \sum_{j=1, j \neq i}^{N} \left[\frac{\partial V_{arij}(l_{ij})}{\partial \mathbf{q}_i} \right]^T - \sum_{k=1}^{M} \left[\frac{\partial V_{aoik}(l_{ik})}{\partial \mathbf{q}_i} \right]^T, \tag{9}$$

where $V_{arij}(l_{ij})$ is APF of the $j^{\,th}$ robot, $V_{aoik}(l_{ik})$ is APF of the $k^{\,th}$ obstacle, $l_{ij} = \|\mathbf{q}_i - \mathbf{q}_j\|$ for i, $j = 1, \dots, N$, $i \neq j$ is the distance between the $i^{\,th}$ and the $j^{\,th}$ robot, $l_{ik} = \|\mathbf{q}_i - \mathbf{q}_k\|$ for $i = 1, \dots, N$ and $k = 1, \dots, M$ is the distance between the $i^{\,th}$ robot and the $k^{\,th}$ obstacle. Graphical interpretation of modified position error is shown in Fig. 4.

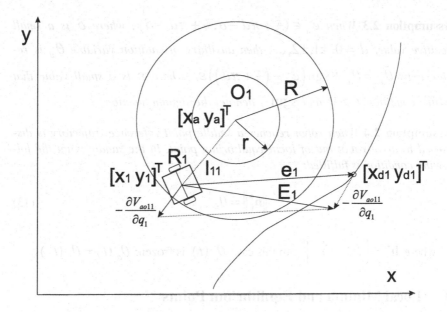

Fig. 4. Robot R_1 is in the repel area of O_1 obstacle. Modified position error is computed as a difference between tracking error and gradient of the obstacle's APF

Auxiliary orientation error is defined as follows: $e_{ai} = \theta_{ai} - \theta_i$. The auxiliary orientation variable θ_{ai} is defined as follows:

$$\theta_{ai} = a\tan 2c(h_{yi}, h_{xi}). \tag{10}$$

Control vector $\mathbf{u}_i = \begin{bmatrix} u_{vi} & u_{\omega i} \end{bmatrix}^T$ for the ith mobile platform is given by the following equation:

$$
\begin{aligned}
u_{vi} &= h_{xi}\cos\theta_i + h_{yi}\sin\theta_i \\
u_{\omega i} &= h_{\theta i}
\end{aligned}
\tag{11}
$$

Assumption 2.1 *Desired trajectories do not intersect APF areas of obstacles and robots do not interact when tracking is executed perfectly.*

Assumption 2.2 *When robot position is in Γ then reference trajectory is frozen:*

$$\mathbf{q}_{di}(t) = \mathbf{q}_{di}(t^-), \tag{12}$$

where t^- is the time value before robot get the area represented by the set Γ. Higher derivatives of $q_{di}(t)$ are kept zero until robot leaves repel area [7].

Assumption 2.3 When $e_{ai} \in (\frac{\pi}{2} + \pi d - \delta, \frac{\pi}{2} + \pi d + \delta)$, where δ is a small positive value, $d = 0, \pm 1, \pm 2, \ldots$, then auxiliary orientation variable θ_{ai} is replaced by $\tilde{\theta}_{ai} = \theta_{ai} + \mathrm{sgn}\left(e_{ai} - \left(\frac{\pi}{2} + \pi d\right)\right)\varepsilon$, where ε is a small value that fulfills condition $\varepsilon > 0$ and $\mathrm{sgn}(\cdot)$ denotes the signum function.

Assumption 2.4 When robot reaches a saddle point reference trajectory is disturbed to drive robot out of local equilibrium point. In the saddle point the following condition is fulfilled:

$$\left\| \mathbf{h}_i^* \right\| = 0, \tag{13}$$

where $\mathbf{h}_i^* = \begin{bmatrix} h_{xi} & h_{yi} \end{bmatrix}^T$. In this case $\theta_{ai}(t)$ is frozen: $\theta_{ai}(t) = \theta_{ai}(t^-)$.

3 Local Minima and Equilibrium Points

Siginificant weakness of APF-based methods is problem of existance of local minima. Especially in case when non-convex obstacles exist in the environment local minima may occur. In this paper it is assumed that only circle shape obstacles exist. Another assumption is that obstacles APF's are separated. In such a case local minima will not appear, however, there will exist local equilibrium points with quantity equal to noumber of obstacles [11] when a set point control is analyzed. These equilibrium points are not attracting and they are also not stable. They appear in positions where tracking error is balanced with repel vector caused by obstacles APF.

Small disturbance is added to the robot position or desired position to prevent robot to get stuck in equilibrium point [14].

4 Stability Analysis

Stability proof gathers three steps: $\lim_{t \to \infty} \begin{bmatrix} x_i & y_i \end{bmatrix}^T = \begin{bmatrix} x_{di} & y_{di} \end{bmatrix}^T$ - proof of stability and asymptotic convergence of robots position to reference position, $\lim_{t \to \infty} \theta_i = \theta_{ai}$ - proof of convergence of orientation to auxiliary orientation variable, $\lim_{t \to \infty} \theta_{ai} = \theta_{di} + \varepsilon_\theta$, where ε_θ is small value - proof of convergence of auxiliary orientation variable to near surroundings of the desired orientation.

4.1 Collision Avoidance with Single Obstacle

In this Subsection for simplicity indexes of robots numbers are omitted. Lyapunov function candidate is given by the following equation:

$$V_l = \frac{1}{2}\mathbf{e}^T\mathbf{e} + V_a(l) = \frac{1}{2}\left(e_x^2 + e_y^2\right) + V_a(l), \tag{14}$$

where V_a is a potential given by (3). Parameter l represents distance between robot and obstacle: $l = \left\| [x \quad y]^T - [x_a \quad y_a]^T \right\|$. Time derivative of (14) is as follows:

$$\frac{dV_l}{dt} = e_x\dot{e}_x + e_y\dot{e}_y + \frac{\partial V_a}{\partial x}\dot{x} + \frac{\partial V_a}{\partial y}\dot{y}. \tag{15}$$

Above equation can be transformed to the form:

$$\frac{dV_l}{dt} = e_x\dot{x}_d + e_y\dot{y}_d - k_p E_x^2 \cos^2(e_a) - k_p E_y^2 \cos^2(e_a) - E_x\dot{x}_d \cos^2(e_a)$$

$$- E_y\dot{y}_d \cos^2(e_a) - E_x\dot{y}_d \sin(e_a)\cos(e_a) + E_y\dot{x}_d \sin(e_a)\cos(e_a). \tag{16}$$

Two cases will be investigated separately:

    1 Robot belongs to set $D^c = \mathbb{R}^2 \setminus D$,

    2 Robot belongs to set Γ.

In case 1 $E_x = e_x$ and $E_y = e_y$ because $\frac{\partial V_a}{\partial x} = \frac{\partial V_a}{\partial y} = 0$. Time derivative of the Lyapunov function simplifies to:

$$\frac{dV_{l1}}{dt} = e_x\dot{x}_d + e_y\dot{y}_d - k_p e_x^2 \cos^2(e_a) - k_p e_y^2 \cos^2(e_a) - e_x\dot{x}_d \cos^2(e_a)$$

$$- e_y\dot{y}_d \cos^2(e_a) - e_x\dot{y}_d \sin(e_a)\cos(e_a) + e_y\dot{x}_d \sin(e_a)\cos(e_a). \tag{17}$$

Additional index denotes case number. Using definition of vector multiplication and introducing matrix:

$$\mathbf{Q} = \begin{bmatrix} k_p \cos^2(e_a) & 0 \\ 0 & k_p \cos^2(e_a) \end{bmatrix} \tag{18}$$

the following inequality is fulfilled:

$$\frac{dV_{l1}}{dt} \leq -\mathbf{e}^T \mathbf{Q} \mathbf{e} + \|\mathbf{e}\| \|\dot{\mathbf{q}}_\mathbf{d}\| \sin^2(e_a) - \|\mathbf{e}\| \|\dot{\mathbf{q}}_\mathbf{d}\| \sin \alpha \sin(e_a) \cos(e_a) \quad (19)$$

$$\leq -\|\mathbf{e}\|^2 \lambda_{\min}(\mathbf{Q}) + 2\|\mathbf{e}\| \|\dot{\mathbf{q}}_\mathbf{d}\|, \quad (20)$$

where $\lambda_{\min}$ is the smallest eigenvalue of matrix $\mathbf{Q}$. Considering $V_{al2} = 0$ one can write:

$$V_{l1} = \frac{1}{2}\|\mathbf{e}\|^2. \quad (21)$$

Substituing $\|\mathbf{e}\|^2 = 2V_{l1}$ into (20) one obtains:

$$\frac{dV_{l1}}{dt} \leq -2\lambda_{\min}(\mathbf{Q})V_{l1} + 2\sqrt{2} \|\dot{\mathbf{q}}_\mathbf{d}\|_{\max} \sqrt{V_{l1}}. \quad (22)$$

Introducing new variables: $\lambda = 2\lambda_{\min}(\mathbf{Q})$ and $\varepsilon = 2\sqrt{2} \|\dot{\mathbf{q}}_\mathbf{d}\|_{\max}$ the above inequality can be transformed to the form:

$$\frac{dV_{l1}}{dt} \leq -\lambda V_{l1} + \varepsilon \sqrt{V_{l1}}, \quad (23)$$

where $\lambda, \varepsilon > 0$ are constants. Solution of (23) is as follows:

$$V_{l1} \leq (V_{l1}(0) - 2\frac{\varepsilon}{\lambda}\sqrt{V_{l1}(0)} + \frac{\varepsilon^2}{\lambda^2})e^{-\lambda t} + 2(\frac{\varepsilon}{\lambda}\sqrt{V_{l1}(0)} - \frac{\varepsilon^2}{\lambda^2})e^{-\frac{1}{2}\lambda t} + \frac{\varepsilon^2}{\lambda^2}. \quad (24)$$

Above equation implies that the system is stable and error is reduced to $\frac{\varepsilon^2}{\lambda^2}$. Substituting (21) into the left side of equation 24 one obtains:

$$\frac{1}{2}\|\mathbf{e}\|^2 \leq (V_{l1}(0) - 2\frac{\varepsilon}{\lambda}\sqrt{V_{l1}(0)} + \frac{\varepsilon^2}{\lambda^2})e^{-\lambda t} + 2(\frac{\varepsilon}{\lambda}\sqrt{V_{l1}(0)} - \frac{\varepsilon^2}{\lambda^2})e^{-\frac{1}{2}\lambda t} + \frac{\varepsilon^2}{\lambda^2}.$$

$$(25)$$

Maximum value of module of error for $t \to \infty$ is as follows:

$$\|\mathbf{e}\|_{max} = \sqrt{2}\,\frac{\varepsilon}{\lambda} = 2\,\frac{\|\dot{\mathbf{q}}_d\|_{max}}{\lambda_{min}(Q)}. \tag{26}$$

By increasing $\lambda_{min}(Q)$ term (increasing k_p parameter) one can reduce module of the error.

In case 2 according to 2.2 equation (16) can be written in the simplified form:

$$\frac{dV_{12}}{dt} = -k_p E_x^2 \cos^2(e_a) - k_p E_y^2 \cos^2(e_a), \tag{27}$$

which always fulfills condition $\dot{V}_{l1} \leq 0$ and the system is stable.

Notice 4.1 *As shown in [13] collision avoidance is guaranteed when $\dot{V}_{l1} \leq 0$ and*

$$\lim_{\|\mathbf{q} - \mathbf{q}_a\| \to r^+} V_a = \lim_{l \to r^+} V_a(l) = +\infty, \tag{28}$$

where $\mathbf{q}_a = [x_a \quad y_a]^T$.

Now convergence analysis will be presented. Using inequality:

$$\dot{V}_{12} \leq \lambda V_{12} + \varepsilon, \tag{29}$$

where λ and ε are constants, after transformations one can write:

$$V_{12}(t) \leq (V_{12}(0) - 2\frac{\varepsilon}{\lambda}\sqrt{V_{12}(0)} + \frac{\varepsilon^2}{\lambda^2})e^{-\lambda t} + 2(\frac{\varepsilon}{\lambda}\sqrt{V_{12}(0)} - \frac{\varepsilon^2}{\lambda^2})e^{-\frac{1}{2}\lambda t} + \frac{\varepsilon^2}{\lambda^2}. \tag{30}$$

In the steady state $V_{12}(t) \leq \frac{\varepsilon^2}{\lambda^2}$. $V_{12}(t)$ is bounded to sphere with radius depending on values ε and λ.

4.2 Collision Avoidance between N Robots and M Obstacles

Now case of N robots that tracks desired trajectories $\mathbf{q_{di}} = [x_{1i} \quad y_{1i}]^T$, $i = 1, \ldots, N$ that avoid collisions in the environment with M obstacles is analyzed. Robots positions are given by vectors $\mathbf{q_i} = [x_i \quad y_i]^T$.

Lyapunov function candidate is given by the following equation:

$$V_l = \sum_{i=1}^{N} \left[\frac{1}{2} \mathbf{e_i}^T \mathbf{e_i} + \sum_{j=1, j \neq i}^{N} \frac{1}{2} V_{arij} + \sum_{k=1}^{M} V_{aoik} \right], \quad (31)$$

where V_{arij} is APF of jth robot that acts on i^{th} robot, V_{aoik} is APF of kth obstacle that acts on i^{th} robot. Differentiating (31) with respect to time results in:

$$\frac{dV_l}{dt} = \sum_{i=1}^{N} \left(e_{xi}\dot{e}_{xi} + e_{yi}\dot{e}_{yi} + \sum_{j=1, j \neq i}^{N} \frac{1}{2} \left(\frac{\partial V_{arij}}{\partial x_i}\dot{x}_i + \frac{\partial V_{arij}}{\partial x_j}\dot{x}_j + \frac{\partial V_{arij}}{\partial y_i}\dot{y}_i + \frac{\partial V_{arij}}{\partial y_j}\dot{y}_j \right) + \right.$$

$$\left. + \sum_{k=1}^{M} \left(\frac{\partial V_{aoik}}{\partial x_i}\dot{x}_i + \frac{\partial V_{aoik}}{\partial y_i}\dot{y}_i \right) \right). \quad (32)$$

Taking into consideration symmetry of interactions between robots one can write:

$$\frac{\partial V_{aij}}{\partial x_i} = -\frac{\partial V_{aij}}{\partial x_j} = -\frac{\partial V_{aji}}{\partial x_j} = \frac{\partial V_{aji}}{\partial x_i}, \quad \frac{\partial V_{aij}}{\partial y_i} = -\frac{\partial V_{aij}}{\partial y_j} = -\frac{\partial V_{aij}}{\partial y_j} = \frac{\partial V_{aij}}{\partial y_j} \quad (33)$$

Using 33 equation 32 can be transformed to the form:

$$\frac{dV_l}{dt} = \sum_{i=1}^{N} \left(e_{xi}\dot{e}_{xi} + e_{yi}\dot{e}_{yi} + \sum_{j=1, j \neq i}^{N} \left(\frac{\partial V_{arij}}{\partial x_i}\dot{x}_i + \frac{\partial V_{arij}}{\partial y_i}\dot{y}_i \right) \right.$$

$$\left. + \sum_{k=1}^{M} \left(\frac{\partial V_{aoik}}{\partial x_i}\dot{x}_i + \frac{\partial V_{aoik}}{\partial y_i}\dot{y}_i \right) \right). \quad (34)$$

Taking into account that above equation is the generalization of (16) to case of N robots and M obstacles convergence of the robots position to desired one can be proven in a similar way as in Subsection 4.1.

4.3   Proof of Convergence of Robots Orientation to Auxiliary Orientation Variable

Now proof of convergence of robots orientation θ_i to auxiliary orientation variable θ_{ai} will be presented:

$$\lim_{t\to\infty}\theta_i = \theta_{ai}. \qquad (35)$$

Substituting last row of (8) into second row of equation (11) and next the obtained result to the last row of (1) one gets:

$$\dot{\theta}_i = k_\theta e_{ai} + \dot{\theta}_{ai}. \qquad (36)$$

The above equation can be transformed to:

$$\dot{e}_{ai} = -k_\theta e_{ai}. \qquad (37)$$

Solving this equation one see that e_{ai} decreases exponentially to zero. Robot orientation converges exponentially to an auxiliary orientation variable θ_{ai}.

4.4   Proof of Convergence of Auxiliary Orientation Variable to Reference Orientation

Now proof of convergence of auxiliary orientation variable θ_{ai} to near surroundings of reference orientation θ_{di} will be presented:

$$\lim_{t\to\infty}\theta_{ai} = \theta_{di} + \varepsilon_\theta, \qquad (38)$$

where ε_θ is a small value.

Comparing (6) and (10), where h_{xi} and h_{yi} are given by (8) one can state that for $k_p E_{xi} \to 0$ and $k_p E_{yi} \to 0$ auxiliary orientation variable converges to reference orientation. When robots are outside interaction area convergence holds for $k_p e_{xi} \to 0$ and $k_p e_{yi} \to 0$. As shown in Subsection 4.1 position error can be reduced to a small value (25) by increasing k_p. By reducing position error auxiliary orientation variable can be bounded to near neighbourhood of the reference orientation.

5 Simulation and Experimental Results

In this section simulation and experimental results are presented. Simulations were performed using Matlab 7.0. In experiments Minitracker robots (Fig. 6) were used [4].

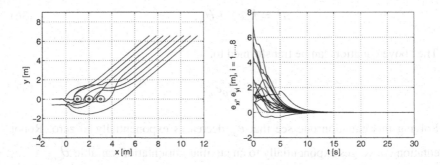

Fig. 5. Simulation results for formation of eight robots tracking linear desired trajectories

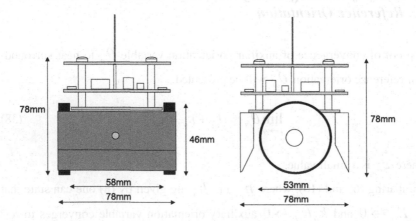

Fig. 6. Minitracker robot

5.1 Simulations

In Fig. 5 simulation results for a group of eight robots that tracks linear desired trajectories are shown. In Fig. 5(a) robots coordinates in cartesian plane are presented. Initial positions of robots were chosen to cause interaction between them and static obstacles. Robots reach their desired trajectories after 15 seconds. Position errors reach values near zero Fig. 5(b). In Fig. 7 Screenshot from the visualization tool is presented. This software was developed by the authors to investigate algorithms before hardware implementation.

Fig. 7. Screenshot form the 3d visualization software tool used to verify effectiveness of proposed algorithm

5.2 Experiments

Experimental results include test for two robots and single static obstacle in the environment. Geomtrical parameters of robots (Fig. 1) are as follows: dostance between robots wheels $2\,b = 65.50mm$, radius of robot wheel $r = 26.5mm$.

Robots have to bypass obstacle avoiding collision at the same time. Robot 2 (Fig. 8(a)) gets into repel area of the obstacle. Obstacles APF acts on it and robot is repelled. Robot 1 avoids collision with robot 2. It changes direction of motion due to some delays in the real system. Position errors of both robots converge to values near zero. Their values for robot 1 are shown in Fig. 8(b) and for robot 2 in Fig. 8(d). In Figs. 8(c) and 8(e) orientation errors of robot 1 and 2 are shown. They also converge to values near zero.

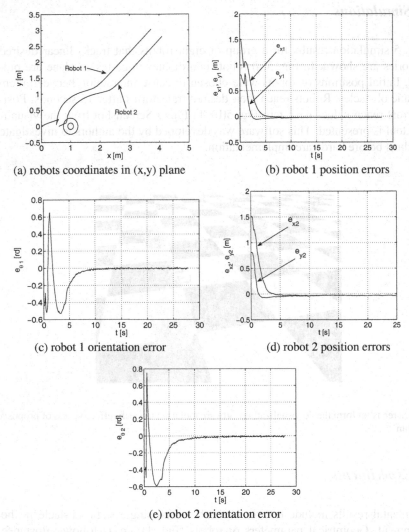

(a) robots coordinates in (x,y) plane

(b) robot 1 position errors

(c) robot 1 orientation error

(d) robot 2 position errors

(e) robot 2 orientation error

Fig. 8. Experimental results for two robots tracking linear desired trajectories

Conclusion

The new control algorithm for formation of mobile differentially-driven robots in the environment with static convex obstacles was presented. Stability proof for the system was shown. Simulation and experimental results illustrate effectiveness of the presented method. Authors plan further experiments with greater number of robots. Control method will be also expanded to be suitable for more complex environments.

References

[1] Do, K.D.: Formation Tracking Control of Unicycle-type Mobile Robots with Limited Sensing Range. IEEE Transactions on Control Systems Technology 16(3), 527–538 (2008)

[2] Egerstedt, M., Hu, X.: Formation Constrained Multi-Agent Control. IEEE Transactions on Robotics and Automtion 17(6), 947–951 (2001)

[3] Fierro, R., Das, A.K., Kumar, V., Ostrowski, J.P.: Hybrid Control of Formations of Robots. In: International Conference on Robotics and Automation, Seoul, Korea, May 21-26, 2001, pp. 3672–3677 (2001)

[4] Jedwabny, T., Kowalski, M., Kieczewski, M., Lawniczak, M., Michalski, M., Michaek, M., Pazderski, D., Kozowski, K.: Nonholonomic Mobile Robot MiniTracker 3 Research and Educational Purposes. In: 35th International Symposium on Robotics, Paris (2004)

[5] Kowalczyk, W.: Control Algorithms for Formation of Mobile Robots, PhD Thesis (in Polish), Poznan University of Technology, Poznan (2008)

[6] Lawton, J.R., Young, B.J., Beard, R.W.: A Decentralized Approach to Elementary Formation Maneuvers. In: Proceedings of the 2000 IEEE International Conference on Robotics and Automation, San Francisco, California, April 2000, pp. 2728–2733 (2000)

[7] Mastellone, S., Stipanowic, D.M., Graunke, C.R., Intlekofer, K.A., Spong, M.W.: Formation Control and Collision Avoidance for Multi-Agent Nonholonomic Systems: Theory and Experiments. International Journal of Robotics Research 27(1), 107–126 (2008)

[8] Michalek, M.: Vector Fields Orientation Control Method for Subclass of Nonholonomic Systems, PhD. Thesis, Chair of Control and Systems Engineering, Poznan University of Technology, Poland (2006) (in Polish)

[9] Ogren, P., Egerstedt, M.: A Control Lyapunov Function Approach to Multiagent coordination. IEEE Transactions on Robotics and Automation 18(5), 847–851 (2002)

[10] Ogren, P., Leonard, N.: Obstacle Avoidance in Formation. In: Proceedings of IEEE International Conference on Robotics and Automation, Taipei, Taiwan, September 14-19, 2003, vol. 2, pp. 2492–2497 (2003)

[11] Rimon, E., Koditschek, D.: Exact Robot Navigation Using Artificial Potential Function. IEEE Transactions on Robotics nad Automation 8, 501–518 (1992)

[12] Speletzer, J., Das, A.K., Fierro, R., Taylor, C.J., Kumar, V., Ostrowski, J.P.: Cooperative Localization and Control for Multi-Robot Manipulation. In: IEEE/RSJ International Conference on Intelligent Robots and Systems, Maui, Hawaii, USA, 29 October-3 November 2001, vol. 2, pp. 631–636 (2001)

[13] Stipanovic, D.M., Hokayem, P.F., Spong, M.W., Siljak, D.D.: Cooperative Avoidance Control for Multiagent Systems. Journal of Dynamic Systems, Measurement and Control 127, 699–707 (2007)

[14] Urakubo, T., Okuma, K., Tada, Y.: Feedback Control of a Two Wheeled Mobile Robot with Obstacle Avoidance Using Potential Functions. In: Proceedings of 2004 IEEE/RSJ International Conference on Intelligent Robots and Systems, Sendai, Japan, 28 October - 2 November 2004, vol. 3, pp. 2428–2433 (2004)

[15] Yamaguchi, H.: Adaptive Formation Control For Distributed Autonomous Mobile Robot Groups. In: Proceedings of the 1997 IEEE International Conference on Robotics and Automation, Albuquerque, New, Mexico, USA, April 1997, pp. 2300–2305 (1997)

Robot Systems for Play in Education and Therapy of Disabled Children

Gernot Kronreif

PROFACTOR GmbH
Im Stadtgut A2, A-4407 Steyr-Gleink, Austria
gernot.kronreif@profactor.at

Abstract From a developmental and educational perspective, play is a "natural" way in which children learn in an enjoyable manner. It can be relaxing, exciting - children can play a role and it is an important possibility to get in touch with other children. On the other hand, disabled children have limited possibilities for inter-action with social and material environment. This paper is focusing on two par-ticular projects coordinated by the author. The PlayROB system is a robot which supports severe handicapped children for playing with LEGO™ bricks. For a long-term field trial six PlayROB systems have been realized and installed at se-lected Austrian education institutions. The user tests revealed that the goal to make autonomous play for children with physical handicaps possible has been ful-ly achieved. The second presented project – the EC-funded project IROMEC – is dealing with an interactive robot system for use in education and therapy. A novel framework for this application area is being developed and evaluated by means of a dedicated robot setup. The main research focus of IROMEC is on the user ori-ented definition of appropriate play scenarios, development of evaluation meth-ods, and finally on the definition of robot behaviors and interaction modes. Ro-bustness, dependability as well as "plug&play" operation of the robot system are specifically addressed.

1   Introduction

In the past "child's play" has often been neglected in comparison to educational objectives such as developing mathematical or language skills. However, state of the art research emphasises the important role of play in children's development, a crucial vehicle to learn about themselves, the environment, and to develop social relationships. The research activities described in this paper target children who are prevented from playing, either due to cognitive or multiple impairments which affect their playing skills, leading to general impairments in their learning poten-tial and more specifically resulting in isolation from the social environment,

I.J. Rudas et al. (Eds.): Towards Intelligent Engineering & Information Tech., SCI 243, pp. 221–234.
springerlink.com   © Springer-Verlag Berlin Heidelberg 2009

including family members and peers. The underlying assumption is that providing tailored means to encourage play using a robotic toy will break down barriers for development through play, fostering individual development up to the person's full potential.

1.1 Play for Enjoyment and Learning

Play in humans is a complex phenomenon, and its roles in development are diverse. Through play, juveniles interact with their physical and social worlds and 'construct' their mental worlds. Thus, play is more than merely 'having fun' or 'practising skills for adulthood'. Opportunities to play e.g. with peers or family members benefit social competence and confidence. Skills acquired through social play behaviour are instrumental in developing and maintaining social relationships and bonds that may last a life-time. Also, from an educational perspective, play is a "natural" way in which children learn in an enjoyable manner. In play, the complexity of stimuli and activities can be gradually increased, thereby guiding the child through a series of experiences that can be designed according to the children's cognitive, emotional, individual and – when applicable – therapeutic needs and possibilities.

Research in the field of educational technology and special education is only recently approaching the right of children with disabilities to play like all their peers. Robots could contribute to change this situation: their nature itself is joyful, they can possess a potentially huge variety of interaction skills, and their power to inspire identification and empathy has been clearly demonstrated. The new technological possibilities offered by robotics have strongly shed light on the early years of children with disabilities, however the appropriate use of technological toys and play activities is still widely unexplored – a situation that the two projects described here aim to remedy. Both projects are an interdisciplinary initiative combining robotics, ICT and other disciplines like cognitive sciences, developmental psychology, pedagogy, human-machine interface in order to demonstrate a novel potential role of advanced robots in society – a role where playing setups for cognitive or multiple disabled children supported by robotics technology finally contribute to enhancement of the following three aspects.

(1) "Quality of Life":
Playing is a substantial and joyful part in the life of children, it can be relaxing, exciting, children can play a role and it is an important possibility to get in touch with other children. In the very recently published "children version" of the ICF (International Classification of Functioning and Disability) the World Health Organisation has carefully considered and described playing activities, both under the "Activities and Participation" and the "Environmental Factors" [1]. Play is then considered as one of the most important aspects in a child's life, a parameter to be considered for assessment of children's quality of life (QoL).

(2) "Social Inclusion":

Children with a disability not only have to deal with the physical and psychological consequences of this impairment itself; the disability also will affect the development of their social roles. For participation in society communication and interaction skills are necessary – functions which could be improved during the play with the robot. Facilitating this development despite of the disability has a life long positive effect on the individual. It will enhance the social inclusion with other children and adults, and this will continue while growing up and in their adult life.

(3) "Learning and Therapy":

There is considerable evidence and recognition of the important role of play activity which validates the high didactic and educational value of play at every stage of life. The most innovative, psychological and educational trends have highlighted the importance of active teaching and the use of a didactic methodology based on the constructivist concept of learning for the correct development of learning processes. Playing and acting in playful environment is the basic key for children's learning; the role of social relationships to increase the children's cognitive skills has been demonstrated firstly in the educational psychology field by Lev Vygotskij [2] and has been confirmed in subsequent studies. More recently, scientific literature in the field has shown the role that playing activities and interactions in technological environments can have and it's positive influence on children's learning skills. This has been demonstrated also for children with cognitive and with physical impairment.

1.2 State-of-the-Art in 'Robot Assisted Play'

Due to the many related activities in different application fields an analysis of the state-of-the-art must observe several areas, like toy market, toy adaptation research (plus assistive devices) as well as activities in the area of 'personal robotics'. For the (robot) toy market several systems – from simple and cheap devices up to very complex and expensive ones – are (or have been) commercially available[1], e.g. AIBO robot dog from Sony Inc., MyRealBaby from Hasbro, or MINDSTORMS from LEGO Inc. Previous research from several groups worldwide however has shown that these kinds of systems are rather limited for the desired purpose of a playing assistant for disabled children.

Other ongoing research projects are investigating different setups and interaction possibilities between robot and human(s) in the framework of "personal robots", e.g. NEC Research laboratories are developing the personal robot PaPeRo to become "family member". Similar work – but more related to Human-Robot-Interaction (HRI) – can be observed in different research laboratories world-wide.

---

[1] See also http://www.robotoys.com/ or http://www.robotshop.ca/robot-toys.html

The IST-FET project Cogniron studies a cognitive robotic companion to be used in domestic scenarios [3]. MIT Media Lab (and other groups) is working on the interaction aspects for sociable robot systems in a laboratory setting; one study is aimed for weight management for people who have lost weight and want to keep it of. The published results demonstrate that this kind of HRI work with typically developing children or adults but cannot be directly applied to the area of assistive technology. For example, work at ATR with Robovie, as well as other work, has shown that interaction levels with children decrease over repeated exposure, while e.g. two independent studies with autistic children by Robins et al. [4] and children with developmental disorders by Kozima et al. [5] have shown that interaction levels increase in a longitudinal study. Other related research by Takanori Shibata at AIST and collaborators (seal type robot PARO) has shown first promising results in using an interactive robot in therapy for children and support for the elderly [6] – a similar approach is by Omron with their NeCoRo robot system, and Michaud et al. [7] who are designing robots for child-development studies.

In the area of robot-assisted playing early research was done by A. Cook and his collaborators where a small industrial robot was used for a particular play setup for children with physical disabilities [8]. Howell et al. [9] presented a robotic system installed at an elementary school utilized for science instruction. Davies described a prototype for a "playing robot" which aims to give assistance during either a painting or a building scenario [10]. The common theme for all of these scenarios is to use the robot for improved interaction with and exploration of 3D objects. The Adaptive Systems Research Group of the University of Hertfordshire has investigated since 1998 the role of robotic toys in therapy and education of children with autism demonstrating that a robot can potentially play a useful therapeutic role encouraging basic social interaction skills (e.g. joint attention and imitation), as well as using the robot as a social mediator facilitating interaction with peers and adults [11], [12].

2   PlayROB and IROMEC – Two Selected Examples for 'Robot Assisted Play'

In the following chapter two selected examples for application of robotics technology as playing assistant for disabled children will be described in more detail. The first described setup – robot system PlayROB – has been designed as assistive system helping severe disabled children in playing with LEGO™ bricks. In this setup the robot is not the toy – but the robot helps to use the toy. On the other hand, the second system described – robot system IROMEC – is representing a special designed robotic toy serving as a mediator for playing activities for cognitive and physically disabled children.

2.1 Playing Assistant PlayROB

"PlayROB" is a remote controlled robot system which aims to assist the severe physically disabled child during play. The main idea behind this system is that – different to many other approaches – the robot should serve as an assistant only. The way of playing is defined by the child, which ensures maximum autonomy. The robot is not the toy – but the robot assists in using the toy, which leads to a "Robot Assisted Play" setup. Using the functionality of such a robot system, the user is now in the position to manipulate real objects (toys) in the real world, despite of her/his impairment.

2.1.1   First Prototype

In a first feasibility study, a dedicated 3DOF[2] Cartesian robot system has been designed for manipulation of LEGO™ bricks (Fig. 1). The design is following a "low cost" approach by using standard components for the entire system.

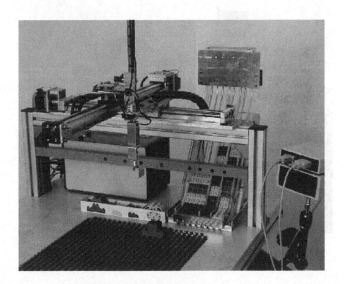

Fig. 1. First prototype of the robot assistant PlayROB – a 3DOF robot with special gripper device and storage system for different kinds of toy bricks

System evaluation, User Trials

The prototype mentioned above was finalized late 2003. In the following, first user trials with this robot prototype were composed of different steps. In a starting series of expert tests the concept per se as well as the functionality and stability of

[2] DOF – Degrees of Freedom

the system could be evaluated. It should be mentioned here that these expert tests and system refinement before starting the tests with end users turned out as very important in order to avoid frustration for the children if the system breaks down.

For next stage of user tests six children were invited to use the robot prototype. For first trials three able-bodied children (between 5 and 7 years old) were confronted with the system for three playing sessions each. Beside further evaluation of the system concept and the user interface (all children have used a 5-key input device) there was also an evaluation of different playing setups (i.e. "free playing", reconstruction of pre-defined figures). In a second series, three disabled children (between 9 and 11 years old; child 1 – multiple physical impairments; child 2 – tetra paresis; child 3 – transverse spinal cord syndrome) were asked to use the robot in the same playing setups as used in the previous series. For both series, no quantitative criteria were used (e.g. time per inserted brick, number of "wrong placements", etc.) – main interest was on acceptance of the system and intuitiveness of the user interface concept.

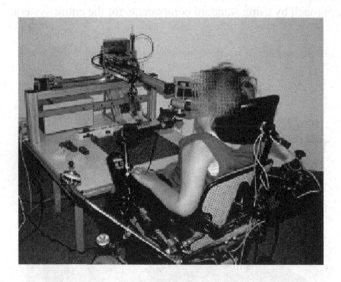

Fig. 2. Robot prototype during user tests

The user tests (Fig. 2) revealed that the chosen approach for a robot system can be an attractive device for children with physical disabilities [13]. The very positive feedback from children, but also from parents and teachers, have encouraged to perform a redesign of the system and the realization of a small series of the robot assistant for a multi-center evaluation study.

2.1.2   Multi-center Evaluation with Redesigned PlayROB System

An important research question for the proposed "Robot Assisted Play" setup is to investigate possible and estimated learning effects. A multi-center study should help to get reasonable answers to that question. Thus, six redesigned systems were installed at selected schools and therapy institutions in Austria since winter semester 2004 in order to introduce the system to many children. Playing with the robot has being included into the regular therapy plan in order to support the evaluation of learning effects.

Main criteria for the redesign were further reduction of the system costs as well as improvement of system safety. The robot system is now completely integrated into a mobile rack – most of the moving parts are covered by the robot housing made from perspex (Fig. 3). Depending on the activity level of the particular user, the system can also be used with locked doors (acrylic glass) in order to avoid any manual intervention during robot operation. Redesign was subject to all system components, i.e. gripper system, storage as well as control system [14]. For efficient execution of the long-term user evaluation, each single playing session is being recorded into a "log-file" in any detail – including name of the player, duration of the playing session and each particular playing sequence.

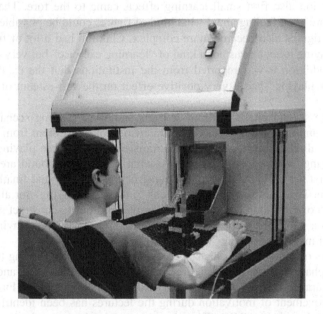

Fig. 3. Redesigned system "PlayROB"

Multi-center System Evaluation

For investigation of possible learning effects in a multi-center study, six PlayROB systems were installed at selected schools and therapy institutions in winter semester 2004 and in summer semester 2005 in order to introduce the system to as many children as possible.

For the desired evaluation of learning effects the following parameters have been recorded for every playing session:

- Duration of playing session
- Number of used bricks and number of different brick types
- Time required for brick placement (bricks/min)
- Utilization of the playground area (%)

The six systems were installed at three institutions in Austria (Waldschule[3], Institut Keil[4], Vereinigung zu Gunsten körper- und mehrfachbehinderter Kinder und Jugendlicher[5]). At each of the three sites, about 5-10 children have used the PlayROB system on regular basis. All of the users are showing significant physical disabilities – in most cases together with different degree of mental retardation. Most of the pupils are not able to speak.

In the first stage of the study (until March 2006) no instructions about what to build were given to the children ("free playing"). Main goal for this first phase of user trials was to evaluate the impact of the redesign measures. Results show that the new system design and the new functions – like laser guidance during brick insertation, new interfaces for children and teacher, new starting procedure, etc. – result in an enhanced acceptance at children and teacher side. Beside of this functional evaluation also first small learning effects came to the fore. The children more and more got a feeling about what kind of figures could be possible by using the bricks – figures also became more complex. Children had a lot of fun during playing – playing to them was not a kind of "learning exercise" but very enjoyable activity. In addition it was reported from the institutions that the experience of "autonomous playing" had a very positive effect on the self-esteem of the children.

During the second phase of the user trials additional playing scenarios have been defined and improved together with the institutions. Different from the original plan, not all children finally could be transferred from "free playing" to "instructed playing" where instructions about what they have to build are given to each child first verbally, then as sample constructions to copy and finally as construction plan. The main reason for not using "instructed playing" for all children was that most of the children had not the cognitive abilities to construct something according to a plan. For some children the proposed "instructed playing"-mode was not that fun as "free playing".

The results obtained during this stage of user trials were confirming the results of previous phases. Most of the children have shown significant advancement in terms of endurance and concentration, but also of spatial perception. Furthermore general improvement of motivation during the lectures has been identified as result of the work with PlayROB. The robot system also has turned out as optimal tool for training with input devices – children were learning different features of the particular input device in a playful environment and with high motivation. In depth analysis has shown a considerable improvement of the recorded parameters for many children [15].

---

[3] Waldschule/ Wr. Neustadt is a school for children with multiple disabilities.

[4] Institut Keil/Vienna is a special therapy institution for children with cerebral palsy.

[5] Vereinigung zu Gunsten körper- und mehrfachbehinderter Kinder und Jugendlicher/Vienna is a parents association for ambulant care for children with physical and or multiple disabilities.

- Placement of bricks has being optimized in terms of time and accuracy.
- Entire area of the playground has being used after some playing sessions.
- "Distance" between selected and "optimal" brick placement has being reduced after some playing sessions.
- With each playing session the number of different bricks used by the player has increased.

Aside of this quantitative analysis there also has been a qualitative evaluation by the teachers/therapists from each involved institute. Teachers/therapists were interviewed about the progress of the children. For example one institute reported that after a 6 month evaluation period – from 7 children playing regularly – one child finally was able to play without any manual or verbal intervention, one other child was able to play with only needing minor verbal intervention. One child has finally used the entire playground area and has created rather complex constructions. All three institutes were reporting that the children are playing with high concentration and fun – also over a longer period of time. There was no significant reduction of interest in playing with PlayROB in course of this long term evaluation. Using the robot has been recognized as "learning with great fun". In addition they reported that the PlayROB was also an attraction for children who were normally able to play with bricks.

Tests on the other hand also have demonstrated that – even if the robot system allows autonomous playing and even if the setup time for the robot could be reduced – introduction of such a robot system to the regular therapy plan also results in an additional working load for the already overloaded teachers and therapists. As a consequence the utilization of the robot systems was a little behind the expectations. Other problems which were significant during the evaluation phase include organizational matters. In all three institutions it was difficult to find a dedicated place/room for the PlayROB continuously. Another problem was the lack of personal. During the evaluation phase it became evident that the two institutions with the therapeutic focus could use the PlayROB more often than it was possible in the school. The reason therefore is – as mentioned before – mainly the lack of personal and the fixed day structure in the school. For the desired "routine use" of the PlayROB system such limiting factors need to be considered and an appropriate framework for efficient use must be developed.

2.2 Toy Robot System IROMEC

Similar to the PlayROB project described above, IROMEC targets children who are prevented from playing, either due to cognitive, developmental or physical impairments which affect their playing skills, leading to general impairments in their learning potential and more specifically resulting in isolation from the social environment. A novel framework for robotic social mediators is being developed and evaluated by means of a dedicated robot setup in the context of therapy and education. The IROMEC project aims on a user centered definition of appropriate play scenarios, development of evaluation methods, and finally on the realisation of a dedicated interactive robot system. IROMEC investigates how robotic toys can provide opportunities for learning and enjoyment. The developed robotic system is tailored towards becoming a social mediator, empowering children with disabilities to discover the range of play styles from solitary to social and cooperative play. Robustness, dependability as well as ease of use are specially addressed.

One of the major aspects of IROMEC is the study of the role of a robot as an enjoyable toy and a social mediator which is widely unexplored and has so far only be demonstrated in very small-scale pilot studies. Further, the project is emphasizing on the development of a dedicated framework encompassing a wide range of children with different kinds of disabilities, rather than purely focusing on specific user groups. Results of IROMEC aim to generalize research on robot mediated play in a social context across different scenarios and user groups. The research focus of IROMEC is on participative design, development of play scenarios which cover all phases of play, definition of robot behaviors and interaction modes resulting from these scenarios, integration of appropriate communication and control technology, and consequent application of a "plug&play" strategy.

2.2.1 IROMEC Play Scenarios

Scenarios serve as central representations throughout development cycles, first describing the goals and concerns of current use, and then being successively transformed and refined through iterative design and evaluation processes [16]. In the IROMEC project the concept of scenarios has been adopted and used for an additional purpose. Here, scenarios are seen as higher level conceptualizations of the 'use of the robot in a particular context'. Scenarios are used not only as intermediary steps or tools in the design and development process of the robot, but more importantly as play contexts which allow users to evaluate specifically implemented functionalities of the IROMEC robot.

Development of IROMEC scenarios has been performed in several phases. The preliminary concepts for play scenarios were based on a detailed literature review as well as experimental investigations and were related to existing technology used in play activities by the various target user groups. The results from the experimental investigation of various concepts of play scenarios together with outcome of the consultation with the panel of expert users (different panels of teachers, therapists, parents related to the different target user groups) were then merged to form "Outline Play Scenarios" that reflect the user requirements and are not related to any specific technological solution/robot. During the final phase of scenario development, these "Outline Play Scenarios" have further been developed, against specific therapeutic and educational objectives, and finally reflect the specific functionalities to be implemented in the IROMEC robot and its various modules. After final discussion with the user panel the following IROMEC Play Scenarios (IS) have been selected for implementation (Fig. 4).

No.	Title of scenario	Characterization	User groups			Social mediation*	Play type				Solitary play	Collaborative play
			AUT	SMI	MMR		EX	SY	AS	GR		
IS01	Turn taking	Turn taking game with a mobile robot	●	◇	◇	H	✓			✓		✓
IS02	Sensory reward	Turn taking game for sensory reward	●	◇	◇	H	✓			✓		✓
IS03	Imitation game	Imitation game	●	◇	○	H	✓					✓
IS04	Make it moves	Cause and effect game	◇	●	◇	M	✓			✓	✓	✓
IS05	Follow-Me	Coordination game	◇	◇	●	H	✓	✓				✓
IS06	Dance with me	a composition game		●	◇	L	✓			✓	✓	
IS07	Build a tower	Solitary constructive game		●	◇	L			✓		✓	✓
IS08	Bring me the ball	Cause and effect game	◇	●	◇	M	✓				✓	✓
IS09	Get In contact	Sensory stimulation game	◇	◇	●	M	✓	✓			✓	✓
IS10	Pretending to be a character	Pretend play		●	◇	H		✓				✓

Fig. 4. Selected IROMEC Play Scenarios with relevant user groups and play type [17]. Definition of play types is using the ESAR classification from IROMEC partner AIJU [18]

IROMEC project claims to strictly adhere to a user-centered design approach. Different kinds of users, therapists, care-givers, children and parents have been iteratively involved in the design of the robot. However, transfer of the play scenarios and the requirements collected during the user panels into a robotic design has been a very challenging task. Problems range from the difficulty of reconciling conflicting needs and different expectations about the final system to the problems of a more direct user involvement into the design process. The final design solution includes a modular and configurable robot platform which allows addressing the very specific needs of each user group and leaves room for the further investigation of consolidated play activities as well as definition and implementation of new scenarios at a later stage. Main components of the robot include *a mobile robot platform*, a dedicated *interaction module* and a *teacher console*. The interaction module mainly consists of a *body* with GUI elements, a *head* with a digital display for both expression and orientation, and *manipulator arms* to guarantee basic manipulation features. Some *add-on components* – like exchangeable coatings for the body with different effects, physical face mask, fur elements, etc – provide additional means for personalization and customization of the robot.

The robot has two main configurations: horizontal and vertical (Fig. 5). In both configurations the body of the robot has a bilateral symmetry. Furthermore, in both configurations the position of the head clearly shows the front of the robot. In vertical configuration the interaction module can be used in "stand-alone" mode – i.e. without need to be connected to the mobile platform and resembles the shape of a human body. In horizontal configuration – in connection with the mobile

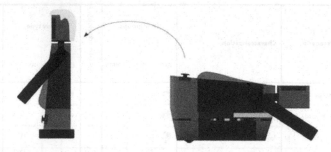

Fig. 5. Basic design of the IROMEC robot in vertical and horizontal configuration

platform—the robot supports a complete set of activities requiring wider mobility and dynamics. In this configuration the robot has a vehicle-like appearance that suits the requirements of Action and Coordination Games. [19]

2.2.2   Prototype and First User Trials

Based on the aforementioned scenario descriptions and the design concept a first prototype of the IROMEC robot has been realized by end of 2008. An evaluation phase has been started in order to assess the prototype's usability, taking into consideration the robot's general usability, the valuation of the play scenarios and, finally, checking the enjoyment and motivation levels experienced by users with regard to the robot.

During this phase the robot has been at six different centres (all over Europe) between February and April 2008. Over the period of evaluation, the robot has been used in a number of trials that have provided some important results for redesign of the robot and implementation of new functions.

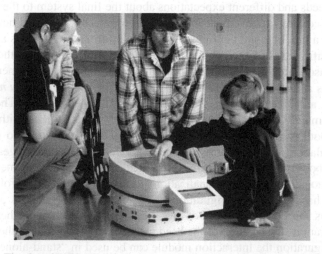

Fig. 6. IROMEC robot (in horizontal configuration) during user trials

37 users participated in the robot's first assessment phase. This includes autistic users with several motor impairments and children with mildly mental retardation. Concluding the trials, the robot has been very positively valued by all the experts who took part in the evaluation. The feedback collected from the secondary users shows that the robot found high interest. Overall assessment is very positive as regards of usability and playability. Another result of the trials is a set of proposals and requirements for redesign and extended functionality which will further improve the performance of the IROMEC robot.

Conclusion

This paper reports on a new research topic dealing with "Robot Assisted Play" for disabled children. Two robot systems are under development coordinated by the author. The PlayROB system aims to assist in manipulation of standard toys and thus allows autonomous playing. A first prototype system as well as a small series of six robots for playing with LEGO™ bricks have been developed and successfully evaluated during a couple of user trials. The second system described here – the IROMEC robot system – is designed to serve as a mediator during playing, increasing the interaction between disabled children and addressing basic objectives according to the ICF-CY classification. Also in this case a first prototype has been developed and could already demonstrate its appropriateness and possibilities to users.

Concluding this paper it should be accentuated that disabled children should get improved access to toys to play with and – besides learning – to simply have great fun. Up-to-date technology can be a useful tool to realize adapted toys for any kind of disabled children.

Acknowledgments. Realization of the six "PlayROB" systems mentioned in this paper was partly funded by Austrian charity organisation "Licht-ins-Dunkel". The author also acknowledges the contributions of Martin Kornfeld, Martin Fürst, Wolfgang Ptacek, Andreas Hochgatterer, Barbara Prazak, Michael Meindl and Stefan Mina as well as of the three participating institutes "Waldschule", "Institut Keil" and "Vereinigung zu Gunsten körper- und mehrfachbehinderter Kinder und Jugendlicher".

EU project "IROMEC" (Interactive Robotic Social Mediators as Companions) is being co-funded by the European Commission in the 6th Framework Programme under contract IST-FP6-045356. Partners in IROMEC project are PROFACTOR GmbH (AT), University of Hertfordshire (UK), Robosoft SA (FR), VILANS (NL), University of Siena (IT), University della Valle d'Aosta (IT), Toy Research Institute (ES), Risoluta SLL (ES) and Austrian Research Centers GmbH (AT). The author specially acknowledges the contributions of all members from the IROMEC team. More information about IROMEC is available at the project homepage www.iromec.eu.

Finally the author expresses his thanks to all children actively participating to the various test phases.

References

[1] World Health Organisation – WHO,
     http://www.who.int/classifications/icf/en/
[2] Vygotskij, L.S.: Play and Its Role in the Mental Development of the Child. Soviet Psychology (Armonk, NY, USA) 5(3), 6–18 (1967)
[3] COGNIRON Project Homepage, http://www.cogniron.org/final/Home.php
[4] Robins, B., Dautenhahn, K., te Boekhorst, R., Billard, A., Keates, S., Clarkson, J., Langdon, P., Robinson, P.: Effects of Repeated Exposure of a Humanoid Robot on Children with Autism. In: Designing a More Inclusive World, pp. 225–236. Springer, Heidelberg (2004)
[5] Kozima, H., Nakagawa, C., Yasuda, Y., Kosugi, D.: A Toy-like Robot in the Playroom for Children with Developmental Disorder. In: Proceedings of the International Conference on Development and Learning, ICDL 2004, San Diego, USA, CD-ROM (2004)
[6] Inoue, K., Wada, K., Ito, Y.: Effective Application of PARO: Seal Type Robots for Disabled People in According to Ideas of Occupational Therapists. In: Miesenberger, K., Klaus, J., Zagler, W.L., Karshmer, A.I. (eds.) ICCHP 2008. LNCS, vol. 5105, pp. 1321–1324. Springer, Heidelberg (2008)
[7] Michaud, F., Theberge-Turmel, C.: Mobile Robotic Toys and Autism. In: Bond, A., Canamero, L., Dautenhahn, K., Edmonds, B. (eds.) Socially Intelligent Agents - Creating Relationships with Computers and Robots, Kluwer Academic Publishers, Dordrecht (2002)
[8] Cook, A.M., Howery, K., Gu, J., Meng, M.: Robot Enhanced Interaction and Learning for Children with Profound Physical Disabilities. Technology and Disability 13, 1–8 (2000)
[9] Howell, R.D., Damarin, S.K., Post, E.P.: The Use of Robotic Manipulators as Cognitive and Physical Prosthetic Aids. In: Proc. RESNA, San Jose, pp. 770–772 (1987)
[10] Davies, R.C.: The Playing Robot: Helping Children with Disabilities to Play. In: Proc. DARS 1995, Vienna, Austria, pp. 71–76 (1995)
[11] Robins, B., Dickerson, P., Stribling, P., Dautenhahn, K.: Robot-mediated Joint Attention in Children with Autism: A Case Study in a Robot Human Interaction. Interaction Studies: Social Behaviour and Communication in Biological and Artificial Systems 2, 161–198 (2004)
[12] Dautenhahn, K., Werry, I., Rae, J., Dickerson, P., Stribling, P., Ogden, B.: Robotic Playmates: Analysing Interactive Competencies of Children with Autism Playing with a Mobile Robot. In: Dautenhahn, K., Bond, A., Canamero, L., Edmonds, B. (eds.) Socially Intelligent Agents - Creating Relationships with Computers and Robots, pp. 117–124. Kluwer Academic Publishers, Dordrecht (2002)
[13] Prazak, B., Kronreif, G., Hochgatterer, A., Fürst, M.: A Toy Robot for Physically Disabled Children? Special issue on Children and Technology, Journal Technology and Disability 16, 131–136 (2004)
[14] Kronreif, G., Prazak, B., Mina, S., Kornfeld, M., Meindl, M., Fürst, M.: PlayROB – Robot-Assisted Playing for Children with severe Physical Disabilities. In: Proc. of the International Workshop on Human-friendly Welfare Robotics Systems - HWRS 2006, Korea, CD-ROM (2006)
[15] Kronreif, G., Prazak, B., Kornfeld, M., Mina, S., Fürst, M.: Robot Assistant "PlayROB" – User Trials and Results. In: Proc. of the 16th IEEE International Symposium on Robot and Human Interactive Communication - RO-MAN 2007, Korea, pp. 113–117 (2007)
[16] Carroll, J.M.: HCI Models, Theories, and Frameworks - Towards a Multidisciplinary Science. Morgan Kaufmann Publishers, San Francisco (2003)
[17] IROMEC Consortium, Final IROMEC Scenarios for Robot Assisted Play, Deliverable D2.1a, internal document, IROMEC (2008)
[18] Garon, D., Filion, R., Doucet, M.: El sistema ESAR: Un método de análisis psicológico de los juguetes (The ESAR System. A Psychological Method of Toy Analysis), AIJU, Ibi (1996)
[19] IROMEC Consortium, Physical and Visual Interface Design of the IROMEC Robot, Deliverable D3.2a, internal document, IROMEC (2008)

Recent Advances in Intelligent Robots at J. Stefan Institute

Jadran Lenarčič, Leon Žlajpah, Bojan Nemec, Aleš Ude, Jan Babič,
Damir Omrcen, and Igor Mekjavič

J. Stefan Institute, Ljubljana, Slovenia
jadran.lenarcic@ijs.si

Abstract. This paper presents an overview of the recent robotics research and technology development of advanced robot systems at the J. Stefan Institute in Ljubljana, Slovenia. The presented projects integrate knowledge from different fields of robotics, automation and biocybernetics bringing into life completely new design and control concepts and approaches. Applications are seen in industrial and service robotics and in particular in the development of future humanoid robots.

1 Introduction

The new robots combine the latest developments in robot mechanism design, new materials, control software, sensors and actuators technology with an enormous amount of new knowledge in human motion and intelligence. We believe that robots will not only be able to clean the house, do the dishes and take out the garbage, but also to play with children, help the elderly and even explore the farthest reaches of space and perform repairs or search-and-rescue missions in hazardous sites. A new generation of industrial production technologies will arise with the introduction of intelligent and robots inspired by human and animal behavior and motion properties. Moreover, the fast development promises that these machines might soon be able to carry out tasks that are beyond human capabilities.

The first developments of robots and intelligent machines at J. Stefan Institute in Ljubljana (Slovenia) date back to the early eighties. Initial investigations and technologies were related to the development of industrial robots and their applications in industrial processes in context to increase the level of automation in factories. Resulting from the strong biocybernetics tradition at the Institute, the robotics research soon adopted new topics in the area of humanoid motion. The first results were related to the human arm motion characteristics, in particular in the determination of the human arm workspace. These investigations were enlarged with new directions in sensor-based adaptive control, motion redundancy and synthesis of humanoid robot kinematics, in particular in manipulation. Later it

I.J. Rudas et al. (Eds.): Towards Intelligent Engineering & Information Tech., SCI 243, pp. 235–245.
springerlink.com   © Springer-Verlag Berlin Heidelberg 2009

expanded to new problems in connection to humanoid shoulder design, jumping, combination of mobile robot platform and a robot manipulator, humanoid stereo vision, humanoid periodical motion, skiing and other related technologies, such as the manikin development and synthesis of humanoid motion in unstructured environments.

The J. Stefan Institute intensively collaborates in the area of intelligent robotics and mechatronics in European framework projects and in bilateral projects with a number of institutes around the world, in particular in Europe, Japan and USA. The aim of this paper is to present an overview of the recent research activities in the Department of Automatics, biocybernetics and robotics at the J. Stefan Institute with the emphasis on bio-inspired robotics and intelligent machines.

2 Human Arm and Shoulder Motion

Investigations are carried out in the area of human arm kinematics in order to determine its positioning ability. A number of models were synthesized on the basis of complex electro-optical measurements and motion recordings of human subjects performing bilateral and unilateral humeral elevation. The obtained serial model of the arm consists of three joints, the inner shoulder joint (sternoclavicular joint and scapulotoracic joint), the outer shoulder complex (glenohumeral joint) and the elbow joint. Based on this model, the reachable workspace was obtained. It was used for the evaluation of the reachable workspace of the human arm. Numerical and graphical representations of the workspace were used for the evaluation of the arm functionality in rehabilitation and ergonomics. The model offers a possibility of a direct comparison of the motion of an impaired and a healthy person. The results are also important for the design, construction and control of the humanoid robots [3].

A further investigation focuses on a humanoid robotic shoulder complex and on the kinematics of humanoid humeral pointing as performed by this complex. The humanoid shoulder complex is composed of two subsystems, a parallel mechanism which serves as the innermost shoulder girdle and a serial mechanism which serves as the outermost spherical glenohumeral joint. These two subsystems are separated by an offset distance and a twist angle. The subsystems operate cooperatively as an offset double pointing system. Humanoid humeral pointing is defined as a configuration in which the displacement of the shoulder girdle and the humerus are co-planar and in which a ratio between an inclination angle in each subsystem achieves a constant value consistent with human humeral pointing. One redundant degree of freedom remains in the humanoid shoulder girdle and it can be used to optimize system configuration and operating criteria, such as avoiding the singular cones of the humanoid glenohumeral joint [1].

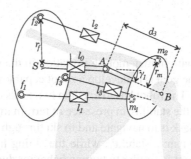

Fig. 1. The parallel humanoid shoulder (frontal view)

3   Humanoid Vertical Jump

The purpose of the research was to propose a new human inspired structure of the lower extremity mechanism by which a humanoid robot is able to efficiently perform fast movements such as running and jumping. We built an efficient dynamic model of the humanoid robot which includes an elastic model of the biarticular muscle gastrocnemius. We determined the role of the biarticular muscles and the elastic tendons in performing the vertical jump. We demonstrated that biarticular links contribute a great deal to the performance of the vertical jump. For the first time in the area of robotics we showed that timing of the biarticular link activation and stiffness of the biarticular link considerably influence the height of the jump. We designed and built a humanoid robotic mechanism that is, by its characteristics, unique in the world. It includes elastic biarticular links which enable the execution of fast dynamic motions. The described robotic mechanism enables in depth research into the field of humanoid robotics and fast motions such as running and jumping. The results are important for the design, construction and control of the humanoid robots and medicine, ergonomics and sports [2].

Fig. 2. A special humanoid robot performing vertical jump

4   Skiing Robot

Projects in this area deal with human motion imitation in the unstructured dynamic environment od skiing. The goal is to develop a robot capable of autonomous skiing on the unknown ski slope. The robot has to execute two tasks during the skiing. The first one is maintaining of the stability in presence of terrain irregularities in order to prevent falls. The second task is to navigate and to ski through the race. The primary task is maintaining the dynamic stability, while the skiing trough the gates is the secondary task. This concept is similar to human behavior during the alpine skiing. The dynamic stability is achieved by appropriate legs pose and movement and is based on zero moment point. For the navigation, a model of the human acceleration is also used. Algorithms were tested in the virtual environments as well as on the real robot on the ski slope [7].

Fig. 3. The skiing robot in the laboratory

5   Robot Manikins

The flame manikin system was designed to test fire protective clothing ensembles in compliance with the draft international standard. The manikin is instrumented with 128 temperature sensors. The system for simulating a "flash fire" comprises 12 gas burners circling the manikin in two rows. During a flame test the manikin is suited with the test clothing ensemble. The skin burn model incorporated in the flame manikin system predicts the severity of burn injury from the skin temperature data. The flame manikin test results are presented as total body surface area with a predicted burn injury.

We developed a sweating thermal foot manikin with a gait simulator. In parallel to this development we investigated the pattern and magnitude of sweating at the foot, incorporated this knowledge in the thermal manikin regulatory system responsible for the simulation of sweating. The system enables the determination of the thermal and evaporative resistance of footwear, as well as other functional characteristics. The system is now manufactured under license by a company, who has already delivered systems to different manufacturers, research institutes, and defence establishments.

Fig. 4. The flame manikin during the experiment

6 Intelligent Robot Industrial Application

Finishing operations in shoe manufacturing processes comprise operations such as application of polishing wax, polishing cream and spray solvents, and brushing in order to achieve high gloss. These operations require skilled workers and are generally difficult to automate due to the complex motion trajectories. We developed a robotic cell for finishing operations in a custom shoe production plant. Such customization of shoe production should allow production of small batches of shoes of the same type. It requires automatic setup and adaptation of the production line. To meet the requirements, a CAD system for automatic generation, optimization and validation of motion trajectories was integrated into a robotic cell. In automatic trajectory generation some of the major problems are the limitations posed by robot joint limits, robot singularities and environment obstacles. These problems were solved by introducing the kinematic redundancy of the robot manipulator.

In collaboration with industry we implemented many work cells for testing and production of footwear. The goal of these applications was to automate production processes in shoe industry and to increase the usage of CAD systems for shoe design in the production processes. We implemented a new robotized cell for distributing the glue to the sole of a shoe. To control the cell an advanced controller was used together with different sensors. We developed an expert system for trajectory planning. The robot trajectories are generated automatically based on the CAD models of the shoe lasts and are optimized regarding the desired accuracy. The generated trajectories are the transferred to the cell controller, where the robot uses these trajectories to apply the glue to the shoe sole. The robotized cell is flexible and suitable for small batch production. This application has shown how in modern production systems we can integrate the design of products and the manufacturing process.

7 Robot Testing

One of our activities is to technologically support the sports equipment manufacturers. This consists of two phases. First is to capture the characteristic motion trajectories during the sports activity, and second is the reproduction of the trajectories using an industrial robot. During the second phase, we detect the ground reaction forces and torques, bending characteristics and tensions in various parts of the sports equipment. This approach enables efficient testing of sports equipment and its subsystems and enables a validation of new materials and technological solutions during the design cycle of a new product.

One of the projects is to determine the exact motion of the human knee to better understand the knee motion and to improve the quality of operations of the knee ligaments. We use an industrial robot as a measuring device, which should not affect the natural knee movement. The robot has to bend a knee and to measure positions of the knee during motion. To measure the loaded or unloaded knee the robot is force controlled. To increase accuracy of the measurement we developed autonomous force sensor gravity and offset compensation. The project was realized in collaboration with medical doctors from university medical center.

8 Adaptive Redundant Robots

We compared and emphasized the conceptual difference between pseudo-inverse and minimal null-space based control algorithms. We demonstrated that minimal null space based control algorithms might become computational unstable and proposed an efficient solution to this problem. Furthermore, we investigated the problem and benefits of vision, tactile, proximity and force sensors interaction in the control loop. Based on the sensors we proposed a method for on-line obstacle avoidance applying the force control to redundant robots. Furthermore, we proposed an obstacle avoidance algorithm by reducing inertia in the null-space and proposed alternative solutions for obstacle avoidance using a vision system. It is well known that the majority of modern control algorithms become inefficient or even unstable in the presence of limits imposed by the real mechanism. Therefore we developed algorithm for the compensation of velocity and/or acceleration limits for redundant robots. The proposed algorithms improves the performance of robots in applications executing actions that require high dynamic response, such as jumping, running, throwing, etc.

We also developed a unique robot controller which combines the velocity and torque control. It is appropriate for robot systems, which are composed of two systems, where one enables only velocity control and the other enables also torque control. An example of such a system is a service robot which is composed of a velocity controlled mobile platform and a torque controlled robot arm. We implemented this control to a mobile manipulator system, which can be used as a service

Fig. 5. The humanoid robot head with stereovision

robot in human environments. The mobility gives large workspace to the system while the manipulator arm enables a more precise object manipulation [8].

9 Service Robots in Humanoid Applications

The aim of the project is to build a service robot system that should help a human in everyday life. We integrated mobile manipulator, robot arm, robot vision and complex sensory system consisting of force, tactile and proximity sensors. We developed a new type of control, which is suitable for controlling redundant systems composed of two or more subsystems. As an example of a human task we have realized the task of pouring a drink to a glass with a robot. For this application we combined the industrial robot manipulator and a mobile robot platform.

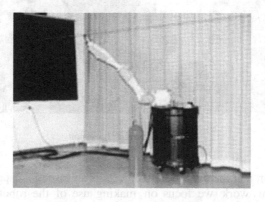

Fig. 6. A combination of a mobile platform and a robot manipulator

The platform was used as a tour guide robot in Modern gallery in Ljubljana. The robot guided visitors in the gallery and commented the exhibition. To navigate in the gallery the robot used preprogrammed map of the gallery including major exhibits. The robot use ultrasonic sensors to locate in the space and to avoid the obstacles (visitors). The robot successfully "worked" in the gallery for a week and guided few hundred visitors.

10   Humanoid Vision and Learning

Complex visual processes such as visual attention are often computationally too expensive to allow real-time treatment on a single computer. To solve this problem we investigated distributed computer architectures that enable us to divide tasks into several smaller problems. We demonstrated how to implement distributed visual attention system on a humanoid robot to achieve real-time operation at relatively high resolutions and frame rates. We started from a popular theory of up visual attention that assumes that information across various modalities is used for the early encoding of visual system uses five different modalities including color, intensity, edges, stereo, and motion. The system was fully implemented on a workstation cluster comprised of eight PCs. It was used to drive the gaze of a humanoid head towards potential regions of interest. We also developed a novel way to integrate top-down and bottom-up processing to filter out level unimportant information while attending to features indicated as important by higher-level processes by a way of top-modulation [4], [9].

Fig. 7. Learning by vision

The exploration and learning of new objects is an essential capability of a cognitive robot. In this work we focus on making use of the robot's manipulation capabilities to learn complete object representations suitable for 3-D object recognition. Taking control the object allows the robot to focus on relevant parts of the object. We propose a systematic method to control a robot in order to achieve a maximum range of motion across the 3-D view sphere. This is done by exploiting the task redundancies typically on a humanoid arm and by avoiding joint limits and self-collisions of the robot. The proposed method enables us to acquire a range of snapshots without re-grasping the object [5].

11 Performing Periodic and Rhythmic Tasks

One of the principal challenges in nowadays robotics is to generate human like motion solving different type of tasks. In this field we concentrated in control algorithms that are capable of performing humanoid rhythmic tasks. The aim is to achieve human-like behavior by using unconventional control algorithms. Common to such tasks is that they are usually more or less an easy task for a human, but a complex task for a robot. Namely, the dexterity of the system and the synchronization are required. A human can use his senses to learn how to perform such tasks. However, developing a robotic system that can perform the same job requires complex sensory systems and advanced control strategies. As the object we have selected a yo-yo and a gyroscopic device. To understand the system we analyzed how a human operates these objects and then we developed models which capture all features important for the control design. We proposed two control strategies: one is based on predefined hand motion patterns (e.g. using some learning methods) and the other generates the motion on-line depending on the current situation. Both methods assure the desired task execution quality [10].

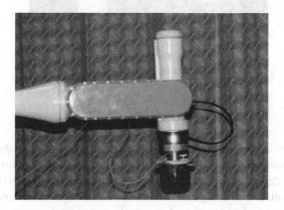

Fig. 8. A robot rotating a gyroscopic device

12   Simulation of Complex Robotic Tasks

Simulation has become a strategic tool in many fields, used by many researchers, developers and by many manufacturers. Robotics is a modern technological branch and simulation plays a very important role, perhaps more important than in many other fields. Modeling and simulation of robot mechanisms and systems has been one of our research topics for several years and we have developed several packages for simulation of robot systems. We developed an integrated environment for the design of robotic controllers. It is implemented on a PC and is based on the Planar Manipulators Toolbox for dynamic simulation of redundant planar manipulators. The tools are fully integrated in the MATLAB/SIMULINK and hence, a lot of standard tools are available for the analysis and control design. Using the real-time simulation it is possible to apply the developed controllers to a real robot manipulator, which can be included in the system via corresponding interfaces, without any additional coding. The main advantage is the flexibility in fast prototyping of different algorithms in the field of control of robotic systems, especially for redundant manipulators. The SYSTEM has been used by more than 400 users all over the world for research and educational purposes [6].

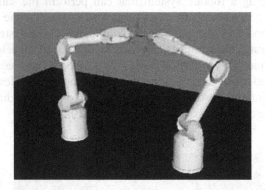

Fig. 9. Collaborating robot manipulators

Conclusions

A brief overview of the recent robotics research and technology development of advanced robot systems at the J. Stefan Institute in Ljubljana, Slovenia, is presented in this paper. It can be seen that the new knowledge combines different fields of robotics, automation and biocybernetics which enables to bring into life entirely new design and control concepts and approaches. Applications are foreseen in industrial and service robotics and in particular in the development of future humanoid robots. The group at the Institute is with its expertise involved in many international projects.

Acknowledgments. · The authors are grateful to all members of the Departments of Automatics, Biocybernetics and Robotics at the J. Stefan Institute who have contributed in the mentioned investigations.

References

[1] Lenarcic, J., Stanisic, M.M.: A Humanoid Shoulder Complex and the Humeral Pointing Kinematics. IEEE Trans. Robot. Autom. 19, 499–507 (2003)

[2] Babic, J., Lenarcic, J.: Optimization of Biarticular Gastrocnemius Muscle in Humanoid Jumping Robot Simulation. Int. J. of Humanoid Robotics 3, 219–234 (2006)

[3] Klopcar, N., Tomsic, M., Lenarcic, J.: A Kinematic Model of the Shoulder Complex to Evaluate the Arm-Reachable Workspace. J. biomech. 40, 86–91 (2007)

[4] Ude, A., Gaskett, C., Cheng, G.: Foveated Vision Systems with Two Cameras per Eye. In: Proc. IEEE Int. Conf. Robotics and Automation, Orlando, Florida, pp. 3457–3462 (2006)

[5] Ude, A., Omrcen, D., Cheng, G.: Making Object Learning and Recognition an Active Process. Int. J. of Humanoid Robotics 2 (2008)

[6] Zlajpah, L.: Integrated Environment for Modelling, Simulation and Control Design for Robotic Manipulators. J. of Intelligent and Robotic Systems 12, 219–234 (2001)

[7] Lahajnar, L., Kos, A., Nemec, B.: Modelling and Control of Autonomous Skiing Robot. In: Proc. EUROSIM (2007)

[8] Nemec, B., Zlajpah, L., Omrcen, D.: Comparison of Null-Space and Minimal Null-Space Control Algorithms. Robotica 25, 511–520 (2007)

[9] Moren, J., Ude, A., Koene, A., Cheng, G.: Biologically-based Top-Down Attention Modulation for Humanoid Interactions. Int. J. of Humanoid Robotics (5), 3–24 (2008)

[10] Gams, A., Zlajpah, L., Lenarcic, J.: Imitating Human Acceleration of a Gyroscopic Device. Robotica 25, 501–509 (2007)

New Trends in Robotic Reinforcement Learning: Single and Multi-robot Case

Duško Katić

Robotics Laboratory, Mihailo Pupin Institute
Volgina 15, 11060 Belgrade, Serbia
dusko@robot.imp.bg.ac.yu

Abstract. A rather general approach to learning control is the framework of Reinforcement Learning, described in this chapter. Reinforcement learning offers one of the most general framework to take traditional robotics towards true autonomy and versatility. Single robot reinforcement learning as well as Multi-robot reinforcement learning are a very challenging areas due to several issues, such as large state spaces, difficulty in reward assignment, nondeterministic action selections, and difficulty in merging learned experiences from other robots. There are still many difficulties in application iof robotics reinforcement learning and in scaling up the multi agent reinforcement learning to multi-robot systems. After reviewing important approaches in this field, some problems and promising research directions will be given.

1 Introduction

Reinforcement learning (RL) is an active area of machine learning research which has received considerable and increasing attention in the last two decades. It is prospective since it assumes that the state space as well as the action space in a given scenario can be divided into discrete ones, and agents in that scenario can learn the optimal policy through reward perceiving. This idea outperforms the traditional control theory which needs to know the model of the environment first. The objective of RL is to learn how to act in a dynamic environment from experience by maximizing some payoff functions or minimizing some cost functions equivalently. In RL, the state dynamics and reinforcement function are at least partially unknown. Thus the learning occurs iteratively and is performed only through trial-and-error methods and reinforcement signals, based on the experience of interactions between the agent and its environment. Reinforcement learning typically requires an unambiguous representation of states and actions and the existence of a scalar reward function. For a given state, the most traditional of these implementations would take an action, observe a reward, update the value

I.J. Rudas et al. (Eds.): Towards Intelligent Engineering & Information Tech., SCI 243, pp. 247–262.
springerlink.com © Springer-Verlag Berlin Heidelberg 2009

function, and select as the new control output the action with the highest expected value in each state (for a greedy policy evaluation). Updating of value function and controls is repeated until convergence of the value function and/or the policy. This procedure is usually summarized under "value update – policy improvement" iterations. The reinforcement learning paradigm described above has been successfully implemented for many well-defined, low dimensional and discrete problems [1-2] and has also yielded a variety of impressive applications in rather complex domains in the last 20 years. In many situations the success or failure of the controller is determined not only by one action but by a succession of actions. The learning algorithm must thus reward each action accordingly. This is referred to as the problem of delayed reward. There are two basic methods that are very successful in solving this problem, *TD learning* [1] and *Q learning* [3]. Both methods build a state space value function that determines how close each state is to success or failure. Whenever the controller outputs an action, the system moves from one state to an other. The controller parameters are then updated in the direction that increases the state value function. Q-learning is a value learning version of RL that learns utility values (Q values) of state and action pairs.

The general purpose of RL is to find a "good" mapping that assigns "perceptions" to "actions". In theory, the formalism and methods of RL can be applied to address any optimal control task, yielding optimal solutions while requiring very little a priori information on the system itself. However, in practice, RL methods suffer from the curse of dimensionality and exhibit limited applicability in complex control problems. Unfortunately, many actual control problems are inherently infinite, described in terms of continuous state variables. And, in the particular case of robotic applications, there is often some degree of uncertainty regarding the state of a system (due to noisy sensors, etc.), requiring a robot to decide upon a (real-valued) belief that describes some probability distribution. The attractiveness of the RL framework and this abundance of interesting but complex control problems emphasize the need to develop more powerful RL methods.

2   Reinforcement Learning for Single Robotic System

The robot learning is essentially concerned with equipping robots with the capacity of improving their behavior over time, based on their incoming experiences. For instance, it could be advantageous to learn dynamics models, kinematic models, impact models, for model-based control techniques. Imitation learning could be employed for the teaching of gaits patterns, and reinforcement learning could help tuning parameters of the control policies in order to improve the performance with respect to given cost functions.

Starting with the pioneering work of Franklin and Benbrahim [4] in the early 1990s, these methods have been applied to a variety of robot learning problems ranging from simple control tasks (e.g., balancing a ball-on a beam, and pole-balancing) to complex learning tasks involving many degrees of freedom such as learning of complex motor skills and locomotion. It is important to notice that RL algorithms are efficiently applied in different robotics fields: manipulation robotics [5-6], mobile robotics [7-8] and humanoid robotics. Reinforcement learning is well suited to training mobile robots, in particular teaching a robot a new behavior (e.g. avoid obstacles) from scalar feedback.

In area of humanoid robotics, there are several approaches of reinforcement learning [4] [9-18] with additional demands and requirements because high dimensionality of the control problem. In paper [13] authors described a learning framework for a central pattern generator (CPG)-based biped locomotion controller using a policy gradient method. The goals in this study are to achieve CPG-based biped walking with a 3D hardware humanoid and to develop an efficient learning algorithm with CPG by reducing the dimensionality of the state space used for learning. It was demonstrated that an appropriate feedback controller can be acquired within a few thousand trials by numerical simulations and the controller obtained in numerical simulation achieves stable walking with a physical robot in the real world. The results suggest that the learning algorithm is capable of adapting to environmental changes. Furthermore, an online learning scheme with an initial policy for a hardware robot is presented to improve the controller within 200 iterations.

Lee and Oh in their research [14], generated a stable biped walking pattern using reinforcement learning. The biped walking pattern is chosen as a simple third order polynomial. To complete the walking pattern, four boundary conditions are needed. Initial position and velocity and final position and velocity of the joint are selected as boundary conditions. In order to find the proper boundary condition value, a reinforcement learning algorithm is used. Also, desired motion or posture can be achieved using the initial and final position. The final velocity of the walking pattern is chosen as a learning parameter. To test the algorithm, a simulator that takes into consideration the reaction between the foot of the robot and the ground was developed. The algorithm is verified through a simulation.

In article [15], Shibata et al. proposed a reinforcement learning (RL) method to train a controller designed for Quasi-PDW of a biped robot which has knees. It is more difficult for biped robots with knees to walk stably than for ones with no knees. The computer simulation shows that a good controller which realizes a stable Quasi-PDW by such an unstable biped robot can be obtained with as small as 500 learning episodes, whereas the controller before learning has shown poor performance. The robot model has closer dynamics to humans in the sense that there are smaller feet whose curvature radius is one-fifth of the robot height, and knees. The reward is simply designed so as to produce a stable walking trajectory, without explicitly specifying a desired trajectory. Furthermore, the controller performs for a short period especially when both feet touch the ground, whereas the existing study above employed continuous feedback control.

Katic et al [16-18], used a non-policy-gradient and policy-gradient methods for learning efficient biped motion. The new integrated dynamic control structure for the humanoid robots is proposed, based on model of robot mechanism. Our approach consists in inclusion of reinforcement learning part only for compensation joints. The basic part of control algorithm represents computed torque control method. The external reinforcement signal was simply defined as fuzzy measure of Zero-Moment-Point (ZMP) error). Internal reinforcement signal is generated using external reinforcement signal and appropriate Critic network. The Critic network provides policy evaluation and can be used to perform policy improvement. For the Critic network, the two layer neural network is proposed. The critic is trained to produce the expected sum of future reinforcement that will be observed given the current values of deviation of dynamic reactions and action.

In order to apply reinforcement learning into the domain of human movement learning, two deciding components need to be added to the standard framework of reinforcement learning: first, we need a domain-specific policy representation for motor skills, and, second, we need reinforcement learning algorithms which work efficiently with this representation while scaling into the domain of high-dimensional mechanical systems such as humanoid robots. Traditional representations of motor behaviors in robotics are mostly based on desired trajectories generated from spline interpolations between points, i.e., spline nodes, which are part of a longer sequence of intermediate target points on the way to a final movement goal. While such a representation is easy to understand, the resulting control policies, generated from a tracking controller of the spline trajectories, have a variety of significant disadvantages, including that they are time-indexed and thus not robust towards unforeseen disturbances, that they do not easily generalize to new behavioral situations without complete recomputing of the spline, and that they cannot easily be coordinated with other events in the environment, e.g., synchronized with other sensory variables like visual perception. In the literature, a variety of other approaches for Dynamic Motor Primitives have been suggested to overcome these problems (see [19]). The resulting framework was particularly well suited for supervised imitation learning in robotics, exemplified by examples from humanoid robotics where a full-body humanoid learned tennis swings or complex polyrhythmic drumming pattern.

Hence, thete are two different application of reinforcement learning: first for traditional spline-based representations, and the second for dynamic motor ptimitives. However, most of the RL methods are not applicable to high-dimensional systems such as humanoid robots as these methods do not scale beyond systems with more than three or four degrees of freedom and/or cannot deal with parameterized policies. Policy gradient methods are a notable exception to this statement.

The advantages of policy gradient methods for parameterized motor primitives are numerous. Among the most important ones are that the policy representation can be chosen such that it is meaningful for the task, i.e., we can use a suitable motor primitive representation, and that domain knowledge can be incorporated, which often leads to fewer parameters in the learning process in comparison to traditional value-function based approaches. Moreover, there exists a variety of different algorithms for policy gradient estimation in the literature, which have a rather strong theoretical underpinning. Additionally, policy gradient methods can be used model-free and therefore also be applied to problems without analytically known task and reward models. Nevertheless, many recent publications on applications of policy gradient methods in robotics overlooked the newest developments in policy gradient theory and its original roots in the literature. Thus, a large number of heuristic applications of policy gradients can be found, where the success of the projects mainly relied on ingenious initializations and manual parameter tuning of algorithms. A closer inspection often reveals that the chosen methods might be highly biased, or even generate infeasible policies under less fortunate parameter settings, which could lead to unsafe operation of a robot. In paper [20] Peters and Schaal have presented an extensive survey of policy gradient methods. All three major ways of estimating first order gradients, i.e., finite difference gradients, vanilla policy gradients and natural policy gradients are discussed in this paper and practical algorithms are given.

3   Reinforcement Learning for Multi-robot Systems

Multi-Robot Systems are distributed systems which consist of a multitude of networked robots and other devices and which, as a whole, are capable of interacting with the environment through the use of perception and action. The main feature of these systems is the use of robust distributed intelligence. The motivations and ideas for developing multi-robot system solutions include: 1) the high complexity of the tasks that a single robot cannot accomplish; 2) the tasks are inherently distributed; 3) building several restricted robots is easier than having a single powerful robot; 4) multiple service robots can solve control problems faster using parallelism; and 5) with the introduction of multi-robot teams, robustness and fault tolerance are increased based on redundancy. Many practical robot applications, such as unmanned aerial vehicles, spacecraft, autonomous underwater vehicles, ground mobile robots, and other robot-based applications in hazardous and/or unknown environments can benefit from the use of multi-robot systems. Based on this facts multi-robot systems have received considerable attention during the last decade [21].

However, there have still been many challenging issues in this field. These challenges often involve the realization of behaviour-based control, or allocating tasks, communication, coordinating actions, team reasoning, etc. Currently, there has been a great deal of research on multi-agent RL [22]. Multi-agent RL allows participating robots to learn mapping from their states to their actions by rewards or payoffs obtained through interacting with their environment. Multi-Robot systems can benefit from RL in many aspects. Robots are expected to co-ordinate their behaviours to achieve their goals. Explicit presentation of an emergent idea of cooperative behaviours through an individual Q-learning algorithm can be found in [23]. Although Q-learning has been applied to many multi-robot systems, such as foraging robots, soccer robots, prey-pursuing robots, prey pursuing robots, patrolling robots and moving target observation robots, etc., most research work in these applications has only focused on tackling large learning spaces. For example, modular Q-learning approaches advocate that a large learning space can be separated into several small learning spaces to ease exploration. Normally, a special selection mechanism is needed in these approaches to select optimal policies generated from different modules. Theoretically, the multi-robot environment is not stationary. Thus the basic assumption for traditional Q-learning working will be violated. Therefore, to solve the scalability problem in multi-Q learning method, it is desirable to develop a distributed multi-Q learning algorithm, where each agent makes its own decision (instead of a central unit) based on its own Q value as well as its neighbors' Q values, instead of all Q values in centralized systems or only its own Q values in single-agent Q learning.

Over the last decade there has been increasing interest in extending the individual RL to multiagent systems, particularly Multi-Robot systems [22], [24]. From a theoretic viewpoint, this is a very attractive research field since it will expand the range of RL from the realm of simple single-agent to the realm of complex multiagents where there are agents learning simultaneously.

Currently multi-agent learning has focused on the theoretic framework of Stochastic Games or Markov Games. Stochastic Games extend one-state Markov Games to multi-state cases by modeling state transitions with Markov Decision Process. Each state in a Stochastic Games can be viewed as a Matrux Game and a Stochastic Game with one player can be viewed as a Markov Decision Process. Stochastic Games have been well studied in the field of multiagent reinforcement learning and appear to be a natural and powerful extension of Markov Decision Process tomulti-agent domains. In the framework of Stochastic Games Nash equilibria is an important solution concept for the problem of simultaneously finding optimal policies in the presence of other learning agents. At a Nash equilibrium each agent player is playing optimally with respect to the others under a Nash equilibrium policy. If all the agents are playing a policy at a Nash equilibrium rationally, then no agent could learn a better policy.

There are several theoretic frameworks for Multi-Robot RL as: Stochastic Games-based, Fictitious Play, Bayesian, Policy Iteration. Multi-robot learning is a challenge for learning to act in a non-Markovian environment which contains other robots. One research on scaling reinforcement learning toward RoboCup soccer has been reported [25]. The most challenging issues also appear in the RoboCup soccer, such as the large state/action space, uncertainties, etc. In [25], an approach using episodic SARSA(λ) with linear tile-coding function approximation and variable λ was designed to learn higher-level decisions in a keepaway subtask of RobotCup soccer. Since the general theory of RL with function approximation has not yet been well understood, the linear SARSA(λ) which could be the best understood among current methods [6] was used in the scaling of reinforcement learning to RoboCup soccer.

Mataric [26] presented a formulation of RL that enables learning in the concurrent multi-robot domain. The methodology adopted in that study makes use of behaviours and conditions to minimize the learning space. The credit assignment problem was dealt with through shaped reinforcement in the form of heterogeneous reinforcement functions and progress estimators. In [27] a variety of methods were reviewed and used to demonstrate for learning in multi-robot domain. In that study behaviours were thought as the underlying control representation for handling scaling in learning policies and models, as well as learning from other agents. Although there have been a variety of RL techniques that are developed for multiagent learning systems, very few of these techniques scale well to Multi-Robot systems. On the one hand, the theory itself on multi-agent RL systems in the finite discrete domains are still underway and have not been well established.

In [28] a modular-fuzzy cooperative algorithm for multi-agent systems was presented by taking advantage of modular architecture, internal model of other agent, and fuzzy logic in multiagent systems. In this algorithm, the internal model is used to estimate the agent's own action and evaluate other agents' actions. To overcome the problem of huge dimension of state space, fuzzy logic was used to map from input fuzzy sets representing the state space of each learning module to output fuzzy sets denoting action space. A fuzzy rule base of each learning module was built through the Q-learning, but without providing any convergence proof. Therefore one can find that the proving techniques and outcomes will be very difficult to extend to the domains of multi-agent reinforcement learning with fuzzy logic generalizations.

In paper [29], authors propose a dynamic correlation matrix based multi – Q learning (DCM-MultiQ) method for a distributed multi-robot system, which focuses on the cooperation between agents. A novel dynamic correlation matrix is proposed, which not only handles each agent's Q value. Furthermore, a theoretical proof of the convergence of the proposed DCM-MultiQ algorithm is also provided using a feedback matrix control theory. To evaluate the efficiency of the proposed DCM-MultiQ method, several case studies of a multi-robot system in forage tasks have been conducted. The simulation results show the efficiency and convergence of the proposed method.

In paper [30], authors propose a reinforcement learning approach to address multi-robot cooperative navigation tasks in infinite settings. The proposed algorithm extends those existing in the literature, allowing to address simultaneous learning and coordination in problems with an infinite state-space. The authors combine an approximate version of Q-learning with an approximate coordination mechanism dubbed approximate biased adaptive play. It is assumed that no communication among the robots and do not require all robots to follow the same decision / coordination algorithm. This is an important advantage: in the presence of a heterogeneous group of robots, this algorithm is still able coordinate to the best decision-rule possible if, for some reason, the other robots act sub-optimally.

In paper [31], the use of RL techniques for solving cooperative problems in teams of homogeneous robots is presented. As an example, the problem of maintaining a mobile robots formation is studied. Authors have formally presented the Line-up problem where a team of robots has to reach a line-up configuration. The experimental results show the viability of the proposed strategy, design and implementation of the Line-up problem. Further research is guided by the idea of experimenting with physical, real robots in a set of more general and harder coordination problems such as coordinated maneuvers, robot interaction and competition.

Multi robot systems often have all of the challenges for multi-agent learning systems, such as continuous state, and action spaces, uncertainties, and nonstationary environment It is unfeasible for robots to completely obtain other robots' information, especially for competitive games since opponents do not actively broadcast their information to share with the other robots. Taking into account the state of the art for multiagent learning system, there is particular difficulty in scaling the established (or partially recognized at least) multi-agent RL algorithms, such as Minimax- Q learning, Nash-Q learning, etc., to Multi-Robot Systems with large and continuous state and action spaces. On the one hand, most theoretic works on multiagent systems merely focus on the domains with small finite state and action sets. As a result, the learning performance (such as convergence, efficiency, and stability, etc.) cannot be guaranteed when approximation and generalization techniques are applied. The most of the multi-agent RL algorithms, is value-function based iteration method. Hence, for applying these techniques to a continuous system the value-function has to be approximated by either using discretization or general approximators (such as neural networks, polynomial functions, fuzzy logic, etc.).

Due to the aforementioned difficulties, a possible opportunity of RL in Multi-Robot systems is to learn robot's coordination. Robots in a team may learn to work cooperatively or share their learned experience to accelerate their learning processes through their limited physical communication or observation abilities. These robots have common interests or identical payoffs. Therefore, more research needs to be performed to gain a clear understanding of Q-learning convergence in the coordination games. Incomplete information, large learning space, and uncertainty are major obstacles for learning in Multi-Robot systems. Learning in Behaviour-Based Robotics can effectively reduce the search space in size and dimension and handle uncertainties locally. The action space will be transformed from continuous space of control inputs into some limited discrete sets. However, the convergence proof for the algorithms using state and action abstraction of Multi-Robot systems will be a very challenging problem. When the state and action spaces of the system are small and finite discrete, the lookup table method is generally feasible. However, in Multi-Robot systems, the state and action spaces are often very huge or continuous, thus the lookup table method seems inappropriate. To solve this problem, besides the state and action abstraction, function approximation and generalization appears to be another feasible solution. The continuous Q-learning method still focuses on single-agent systems. Hence a version of continuous Q-learning method for multi-agent systems is expected accordingly. Another ongoing research for solving continuous cases is continuous RL for feedback control systems [32]. In [32] a continuous RL algorithm was developed and applied to the control problem involving the refinement of a Proportional-Integral (PI) controller.

4   Example

As example of RL application in robotics, policy-gradient method for learning efficient biped motion is chosen. The kinematic scheme of the biped locomotion mechanism whose spatial model that will be considered is shown in Fig. 1a. The mechanism possesses 18 powered DOFs, designated by the numbers 1-18, and two underactuated DOFs (1' and 2') for the footpad rotation about the axes passing through the instantaneous ZMP position. The overall dynamic model of the locomotion mechanism is represented in the following vector form:

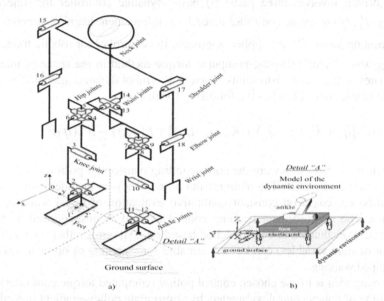

Fig. 1. Model of the biped locomotion mechanism

$$P + J^{T}(q)F = H(q)\ddot{q} + h(q,\dot{q}) \tag{1.1}$$

where: P is the vector of driving torques at the humanoid robot joints; F is the vector of external forces and moments acting at the particular points of the mechanism; H is the square matrix that describes 'full' inertia matrix of the mechanism shown in Fig. 1a; h is the vector of gravitational, centrifugal and Coriolis moments acting at n mechanism joints; J is the corresponding Jacobian matrix of the system; $n = 20$, is the total number of DOFs; q is the vector of internal coordinates; $\dot{q}$ is the vector of internal velocities. The robot's bipedal gait consists of several phases that are periodically repeated. Hence, depending on whether the system is supported on one or both legs, two macro-phases can be distinguished: (i) single-support phase (SSP) and (ii) double-support phase (DSP). Double-support phase has two micro-phases: (i) weight acceptance phase (WAP) or heel strike, and (ii) weight support phase (WSP). The indicator of the degree of dynamic balance is the ZMP, i.e. its relative position with respect to the footprint of the supporting foot of the locomotion mechanism. The ZMP is defined as the specific point under the robotic mechanism foot at which the effect of all the forces acting on the mechanism chain can be replaced by a unique force and all the rotation moments about the x and y axes are equal zero. The ZMP position is calculated based on measuring reaction forces under the robot foot.

In order to enable a balancing controller, the application of the so-called integrated dynamic control was proposed. Based on the above assumptions, the control algorithm involves three parts: (i) basic dynamic controller for trajectory tracking P_1, (ii) dynamic controller tuned by reinforcement learning structure for compensation joints P_2. As applied approach, the controller for robotic trajectory tracking was adopted using the computed torque method in the space of internal coordinates of the mechanism joints based of the robot dynamic model. The proposed dynamic control law has the following form:

$$P_1 = H(q)[\ddot{q}_0 + K_V(\dot{q} - \dot{q}_0) + K_p(q - q_0)] + h(q, \dot{q}) - J^T(q)F \qquad (1.2)$$

The matrices Kp and Kv are the corresponding matrices of position and velocity gains of the controller. The main intention and idea is to include learning control component based on constant qualitative evaluation of biped walking performance. The reinforcement learning control as kind of unsupervised learning environment (evaluation of control action based on ZMP error rather than numerical error of state variables) can be very suitable for searching of optimal and balanced biped walking.

The main idea is to use chosen control policy (computed torque controller) but with tuning of policy control parameters by appropriate policy-gradient procedure. This reinforcement control part P_2 is realized only for special compensation joints. P_2 is the vector of compensation control torques at the selected compensation joints The control torques P_2 has to be 'displaced' to the other (powered) joints of the mechanism chain. Considering the model of locomotion mechanism presented in Fig. 1, the compensation was carried out using the following mechanism joints: 1, 6 and 14 to compensate for the dynamic reactions about the x-axis, and 2, 4 and 13 to compensate for the moments about the y-axis. The proposed Reinforcement learning structure is based on policy gradient Methods [20].The policy-gradient method is a stochastic gradient-descent method. The policy can therefore be improved upon every update. In this case, control policy represents computed torque controller structure with aim to select/tune the best control parameters. Exactly, the control policy in this case, represents the set of control algorithms with different control parameters. The input to control policy is state of the system, while the output is control action (signal). The general aim of policy optimization in reinforcement learning is to optimize the control parameters policy κ in this way that the expected return

$$J(\kappa) = E\left(\sum_{i-0}^{L} \gamma^i r_i\right) \tag{1.3}$$

is optimized (where $\gamma^i \in [0, 1]$ is a discount factor; r_i is reward or reinforcement signal. It is important to notice that for biped motion, drastic change of control parameter is not valid and and smooth parameter change is required. Hence, policy gradient method based on steepest descent is chosen. The control parameter policy is updated according to the following rule:

$$\kappa_{t+1} = \kappa_k + \alpha \nabla_k J(\kappa) \tag{1.4}$$

where α is learning rate; $k = (0; 1; 2;...)$. In this case, policy parameter vector is is defined as $\kappa = [Kp \ Kv \ \sigma \ ]^T$, while control policy is defined as

$$\pi = \frac{1}{\sqrt{2\pi\sigma}} \left(\frac{-(P_2 - \psi(x,\kappa))^2}{2\sigma^2} \right) \tag{1.5}$$

where $\psi(x,\kappa) = \kappa^T \phi(x); x$ is the state of the system; $\phi(x)$ is the Gaussian basis function. There are various methods for gradient estimation, but following algorithms is chosen:

$$\kappa_{t+1} = \kappa_t + \alpha B_t \delta_t \tag{1.6}$$

$$B_t = \beta B_{t-1} + \nabla \log \pi \tag{1.7}$$

where β is a constant factor; defined by state value function.V

The estimated value function represents a *Critic*, because it criticizes the control actions made by the basic controller. Critic network maps position and velocity tracking errors and external reinforcement signal R in scalar value which represent the quality of given control task. The output scalar value of Critic is important for calculation of internal reinforcement signal R. Critic constantly estimate internal reinforcement based on tracking errors and value of reward. Critic is standard 2-layer feedforward neural network (perceptron) with one hidden layer. The activation function in hidden layer is sigmoid, while in the output layer there is only one neuron with linear function. The input layer has a bias neuron. The output scalar value v is calculated based on product of set C of weighting factors and values of neurons in hidden later plus product of set A of weighting factors and input values and bias member. There are also one more set of weighting factors A between input layer and hidden layer. The number of neurons on hidden later is determined as 5. The internal reinforcement is defined as TD(λ) error by the following equations:

$$e_t = 1 + \gamma \lambda e_{t-1} \qquad \qquad if \quad x = x_t \qquad (1.8)$$

$$e_t = \gamma \lambda e_{t-1} \qquad \qquad otherwise \qquad (1.9)$$

$$R_{t+1} = \delta_t = (R_t + \mathcal{W}_{t+1} - v_t) e_t \qquad (1.10)$$

where γ is a discount coe±cient between 0 and 1 (in this case γ is set to 0.9).; $\lambda, o \prec \lambda \prec 1$, is a new parameter. The learning process for value function is accomplished by step changes calculated by products of internal reinforcement, learning constant and appropriate input values fromprevious layers.

The corresponding experiments were carried out in a caption motion studio [33]. For this purpose, a middle-aged (43 years) male subject, 190 [cm] tall, weighing 84.0728 [kg], of normal physical constitution and functionality, played the role of an experimental anthropomorphic system whose model was to be identified. The selected subject, whose parameters were identified, performed a number of motion tests (walking, staircase climbing, jumping), whereby the measurements were made under the appropriate laboratory conditions. We assumed that it is possible to design a bipedal locomotion mechanism (humanoid robot) with defined parameters (same as in Fig. 1a). On the basis of the measured values of positions (coordinates) of special markers in the course of motion it was possible to identify angular trajectories of the particular joints of the bipedal locomotion system. Some special simulation experiments were performed in order to validate the proposed reinforcement learning control approach. The tracking errors converge to zero values in the given time interval. It means that the controller ensures good tracking of the desired trajectory. Also, the application of reinforcement learning structure ensures a dynamic balance of the locomotion mechanism. In the simulation example, it was shown how the basic dynamic controller together with reinforcement learning control structure is able to compensate the deviations of dynamic reactions even in the presence of uncertainty of the ground surfaceinclination. 100 basis functions $\phi(x)$ are allocated to represent mean of the policy $\phi(x)$. In Figs. 2 and 3 the comparison of the simulation results for ZMP errors in coordinate directions are shown. In Fig. 4 value of reward or internal reinforcement through process of walking is presented. It is clear that task of walking within desired ZMP tracking error limits is achieved in a good fashion.

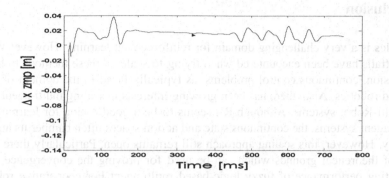

Fig. 2. ZMP error in x-direction

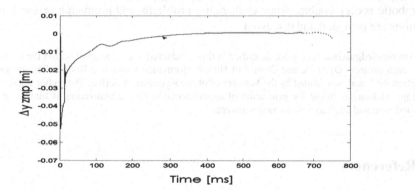

Fig. 3. ZMP error in y-direction

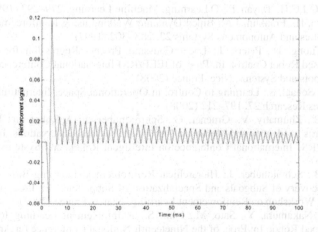

Fig. 4. Acquired reward

Conclusion

Robotics is a very challenging domain for reinforcement learning, However various pitfalls have been encountered when trying to scale up these methods to high dimension, continuous control problems, as typically faced in the domain of humanoid robotics. Also, there has been growing interests in scaling multi-agent RL to Multi-Robot systems. Although RL seems to be a good option for learning in multi-agent systems, the continuous state and action spaces often hamper its applicability. However, this scaling approach still remains open. Particularly there is a lack of theoretical grounds which can be used for proving the convergence and predicting performance of fuzzy logic-based multi-agent For cooperative robots systems, although some research outcomes in some special cases have been available now, there are also some difficulties (such as multiple equilibrium and selecting payoff structure, etc.) for directly applying them to a practical systems, e.g., robotic soccer system. Some challenging problems and promising research directions are provided and discussed.

Acknowledgments. The work described in this conducted was conducted within the national research project "Dynamic and Control of High-Performance Humanoid Robots: Theory and Application". and was funded by the Ministry of of the Republic of Serbia. The author thanks to Dr. Ing. Aleksandar Rodić for generation of experimental data and realization of humanoid robot modeling and trajectory generation software.

References

[1] Sutton, R.S., Barto, A.G.: Reinforcement Learning: An Introduction. The MIT Press, Cambridge (1998)
[2] Bertsekas, D.P., Tsitsiklis, J.N.: Neuro-Dynamic Programming. Athena Scientific, Belmont (1996)
[3] Watkins, C.J.C.H., Dayan, P.: Q Learning. Machine Learning, 279–292 (1992)
[4] Benbrahim, H., Franklin, J.A.: Biped Dynamic Walking using Reinforcement Learning. Robotics and Autonomous Systems 22, 283–302 (1997)
[5] Nguyen-Tuong, D., Peters, J.: Local Gaussian Process Regression for Real-time Model-based Robot Control. In: Proc. of IEEE/RSJ International Conference on Intelligent Robots and Systems, Nice, France (2008)
[6] Peters, J., Schaal, S.: Learning to Control in Operational Space. International Journal of Robotics Research 27, 197–212 (2008)
[7] Bakker, B., Zhumatiy, V., Gruener, G., Schmidhuber, J.: A Robot that Reinforcement-Learns to Identify and Memorize Important Previous Observations. In: Proc. of the IEEE/RSJ International Conference on Intelligent Robots and Systems, pp. 430–435 (2003)
[8] Bakker, B., Schmidhuber, J.: Hierarchical Reinforcement Learning Based on Automatic Discovery of Subgoals and Specialization of Subpolicies. In: Proc. of the 2003 European Workshop on Reinforcement Learning, Nancy, France (2003)
[9] Mori, T., Nakamura, Y., Sato, M., Ishii, S.: Reinforcement Learning for a CPG-driven Biped Robot. In: Proc. of the Nineteenth National Conference on Artificial Intelligence (AAAI), pp. 623–630 (2004)

[10] Nakamura, Y., Sato, M., Ishii, S.: Reinforcement Learning for Biped Robot. In: Proc. of International Symposium on Adaptive Motion of Animals and Machines (2003)

[11] Peters, J., Vijayakumar, S.M., Schaal, S.: Reinforcement Learning for Humanoid Robotics. In: Proc. of Third IEEE-RAS International Conference on Humanoid Robots, Karlsruhe, Germany (2003)

[12] Tedrake, R., Zhang, T.W., Seung, H.S.: Stochastic Policy Gradient Reinforcement Learning on a Simple 3d Biped. In: Proc. of the 2004 IEEE/RSJ International Conference on Intelligent Robots and Systems (2004)

[13] Endo, G., Morimoto, J., Matsubara, T., Nakanishi, J., Cheng, G.: Learning CPG-based Biped Locomotion with a Policy Gradient Method: Application to a Humanoid Robot. International Journal of Robotics Research 27, 213–228 (2008)

[14] Lee, J., Oh, J.H.: Walking Pattern Generation for Planar Biped Walking Using Q-learning. In: Proc. of the 17th World Congress The International Federation of Automatic Control, Seoul, Korea, pp. 3027–3032 (2008)

[15] Shibata, T., Hitomoi, K., Nakamura, Y., Ishii, S.: Reinforcement Learning of Stable Trajectory for Quasi-Passive Dynamic Walking of an Unstable Biped Robot. In: Hackel, M. (ed.) Humanoid Robots: Human-like Machines, Itech, Vienna, Austria, pp. 211–226 (2007)

[16] Katić, D., Vukobratović, M.: Reinforcement Learning Algorithms in Humanoid Robotics. In: de Pina Filho, A.C. (ed.) Humanoid Robots: New Developments, Advanced Robotic Systems International and I-Tech, Vienna, pp. 367–400 (2007)

[17] Katic, D., Rodic, A., Vukobratovic, M.: Hybrid Dynamic Control Algorithm For Humanoid Robots Based on Reinforcement Learning. J. of Intelligent and Robotic Systems 51, 3–30 (2008)

[18] Katic, D., Rodić, A.: Dynamic Control Algorithm for Biped Walking Based on Policy Gradient Fuzzy Reinforcement Learning. In: Proc. of the 17th IFAC World Congress, Seoul, Republic of Corea (2008)

[19] Nakanishi, J., Morimoto, J., Endo, G., Cheng, G., Schaal, S., Kawato, M.: A Framework for Learning Biped Locomotion with Dynamic Movement Primitives. In: Proc. of IEEE-RAS/RSJ International Conference on Humanoid Robots, Los Angeles, USA (2004)

[20] Peters, J., Schaal, S.: Policy Gradient Methods for Robotics. In: Proc. of the IEEE International Conference on Intelligent Robotics Systems, Beijing, China (2006)

[21] Parker, L.E.: Distributed Intelligence: Overview of the Field and its Application in Multi-Robot Systems. J. of Physical Agents 2, 5–14 (2008)

[22] Yang, E., Gu, D.: Multiagent Reinforcement Learning for Multi-Robot Systems: A Survey. Technical Report CSM-404, Department of Computer Science, University of Essex (2004)

[23] Park, K.H., Kim, Y.J., Kim, J.H.: Modular Q-Learning-based Multi-Agent Cooperation for Robot Soccer. Robotics and Autonomous Systems 35, 109–122 (2001)

[24] Touzet, C.F.: Distributed Lazy Q-Learning for Cooperativemobile Robots. International Journal of Advanced Robotic Systems 1, 5–13 (2004)

[25] Stone, P., Veloso, M.: Multiagent Systems: a Survey from a Machine Learning Perspective. Autonomous Robots 8, 345–383 (2000)

[26] Mataric, M.J.: Reinforcement Learning in the Multi-Robot Domain. Autonomous Robots 4, 73–83 (1997)

[27] Mataric, M.J.: Learning in Behavior-based Multi-Robot Systems: Policies, Models, and Other Agents. J. of Cognitive Systems Research 2, 81–93 (2001)

[28] Gultekin, I., Arslan, A.: Modular-Fuzzy Cooperative Algorithm for Multi-Agent Systems. In: Yakhno, T. (ed.) ADVIS 2002. LNCS, vol. 2457, pp. 255–263. Springer, Heidelberg (2002)

[29] Guo, H., Meng, Y.: Dynamic Correlation Matrix-based Multi-Q Learning for a Multi-Robot System. In: Proc. of IEEE/RSJ International Conference on Intelligent Robots and Systems, Nice, France, pp. 840–845 (2008)

[30] Melo, F.S., Ribeiro, M.I.: Reinforcement Learning with Function Approximation for Cooperative Navigation Tasks. In: Proc. of the 2008 IEEE International Conference on Robotics and Automation, Pasadena, USA, pp. 3321–3327 (2008)

[31] Sanz, Y., de Lope, J., Martín, J.A.H.: Applying Reinforcement Learing to Multi-Robot Team Coordination. In: Corchado, E., Abraham, A., Pedrycz, W. (eds.) HAIS 2008. LNCS, vol. 5271, pp. 625–632. Springer, Heidelberg (2008)

[32] Tu, J.: Continuous Reinforcement Learning for Feedback Control Systems. Master's thesis, Computer Science Department, Colorado State University, Fort Collins, USA (2001)

[33] Rodić, A., Vukobratović, M., Addi, K., Dalleau, G.: Contribution to the Modeling of Non-smooth, Multi-point Contact Dynamics of Biped Locomotion – Theory and Experiments. Robotica 26, 157–175 (2008)

Points of View on Building an Intelligent Robot

Claudiu Pozna[1], Radu-Emil Precup[2], Stefan Preitl[2], Fritz Troester[3], and József K. Tar[4]

[1] Department of Product Design and Robotics, Transilvania University of Brasov
Bd. Eroilor 28, 500036 Brasov, Romania
cp@unitbv.ro

[2] Department of Automation and Applied Informatics, "Politehnica" University of Timisoara
Bd. V. Parvan 2, 300223 Timisoara, Romania
radu.precup@aut.upt.ro, stefan.preitl@aut.upt.ro

[3] Department of Mechanical and Electrical Eng., University of Applied Science Heilbronn
Max Planck Str. 39, 74081 Heilbronn, Germany
troester@fh-heilbronn.de

[4] Institute of Intelligent Engineering Systems, Budapest Tech Polytechnical Institution
Bécsi út 96/B, H-1034 Budapest, Hungary
tar.jozsef@nik.bmf.hu

Abstract. Aspects concerning the building of an intelligent robot are discussed. The intelligent robot belongs to a class of autonomous robots. The hardware and software architectures of the robot are analyzed. They are part of the new three-level intelligent control system architecture. The mathematical model of the merged robot and trajectory tracking is derived in order to be used as controlled plant. The pole placement approach is applied in the design of the state feedback controller. Real-time experimental results done in trajectory tracking validate the architectures, models and design method.

Keywords: artificial intelligence, driving robot, intelligent robot.

1 Introduction

One spectacular application of the intelligent robots concerns the autonomous cars. An autonomous care is able to drive by itself. Therefore the car copies the behavior and performance of the human driver, thus being viewed as an intelligent robot. The autonomous cares have attracted the interest of researchers from both industry and academia.

I.J. Rudas et al. (Eds.): Towards Intelligent Engineering & Information Tech., SCI 243, pp. 263–277.
springerlink.com   © Springer-Verlag Berlin Heidelberg 2009

One example of autonomous car projects produced by the automotive industry projects is the Autonomous Driving Project managed by Volkswagen. The purpose of that project was to develop an autonomous vehicle with the options of accidents avoidance and automatic driving. The project partners were the Brunswick Technical University, Robert Bosch GmbH, Kasprich-IBEO GmbH and Sondermachinen GmbH. Accordingly, up to ten vehicles were simultaneously driven automatically by robot-drives. To transform a VW in mobile robot a driving robot was implemented in the car. That driving robot has three "legs" (which allow it to manipulate the gas, clutch and brake pedals) and two "arms" (which manipulate the steer and the gearbox lever). The environment recognition was possible by means of radar sensors, one laser scanner, and two video cameras. All those systems enable the vehicle guidance based on the computation of the desired trajectory; vehicle regulation, sensors functions, etc. Another project managed by the automotive industry in the framework of the European Prometheus Project was VITA II. Daimler-Benz has presented a vehicle which is capable of driving autonomously on highways and doing overtaking maneuvers without interaction.

One of the autonomous car projects produced by the academia is the Stanley mobile robot developed by Stanford University. Another example is the SafeMove France-Korean project which developed the CyCab robot designed at INRIA and the pi-Car prototype of IEF. The NAVLAB robot has been developed by the Carnegie Mellon University Robotics Institute. An analysis of those developments has been done in (Pozna and Troester 2007) with focus on the Automotive Competence Center (ACC) autonomous car built at the University of Applied Science Heilbronn. The current research directions concerning the autonomous cars include the car tracking (Sadoghi Yazdi et al. 2006), (Klein et al. 2008), vision-based navigation (Giovannangeli and Gaussier 2008), (Kubota et al 2009), sensor fusion and parking (Shunguang et al 2008), safety assessment (Althoff et al 2007), quasi optimal path generations (Laszka et al 2009), the use of topological primitives (Rawlinson and Jarvis 2008), sliding mode and fuzzy logic control (Shahmaleki and Mahzoon 2008), (Hwang et al 2009), etc. A good overview on those research directions dedicated to mobile robots has been done in (Bensalem et al 2009).

The main contribution of this paper deals with extending the previous work accomplished in the framework of the ACC project. A special attention will be given to the elements of intelligence in the control system of the autonomous car.

The paper treats the following topics: the description of the construction and hardware architecture of the intelligent robot accepted as autonomous car, the brief presentation of several aspects concerning the implementation of the control system, and the software architecture of the control system focused on the control program. Results corresponding to real-time trajectory tracking tests are included.

2   Construction and Hardware Architecture of Intelligent Robot

The intelligent robot accepted here consists of the following systems: the car, the driving robot, the control system and the extra sensory system. The ACC project has been organized in terms of the functional design concept presented in Fig. 1. The three major levels in this procedure are: construct, implement and test the driving robot in the car, construct the control system, integrate the sensors needed for the environment recognition and finally test the autonomous car playing the role of intelligent robot.

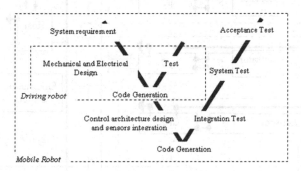

Fig. 1. Functional design concept

The hardware architecture of the driving robot is presented in Fig. 2. Its five subsystems are: steering, acceleration, turning the ignition key, turning the gearbox lever and braking the car. Each subsystem copies a certain action of the human driver and transfers the intelligence to the robot. Each subsystem consists of actuators, sensors and microcontrollers which solve the local control problems. A main feedback control loop is designed over the five local feedback loops. The hardware support of that loop consists of the environment sensors, the target PC, and the CAN data transfer network. The program which runs on the target computer implements the control algorithms as part of the control system.

The mechanical structure of the robot is presented in Fig. 3a. The integration of the driving robot in the autonomous car is shown in Fig. 3b to illustrate its capabilities.

The indoor tests must be done after the driving robot integration as it is outlined in the project management workflow (Fig. 1). With this regard the test programs have been coded in Matlab using the xPC Toolbox. They have been developed in the host computer (Fig. 2) and downloaded onto the target computer.

The driving robot is an operational model of the human driver. All human driver actions can be approximated in terms of the driving robot.

Several tests have been done making use of the hardware architecture illustrated in Fig. 2. They were successful and justified the transfer to the next level of the project (Fig. 1) represented by the functional design of the control system to be presented in the next Section.

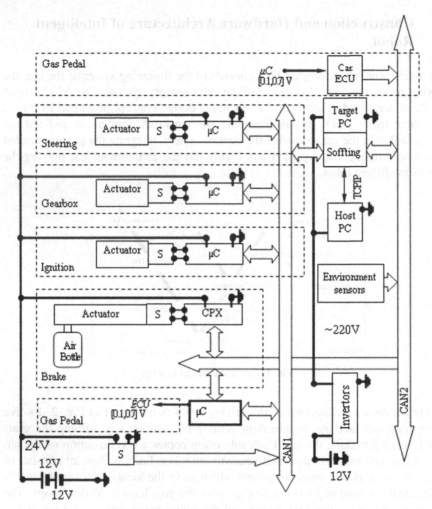

Fig. 2. Hardware architecture of driving robot

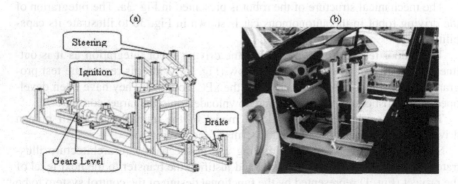

Fig. 3. a Mechanical structure of driving robot; **b** integration of driving robot in autonomous car

3   Control System Structure and Software Architecture

The control system structure is based on the concept of human driver behavior model. It incorporates the ideas organized as longitudinal behavior models (Bengtsson et al 2001), lateral behavior models (Ungoren and Peng 2005), brake behavior models (Goodrich et al 1999), (Wada et al 2008), etc. The hierarchical control system structure used here is inspired from the reference model architecture used in the design of intelligent systems (Meystel and Albus 2002). It consists of a control system structure organized as a three-level architecture consisting of a strategic level, which establishes the goal of the driving, a tactical level, which finds the solution to accomplish the goal, and an operational level, which implements this solution on low level control of the vehicle. The software architecture corresponding to the control system structure is presented in Fig. 4.

It has been accepted that it is more suitable to model and implement the "human driver decisions act", than the "human driver actions". This idea transfers the approximation of the human driver behaviors from a mechanical to an artificial intelligence problem. That type of problem involves a preliminary analysis which must provide answers to the following questions: "What are the driving behaviors?"; "Can we obtain some fundamental true about these behaviors and use them in our construction?".

A phenomenological research has been conducted. It starts with the semantic characterization of the "driving behavior". First, it is important to establish the category tree of that expression resulting in act → activity → behavior, practice, ... (Pozna and Troester 2007). Consequently the behavior is "an action or a set of actions performed by a person under specified circumstances that reveal some skill, knowledge or attitude". Therefore the driving behavior has a special feature. To describe it, the focus was on the word "custom" which belongs to the same category tree i.e. act → activity → practice → habit, and which is defined as "accepted or habitual practice". Those habits have a certain special nature in many situations such as the automatism standing for any reaction that occurs automatically without conscious thought or reflection. Concluding, the driving behavior is an action or a set of actions performed by a person under driving circumstances, actions which tend to be transformed in habits and even in automatisms. In fact the driving behavior is composed of a series of behaviors (the driver's behavior when he makes the ignition, the driver's behavior when he stops the car, etc.). Therefore the fundamental truth of the driving behavior can be set in terms of the following synthetic propositions:

- A priori, the driver establishes the current driving goal.
- A behavior is a set of actions.
- The behaviors are linked together, creating a system which allows the driver to obtain solutions in the driving circumstance.
- The translation from one behavior to another is triggered by the occurrence of an event.
- The system is developed by learning and experience.

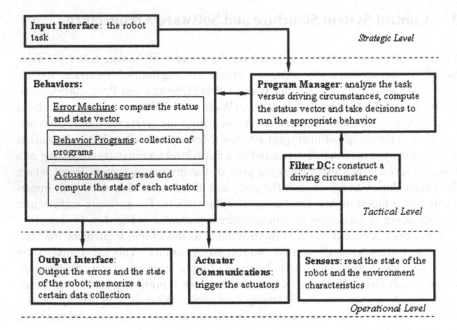

Fig. 4. Software architecture

- The behaviors presume decisions based on incomplete sets of information.
- The sets of actions tend in time to be transformed in habits and automatisms.

The above propositions agree with the three-level architecture. Using the propositions, for the design of the tactical level it is important to model (approximate) the human driving behaviors by a collection of high linked programs (behaviors) which are stored in a memory. The decision to run a certain program is made by a manager program. That decision is based on the driving goal and on the environment understanding (driving circumstance). Each program (behavior) is a set of instructions (actions) which impose parameters and trigger actuators. The propositions enable also the possibility to use the state machines to handle the behaviors, fuzzy logic to model the decisions or describe the environment, and neuronal networks to implement the learning processes.

The following comments are important in relation with Fig. 4:

- The strategic level, where the robot receives its task (goal) is an interface which helps the human operator to impose the goal.
- The "Program Manager" analyzes the goal versus the driving circumstances which are obtained from the sensors; the result of this process is the status vector of the robot (the desired position, velocity, etc.) and also the decision to run a certain program from the "Behaviors" subsystem.
- The "Behaviors" consists of three parts:
  - The "Error Machine" which compares the status vector with state vector (the positions, velocity, etc. obtained from the sensors).

- o The "Behavior Programs" which is a collection of programs (behaviors); each program is able to solve a special environment situation (ignition, emergency stop, zero position, errors, etc.).
  - o The "Actuators Manager" which manages the actuators of the robot.
- The "Output Interface" allows the states and the errors reading and also the robot state history memorization.
- The "Actuators Communications" outputs the data to the microcontrollers attached to each actuator.

In order to build the "Behaviors" subsystem it is important to set the structure of the programs as part of the "Behaviors Programs". Three different structures are presented in Fig. 5. They are referred to as basic behaviors, error behaviors and simple behaviors. The main differences between the programs are the connection types belonging to the set {P=previous, N=next, E=error, QI=quick in, QO=quick out} and the direction of information flow.

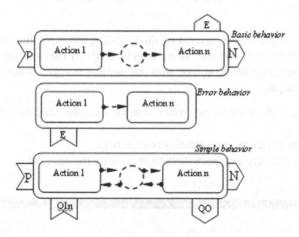

Fig. 5. Structure of "Behavior Programs"

The "Error Machine" program is part of the "Behaviors" subsystem. The aim of that program is to compare the status vector (the reference inputs of the robot: desired car speed, desired steering angle, etc.) specifying the robot goal with the state vector (the variables measures by means of the sensors: car speed, steering angle, etc.). The decisions on the program to be run are made by the "Program Manager" illustrated in Fig. 6.

The "Program Manager" compares the goal of the robot with the driving circumstance, calculates the status vector and enables the program running. After these decisions, the program continues to compare the robot goal with the driving circumstance. If the result is acceptable, nothing is changed (the same program is run); otherwise a "Crisis" or a "Failure" event is signaled. A "Crisis" means that a new behavior is needed, so the status vector as well as the program will be changed. A "Failure" means that no solutions (behaviors) are available to solve the problem, so the robot must be stopped safely.

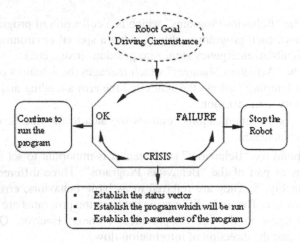

Fig. 6. Structure of "Program Manager"

The control program is composed of three levels, the input interface (Fig. 7), the tactical level (Fig. 8) and the output interface which runs on the target computer. The navigation problem solved with the intelligent robot is defined in terms of the next procedure:

- Define the task of the mobile robot to track a certain trajectory between two points.
- Define the trajectory mathematically on a map.
- Avoid the initially unexpected obstacles during the navigation.

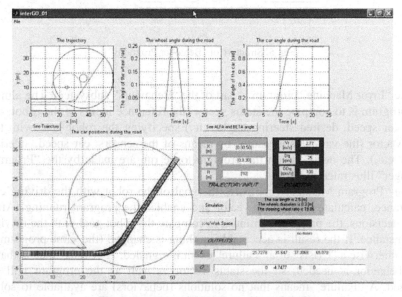

Fig. 7. Graphical user interface of input interface

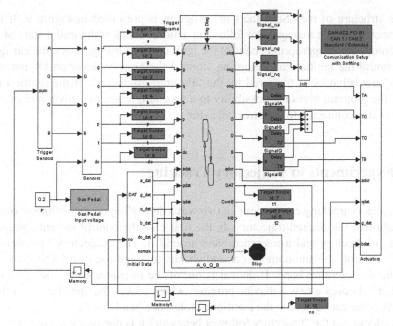

Fig. 8. Structure of implemented tactical level

The tactical level has been designed to fulfill the above procedure. The interface transforms the robot desired trajectory which is defined in the Cartesian coordinates into the desired trajectory which refers to the steering actuator and the car speed. Therefore the interface can simulate the working volume of the mobile robot in a desired map. The interface gives also the possibilities to verify if the dynamical characteristics of the driving robot (the maximum steering torque, the braking force, etc.) admit the kinematics of the mobile robot (the possibility to track a certain trajectory within a specified time frame).

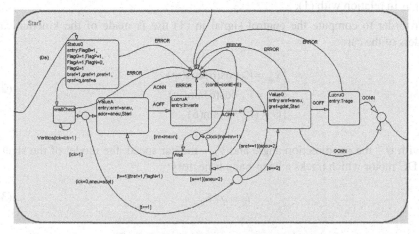

Fig. 9. Structure of "Start the Car" program

The structure of the "Start the Car" program is presented in Figure 9. It is a state machine which manages the following actions: check if the initial state of the autonomous car is appropriate, control the ignition key turning, check the car ignition, control the ignition key returning, control the gearbox lever on D, and wait until a new behavior is triggered by the "Program Manager". If the ignition has failed, the program gives the possibility to a second ignition maneuver and in case of errors it is connected to the error behavior.

4 Experiments in Trajectory Tracking

The trajectory tracking can be done in two ways depending on the offline or online calculation of the desired trajectory. In the first case the control system computes offline, a trajectory and a tolerance band around it. The "trajectory follower behavior" controls the autonomous car and checks (making use of a GPS) if the car is inside the tolerance band. If the car is out of the tolerance band, the "Program Manager" chooses a new behavior program which makes the specific corrections and drives the car back onto the mentioned tolerance band.

In order to run the "trajectory follower behavior" it is necessary to compute offline the control signal vector

$$\begin{Bmatrix} L \\ \delta \end{Bmatrix} = \begin{Bmatrix} L_1 & L_2 & ... & L_n \\ \delta_1(t) & \delta_2(t) & ... & \delta_n(t) \end{Bmatrix}, \tag{1}$$

where L_i is the length of the trajectory that must be covered with the steering angle $\delta_i(t)$. Since the desired trajectory consists of linear trajectories which are connected by circular trajectories (Fig. 10) and the car is a nonholonomic vehicle, imposing an appropriate trajectory means finding a good compromise to minimize the tracking errors in the rectilinear zone. The total number of trajectories is n, and $i = \overline{1, n}$ in relation with (1).

In order to compute the control signal in (1) use is made of the kinematical models of the car:

$$\begin{cases} \dot{x}_m = V_R \cos(\delta) \cos(\psi) \\ \dot{y}_m = V_R \cos(\delta) \sin(\psi) \\ \dot{\psi} = (V_R / L) \sin(\delta) \end{cases}, \tag{2}$$

with ψ – the car direction angle and V_R – the car speed, the model of the steering DC motor which tracks a bang-bang trajectory:

$$i_R = q / \delta, \tag{3}$$

and the model of the mechanical transmission between the steering wheel and the car wheel:

$$q = q(q_0, q_d, \dot{q}_d, \ddot{q}_d, t),$$ (4)

where q_d is the desired position, $\dot{q}_d$ is the maximum angular velocity, and $\ddot{q}_d$ stands for the acceleration / deceleration.

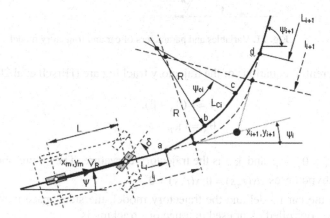

Fig. 10. Transformation of desired trajectory

The approximation implemented here replaces the circular trajectories by clotoidal ones (Pozna and Troester 2007). The value of δ_d has been computed off-line by the numerical integration of (4).

The online calculation of the desired trajectory should be used in the avoiding obstacles maneuvers. If the car recognizes an obstacle, the on line procedures compute the avoiding trajectories and control the car on those trajectories.

In order to design the control algorithm the dynamic model of the car is divided in the longitudinal and the lateral dynamics. Use is made of the following model proposed in (Isermann 2001), based on the assumptions that the car velocity is constant during the locomotion and the angles are small:

$$\begin{bmatrix} \dot{\beta}(t) \\ \ddot{\psi}(t) \end{bmatrix} = \begin{bmatrix} a_{11} & a_{12} \\ a_{21} & a_{22} \end{bmatrix} \cdot \begin{bmatrix} \beta(t) \\ \dot{\psi}(t) \end{bmatrix} + \begin{bmatrix} b_1 \\ b_2 \end{bmatrix} \cdot \delta,$$

$$a_{11} = -2c/(mV), a_{12} = (cl_s - cl_f)/(mV^2), a_{21} = c(l_s - l_f)/J_z,$$ (5)

$$a_{22} = -c(l_s^2 + l_f^2)/(J_z V), b_1 = c/(mV), b_2 = cl_f/J_z,$$

where (Fig. 11a) β is the angle between the car velocity V and the car direction, m and J_z are the mass and inertia momentum of the car, c is the rotational stiffness of the wheels, and l_f and l_s are the lengths from the mass center to the front and back wheels.

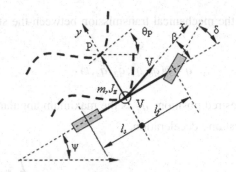

Fig. 11. Variables and parameters of car and trajectory model

The differential equations of the trajectory tracking are (Hirsch et al 2005):

$$y = V(\theta_\Delta + \beta), \tag{6}$$
$$\dot{\theta}_\Delta = V\kappa_P - \dot{\psi},$$

where $\theta_\Delta = \theta_P - \psi$ and κ_P is the trajectory curvature. They were obtained accepting the hypotheses $dx(r,s) = 0, x(r,s) = 0$.

Merging the car model and the trajectory model, the state-space mathematical model of the controlled plant used in trajectory tracking is

$$\begin{bmatrix} \dot{\beta} \\ \ddot{\psi} \\ \dot{\theta}_\Delta \\ \dot{y} \end{bmatrix} = \begin{bmatrix} a_{11} & a_{12} & 0 & 0 \\ a_{21} & a_{22} & 0 & 0 \\ 0 & -1 & 0 & 0 \\ V & 0 & V & 0 \end{bmatrix} \cdot \begin{bmatrix} \beta \\ \dot{\psi} \\ \theta_\Delta \\ z \end{bmatrix} + \begin{bmatrix} b_1 \\ b_2 \\ 0 \\ 0 \end{bmatrix} \cdot \delta + \begin{bmatrix} 0 \\ 0 \\ V \\ 0 \end{bmatrix} \cdot \kappa_P, \tag{7}$$

$$y = \begin{bmatrix} 0 & 0 & 0 & 1 \end{bmatrix} \cdot \begin{bmatrix} \beta & \dot{\psi} & \theta_\Delta & y \end{bmatrix}^T + \begin{bmatrix} 0 \end{bmatrix} \cdot \delta.$$

The transfer function representation of the controlled plant is

$$Y(s) = G_{y/\delta}\Delta(s) + G_{y/\kappa}K_P(s), \tag{8}$$

where:

$$G_{y/\delta}(s) = V \frac{b_1 s^2 + ((a_{12}-1)b_2 - a_{22}b_1)s - (a_{21}b_1 - a_{11}b_2)}{s^4 - (a_{11}+a_{22})s^3 + (a_{11}a_{22} - a_{12}a_{21})s^2}, \tag{9}$$

$\Delta(s)$ is the Laplace transform of $\delta(t)$,

$$G_{y/\kappa}(s) = \frac{V^2}{s^2}, \tag{10}$$

$K_P(s)$ is the Laplace transform of $\kappa_P(t)$, and $Y(s)$ is the Laplace transform of the controlled output $y(t)$.

The pole placement approach has been applied in the design of the control algorithm. It has resulted in a state feedback controller. The implemented control system structure is presented in Fig. 12. It consists of several blocks: the car model (7), the controller, the observer, the curvature source and the disturbance transfer function (10).

Several real-time experiments have been performed to track linear or circular trajectories making use of one real robot acting in a certain test field. Part of the results of those experiments is presented in Fig. 13. The good behavior of the intelligent robot has been proved.

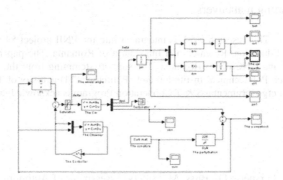

Fig. 12. Implemented control system structure

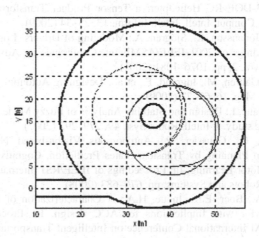

Fig. 13. Real-time tracking of two trajectories

Conclusions

Several points of view concerning the building of an intelligent robot as autonomous car have been discussed. A low-cost driving robot was designed and integrated in the car employing the three-level control system architecture.

The human behavior model was integrated. It ensures a certain degree of intelligence of the control system.

The first direction of future research will deal with the modeling and model-based design of the intelligent control system based on fuzzy logic or conventional modeling and design techniques (Horváth and Rudas 2004), (Johanyák et al 2006), (Blažič and Škrjanc 2007), (Vaščák 2008), (Tanaka et al 2009), (Baranyi et al 2009). Another direction concerns the development of the driving robot to fulfill the obstacle avoiding maneuvers.

Acknowledgments. This research work is integrated into the PNII project 842/2008 supported by the Ministry of Education, Research and Innovation of Romania. The paper was supported also by the CNMP and CNCSIS of Romania. The support stemming from the cooperation between Budapest Tech Polytechnical Institution and "Politehnica" University of Timisoara in the framework of the Intergovernmental S & T Cooperation Program is acknowledged.

References

[1] Althoff, M., Stursberg, O., Buss, M.: Safety Assessment of Autonomous Cars Using Verification Techniques. In: Proceedings of 2007 American Control Conference, pp. 4154–4159 (2007)

[2] Baranyi, P., Korondi, P., Tanaka, K.: Parallel Distributed Compensation-based Stabilization of a 3-DOF RC Helicopter: a Tensor Product Transformation-based Approach. J. Adv. Comput. Intell. Intell. Inform. 13, 25–34 (2009)

[3] Bengtsson, J., Johansson, R., Sjögren, A.: Modeling of Drivers' Longitudinal Behavior. In: Proceedings of 2001 IEEE/ASME Int. Conference on Advanced Intelligent Mechatronics, vol. 2, pp. 1076–1081 (2001)

[4] Bensalem, S., Gallien, M., Ingrand, F., et al.: Designing Autonomous Robots. IEEE Robot. Autom. Mag. 16, 67–77 (2009)

[5] Blažič, S., Škrjanc, I.: Design and Stability Analysis of Fuzzy Model-based Predictive Control - a case study. J. Intell. Robot Syst. 49, 279–292 (2007)

[6] Giovannangeli, C., Gaussier, P.: Autonomous Vision-based Navigation: Goal-oriented Action Planning by Transient States Prediction, Cognitive Map Building, and Sensory-Motor Learning. In: Proceedings of IEEE/RSJ International Conference on Intelligent Robots and Systems, pp. 676–683 (2008)

[7] Goodrich, M.A., Boer, E.R., Inoue, H.A.: A Characterization of Dynamic Human Braking Behavior with Implications for ACC design. In: Proceedings of 1999 IEEE/IEEJ/JSAI International Conference on Intelligent Transportation Systems, pp. 964–969 (1999)

[8] Hirsch, K., Hilgert, J., Lalo, W., et al.: Optimization of Emergency Trajectories for Autonomous Vehicles with Respect to Linear Vehicle Dynamics. In: Proceedings of 2005 IEEE/ASME International Conference on Advance Intelligent Mechatronics, pp. 528–533 (2005)

[9]   Isermann, R.: Diagnosis Methods for Electronic-controlled Vehicles. Veh. Syst. Dyn. 36, 77–117 (2001)

[10] Horváth, L., Rudas, I.J.: Modeling and Problem Solving Methods for Engineers. Academic Press, Elsevier, Burlington (2004)

[11] Hwang, C.L., Wu, H.M., Shih, C.L.: Fuzzy Sliding-Mode Underactuated Control for Autonomous Dynamic Balance of an Electrical Bicycle. IEEE Trans. Control Syst. Technol. 17, 658–670 (2009)

[12] Johanyák, Z.C., Tikk, D., Kovács, S., Wong, K.K.: Fuzzy Rule Interpolation Matlab Toolbox - FRI Toolbox. In: Proceedings of 15th IEEE International Conference on Fuzzy Systems, pp. 1427–1433 (2006)

[13] Klein, J., Lecomte, C., Miche, P.: Preceding Car Tracking Using Belief Functions and a Particle Filter. In: Proceedings of 19th International Conference on Pattern Recognition, vol. 4 (2008) doi:10.1109/ICPR.2008.4761008:4

[14] Kubota, T., Hasimoto, T., Kawaguchi, J., et al.: Vision-based Navigation by Landmark for Robotic Explorer. In: Proceedings of IEEE International Conference on Robotics and Biomimetics, pp. 1170–1175 (2009)

[15] Laszka, A., Várkonyi Kóczy, A.R., Pék, G., Várlaki, P.: Universal Autonomous Robot Navigation Using Quasi Optimal Path Generation. In: Proceedings of 4th International Conference on Autonomous Robots and Agents, pp. 458–463 (2009)

[16] Meystel, A., Albus, J.S.: Intelligent Systems. John Wiley and Sons, New York (2002)

[17] Pozna, C., Troester, F.: Research on the ACC Autonomous Car. J. Autom. Mob. Robot Intell. Syst. 1, 32–40 (2007)

[18] Rawlinson, D., Jarvis, R.: Ways to Tell Robots Where to Go - Directing Autonomous Robots Using Topological Instructions. IEEE Robot Autom. Mag. 15, 27–36 (2008)

[19] Sadoghi Yazdi, H., Lotfizad, M., Fathy, M.: Car Tracking by Quantised Input LMS, QX-LMS Algorithm in Traffic Scenes. IEE Proc. Vis. Image Signal Process 153, 37–45 (2006)

[20] Shahmaleki, P., Mahzoon, M.: Designing a Hierarchical Fuzzy Controller for Backing-up a Four Wheel Autonomous Robot. In: Proceedings of 2008 American Control Conference, pp. 4893–4897 (2009)

[21] Shunguang, W., Decker, S., Chang, P., et al.: Collision Sensing by Stereo Vision and Radar Sensor Fusion. In: Proceedings of 2008 IEEE Intelligent Vehicles Symposium, pp. 404–409 (2008)

[22] Tanaka, K., Yamaguchi, K., Ohtake, H., Wang, H.O.: Sensor Reduction for Backing-up Control of a Vehicle with Triple Trailers. IEEE Trans. Ind. Electron. 56, 497–509 (2009)

[23] Ungoren, A.Y., Peng, H.: An Adaptive Lateral Preview Driver Model. Veh. Syst. Dyn. 43, 245–259 (2005)

[24] Vaščák, J.: Fuzzy Cognitive Maps in Path Planning. Acta Tech. Jaurinensis Ser. Intell. Comput. 1, 467–479 (2008)

[25] Wada, T., Doi, S., Tsuru, N., et al.: Formulation of Deceleration Behavior of an Expert Driver for Automatic Braking System. In: Proceedings of International Conference on Control, Automation and Systems, pp. 2908–2912 (2008)

Autonomous Locomotion of Humanoid Robots in Presence of Mobile and Immobile Obstacles

Trajectory Prediction, Path Planning, Control and Simulation

Gyula Mester[1] and Aleksandar Rodić[2]

[1] Department of Informatics, University of Szeged, Árpád tér 2, H-6720 Szeged, Hungary
gmester@inf.u-szeged.hu
[2] Robotics Laboratory, Mihajlo Pupin Institute, Volgina 15, 11060 Belgrade, Serbia
roda@robot.imp.bg.ac.yu

Abstract. The article is addressed to the control synthesis of an intelligent autonomous locomotion (artificial gait) of biped robots operating in unknown and unstructured dynamic environments, based on perception, spatial reasoning, and learning the skill of human locomotion. Focusing the research activities to the embodied cognition and computational intelligence, this paper contributes to the extension of the intelligent robot behavior through building advanced algorithms for dynamic environment understanding, simultaneous localization, trajectory prediction, path planning, obstacle avoidance, collision avoidance and scenario-driven behavior. The article includes some characteristic simulation results to demonstrate the efficiency of the developed control algorithms and verify the obtained results.

1 Introduction

Nowadays there are humanoid robots capable to walk, maintain a dynamic balance and perform lack of manipulation tasks. Most of them are remote controlled by a human operator or by a remote computer and have no cognitive capabilities to do the imposed tasks quite autonomously. The considerations in this article are addressed to the advancements of embodied cognition with humanoid robots through building of intelligent cognitive algorithms for extended autonomy, better environment understanding, spatial reasoning and bio-inspired adaptation to the external world (Fig. 1). Main objectives of the paper are addressed to synthesis of intelligent algorithms for trajectory prediction, path planning, avoidance of static and

I.J. Rudas et al. (Eds.): Towards Intelligent Engineering & Information Tech., SCI 243, pp. 279–293.
springerlink.com
© Springer-Verlag Berlin Heidelberg 2009

mobile obstacles in robot environment and human-like (anthropomorphic) locomotion. The paper concerns with the synthesis of an intelligent, autonomous, bio-
inspired locomotion (artificial biped gait) in unknown and unstructured dynamic
environment (Fig. 1). Advanced cognitive performances, that enable free autonomous biped locomotion, can be reached through integration of sensory perception,
learning and imitation of anthropomorphic psycho-physical (mental and locomotive) reactions, building algorithms for objects recognition, bio-inspired localization and mapping in unknown environments and scenario-driven behaviour.

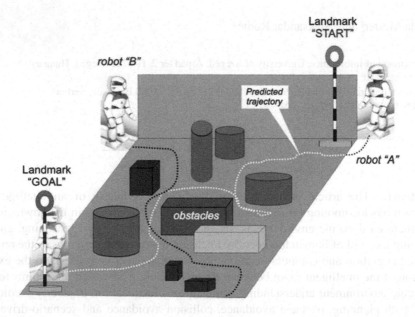

Fig. 1. An example of the typical obstacle avoidance and collision avoidance dynamic scenario

From the infant psychology it is well known that a child rapidly increases its
intellectual abilities in the moment when it stands on the legs and begin walking,
getting in such a way necessary numerous information about 3D workspace. In
this period of life children intensively learn their locomotive, manipulative and
cognitive skills through training, learning by imitation and learning by trial-and-
error actions. This learning process is much effective if a child is able to walk
autonomously, extending his or her space of exploration in the external world.
These biological principles can be applied to the humanoid robots to learn about
the 3D-world by changing their positions autonomously and doing the imposed
tasks depending on real situations.

Free locomotion is one of the most important factors that enable robotic system
to develop its advanced cognitive characteristics – artificial intelligence capable to
manage robot behavior in unknown environment. Nowadays, there are few high-
tech humanoid robot platforms (Asimo/Honda [1], Hoap-2/Fujitsu [2],
QRIO/Sony [3], iCub [4]) with the advanced technical performances. These biped
robots are dedicated to the R/D purposes predominantly. Among the research chal-

lenges to be considered in the paper, too, is to propose an appropriate biped robot control system that is capable to re-product and to imitate human behavior with a large movement repertoire, variable speed, various constraints and many uncertainties in dynamic environments in a fast and reactive manner. Hence, the control structure to be proposed in this paper involves learning from experience and creating adaptive intelligent control architecture. For that purpose, the advanced algorithms for simultaneous localization and environment understanding as well as approach for learning locomotion and navigation in the space of static and mobile obstacles through experimentation (simulation) will be developed. The learning models allow to the biped robot to predict and plan its own motions as well as to interpret the motions of other subjects (e.g. human models). Modeling and simulation software developed for the purpose of researches conducted in this paper provides a good research platform for design and validation of Reinforcement learning algorithms and fuzzy logic algorithms to be applied in control locomotion and verification of control performances.

2   Autonomous Locomotion of Biped Robots

In this section, some aspects of biologically inspired autonomous locomotion with biped robots will be considered. In that sense, the natural principles of human localization and navigation will be studied. The results of one such analysis will be used to develop bio-inspired algorithms for robot autonomous locomotion that includes a way of trajectory prediction and path planning in presence of mobile and immobile obstacles.

2.1   Bio-inspired Localization and Principles of Natural Navigation

Healthy adults move free, quite autonomously in the 3D workspace based on their natural perception (visual, sound, vestibular, scent, etc.), knowledge, experience and skill of logical reasoning. Infants learn skill of navigation and walking in free space through training (trial-and-error) and natural instinct. In such process of learning they have only their perception and still entirely non-developed intellectual capabilities. They have no significant experience about terrain topology and dimensional relationships between existing objects in surrounding. In spite of that, infants learn quickly by exploration of the space around themselves.

Human beings use the visual feedback to determine the exact position in 3D workspace as well as to plan their motion. Object-based localization is the crucial natural principle enables humans to guide themselves in the unknown environment. In the environment, humans determine their relative position and find their direction of motion with respect to the characteristic, well-visible object(s) as it is shown in Fig. 2. A lantern (light) represents a landmark object in this example shown in Fig. 2. Under the notion landmark object or marker we assume a real object or shape that dominates in a certain way comparing with other objects/shapes in surrounding by its size (length, height, width), brightness, color, etc. Conventionally, biped robots are equipped by a stereo-vision system (two cameras). The role of cameras is to identify the relative position d and direction of motion (azimuth) β of biped robot with respect to the chosen landmark object (Fig. 2) with a satisfactory accuracy.

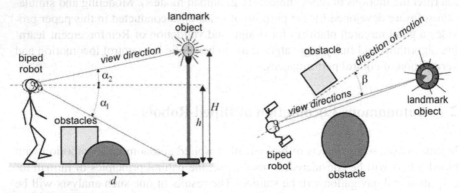

Fig. 2. Biologically inspired localization of a biped robot: a) identification of the relative position and size of different landmark objects in the environment, b) determination of the azimuth angle with respect to the target landmark object

The choice of the appropriate marker can be made by training an appropriate artificial connectionist structure (a kind of robot brain/memory) embedded into the robot's high-level control block. Such network structure is trained off-line to recognize the potential landmark objects, i.e. bright, large, high or colored objects that can be potentially well visible in the unknown scene. An arbitrary indoor scene with a biped robot and obstacles are presented in Fig. 2 in two geometry perspectives – side view and a top view. Elevation angles of robot eyes/cameras α_1 and α_2 as well as their attitude h are known (measurable). Robot relative position (distance d) is calculated from the simple relation $d = h/tg(\alpha_1)$ while the height of the object H can be estimated as $H = d \cdot tg(\alpha_2)$. Elevation angles α_1 and α_2 can be obtained from the encoder sensor attached to the neck's pitch joint (Fig. 2) or by measuring the tilt (pitch) angles of cameras as alternative. Azimuth angle, i.e. angle of direction of motion β is estimated by measuring the relative yaw angle of the neck joint as it is shown in Fig. 2. In such a way, a bio-inspired, simple way of robot Simultaneous Localization and Mapping (SLAM) and advanced navigation algorithms will be realized. Determined geometry values α_1, α_2, β, d and H acquired by corresponding robot sensor system are forwarded to the high-level (cognitive) control block of a humanoid robot. Beside the visual feedback information ensured by a pair of video cameras (see Fig. 2b), additional information about existence of obstacles is necessary for obstacle avoidance as well as robot trajectory prediction. The accurate distance(s) of the obstacle(s) inside the circle of $r \sim 1.00 - 1.50$ $[m]$ can be obtained from the appropriate distance sensors. For that purpose Ultrasonic Range Finder (USRF) or laser Light Detection and Ranging (LIDAR) are commonly used in robotic practice depending on desired accuracy, assembling possibilities to the mechanical structure, price, etc. By implementation of the USRF sensors it is possible to detect existence of the obstacles in a robot collision zone as well as direction of motion of possible mobile obstacles/objects in the robot surrounding. By numerical differentiation of the identified/measured distance(s) between the robot and moving object(s) it is possible to estimate its/their speed(s) and acceleration(s) of motion. These are important indicators to be used for making the strategy of collision avoidance.

During motion in unknown environment people comes in zones close to the obstacles (Fig. 2b). In order to avoid obstacles they make appropriate actions: change the course, i.e. direction of motion ε, vary the forward velocity v, step length s, step period T, foot lifting height h_f, etc. Mentioned variables ε, v, s, T, h_f represents the *gait parameters* G_p. These parameters represent output variables of the new cognitive block for robot trajectory prediction and planning (generation of feet cycloids) that will be integrated in the robot's high-level control structure. During a walk, humans do not know quantitative values how far they are from the closest obstacle or how fast they run. They have a linguistic, i.e. symbolic information in the mind that their relative position is in the range "near-far" i.e.: very near (beside), near, moderately far, far, very far (indefinite far). Similar values gradation is appeared with the forward speed v e.g.: wait, moves very slow, slow, moderately fast, fast, very fast. Concerning the dimensions of the obstacles (height, width, depth) the following descriptive indicators are of importance: very small, small, moderate, large and very large/huge. The mentioned linguistic/symbolic indicators/states can be mathematically formulated using fuzzy functions. A robot can be trained to distinguish the mentioned ranges of symbolic/linguistic indicators in order to predict desired motion in a 3D workspace free of collision. Implementing fuzzy rules and fuzzy reasoning in the scope of the cognitive control block, robot will be capable to understand environment and to make appropriate decisions to response to the real circumstances in surrounding. In that sense, elements of artificial intelligence will be incorporated into the biped robot control structure to extend the existing cognitive system behavior. The fundament of the robot artificial intelligence to be built in the advanced intelligent control structure makes a corresponding cognitive block. It consists of a corresponding artificial connectionist structure as well as a fuzzy system. Both mathematical tools enable robot fast learning, building algorithms for environment understanding as well as decision making capabilities.

2.2 Building Algorithms for Obstacle and Collision Avoidance

To build intelligent control algorithms for avoidance of mobile and immobile environmental obstacles corresponding geometry model of the obstacle avoidance scenario should to be developed. For that purpose, the following geometrical as well as kinematical scenario models of obstacle avoidance as well as collision avoidance are developed. They are presented in Figs. 3 and 4.

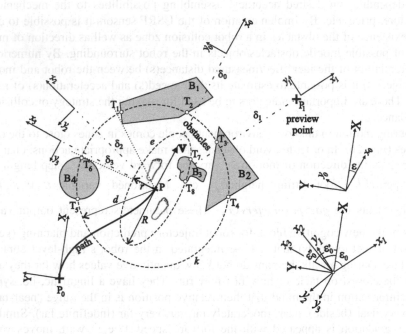

Fig. 3. General model of an obstacle avoidance scenario – the geometrical and kinematical indices used by the algorithm for obstacle avoidance

The point P in the model presented in Fig. 3 denotes the actual relative position of biped robot (i.e. the projection of its mass center to the ground support) with respect to the XOY inertial coordinate system. P_0 is the starting point. Robot moves towards the *preview point*. In general case, the preview point can be situated on some landmark object or part of object (e.g. edge, corner, etc.). Due to the presence of obstacles, biped robot has to change its course of motion to avoid collision and to continue motion towards the preview point. Direction of motion δ_0 is collinear to the sagital direction of biped robot. Course δ_1 corresponds to the direction towards the preview point. Biped robot detects the obstacles in its surrounding. The arrangement of the objects within the range bounded by the circle k causes the direction δ_2 of escaping obstacles. If possible the robot should go forward or in lateral directions. Only in the case of a dead-lock, the robots have to go backward to continue the trip. Direction δ_3 determines the course orthogonal to the δ_0. Robot is obliged to moves along the direction δ_3 only in the case when the proximity sensors indicate that the robot can strike some objects (defined by the points T_7 and T_8) by the swinging leg as presented in Fig. 3. The proximity range is bounded by the ellipse e. A fuzzy block determines, according to the actual situation in every sampling time, the direction δ_1, δ_2 or δ_3 in which the biped robot has to move in order to escape the obstacles. Humans also reason in a similar way trying to optimize their trajectories moving towards the preview points in their environment. In a similar way, humans try to solve the problem of motion in the presence of mobile obstacles as presented by the model in Fig. 4.

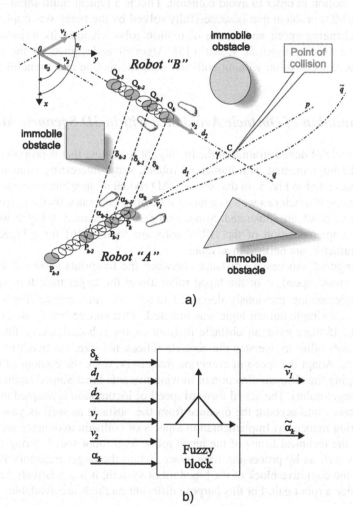

Fig. 4. Model of a collision avoidance scenario – the geometrical and kinematical indices used by the algorithm for avoidance the collision of two walking robots – "A" and "B"

Two robots "A" and "B" move in the vicinity of each other by different forward speeds v_1 and v_2 as depicted in Fig. 4. Their directions of motion cross each other in the collision point C. Depending on the speeds the robots can come in collision if not adapt their speeds and course of trajectories. d_1 and d_2 represent the distances of both robots from the potential collision point in the kth sampling time (Fig. 4). δ_{k-2}, δ_{k-1} and δ_k are the shortest distances between the robots "A" and "B" in three successive sampling times. From the particular triangles presented in Fig. 4 both robots can estimate the current speed of the other one as well as to predict the own distance to the collision point. Corresponding fuzzy block within the control block is designed to determine the referent speed as well as direction of motion in order to avoid collision. This is a typical multi-input – multi-output (MIMO) problem that is successfully solved by the fuzzy sets implementation. By changing speed and course of motion robot changes its trajectory and adapt to the existing dynamic scenario [5]. Algorithms designed for the purpose enables biped locomotion without collision with the mobile and immobile obstacles.

2.3 Simulation of Obstacle Avoidance within 3D Scenario Model

As an example of demonstration of the intelligent reasoning that is proposed in the paper to be implemented with humanoid robots, some interesting simulation results are presented in Fig. 5. In that sense, CAD model of an arbitrary environment with immobile obstacles as well as a target trajectory generated by the corresponding cognitive block are illustrated. Simulation results presented in Fig. 5 were obtained by implementation of the HRSP software toolbox [6] for a biped robot whose parameters are defined in advance.

Step lengths s (successive distances between the footprints shown in Fig. 5b) and the forward speed v of the biped robot along the target trajectory is determined implementing previously described fuzzy inference engine. For creating fuzzy rules, a simple human logic was imitated. That can be briefly described in few words. If there exist an obstacle in front of the robot, decrease the speed, checking possibility to overstep the obstacle (check the size, i.e. height) or move left or right. Adapt the speed of cornering maneuver; track the contour of the obstacle keeping the ultimate direction of moving towards the assumed landmark object (i.e. target point). The actual forward speed of locomotion is adapted in such a way that takes into account the distance from the obstacle as well as yaw rate of the cornering maneuver. Implementation aspects of collision avoidance are conditioned by the technical limits of the biped robot (e.g. robot speed during the maneuver) as well as by processing time. Once, when the target trajectory is determined by the cognitive block of the biped robot system, it is a relatively easy task to synthesize a robot gait. For this purpose different methods are available.

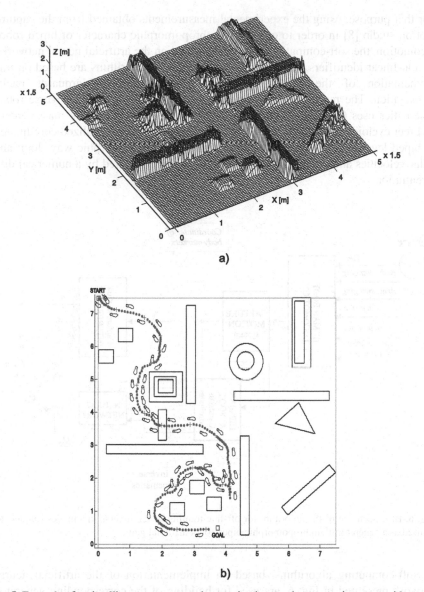

Fig. 5. Example of an intelligent autonomous locomotion in an unknown environment with presence of static obstacles; a) CAD-model of the environment, b) Obstacle avoidance and target trajectory prediction with the footprints depicting gait characteristics

For the purpose of researches presented in this article, the path planning is performed by implementation of the generator of the artificial biped gait developed by the usage of the 36 degrees of freedom (DOFs) biped robot model [6, 7] and based on the experimental results obtained from capture motion system [8]. Joint trajectories of biped robot are determined by calculation of its inverse kinematics.

For that purpose, using the experimental measurements obtained from the capture motion studio [8] in order to ensure an anthropomorphic character of biped robot locomotion, the soft-computing algorithms, based on the artificial neural networks as non-linear identifiers, are derived. The mentioned algorithms are based on implementation of the off-line fast-convergent Levenberg-Marquardt back-propagation. The multi-layer network structure chosen to train the inverse robot kinematics uses Cartesian coordinates of hip joint centers, hip link mass centers and feet cycloids of motion (Fig. 6). At the output it gives generalized coordinates of biped legs that enable robot locomotion in an anthropomorphic way. Joint angular velocities and corresponding accelerations are calculated by a numerical differentiation.

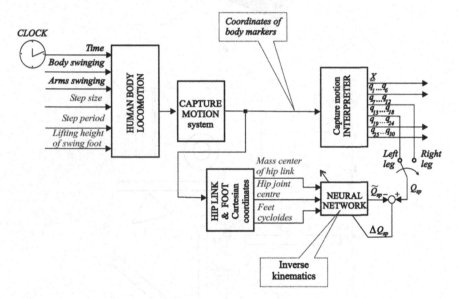

Fig. 6. Block-scheme of the algorithm for off-line training of leg inverse kinematics (model) to be used as a generator of anthropomorphic biped robot artificial gait

Soft-computing algorithms, based on implementation of the artificial neural network presented in Fig. 6, are used for building of the corresponding generator of the artificial biped gait. Generator of biped robot gait serves to generate appropriate robot joint trajectories based on the imposed gait parameters: step size, gait speed and lifting height of the swing foot. In such a way the fore mentioned biped gait generator serves as an appropriate interface between the high-level trajectory prediction and path planning block and corresponding low-level control block (servo-control block). The high-level control block is responsible for the strategy of autonomous locomotion including obstacle avoidance and adaptation to the environment while the low-level control block is charged with the trajectory tracking in the joint space of biped robots.

3   Hierarchy Control Structure of Biped Walking

Joint trajectory tracking, posture stability and dynamic balance will be ensured using *position/velocity feedback* in the joint space, *impedance control* as well as *feedback upon dynamic reactions* [9, 10], i.e. feedback upon the Zero Moment Point (ZMP) deviations [11]. Control of biped robot dynamics will be realized at the low-control level (servo level) using the corresponding sensor system (encoders, tension/torques sensors, gyro, etc.). Additional contact force/torque sensors attached to the foot sole of the biped robot are necessary, too. For that purpose, the industrial force sensing resistors or 6-axial Force-Torques sensors have to be implemented. Overall control system structure is shown in Fig. 7.

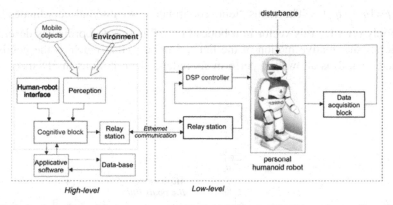

Fig. 7. Control system architecture that supports an intelligent autonomous locomotion of biped robot

Two control blocks represent a brain of the system consisting of *high-level control block* (i.e. cognitive block) and *low-level servo control block* [12, 13]. Control of robot dynamics (biped locomotion) will be designed at the servo-level while the intelligent control algorithms (cognitive behavior) enabling non-restricted autonomous locomotion and advanced reasoning will be synthesized at the high control level. Corresponding data-acquisition block ensure state feedback as well as information (actual relative position, distance/range, obstacle position and relative velocity, etc.) about the World in surrounding. Relay station enables reliable communication between these two hierarchical levels via Ethernet lines. Certain upgrade of the human-robot interface will be done according to the chosen demonstration (simulation) scenarios in order to enable task definition: introducing of the start and goal sites/positions, memorizing of the object image to be found and manipulated, understand of manual and/or sound commands by human operator, etc.

Advanced humanoid robot platforms have both control blocks (high and low) integrated on-board in order to speed up the system response. In that sense, we can speak about full system autonomy where the robot is capable to make decisions, plan its trajectories as well as enable stabile and reliable walking in real-time.

4   Modeling and Simulation of Biped Robot Locomotion

The proposed control strategy of biped robot autonomous locomotion, including on-line trajectory prediction and path planning by use of the generator of artificial biped gait, is evaluated through the corresponding simulation experiments. Human body for its complex motion uses synergy of more than 600 muscles. It has more then 300 DOFs [8]. Some of these particular motions are essential for the human activities while the others give it a full mobility. In this article, a 36 DOFs biped locomotion mechanism of the anthropomorphic structure (Fig. 8) will be considered as an appropriate model of biped locomotion mechanism. Let the joints of the system be such to allow n independent motions. Let these joint motions be described by joint angles forming the vector of the generalized coordinates $q = [q_1 \cdots q_n]^T$. The terms 'joint coordinates' or 'internal coordinates' are commonly used for this vector in robotics [11]. This set of coordinates describes completely the relative motion of the links. The basic link (such as the pelvis in this case, Fig. 8) is allowed to perform six independent motions in 3D-space. Let

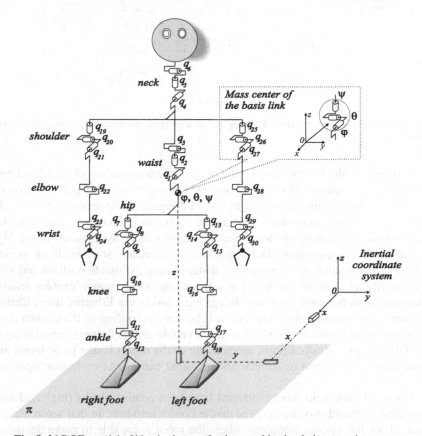

Fig. 8. 36 DOFs model of biped robot mechanism used in simulation experiments

the position of the basic link be defined by the three Cartesian coordinates (x, y, z) of its mass center and the three orientation angles (φ-roll, θ-pitch and ψ-yaw), forming the vector $\underline{X} = [x \; y \; z \; \varphi \; \theta \; \psi]^T$. Now, the overall number of DOFs for the system is $N = 6 + n$, and the system position is defined by

$$Q = [\underline{X} \; q]^T = [x \; y \; z \; \varphi \; \theta \; \psi]^T. \tag{1.1}$$

It is assumed that each joint has an appropriate actuator. This means that each motion q_j has its own drive – the torque τ_j. Note that there is no drive associated to the basic body coordinates $\underline{X}$. The vector of the joint drives is $\tau = [\tau_1 \cdots \tau_n]^T$, and the augmented drive vector (*N*-dimensional) is $T = [\underline{0}_6 \; \tau]^T = [0 \cdots 0 \; \tau_1 \cdots \tau_n]^T$. The kinematical scheme of biped model used in this paper is presented in Fig. 8.

Animation of anthropomorphic biped gait obtained in computer simulation is presented in Fig. 9.

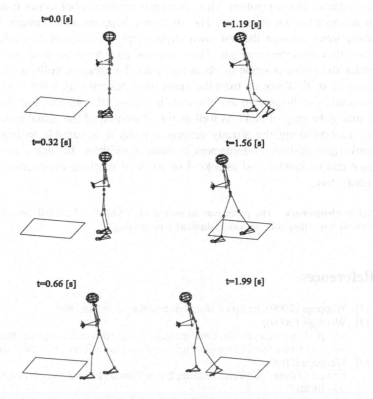

Fig. 9. Snapshots of the computer simulation of a biped robot anthropomorphic gait [5] as a sequence of the autonomous locomotion presented in Fig. 5

Conclusion

During autonomous locomotion human beings use their experience to improve the skill of navigation and to speed up locomotion in a space with obstacles. Robots use their memory and cognitive capabilities to do the same. Every path (i.e. its gait parameters G_p) performed is stored in the memory of the robot controller. After more experimental walking, robot's control system acquires information about different paths. For example, memorized path enables robot to go back along the same trajectory much faster then previously. In such a way, robot's cognitive block creates a kind of topology map of available paths that can be used in the later tasks. Using perception system, robot is able to determine its relative position in the scene in any time instant. Then, high-control level compares robot's current position with the memorized discrete positions on the map of paths. Finding the closest point appearing on the other path saved in the map of paths, robot is able to continue motion along the known trajectory. In that case, robot uses already memorized gait parameters G_p as new trajectory parameters. Using known trajectory parameters as desired path information, robot saves processing time used for localization and navigation. That potentially enables robot to run faster across the in advance known trajectory. The same case happens with humans. Initially, humans move through the unknown environment slowly and carefully. Any other time, they pass the same path faster because they have not to do navigation or to check the global position of obstacles again. To escape a collision with obstacles due to a small deviation from the memorized desired path robot has to check permanently the distance to the obstacles in a zone of collision. For this purpose of learning the map of paths as well as for adaptation of the actual robot position to the motion along the already generated paths it is suitable to implement fast-convergent on-line reinforcement learning algorithms. In such a way, an experience can be synthesized as a kind of artificial machine-experience of humanoid robots, too.

Acknowledgments.   The paper was supported by TÁMOP-4.2.2/08/01 project titled "Sensor network based data collection and information processing".

References

[1] Webpage (2009), http://world.honda.com/ASIMO/
[2] Webpage (2009),
     http://jp.fujitsu.com/group/labs/downloads/en/business/
     activities/actties-4/fujitsu-labs-robotics-005-en.pdf
[3] Webpage (2009),
     http://www.sony.net/SonyInfo/News/Press_Archive/200312/
     03-060E/
[4] Metta, G., Sandini, G., et al.: The RobotCub Project – an Open Framework for Research in Embodied Cognition. In: Proc. of the, IEEE-RAS Int. Conf. on Humanoid Robots (2005)

[5] Mester, G.: Intelligent Engineering Systems and Computational Cybernetics. In: Machado, J.A.T., et al. (eds.) Intelligent Mobile Robot Control in Unknown Environments. Springer, Heidelberg (2009)

[6] Rodic, A.: Humanoid Robot Simulation Platform – HRSP. Matlab/Simulink Software Toolbox for Modeling, Simulation & Control of Biped Robots. Robotics Lab. Mihajlo Pupin Institute (2009),
     http://www.institutepupin.com/RnDProfile/ROBOTIKA/
     comprod.htm

[7] Mester, G.: Simulation of Humanoid Robot Motion. In: Proceedings of the International Kando Konference 2008, Budapest, Hungary, pp. 1–8 (2008)

[8] Rodić, A., Vukobratović, M., Addi, K., Dalleau, G.: Contribution to the Modeling of Non-smooth, Multi-point Contact Dynamics of Biped Locomotion – Theory and Experiments. Robotica 26(2), 157–175 (2008)

[9] Vukobratović, M., Rodić, A.: Contribution to the Integrated Control of Biped Locomotion Mechanisms. International Journal of Humanoid Robotics 4(1), 49–95 (2007)

[10] Rodić, A., Vukobratović, M.: Control of Dynamic Balance of Biped Locomotion Mechanisms in Service Tasks Requiring Appropriate Trunk Postures. Engineering & Automation Problems 5(1), 4–22 (2006)

[11] Vukobratović, M., Potkonjak, V., Rodić, A.: Contribution to the Dynamic Study of Humanoid Robots Interacting with Dynamic Environment. Robotica 22, 439–447 (2004)

[12] Katić, D., Rodić, A., Vukobratović, M.: Reinforcement Learning Control Algorithm for Humanoid Robot Walking. International Journal of Information & Systems Sciences 4(2), 256–267 (2007)

[13] Katić, D., Rodić, A., Vukobratović, M.: Hybrid Dynamic Control Algorithm for Humanoid Robots Based On Reinforcement Learning. Springer Journal of Intelligent and Robotic Systems (1), 3–30 (2008)

Autonomous Advertising Mobile Robot for Exhibitions, Developed at BMF

Péter Kucsera

Budapest Tech
Tavaszmező u. 17, H-1084 Budapest, Hungary
kucsera.peter@kvk.bmf.hu

Abstract. In this article an autonomous advertising mobile robot that has been realized in practice is going to be described. The whole development is based on industrial components. In the controlling and communication part, Phoenix Contact products were used, whereas the sensing was based on a Sick Laser Scanner. These two companies donated the whole project, and the final aim was to build up a working robot, which should advertise the companies on Hungarian and international exhibitions. The robot was successfully built up in the Process Automation Lab of Kandó Kálmán Faculty of Electrical Engineering, Budapest Tech, and was exhibited with great success on Regula 2009, which is the biggest Hungarian automation exhibition.

Keywords: mobile robot, navigation, sensors, industrial control, robotics, mobile robotics.

1   The Aim of the Development

In 2008 the students of the Process Automation Lab of Kandó Kálmán Faculty of Electrical Engineering, Budapest Tech, led by Péter Kucsera took part in the Xplore New Automation Award organized by Phoenix Contact with good results where they built a small experimental mobile robot. This robot could move along a predefined path, and could accomplish docking task with high accuracy. The project was chosen to the finals and it was so successful that Phoenix Contact decided to encourage us to continue the development of the robot and build up an advertising mobile robot of bigger size and more sophisticated mechanics and control. The company together with Sick wanted to have a moving object with a screen for advertising videos running on it, and they also wanted to place some leaflets on the robot for people to take them. It was essential for the robot to be absolutely safe. It is not allowed to run over the visitors, or to move to other companies' stands. The accomplished system can be seen in Figure 1.

I.J. Rudas et al. (Eds.): Towards Intelligent Engineering & Information Tech., SCI 243, pp. 295–303.
springerlink.com   © Springer-Verlag Berlin Heidelberg 2009

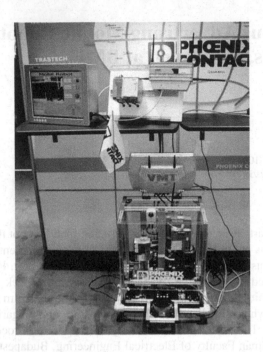

Fig. 1. The accomplished advertising mobile robot system

The mobile robot system can be divided into two parts:

- The moving mobile robot,
- The operator workstation placed on the Phoenix Contact stand.

In the case of a mobile robot system, the following tasks have to be solved. Firstly, information from the environment has to be sensed. Then, this information has to be processed and the proper command has to be sent to the actors. The acting of course changes the environment, so this process has to be done cyclically.

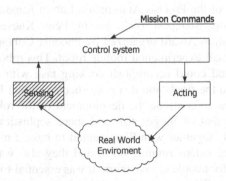

Fig. 2. Scheme of mobile robot systems

The robot has to have some task, given by the operator, sent by a communication channel. Since the robot is moving, only wireless methods can be applied between the operator and the robot. This simple concept can be seen in Figure 2.

2   Sensing

It is really important to have fast distance measurement from the robot to the surrounding obstacles. If we see more, controlling is easier. So a fast 2 or 3 dimensional scanned distance measurement is needed. The typical method to measure distance is based on the time of flight measurement. A wave package is transmitted and after reflecting from the target it gets back to the transmission point and is sensed. Knowing the propagation speed of the wave and the running time, the distance can be calculated. Transmitting sound causes slow and inaccurate measurement, so usually laser light is used. This measuring equipment is called laser scanner. On our platform we used a SICK LMS 100 2 dimensional laser scanner. This is the smallest scanner on the market whit 20 m measuring range and 0.25^O phase resolution in 270° scanning angle. The product was so new on the market, that we were probably the first who could try it in Hungary. The scanner can handle more predefined safety zones, switched by discrete inputs. This was really practical, since when the robot is turning the path is much different than when it moves straight.

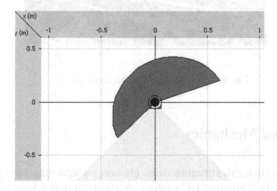

Fig. 3. The defined safety zone when the robot is turning

A web camera was also placed on the mobile part, there was always real time picture information from the robot, which made the operation easier, and the operator screen much more spectacular.

3   Control

The controlling system is based on a Phoenix industrial controller (PLC - Programmable Logic Controller). We used medium class PLC, which can handle Ethernet TCP/IP and RS232, serial communication. There are on board integrated I/O-s and industrial Bus couplers, so with connected extension modules new functions can be integrated. Handling the servos only needs two servo drive modules connected next to the PLC. This results in a really flexible system, which can be easily improved. Our whole controlling algorithm only loaded the CPU with 32%, so there is still opportunity to improve the program. It is really practical that the Ethernet port is integrated, since it makes it easy to communicate with other devices like the laser scanner, and operator panel.

Fig. 4. The controller and the servo drives

4   Actors and Mechanics

The simplest mechanical structure was chosen, which has two independently driven wheels by two brushed DC motors. A third support wheel is used to make the platform stable, and support it. Encoders were used to have feedback from the wheels. The wheel speed is controlled by the drives using PID closed control loop. Applying this structure has a great advantage. From the rotation of the wheels, when they are not slipping, the position can be estimated. This is really useful information in navigation.

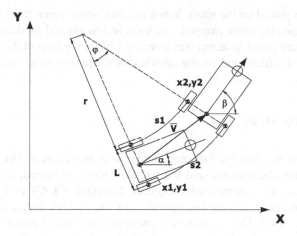

Fig. 5. Calculating the position from the turn of the wheels

5   Controlling Algorithm

The planned algorithm could build up a map from the real environment, and when the map is ready, a root could be defined. The navigation would be based on distance measurement between objects with known position and the robot, in this the position calculation from the wheel movement could help a lot. We had no time to accomplish this algorithm, since on an exhibition there are many dynamic, moving obstacles (the visitors) and it is really hard to make difference between the mapped obstacles and the dynamic ones. We decided to simple down the problem, by let the robot roam, without a predefined root, always looking for free spaces which are enough for it to change its position. When the robot has free area straight ahead, it moves straight, when it is not possible, it starts to rotate the scanning area, and searches the way, closest to the straight, where there is enough space for it to move. Also safety zones are defined in the scanner. If a dynamic object, for example a visitor, steps in the safety zone, the robot stops, and starts searching again for free roots.

It was also requirement not to let the robot to move other companies stands. Fortunately, the carpet of the visitor area had different color, so only this had to be detected. Some of the companies' stands were on an about 10 cm high podium. To detect them we had to place the scanner as low as possible. Placing the scanner to the lowest point, minimum 5 cm high objects could be detected.

Despite these methods, the robot sometimes stuck on some especially difficult locations. For example, if the podium was lower than 5 cm, the robot could not see it, but it was not powerful enough to climb up. It could not detect obstacles which were wider in the upper part, and narrowed down in the lower part. This caused the robot stuck, which meant higher current on the motors, which was detected. A bumper was also placed on the robot. When an error signal came from the bumper or from the motors, the robot stopped, and switched to manual mode, and also on the remote operator panel an alarm was generated. With the help of the IP camera, the operator could actuate, and get the robot out of the problematic situation.

6 Communication

On the whole system, standard Ethernet communication was used. Since the PLC, the Laser Scanner, the operator and the webcam were connected, a switch was needed to connect these components together. Standard WLAN communication was used to communicate with the operator. In the system, industrial Ethernet components were used. This equipment is more expensive than the office components, but they are more reliable, it is easier to handle them, and they also have some extra functions. When the communication between the laser scanner and the PLC was tested, it was possible to mirror one communication port of the switch to another. This way, with a network analyzer program we could see the TCP/IP packages between the two communicating devices, although the packages were not sent to our computer. The WLAN devices have a fast rooming function, so when a large area has to be covered by more than one Access Point, the client can switch faster between them.

7 Power Supply

In the case of a mobile robot providing energy for the electric systems is quite an important problem. On our robot we have two DC motors with 120 W maximum power consumption. We also placed an industrial PC on the robot, which has about 80 W power need. Compared to these two loads the other components power consumption is irrelevant. At least one hour continuous working is needed, so a 12 Ah maintenance-free lead gel rechargeable battery can supply the system properly.

8 Operator Panel

To operate the system, an operator computer is used. On this computer, it is possible to monitor the state of the robot, operate the behavior, switch to manual mode, and give commands manually, and also see the pictures coming from the camera which was placed on the robot. Reading values from the PLC is possible by using a special server program, called OPC server. The OPC server gives us an open source platform, to read the PLC registers.

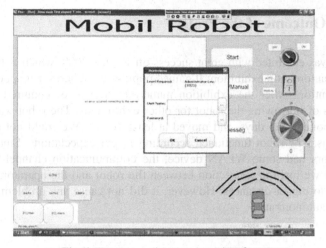

Fig. 6. The accomplished operator interface

Visualizing the read values can be accomplished by an embedded software, but in the industry, there are ready to use software packages, called SCADA software. Since we used a Phoenix Contact controller, it was logical to use the SCADA of the same company which is called Visu+. With Visu+ it was really simple to establish communication with the PLC, reading and writing registers, and also to visualize the picture of the cam.

The whole system, with the main components can be seen on Figure 7.

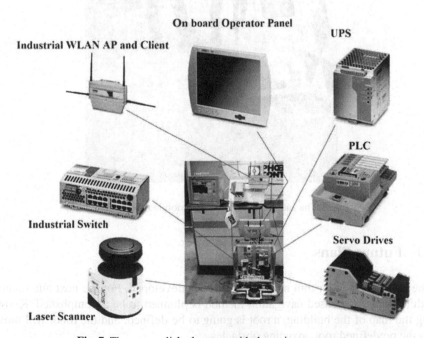

Fig. 7. The accomplished system with the main components

9   The Outcome of the Project

The robot was exhibited with great success on Regula 2009, which is the biggest Hungarian automation exhibition. It gained popularity. Wherever it appeared it attracted attention. After the exhibition numerous companies enquired about the possibilities of borrowing the robot for other exhibitions. The robot was in function for 8 hours for 4 days, and moved at least 10 km. We could not detect any malfunctions. The robot functioned according to our expectations. Since most of the exhibitors had some WLAN device, the communication channel was overloaded, and we lost the connection between the robot and the operator, when the robot went to bigger distances. However, it did not cause any problem, since the platform is autonomous.

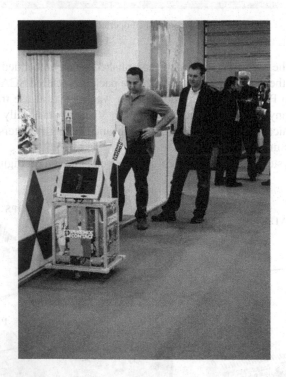

Fig. 8. The autonomous advertising mobile robot in function

10   Future Plans

The controlling algorithm needs to be further developed. For the next automation exhibition a map based navigation method is planned to be accomplished. Knowing the map of the building, a root is going to be defined, and the robot will move on the predefined root, avoiding obstacles.

Conclusion

In this article, an autonomous advertising mobile robot is introduced. The robot was developed in the laboratory of Process Automation at Kandó Kálmán Faculty of Electrical Engineering, Budapest Tech. The aim of the project was to develop a working advertizing mobile robot platform for exhibition purposes, sponsored by Phoenix Contact and Sick Sensor Intelligence. Many students worked on the development, and gained useful practical knowledge in the field of automation, industrial communication, and visualization. On the robot, mainly industrial components were used. The main controller is a PLC, the communication based on industrial WLAN, and Ethernet components, and for the operation an industrial PC with SCADA software is used. The project was exhibited with great success on Regula 2009, which is the biggest Hungarian automation exhibition.

References

[1] Siegwart, R., Nourbahsh, I.R.: Introduction to Autonomous Mobile Robots. The MIT Press Massachutsetts Institute of Technology, Cambridge (2004)
[2] Mester, G.: Modeling of the Control Strategies of Wheeled Mobile Robots. In: Proceedings of The Kandó Conference, Budapest, pp. 1–4 (2006) ISBN 963-7154-42-6
[3] Matijevics, I.: Microcontrollers, Actuators and Sensors in Mobile Robots. In: 4th Serbian-Hungarian Joint Symposium on Intelligent Systems, Subotica, Serbia, September 29-30 (2006)
[4] LMS100 Laser Measurement System application manual,
    http://www.sick.com/home/factory/news/autoident/lms_100/
    en; LMS100_ProductInformation.pdf (downloaded on: 2008.10.10)
[5] User Manual for the WLAN Devices FL WLAN 24 (D)AP 802.11 and FL WLAN 24 EC 802.11, http://eshop.phoenixcontact.com/phoenix/;
    um_en_fl_wlan_ap_ec_7190_en_01.pdf (downloaded on 2008.10.10)
[6] Uninterruptible Power Supply Unit for Universal Use,
    http://eshop.phoenixcontact.com/phoenix/;
    db_en_quint_dc_ups_24dc_10_101998_01_gb.pdf (downloaded on: 2008.10.10)
[7] Installing and Operating the Inline Controllers ILC 330 ETH, ILC 350 ETH, ILC 350 ETH/M and ILC 350 PN, http://eshop.phoenixcontact.com/;
    um_en_ilc_330_350_6959_en_05.pdf (downloaded on: 2008.10.10)

Part IV
Information Technology

High-Level Specification of Games

Jaak Henno

Tallinn University of Technology, Estonia
jaak@cc.ttu.ee

Abstract. Games, MMOG-s (Massively Multiplayer On-Line Games), on-line so-cial sites etc have become a major cultural and economic force. The main distin-guishing feature and attraction of games is their interactivity – participants con-stantly change the state of affairs with their actions. The resulting dynamic flow of events, gameplay is like execution of an algorithm, where elementary actions are defined by game rules, but the logic, the flowchart is composed "on-the-fly" by players.

To many IT study programs have been introduced Game Programming courses. However, usually in these courses is not discussed design of games. Books on game programming and courses on this subject are based on some specific pro-gram language (C, C++), software package, pre-programmed set of classes etc, thus instead of discussing games on general they consider specific features of these programming environments and only through them also something about games. We do not have adequate formal methods for description and specification of games.

On basc of analyze of several popular packages (Gamemaker, Flash AS3, Panda3D) used for programming simple, casual games here is presented a meta-language for object-oriented, structural description of games as event-driven ob-ject-oriented systems. Specifications of games created using this language are easy to transform into implementations using some of these concrete game program-ming environments. This allows to present games on general abstract level, which does not depend on implementation environments. It allows to consider advan-tages and problems with concrete packages and to compare implementations of games in different environments. The specification method is illustrated with three examples.

Keywords: Game, game programming system, specification, event-driven archi-tecture, formal description, emergence.

1 Introduction

Video games (computer and console games) have become a major economic and cultural force. In last ten years, total revenues of game business in USA have grown ten times and for several years are already greater than revenues of cinema

I.J. Rudas et al. (Eds.): Towards Intelligent Engineering & Information Tech., SCI 243, pp. 307–322.
springerlink.com © Springer-Verlag Berlin Heidelberg 2009

business. Even the current economic confusion has not affected the game business – in December 2008 game industry reported the highest single-month revenues in video game history [1] and for 2009 it is predicted that "…videogame industry revenue is set to reach \$41.9 billion this year, having risen from \$27.2 billion in 2004" [2]. More and more people are participating in MMOG-s (Massively Multiplayer On-line Games) and in so-called social networks (Twitter, Slasdot, Digg), which are rather similar to MMOG-s. Games have also changed psychology of young generation, the "post-Pong" (computer game Pong was introduced in 1973) generation, "digital natives" [3], is in many ways different from older generation, "digital immigrants" [4].

Game programming has been introduced to IT curricula in many universities, but the theoretical base for games and game programming is week. We do not have formal methods to describe games and gameplay, to classify games in meaningful way [5]. Description and classification of games is often based on visual appearance (horizontal scroller, maze), on technology used (arcade games, console games) or simply by game size (casual games), but these words do not help to understand "what is there inside", the idea of the game, structure of game and game program. Design and development of games is a complex activity involving participants from several disciplines, who all should first understand, what they are supposed to create, the basic structure of the game. However, we do not have formal methods for representation of games. Game programming courses often do not begin with general introduction and analyze of games, but with discussion of e.g. blitting and video memory buffers, i.e. low-level issues, which are not directly related to game design and game program structure. Books and courses on game programming are based on some concrete game programming system or software package, thus instead of discussing games they actually teach some concrete implementation environment, which often is also very low-level – C, C++ etc. Dependence on implementation issues is very confusing for students and often does not allow considering and understanding the game structure before implementation, game becomes "visible" only after implementation.

2   What Is a Game?

It is rather difficult to define what is a "game", to determine the basic features, common to all/most games. When Ludwig Wittgenstein considered different forms of games (children games, Olympic games, flirtation, Go etc) he concluded that word "game" can not be explained as one concept, but "game" contains several concepts which share a "family resemblance" to one another [6]. Johan Huizinga considered games as an art form: "There is beauty in games. … I am fascinated by the rich gameplay so many games offer using only a handful of rules. … Every session of play becomes a performance; a ritualized dance that is focused and confined by the game's rules and premise, but which is never the same twice. Mine is an aesthetic appreciation of the freedom and possibilities set up within the logical game space within the magic circle of the game"; „play and games are in fact at the very center of what makes us human. Play is older than culture" [7].

Academic community stated to think about games seriously only with the turn of the new century. In 2001 was established the international journal of computer game research "Game Studies" [8]. Year 2001 was declared "year one" of game studies and Gonzalo Frasca made in his article popular the word "ludology" - the discipline that studies games, play, toys and videogames [9] (ludus = game (Latin), logos = science (Greek)).

Game authors use lot of metaphors and poetic language when speaking about games (gameplay is "choreography", "We like games that deliver a sense of an endless world, living environments, and open-endedness" [10]). Academic researches (especially with IT background) consider games as programs, dynamic systems. For dynamic systems we have lot of theory, many types of models and modeling technologies can be used to describe their structure and dynamics. It is not difficult to develop structural models to describe already existing games, but these models provide little help for developers who want to invent new types of games. Games are very different from conventional IT systems. For instance, the structural model of the Portal game from Valve (at least the first levels) is very simple, but this does not tell anything about surprising gameplay. Even simpler is the structure of the game Auditorium [11] – just on input flow of energy, several manipulators which allow to change the direction of the flow and sound organs, the flow of interactions is also quite simple, but what kind of model could describe gameplay, the search and "aha!"-s?

Since attempts to invent a generative device, a "game grammar" [12], [13], [14]] have not been successful, several researchers have tried to describe games by UML and other types of diagrams. Flowchart-type diagrams were used to describe possibilities to advance in simulations or adventure games [15], where gamers have to move around, find objects and fill some tasks; similar approach was used in [16]. These schemas describe the game's static environment – plan of rooms and what is in these rooms. The diagrams are similar to story-beat diagrams [17] which describe the scene location, scene description, the object and cast members in the scene, events game engine listens and what actions they trigger, what happens in response to these events. The UML diagrams used in [18] also describe the structure of a game, i.e. game settings, but not what really happens in gameplay.

Attempts to apply existing system design and specification methods e.g. use case diagrams [19] to games have not been very encouraging. Indications that player can *"View current crew assignments"* or decision block *"[if more data needed] – getMoreData()"* say very little about game idea and structure. In [20] were used colored Petri nets to describe on a very high level (inputs are "skill", "opponents skill" etc) possible flows during gameplay, but this very abstract model does not consider operational rules of the game.

The current state of understanding of game design is sometimes compared with alchemy: "...it was clear to the alchemists that 'something' was generally being conserved in chemical processes, even in the most dramatic changes of physical state and appearance; i.e. that substances contained some 'principles' that could be hidden under many outer forms, and revealed by proper manipulation" [21]. Game designers also are convinced, that in a good game there is "something" what makes it good: "It is my belief that a highly mechanical and predictable heart, built on the foundation of basic human psychology, beats at the core of every single successful game" [22]. Unfortunately, we still have not understood what this "something" is and what are the "principles" which tie the ingredients into a good game.

Gameplay is like execution of a program. It consists of elementary actions (linear code) and logic, which ties these elementary actions together. In gameplay, the elementary actions (what player can do) are fixed by game rules. However, the logic, which ties these actions into gameplay, is invented by players. Rules give only possible operations, decision blocks are from player. Games are programs with variable logic, ballet with fixed pa's, but free choreography, tales, with known character types and plot elements, but the player determines how the story unfolds.

This unpredictable element – player's decisions – is the main difference in modeling and specifying games compared to e.g. Information Systems. Design of an Information System usually begins with considering typical use cases; it is assumed, that the system will always be used in a similar way. In games, we can not predict player's decisions. The main purpose of formal specifications for games is to help to present possibilities for the gameplay/story and the basic structure of the game program, to assure, that we have planned correctly all event-information-action flows, that players always have necessary information, can perform intended actions and game engine (controller) can all the time control the whole process.

3   Game Specification Language

In the following is presented an abstract system (language) for describing game program's structure. The formalism generalizes popular game programming systems: (Gamemaker [23], Pygame [24]) and it is very convenient also for an OO-programming language/system, e.g. Flash programming language Actionscript (the AS3 version), Panda3D [25] etc.

On the logical level the described here formalism is somewhat similar to Game Description Language [26] created in Stanford AI department. The Stanford GDL is a logic language for describing the logical structure of games (with full information) and to check automated game-playing programs. It is not suitable to present a new game, especially to programmers, who do not have previous experience with logic languages, since game description is presented using Prolog rules. For instance, the main game loop is presented by the following rule (the system predicate next declares, that the argument clause will be true in the next game frame/state) [27]:

$$(\Leftarrow (\text{next (cell ?x ?y ?player))}$$
$$(\text{does ?player (mark ?x ?y)))}$$

The presented here game specification language does not use logic language, but describes game as an event-driven system, using a functional style, somewhat is considered also game visual and organizational structure. The main elements of specifications are game objects, game events and objects reactions to events. Game logic is considered only on the level of possible interactions in gameplay – how events are dispatched and what actions they trigger. The formal level is not suitable for proving correctness of game-playing programs, but it allows considering information flows in game program and thus finding inconsistencies in specification very early. The formalism does not require end states and termination, thus it can be used also to describe simulations, e.g. interactions inside a multi-agent system when agents create a common vocabulary/language (example 3).

The following is an informal description of the language.

Game is a finite automaton, which players see as a frame-based animation. Game rules define possible actions of players - how they can change game states. Gameplay is like execution of a program (this automaton); possibly state changes are determined by rules, but execution logic – by players.

Game objects and attributes encode the states of this automaton. Possibilities for state changes appear as game events, actual changes are player's reactions (actions) to events.

3.1 Requirements for Implementation Environment

The main purpose of described here formalism is to help to describe high-level structure of a game (describing actions using pseudocode) and not to deal with low-level programming issues. Created specifications are easy to implement without essential changes in program's structure, if the implementation environment satisfies certain requirements.

It should implement the basic game loop:

Get player's input, i.e. capture IO events (keyboard, mouse) and game-specific (i.e. defined in this particular game) events

Calculate the next state

Create NEXT_FRAME event and other (game-specific) events, i.e. send information to players and create visual output.

For video games the visual side is very essential, thus the environment should implement sprite graphics (drawing sprites on screen, scaling, rotating, flipping, alpha blending and color transform) and also have vector graphics (2D or 3D). Graphic objects - sprites and vector objects - should have coordinates and some attributes to describe movement – direction, speed. It should have collision detecting, at least the simplest version, where the bounding box or sphere is used to detect collisions.

Several povpular game programming environments: Gamemaker, Flash (AS3), Openscript (Toolbook), Panda3D, Pygame etc satisfy (most) of these requirements, thus are suitable as implementation environments for games, specified using this language.

3.2 Game Specification Language

The main categories of game are game objects (classes), events and reactions of objects to events (actions). Objects can be passive (also called non-playing characters or background objects), which do not receive, recognize events and active, which receive events and can perform actions (reactions). Classes (both active and passive) have attributes; e.g. object coordinates on screen (game window). Actions of objects (active objects) change game state.

Games also have some additional structure (rooms, levels) and general attributes (screen/window size, time); in the following, this structure is denoted by S (*Setup*).

Game specification is a half-formal presentation of the game program, which allows considering game before implementation and selecting the most suitable implementation environment. Specification is a 5-typle of disjoint sets

$$G =< Description, O, E, A, S >$$

Below is described meaning of these components and their structure.

Description – a brief description of the game in natural language (with images).

O – classes of game objects. The set of classes O is hierarchically ordered, i.e. a new class can extend some already existing class or some class defined by implementation environment; notation $O_1 \subseteq O_2$ means that class O_2 extends class O_1. Objects of classes (class instances) can get objects or data structures created from other objects and/or their attributes as arguments when new objects are created, e.g. object *ship* needs arrays of missiles, enemies and explosions and on firing passes them to missiles; missile seeks collisions with enemies and starts explosions.

The top object is constant $controller \in O$ - this is the game engine, i.e. the program, running the game.

Objects can receive events E (created by other objects); in response to an event, they perform some actions and/or create new events. In event-driven systems events are usually considered to be properties of the whole class, i.e. they are sent to class, not to some object of the class, but in specifications it is often very convenient to consider events which are sent to some distinguished object of the class and not to whole class. For instance, in turn-based game (Tic-Tac-Toe) event *turn* is sent only to one player and e.g. it does not have sense to send event *next* (next frame) to enemies, who are dead. This behavior is easy to implement using additional attributes in object specifications and is used in following.

Classes (both active and passive) have the set $Attr(O)$ of properties (attributes), i.e. class variables. Attributes have type – *public, private, static* with usual semantics. In most environments (Gamemaker, Flash) default type for all attributes is *private (local* in Gamemaker*)*, but for beginners it is often more convenient to consider all (or most of) attributes *public* – beginners do not always understand issues concerning limited access to attributes.

All attributes have a fixed domain of allowed values. Objects $o \in O$, which have visual representation on screen, have attribute $sprite(o)$ - sprite of the object. Sprite can be one image, but also a set of several images and/or animations, e.g. a DirectX graphical 3D-object can have built-in animation $walk$.

For correct rendering to screen sprites are collected into an ordered *display list*. Adding its sprite to display list is an action, which objects must perform when they are added to game, a reaction to event *Create*. In some implementation environments, this is done automatically (Gamemaker), in some environments (AS3, Panda3D) this should be done in code. Since this depends on implementation environment, the feature is not shown in specifications.

Usually domains of attributes are numeric and are (primitive) types in underlying implementation language/system, i.e. int, $real$, $Array$ etc. There could be also some operations defined on attribute domains (e.g. the negation operation on the domain of marks in example 1). Common properties for visual objects are their coordinates in game space (2D or 3D) and their movement characteristics – speed, direction. While coordinates are usually expressed in Cartesian coordinate system, the movement can have several representations – as a vector (i.e. speed vector coordinates) but can be expressed also as direction angle and linear speed along the direction line (polar representation).

Most classes allow generating several instances of objects from class, but some classes can have only unique object (*controller*).

The objects receive and dispatches events $e \in E$ to active objects (passive objects do not receive events). The set of events contains constant $create \in E$ - this is the event, when an object (instance of a class) is added to game (in object-oriented systems this is executing the constructor function). Another common event is *step* - frame change, i.e. new frame. Sometimes it is convenient to divide the *step* event into several consecutive events: *begin step, step, end step* (e.g., this is used in Gamemaker). The reaction to *step* event can be e.g. change in object coordinates – this creates movement.

Common events are mouse and keyboard events: *click, over, out, keypress* and interaction, *collision* of objects – reaction to this event can make ball to bounce from a pad. The *collision* event is not symmetrical; it occurs only when an object (e.g. object *bullet*) executes testfunction *collision(Other)* and result is passed only to the object who initiated test. Collision has different variations depending on how precisely it is calculated – whether only bounding boxes are used or it is necessary to find exact contact of object surfaces.

Sending an event e to object O is denoted by

$$\Rightarrow (O, e, Attrs)$$

Here $Attrs$ are additional attributes of the event.

Game begins from event

$$\Rightarrow (controller, create)$$

This is a unique event, $create$ (for the object $controller$) can not occur later. This start event may have attributes – e.g. difficulty of level, addresses of servers where the game will be hosted etc.

To receive an event, object should be made event listener. In the following, the listener declarations are omitted. It is assumed that when an event $e \in E$ is sent to some class O or to some specific object $o \in O$ it has reaction (action) $f(O, e, Attrs)$.

It is often convenient to allow objects to have states. Very common are states based on attribute $visible$ – non-visible objects usually do not perform many actions, thus they also do not respond to several events, e.g. to animation event $next$. Structure of object states can be more complex, e.g. a fighter can have states $Creeping, Shooting, Dead$, a rocket can have states $Flying, Landed, Fuelling$ etc. For objects with states, the event actions become case-structures.

The following examples intend to illustrate the main ideas behind the proposed high-level specification method: example 1 – the data/information flow, example 2 – explanation of safe implementation method where possibilities for memory leak are minimized, example 3 – use of more complex data structures (dictionaries) for class attributes.

In the following examples variable identifiers begin with capital letter, constant (object) identifiers – lowercase latter, "_" stands for arbitrary value, prefix Arr denotes an array (actually set, the order and indexing are not used) and actions are described in a pseudocode. Many details are not specified but left for common sense and previous knowledge about specifications and programming.

Example 1: Tic-Tac-Toe

Tic-Tac-Toe=
$\quad < \{controller, Player, Square\}, \{create, turn, move\}, S >$

$Description$: two players take turns to mark a 3 x 3 grid; player who succeeds in marking a whole a row, column, or diagonal wins.

$attr(controller) = \{Arr_{blank}[\,], Arr_x[\,], Arr_o[\,], Arr_{lines}\}$

$attr(Square) = \{r, c\}$

$E(controller) = \{create, move(Sprite, i, j)\}$

$attr(Player) = \{Arr_{blank}, Arr_{marked}, Sprite\}$

$Sprite \in \{x, o\}$

The operator $\neg$ acts on the set of sprites as negation, i.e. $\neg o = x$, $\neg x = o$

$f(controller, create) = \{$

$for(i = 1, 2, 3)\, for(j = 1, 2, 3)$

$\{sq(ij) = Square.create(r = i, c = j), Arr_{blank}.push(sq(ij))\};$

$Arr_{lines} = [[sq(11), sq(22), sq(33)],$

$[sq(21), sq(22), sq(23)],$

$[sq(31), sq(32), sq(33)],$

$[sq(11), sq(21), sq(31)],$

$[sq(21), sq(22), sq(23)],$

$[sq(31), sq(32), sq(33)],$

$[sq(11), sq(22), sq(33)],$

$[sq(13), sq(22), sq(31)]]$

$playerx = Player.create(sprite = x, Arr_{blank}, Arr_{marks} = [\,]);$

$playero = Player.create(sprite = o, Arr_{blank}, Arr_{marks} = [\,]);$

$\Rightarrow (playerx, turn)\}$

$f(controller, mark(Sprite, i, j))\{$

$Arr_{Sprite}.push(sq(i, j))$

$if(Arr_{lines} \cap Arr_{Sprite} \neq \varnothing)\ \{$

$\quad \Rightarrow (won, player(\_, \_, Sprite)),$

$\quad \Rightarrow (loose, player(\_, \_, \neg Sprite)),$

$\quad exit\}$

$else\{$

$\quad Arr_{blank} \leftarrow Arr_{blank} - \{sq(i, j)\}$

$\quad if(Arr_{blank} \neq \varnothing)\{$

$\quad\quad \Rightarrow (turn, player(\_, \_, \neg Sprite)\}\}$

$$E(Player) = \{create, turn, move(\_, i, j), won, loose\}$$
$$f(Player(Arr_{blank}, Arr_{marks}, Sprite), turn) = \{$$
$$(sq(i, j) \in Arr_{blank})\{\Rightarrow (controller, move(Sprite, i, j),$$
$$\Rightarrow (player(\_, \_, \neg Sprite), move(\_, i, j))$$
$$Arr_{blank} \leftarrow Arr_{blank} - \{sq(i, j)\}$$
$$Arr_{marks} \leftarrow Arr_{marks} \cup \{sq(i, j)\}$$
$$draw\_sprite(Sprite, sq(i, j))\}$$

$$f(player(Arr_{blank}, \_, \_), move(\_, i, j))\{$$
$$Arr_{blank} \leftarrow Arr_{blank} - \{sq(i, j)\}\}$$

The last action (for player, who did not make the last move) means, that he also has to remove the marked cell from his list of free cells.

In above actions players do not use the list of already marked cells (attribute Arr_{marks}) for deciding the next move; this is included only for possibility to develop some AI algorithm for deciding the next move.

The setup S consists of game grid (passive background object).

Example 2: Star Track Battle

Fig. 1. Star Track Battle

Description: on the background of moving planets a single battleship should fight with enemies, who fly constantly over the screen from right to left; battleship can move only vertically and shoot. Enemies do not shoot, but if they collide with battleship the ship gets damages (its health decreases). If player can shoot down all enemies (fixed number at the beginning of the game) he wins, but if ships health becomes zero, player looses.

The following specification (partial) demonstrates safe use of memory. It uses the ship and fixed number of all other objects - missiles, enemies, explosions.

Enemies fly in from right to left. If they remain alive, they return to right side of the screen and repeat; if they are hit, they become invisible and are removed from screen (but not from game memory). Missiles become active (visible, moving) when they are fired; when they hit an enemy or leave the screen area they become invisible and do not receive step event, which makes them move. Explosions are small animations (growing and getting darker) and behave the same way – they become active, visible and are shown only when a missile hits an enemy; when they finish playing they become invisible and do not receive step events. This simplifies memory handling in implementation. If on firing new missile objects were created and lost missile objects where removed from game then the program should constantly interact with operating system and memory leaks become very possible.

The Game =

$< \{controller, ship, Star \subset Enemy, Missile, Explosion\},$

$\{E(ship) = \{create, move\_up, move\_down, shoot,$

$\qquad\qquad collision(Enemy)\},$

$E(Enemy) = \{create, step, damaged\},$

$E(Missile) = \{collision(Enemy)\} >$

$E(Star) = \{create, step\}$

$Attr(controller) = \{ship, Arr_{stars}, Arr_{enemies}, Arr_{explosions}, Arr_{missiles}\}$

$Attr(ship) = \{y, shooting\}$

$Attr(Star) = \{x, y, size, speed_x\}$

$Attr(Enemy) = Attr(Star) \cup \{visible, active\}$

$Attr(Explosion) = \{x, y, visible\}$

$Attr(Missile) = \{x, y, speed_x, visible\}$

$f(controller, create) = \{$

$for(i = 1, \text{STARS})(star = Stars.create, Arr_{stars}.push(star))$

$for(i = 1, \text{EXPLOSIONS})(expl = Explosion.create(visible = false),$

$Arr_{explosions}.push(expl))$

$for(i = 1, \text{ENEMIES})(eny = Enemy.create,$

$Arr_{enemies}.push(eny))$

$for(i = 1, \text{MISSILES})(m = Missile.create(visible = false,$

$speed_x = 6),$

$Arr_{missiles}.push(m))\}$

$$f(Star, create) = \{$$
$$x = random(0, screen.width)$$
$$y = random(0, screen.height)$$
$$size = random(size_{min}, size_{max})$$
$$speed_x = -random(speed_{min}, speed_{max})\}$$
$$f(Star, step) = \{x+ = speed_x$$
$$if(x < 0) \; x = screen.width\}$$

The class *Enemy* extends class *Star* and does not need any additional properties concerning movement.

$$f(Enamy, create) = \{visible = true, active = true\}$$
$$f(Enemy, damaged) = \{visible = false, active = false\}$$
$$f(Ship, key_{up}) = \{if(y > 0) \; y--\}$$
$$f(Ship, key_{down}) = \{if(y < screen.height) \; y++\}$$
$$f(Ship, key_{shoot}) = \{$$
$$(\exists m \in Arr_{missiles}, m.visible = false)\{$$
$$m.x = ship.x$$
$$m.y = ship.y$$
$$m.visible = true\}\}$$
$$f(Missile, step) = \{$$
$$if(visible = true)\{$$
$$(for(eny \in Arr_{enemies}, eny.active = true)\{$$
$$if(collision(eny))\{$$
$$eny.active = false$$
$$eny.visible = false$$
$$(\exists expl \ni Arr_{explosions}, expl.visible = false)\{$$
$$expl.x = x$$
$$expl.y = y$$
$$expl.visible = true\}\}\}$$
$$\}$$

In functionality specifications (e.g. in the above $f(Missile, step)$) variables without prefixes always belong to the object whose functionality is explained, i.e. here $x = missile.x$, where $missile$ is the concrete instance of class $Missile$ whose functionality is explained. Thus the last specification for $f(Missile, step)$ says, that if missile has been fired (is visible) and if it collides with active enemy, then enemy becomes non-active, vanishes and the array of explosions is searched for currently "non-playing" (i.e. not visible) explosion, which is shown in the place of collision. Explosions are small animations which themselves quickly vanish (become invisible). Total number of explosions (length of the array of explosions) is set approximately twice higher than the number of missiles, thus for every missile (if it hits an enemy) there is a non-playing explosion, which can be used.

Example 3: Simulation of emergence of common vocabulary

This example is not a game – there is no goal, but it shares many features of MMOG-s and it is easy to transform this to game (e.g. make agents to compete). This example demonstrates use of object states and use of dictionary-type data structures for attributes.

Description. A multiagent society is searching labyrinth for food [28], [29]. When agents meet each other, they try to communicate, tell what they have found (food) or warn that there is a danger (deep hole) or that the passage is a dead end. Thus their messages to each other are about "food", "danger", "dead end". For these semantic concepts they invent words (arbitrary sounds) or use words which they have got earlier from other agents. Their short-time memory buffer is bound to local spatial context - they remember facts only when moving along a corridor without branching (the local spatial environment). At the beginning, they do not have any vocabulary, but in the process emerges common vocabulary.

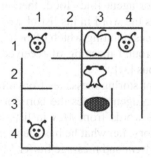

Fig. 2. Simulation of emergence of common vocabulary

Emergence_of_Vocabulary =

$$< \{Agent, Room, Food, Danger, Dead\_end\}, E, A, F\} \cup S >$$

$$Attr(Room) = \{open_{left}, open_{right}, open_{up}, open_{down}, food, danger\}$$

The labyrinth is rectangular grid. Elements of class *Room* are its cells. Boolean attributes *open* indicate directions where the room is open, e.g. on the above picture for the cell (2,2) $open_{left} = true$, $open_{right} = false$, $open_{up} = false$, $open_{down} = true$. Agents perceive these attributes and use them to decide, when they enter a dead end (a room where only one of these four attributes is true) and when they leave the current corridor (this is a room where more than two *open*-attributes are true).

$$attr(Ag) = \{Mem, Dir, Mes, L\}$$

$$Mem = [F, Dan, De]$$

Attribute *Mem* is agent's short-term memory. Boolean variables *F, Dan, De* denote correspondingly presence of food, danger or dead-end in the current local spatial context, i.e. in the corridor where agent is. When agent finds a room with food or danger or a dead end, he stores these facts in corresponding variables of his short-term memory. When agent leaves a corridor (enters a room, which is open in more than two directions), the short-term memory is cleared (all variables become *false*).

Dictionary-type variable *L* is for storing emergent vocabulary. Here for semantic values *f* (food), *dan* (danger), *de* (dead end) are recorded finite lists of possible words (denotations) for these concepts.

At the beginning of simulation all these lists are empty. When agent meets another agent, he creates a message about entities, which are true in his short-term memory. He uses word from his list, which previously has been used most often; if the list for a concept is empty, he creates a new word.

The spatial context (current corridor) restricts possible meaning of received messages. For instance, if another agent's message contained one word and when moving further in the corridor agent finds food, then he knows that the received word denoted food and stores the word in his list of words for concept food. Thus here the spatial context replaces pointing which often has been used in simulations of language emergence to ground meaning of received messages [30], [31] and which is considered ambiguous [32].

Agent's attribute *Mes* is for storing messages from other agents received in current context (corridor). When agent leaves the corridor (in first room with more than two exits) he stores all words from *Mes* in lists of all concepts which are stored in his short-term memory, i.e. what he has found in this context and empties both the short-term memory *Mem* and message buffer *Mes*.

In the attribute *Dir* is stored the direction of his last movement – this is used to avoid random vreturns to previous room.

Conclusions

Because of economic impact of games, game programming has become essential part of many IT curricula, but we do not have established methods for game specification and formal presentation. Games are rather different from "classical" software systems (e.g. Information Systems) and classical methods for specification do not work here. Most books about game programming are tied to implementation environment (C, C++, C# etc) and instead of discussing games are actually teaching these concrete software environments. Here is presented a high-level specification method for of games as event-driven object-oriented software systems. The method allows describing games without using low-level details of programming languages; conversion of specifications to working programs using e.g. Gamemaker, Flash AS3 or Panda 3D environments is rather straightforward.

References

[1] http://www.gamasutra.com/
    php-bin/news_index.php?story=21885 (14.01.2009)
[2] http://gamedeveloper.digitalmedianet.com/articles/
    viewarticle.jsp?id=704781
[3] Prensky, M.: Digital Natives, Digital Immigrants (14.01.2009),
    http://www.marcprensky.com/writing/
[4] Prensky, M.: Do They Really Think Differently? On the Horizon, vol. 9(6). MCB University Press (December 2001),
    http://www.marcprensky.com/writing/ (12.01.2009)
[5] Natkin, S., Vega, L.: A Petri Net Model for Computer Games Analysis. Int. J. Intelligent Games and Simulation 3, 1 (2004)
[6] Wittgenstein, L.: Philosophical Investigations (1953/2002) ISBN 0-631-23127-7
[7] Huizinga, J.: Homo Ludens. A Study of the Play Element in Culture. Beacon Press, Boston (1955)
[8] http://gamestudies.org/0802 (14.01.2009)
[9] http://www.ludology.org/ (14.04.2009)
[10] Raph Koster, Theory of Fun for Game Design. Paraglyph Press (2004) ISBN 10: 1-932111-97-2
[11] http://www.playauditorium.com/ (14.04.2009)
[12] Defining Games: Raph Koster's Game Grammar (14.04.2009),
    http://www.gamasutra.com/view/feature/1979/
    defining_games_raph_kosters_game_.php
[13] Koster, R.: A Grammar of Gameplay. In: Game Developers Conference 2005: Futurevision (2005)
[14] Bura, S.: Game Grammar (2006),
    http://www.stephanebura.com/diagrams/ (14.01.2009)
[15] Bethke, E.: Game Development and Production. Wordware Publishing, Plano, TX (2003)
[16] Siang, A., Rao, G.: Designing Interactivity in Computer Games: A UML Approach. Int. J. Intelligent Games and Simulation 3, 2 (2004)
[17] Onder, B.: Writing the Adventure Game. In: Laramee, F. (ed.) Game Design Perspectives, pp. 28–43. Charles River Media, Hingham (2002)

[18] Dormans, J.: Visualizing Game Dynamics and Emergent Gameplay. Meaningful Play (2008),
     http://meaningfulplay.msu.edu/proceedings2008/
     mp2008_paper_40.pdf (24.04.2009)
[19] Taylor, M.J., Gresty, D., Baskett, M.: Computer Game-Flow Design. ACM Computers in Entertainment 4(1), article 5 (2006)
[20] Bura, S.: Game Grammar (2006),
     http://www.stephanebura.com/diagrams/ (14.01.2009)
[21] http://interpretivealchemy.blogspot.com/ (14.01.2009)
[22] Cook, D.: The Chemistry Of Game Design (14.01.2009),
     http://www.gamasutra.com/view/feature/1524/
     the_chemistry_of_game_design.php
[23] http://www.yoyogames.com/make (14.04.2009)
[24] http://www.pygame.org (14.04.2009)
[25] http://www.panda3d.org/ (14.04.2009)
[26] Game Description Language,
     http://games.stanford.edu/language/language.html
[27] http://games.stanford.edu/language/language.html
[28] Henno, J.: Emergence of Language: Hidden States and Local Environments. In: Jaakkola, H., Kiyoki, Y., Tokuda, T. (eds.) Information Modelling and Knowledge Bases XIX, pp. 170–181. IOS Press, Amsterdam (2007)
[29] Henno, J.: Emergence of Names and Compositionality. Information Modelling and Knowledge Bases, vol. XVIII. IOS Press, Amsterdam (2007)
[30] Steels, L., Vogt, P.: Grounding Adaptive Language Games in Robotic Agents. In: Husbands, C., Harvey, I. (eds.) Proceedings of the Fourth European Conference on Artificial Life, Cambridge MA and London. MIT Press, Cambridge (1997)
[31] Steels, L., Kaplan, F.: AIBO's First Words: The Social Learning of Language and Meaning. Evolution of Communication 4(1) (2001)
[32] Quine, W.V.: Word and Object. MIT Press, Cambridge (1960)

New Trends in Non-volatile Semiconductor Memories

Zsolt J. Horváth[1, 2] and Péter Basa[1]

[1] Research Institute for Technical Physics and Materials Science, Hungarian Academy of Sciences, P.O.Box 49, H-1525 Budapest 114, Hungary

[2] Institute of Microelectronics and Technology, Kandó Kálmán Faculty of Electrical Engineering, Budapest Tech, Tavaszmező u. 15-17, H-1084 Budapest, Hungary
horvath.zsolt@kvk.bmf.hu

Abstract. The construction and the physical background of operation of floating gate, nanocrystal, silicon nitride-based, phase-change, ferroelectric and magnetoresistive memories are breafly summarized.

Keywords: flash memories, nanocrystal memories, SONOS, phase-change memories, FeRAMs, MRAMs.

1 Introduction

Non-volatile memories are those memories that hold the information even after turning off a power supply. Nowadays the most frequent non-volatile memories are flash memories (FMs), which are applied in the most embedded systems. They are used in mobile phones, MP3 players, pocket computers, digital cameras as well as in pen drives, memory cards and hybrid hard disks. They are of high density, updateable, electrically erasable, relatively fast, and reliable [1].

However, scaling down technology node faced difficulties in these memories, because the reduction of their dimensions is limited due to reliability problems [2]. This inspires research for new constructions and for new operation principles of non-volatile memory elements and arrays. Another driving force for research for new solutions in this field is the permanent demand for creating faster and faster memories.

Most of the non-volatile memories are Random Acces Memories (RAMs). In addition to flash memories, recent Non-Volatile Random Acces Memories (NVRAMs) being in production and/or under development are the Phase-Change RAMs (PRAMs), the Ferroelectric RAMs (FeRAMs), and the Magnetoresistive

RAMs (MRAMs). Although the magnetic tunnel junction, which is the memory element in MRAMs, isn't made of semiconductors, they are integrated in a semiconductor memory chip. The compatibility of their preparation with the CMOS technology is a general requirement to memory elements.

Non-volatile memory effect has been demonstrated in devices with carbon nanotubes as well [3]. Establishment of nanoelectronics made on carbon nanotubes – including non-volatile memories – is a new challenge of nowadays solid-state electronics.

In this paper the construction and the physical background of operation of non-volatile memories are breafly summarized.

2 Main Characteristics of Non-volatile Memories

One of the main characteristics of memories is their absolute and areal density, which is strongly related to the cell area. Information in non-volatile memory elements is held by the threshold voltage of Field Effect Transistors (FETs) in FMs and FeRAMs, by the conductivity of the phase-change material (crystalline vs. amorphous) in PRAMs, by electrical polarization of the ferroelectric material in FeRAMs, and by the conductivity of a magnetic tunnel junction connected with its magnetic polarization in MRAMs. So, memory cell consist of a single memory FET in FMs and novel FeRAMs, of a transistor and capacitor in FeRAMs of earlier development, of a transistor and resistor or of two transistors in PRAMs, and of a magnetic tunnel junction and a resistor or a transistor in MRAMs, respectively [4].

Another important parameter is the requirements for writing, erasing and reading. This means the pulse amplitude and width of applied voltage in FMs and FeRAMs, or applied current in PRAMs and MRAMs, which are necessary for Write/Erase (W/E) operations, on one hand, and reading time and voltage, on the other hand. W/E and read speed depends on the memory type, architecture, technology node, actual mechanism, etc.

Memory behaviour is characterized by the difference of the actual parameter holding the information between the two logical states. This is, e.g., the difference between the threshold voltage values attributed to logical states "0" and "1" in FETs, which is called memory window, or the difference in conductivity in PRAMs and MRAMs.

The basic characteristics of reliability are endurance and retention. Endurance means the number of W/E cycles before degradation. Except MRAMs, the degradation is clearly connected to the W/E process in all the non-volatile memories. The speed of information loss is characterized by retention. For non-volatile memories the required retention time is 10 years. As read process is non-destructive in the most non-volatile memories, "read disturb" is another parameter of reliability. It is attributed to the change of the parameter holding the information during read process.

3   Conventional Non-volatile Memories

The conventional non-volatile memories are the Electrically Programmable Read Only Memories (EPROMs), the Electrically Erasable Programmable Read Only Memories (EEPROMs) and flash memories. The main difference between EPROMS and EEPROMs is that the latter ones can be erased electrically, while EPROMS are erased by UV illumination. The main difference between EEPROMS and flash memories is in the organization of memory arrays. In EEPROMs memory cells are erased by bytes, while in FMs by blocks using a "flash", i.e., a common erasing voltage pulse [4].

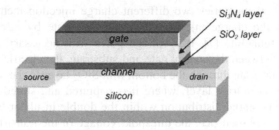

Fig. 1. The MNOS memory transistor

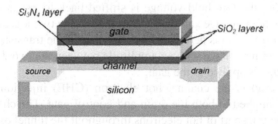

Fig. 2. The SONOS transistor

The first realized memory element was the metal-nitride-oxide-silicon (MNOS) FET [5]. The schematic image of this device is presented in Fig. 1. In these devices the charge is injected to and stored in traps in the nitride layer near the SiO_2/Si_3N_4 interface. Later silicon-oxide-nitride-oxide-silicon (SONOS) devices with polysilicon gates [6] were developed to increase the retention time of charge trapped in the nitride layer. The schematic image of a SONOS transistor is presented in Fig. 2.

The most common memory elements of present non-volatile memories are the floating gate FETs. These devices contain two gates, as shown in Fig. 3. The floating gate is embedded between two SiO_2 layers (called tunnel and control layers, respectively), and has no external contacts. The control gate is on the top of the upper SiO_2 layer. Information storage in these memories is based on changing the threshold voltage of FETs by high voltage pulses. The actual mechanism is injection of charge to and its storage at the floating gate.

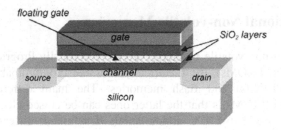

Fig. 3. The floating gate memory transistor

In these types of memories two different charge injection methods are used. One of them is the charge injection from the substrate by direct or Fowler-Nordheim (FN) tunneling [7]. If a voltage pulse with necessary amplitude and width is applied between the control gate and substrate, the electrical charge is injected from the substrate through the bottom thin SiO_2 layer to the floating gate or into the traps in the nitride layer, where it is captured and stored. The captured charge alters the potential distribution within the double insulator layer under the control gate. As a consequence, the threshold voltage of the transistor is changed. Applying a voltage pulse of opposite sign to the gate, either the stored charge can be emitted, or charge with opposite sign can be injected, which compensates the trapped charge. So, the threshold voltage is shifted back to its earlier value. The two different threshold voltage values are attributed to the two logical states.

Reading the information is performed by checking, if the transistor is opened or closed, when a reading voltage with an amplitude between the two different values of threshold voltage is applied to the gate.

The other method is the channel hot electron (CHE) injection. In this case charging pulse is applied to both the drain and control gate. The charging process occurs by ballistic transport of hot electrons throughout the tunnel oxide. Hot electrons are generated by avalanche breakdown of the drain-substrate p-n junction [6].

CHE injection is much faster than FN or direct tunneling. On the other hand, CHE or FN injection requires higher electric fields – and so higher pulse amplitudes – than direct tunneling. As a consequence, tunnel oxide degrades faster in memories using CHE and FN injection, than in memories with direct tunneling [8].

Two different architectures are used in flash memories, the NOR and NAND architecture. In NOR architecture all the cells are parallel connected with a common ground, and the bit lines are directly connected to the drains of memory transistors. In NAND arrays a finite number of transistors (usually 16 or 32) are connected in series between ground and the bit-line. Due to the difference in the organization, the density of NAND arrays is higher, but their speed is much lower than that of NOR arrays. The smallest area of memory cells realized so far in FMs is 0.008 μm^2 using 45 nm technology node [3].

In NOR FMs the information can be read randomly, while in NAND FMs in series only. So, the read time is a few ns or a few μs, respectively. Write speed difference is connected to the different writing mechanisms. In NOR architecture CHE injection is used, while in NAND FN tunneling, which yield write time about 1 μs or 300 μs, respectively [4].

4　Problems with Conventional Non-volatile Memories and the Possible Ways of Solution

As mentioned above, the main part of present flash memories is based on floating gate FETs. The reduction of dimensions is limited in floating gate devices due to defects in the tunnel oxide. The reduction of lateral dimensions requires the reduction of voltage level used, and so the thickness of insulator layers in memory FETs. The main problem in floating gate transistors is that through the defects or weak points of tunnel oxide with reduced thickness the whole amount of stored charge carrying the information can be lost [9].

The other problem of floating gate devices is the drain turn-on effect [9]. It is connected with strong capacitive coupling between the drain and floating gate and between the source and floating gate. So, if the drain voltage is increased, the potential of floating gate increases as well, and the drain current does not saturate.

It has been considered that one of the possible solutions to overcome the above effects is to replace floating gate with separated semiconductor nanocrystals (NCs) [10], which are electrically isolated, as shown schematically in Fig. 4. In this case the loss of information via local defects can be avoided, and the drain turn-on effect is reduced as well [9].

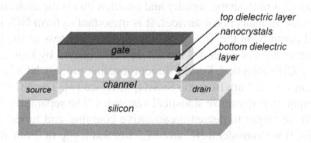

Fig. 4. The nanocrystal memory transistor

Another possible and more perspective way to avoid the above difficulties is the application of Si_3N_4 based devices. In these devices the charge holding the information is stored in the traps of nitride layer, which are electrically isolated by their nature. So, the effect of local defects is reduced significantly [8, 9].

PRAMs, FeRAMs and MRAMs are perspective for solving the above problem and for the further improvement of non-volatile memory performance as well.

5   Nanocrystal and Nitride-Based Memories

As mentioned above, reliability problem of floating gate memories can be over-come, if the floating gate is replaced by a charge trapping layer, i.e., by a properly located sheet of semiconductor or metal NCs [8, 10, 11], or by a Si_3N_4 layer [4, 8]. Another advantage of using a Si_3N_4 layer is that – due to its higher dielectric con-stant – higher electric field will be developed in the tunnel oxide for the same layer thicknesses and voltage pulses, than for a SiO_2 control layer [11]. The higher electric field enhances the charge injection. However, this is partly compensated or even overcompensated by the higher conductivity of Si_3N_4 [12]. Another disad-vantage of nitride based memories in comparison with nanocrystal ones is that the energy level of traps is shallower (1-2 eV), than energy levels in nanocrystals (about 3 eV), which contributes to faster retention behaviour of nitride based structures [7, 8].

In contrast to floating gate FETs, nanocrystal or SONOS transistors make pos-sible information storage of two bits by one memory transistor. Using CHE injec-tion electrons are transferred to and stored in NCs or traps near the drain only. NCs or traps near the source are undisturbed. On the basis of the symmetrical structure of the FET two bits can be stored by changing the polarity of potential between source and drain during charge injection and reading [8, 13]. Multilevel information storage is also possible in these devices, which is similar to analog storage: different threshold voltage values are attributed to the different logical values [14, 15]. However, in the case of nanocrystal memories multilevel storage requires very narrow scatter of NC size and their homogeneous lateral distribution to avoid statistical variations.

The NC size (2-4 nm), shape, density and position inside the dielectric layer are crucial in any case in nanocrystal devices. It is important to form NCs in a sheet at a well-defined depth from the Si/SiO_2 interface. The thickness of the tunnel layer below the sheet of NCs is about 4-8 nm for devices charged by Fowler-Nordheim tunneling or by CHE injection. For the direct tunneling process NCs have to be lo-cated at a distance of 2-3 nm from the Si/SiO_2 interface [3]. The NC density about 10^{12} cm^{-2} is required to minimize statistical variations. The separation of NCs must also be 3-4 nm or larger to decrease capacitive coupling and tunneling between them [3, 8]. So, if we consider a NC size of 2 nm and a gap of 3 nm among them, the channel lenght has to be 30 nm at least (6 NC rows in the channel) for reliable one-bit storage, and about 50 nm (10 NC rows) for two-bit storage [3, 4]. But, degradation is much slower in devices, where the writing process is performed by direct tunneling. However, direct tunneling can be applied for one-bit storage only.

Most common nanocrystals studied for memory transistors are those of Si and Ge [7, 8, 16]. An advantage of Ge NCs in comparison with Si ones is that Ge has narrower band-gap, than Si. So, Ge NCs create deeper energy levels in the dielec-tric layer. However, quantum confiment in Ge is more pronounced, than in Si, due to the difference in permittivity values [3]. From the technological point of view,

Ge can be easily deposited (e.g. by evaporation) [17], but it can evaporate during high temperature processes as well, reacting with SiO_2 and forming GeO, which is volatile [18]. On the other hand, Ge atoms diffuse into the SiO_2 layer easily, in particular in the presence of oxygen vacancies [18]. The other drawback of Ge is that it has not been used in standard CMOS technology [3].

Nitride-based memories seems much better from the point of view of scalability, because the dimensions of potential wells related to traps, which are attributed to dangling bonds, are much smaller, than NCs. Their technological process is simpler either, than that of nanocrystal memories.

Although it seems obvious to merge the advantages of nanocrystals and nitride based memory structures, only very few works has been devoted to nitride based memory structures with embedded nanocrystals [3, 8, 12, 19-25]. It has been obtained in most cases that the presence of nanocrystals improved the charging and/or retention behaviour of the nitride based devices.

6 Phase-Change Memories

Phase-change memories are perspective to replace flash memories in the future. In these memories the memory effect is connected to the change of the conductivity of a chalcogenide layer (mainly GeTe-based ternary or quaternary compounds, as $Ge_2Sb_2Te_5$), when its structure is changed from crystalline to amorphous or back. Amorphous and crystalline states show low and high conductivity, respectively. Amorphous state can be reach by heating the layer above its melting point, and a consecutive fast cooling. For switching to crystalline state the layer has to be heated again to a critical temperature, where the crystallization is going on. The swithching can be performed by Joule heating and cooling by current pulse through the layer. The two states can be reversibly reached. For change to amorphous state short current pulses with high amplitude, while to crystalline phase wider pulses with lower amplitude are required. Read is performed by low current levels [4, 26-28].

Phase change memories have several advantages in comparision with field effect memory transistors discussed above. In phase change memories conductivity values between the two logical states are different by several orders of magnitude. No physical limits of scaling are known. W/E time is about 300 ns and 50 ns, respectively. Their endurance is much better as well [28].

The main disadvantage of phase change memories is the high writing current, which yields high power consumption, and leads to degradation. However, writing current decreases with scaling of technology. The other perspective way for its reduction is to increase the resistance of the chalcogenide layer [28].

7 Ferroelectric Memories

Another perspective alternatives of FMs are the ferroelectric memories. The memory effect in these devices is connected to the electrical polarization of a ferroelectric material under external electric field. As ferroelectric, mainly $PbZr_xTi_{1-x}O_3$ (PZT) is used for memory purposes [4, 29].

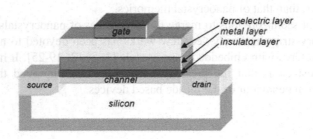

Fig. 5. Floating gate ferroelectric memory transistor

In earlier developments a 1 transistor/1 capacitor cell (similar to a DRAM cell) was used. The information was stored in the capacitor using a destructive reading. Novel 1 transistor cells can be read out non-destructively. In these memories ferroelectric material is replaced between an insulator layer and the metal gate of the FET. The transistor is similar to the MNOS transistor presented in Fig. 1, but the nitride layer is replaced by the ferroelectric layer. The lower insulator layer is necessary to separate the ferrolectric from the substrate during the crystallization of the ferroelectric layer [29].

For the reduction of the polarization W/E voltages and for the improvement of information storage, a floating gate is often inserted between the insulator and the ferroelectric layers, as shown in Fig. 5. In this construction the gate area can be smaller, than the floating gate (see Fig. 5), and so the ratio of floating gate – substrate capacitance to floating gate – control gate capacitance can be increased. The two capacitances act as a capacitive devider during writing and erasing, on one hand. On the other hand, when the gate is grounded, the remanent polarization of the ferroelectric results in the polarization of the bottom insulator layer. This is a depolarization field enhancing information loss. If the capacitance ratio is decreased, the depolarization field is decreased as well. The drawback of this construction is that if there is a leakage current through the bottom insulator layer, it charges up the floating gate, and the information cannot be read out [29].

8 Magnetoresistive Memories

Third future alternatives of FMs are the magnetoresistive memories.The main part of these memories is the magnetic tunnel junction. This means a layered structure of two magnetic materials separated by an insulator layer, which is thin enough for electron tunneling. One of the magnetic layers has a fixed, permanent polarization, while the other can be polarized in parallel or opposite direction to the fixed polarization of the first magnet. The conductivity of the magnetic tunnel junction depends on the direction of polarization of the free magnet [4, 30-32].

The information can be written by different methods. One of them is direct write by currents flowing in word and bit lines at the same time. This method requires high current levels and can lead to false writes. Another method is the toggle mode, which can be applied in more complicated cells only, which contains synthetic antiferromagnets, where two or more magnetic layers are antiparallel coupled [31]. The writing process is acomplished by ceating a rotating magnetic field. The third method is the spin-torque-transfer, which means the polarization by spin-aligned electrons [30]. Another method to decrease writing current is the thermomagnetic writing, when the magnetic material is heated to reduce the energy necessery for polarization [32].

Magnetoresistive memories are much faster, than FMs, and do not degrade. Their retention time is over 10 years [30].

Conclusions

Although phase-change, ferroelectric and magnetoresistive memories face many challenges yet, their excellent properties (speed, scalability, retention and/or redundance) make them real perspective alternatives of flash memories.

References

[1] Parat, K.K.: Flash Memory Technology - Recent Advances and Future Outlook. In: Bath, K.N., DasGupta, A. (eds.) Physics of Semiconductor Devices, pp. 433–438. Narosha Publishing House, New Delhi (2004)
[2] Cappelletti, P.: Flash Memory Reliability. Microelectron. Reliab. 38, 185–188 (1998)
[3] Horváth, Z.J., Basa, P.: Nanocrystal Memory Structures. In: Torchinskaya, T.V., Vorobiev, Y. V. (ed.) Nanocrystals and Quantum Dots of Group IV Semiconductors, ch. 5. American Scientific Publishers (in press)
[4] Bez, R., Camerlenghi, E., Pirovano, A.: Materials and Processes for Non-Volatile Memories, Mater. Sci. Forum (2009) doi:10.4028/www.scientific.net/MSF.608.111
[5] Frohman-Bentchkowsky, D., Lenzlinger, M.: Charge Transport and Storage in Metal-Nitride-Oxide-Silicon (MNOS) Structures. J. Appl. Phys. 40, 3307–3319 (1969)

[6] Lusky, E., Shacham-Diamand, Y., Mitenberg, G., Shappir, A., Bloom, I., Eitan, B.: Investigation of Channel Hot Electron Injection by Localized Charge-Trapping Nonvolatile Memory Devices. IEEE Trans. Electron. Dev (2004) doi:10.1109/TED.2003.823245

[7] Horváth, Zs. J., Basa, P.: Nanocrystal Non-Volatile Memory Devices. Mater. Sci. Forum (2009) doi:10.4028/3-908454-02-6

[8] Rao, R.A., Steimle, R.F., Sadd, M., Swift, C.T., Hradsky, B., Straub, S., Merchant, T., Stoker, M., Anderson, S.G.H., Rossow, M., Yater, J., Acred, B., Harber, K., Prinz, E.J., White Jr., B.E., Muralidhar, R.: Silicon Nanocrystal-based Memory Devices for NVM and DRAM Applications. Solid-State Electron (2004) doi:10.1016/j.sse, 03.021

[9] Compagnoni, C.M., Ielmini, D., Spinelli, A.S., Lacaita, A.L.: Extraction of the Floating-Gate Capacitive Couplings for Drain Turn-On Estimation in Discrete-Trap Memories. Microel. Eng (2006) doi:10.1016/j.mee.2005.09.005

[10] Tiwari, S., Rana, F., Hanafi, H., Hartstein, A., Crabbé, E.F., Chan, K.: A Silicon Nanocrystals-based Memory. Appl. Phys. Lett. 68, 1377–1379 (1996)

[11] Horváth, Z.J.: Semiconductor Nanocrystals in Dielectrics: Optoelectronic and Memory Applications of Related Silicon-based MIS Devices. Current Appl. Phys. (2006) doi:10.1016/j.cap.2005.07.028

[12] Basa, P., Horváth, Z.J., Jászi, T., Pap, A.E., Dobos, L., Pécz, B., Tóth, L., Szöllősi, P.: Electrical and Memory Properties of Silicon Nitride Structures with Embedded Si Nanocrystals. Physica E (2007) doi:10.1016/j.physe, 12.016

[13] Eitan, B., Pavan, P., Bloom, I., Aloni, E., Frommer, A., Finzi, D.: NROM: A Novel Localized Trapping, 2-bit Nonvolatile Memory Cell. IEEE Electron Dev. Lett. 21, 543–548 (2000)

[14] Eitan, B., Cohen, G., Shappir, A., et al.: 4-bit per Cell NROM Reliability. In: IEEE International Electron Devices Meeting, Technical Digest, pp. 547–550 (2005)

[15] Nagel, N., Muller, T., Isler, M., et al.: A New Twin Flash (TM) Cell for 2 and 4 bit Operation at 63 nm Feature Size. In: 2007 International Symposium on VLSI Technology, Systems and Applications (VLSI-TSA), Proceedings of Technical Papers, pp. 90–91 (2007)

[16] Normand, P., Dimitrakis, P., Kapetanakis, E., Skarlatos, D., Beltsios, K., Tsoukalas, D., Bonafos, C., Coffin, H., Benassayag, G., Claverie, A., Soncini, V., Agarwal, A., Sohl, C., Ameen, M.: Processing Issues in Silicon Nanocrystal Manufacturing by Ultra-Low-Energy Ion-Beam-Synthesis for Non-Volatile Memory Applications. Microelectron. Eng. (2004) doi: 10.1016/j.mee.2004.03.043

[17] Basa, P., Molnár, G., Dobos, L., Pécz, B., Tóth, L., Tóth, A.L., Koós, A.A., Dózsa, L., Nemcsics, Á., Horváth, Z.J.: Formation of Ge Nanocrystals in SiO2 by Electron Beam Evaporation. J. Nanosci. Nanotechnol. (2008) doi:10.1166/jnn.2008.A122

[18] Beyer, V., von Borany, J.: Elemental Redistribution and Ge Loss during Ion-Beam Synthesis of Ge Nanocrystals in SiO2 films. Phys. Rev. B (2008) doi: 10.1103/PhysRevB.77.014107

[19] Dai, M., Chen, K., Huang, X., Wu, L., Zhang, L., Qiao, F., Li, W., Chen, K.: Formation and Charging Effect of Si Nanocrystals in a-SiNx/a-Si/a-SiNx Structures. J. Appl. Phys. (2004) doi: 10.1063/1.1633649

[20] Ammendola, G., Ancarani, V., Triolo, V., Bileci, M., Corso, D., Crupi, I., Perniola, L., Gerardi, C., Lombardo, S., DeSalvo, B.: Nanocrystal Memories for FLASH Device Applications. Solid-State Electron (2004) doi:10.1016/j.sse2004.03.012

[21] Choi, S., Yang, H., Chang, M., Baek, S., Hwang, H., Jeon, S., Kim, J., Kim, C.: Memory Characteristics of Silicon Nitride with Silicon Nanocrystals as a Charge Trapping Layer of Nonvolatile Memory Devices. Appl. Phys. Lett. (2005) doi: 10.1063/1.1951060

[22] Huang, S., Oda, S.: Charge Storage in Nitrided Nanocrystalline Silicon Dots. Appl. Phys. Lett. (2005) doi: 10.1063/1.2115069

[23] Tu, C.-H., Chang, T.-C., Liu, P.-T., Weng, C.-F., Liu, H.-C., Chang, L.-T., Lee, S.-K., Chen, W.-R., Sze, S.M., Chang, C.-Y.: Formation of Germanium Nanocrystals Embedded in Silicon-Oxygen-Nitride Layer. J. Electrochem. Soc. (2007) doi: 10.1149/1.2717494

[24] Chen, W.-R., Chang, T.-C., Hsieh, Y.-T., Sze, S.M., Chang, C.-Y.: Appl. Phys. Lett. 91, 102106 (2007) doi: 10.1063/1.2779931

[25] Horváth, Z. J., Basa, P., Jászi, T., Pap, A. E., Dobos, L., Pécz, B., Tóth, L., Szöllősi, P., Nagy, K.: Electrical and Memory Properties of Si3N4 MIS Structures with Embedded Si Nanocrystals. J. Nanosci. Nanotechnol. (2008) doi:10.1166/jnn.2008.A120

[26] Hosaka, S., Miyauchi, K., Tamura, T., Sone, H., Koyanagi, H.: Proposal for a Memory Transistor Using Phase-Change and Nanosize Effects, Microelectron. Eng. (2004) doi: 10.1016/j.mee, 03.044

[27] Wuttig, M., Yamada, N.: Phase-Change Materials for Rewriteable Data Storage. Nature Mater. (2007) doi: 10.1038/nmat2009

[28] Kim, K., Jeong, G.: The Prospects of Non-Volatile Phase-Change RAM, Microsyst. Technol. (2007) doi: 10.1007/s00542-006-0150-y

[29] Ishiwara, H.: Current Status of Ferroelectric-Gate Si Transistors and Challenge to Ferroelectric-Gate CNT Transistors, Current Appl. Phys. (2008) doi: 10.1016/j.cap.2008.02.013

[30] Braganca, P.M., Katine, J.A., Emley, N.C., Mauri, D., Childress, J.R., Rice, P.M., Delenia, E., Ralph, D.C., Buhrman, R.A.: A Three-Terminal Approach to Developing Spin-Torque Written Magnetic Random Access Memory Cells. IEEE Trans. Nanotechnol. (2009) doi: 10.1109/TNANO.2008.2005187

[31] Fukumoto, Y., Nebashi, R., Mukai, T., Tsuji, K., Suzuki, T.: Toggle Magnetic Random Access Memory Cells Scalable to a Capacity of Over 100 Megabits. J. appl. Phys. (2008) doi: 10.1063/1.2826744

[32] You, L., Kato, T., Tsunashima, S., Iwata, S.: Thermomagnetic Writing on Deep Submicron-Patterned TbFe Films by Nanosecond Current Pulse. J. Magnetism Magnetic Mater. (2009) doi: 10.1016/j.jmmm.2008.10.026

Biometric Motion Identification Based on Motion Capture

Ryszard Klempous

Institute of Computer Engineering, Control and Robotics, Wroclaw University of Technology
27 Wybrzeze Wyspianskiego Street, 50-370 Wroclaw, Poland
ryszard.klempous@pwr.wroc.pl

Abstract. Motion tracking, motion capture (or Mocap) is a computer technique of digitally recording movements for sport, entertainment, security, military and medical applications. Biometrics (Greek: *bios* ="life", *metron* ="measure") is the field of study that involves methods of unique recognition of humans based on one or many intrinsic physical or/and behavioral characteristics. In ICT, biometric authentication refers to technologies that analyze and measure human physical and behavioral features for the purpose of authentication. Identification of a person involves extracting of biometric info from a person and then comparing this info with samples stored in database. Identification is more far more complex than verification as the biometric information has to be compared with all other entries in the database in order to provide a definite answer.

Keywords: motion analysis, motion capture, motion representations and analysis, gait recognition.

1   Introduction

The biometrics is the science that studies occurrence of the individual features of live organisms by using mathematical methods. Most commonly it is used in anthropology, medicine, palaeontology, physiology, genetics and farming. The methods based on the biometric identification belong to the most rapidly developing computer technologies. The biometric methods deliver many fast, effective and often non-invasive ways of user authentication. The development of statistical methods and computer system technologies contribute to ever increasing and widespread use of biometrics in the access control systems. According to reports of the International Biometric Group, on the global scale, the total value generated by the biometric systems industry reached almost 1.3 billion Euros. The largest part in this market segment belongs to USA which in 2008 alone reached 1.4 billion dollars [33]. The field of biometrics can be is made of two sections:
- The static biometrics -analysing the man physical and biological features in the aim of his identification,
- The dynamic biometrics (behavioural) -analysing way and the features of behaviour oneself the given individual.

I.J. Rudas et al. (Eds.): Towards Intelligent Engineering & Information Tech., SCI 243, pp. 335–348.
springerlink.com   © Springer-Verlag Berlin Heidelberg 2009

According to the Department of Biometrics at Cornell University, "Biometry is the application of statistics, probability, mathematics, systems analysis, operations research, engineering, computer science, and other areas to studies of the life sciences" [34]. Examples of physical (or physiological or biometric) characteristics include fingerprints, eye retinas and irises, facial patterns and hand measurements, while examples of mostly behavioral characteristics include signature, gait and typing patterns. Practical applications of biometrics include:

- Access control and privileges checking
- Intruder/suspect identification
- Security systems (personal tracking)
- Medical data analysis (diagnosis/treatment

All behavioral biometric characteristics have a physiological component, and, to a lesser degree, physical biometric characteristics have a behavioral element. In some papers, authors have coined the term behaviometrics for behavioral biometrics such as typing rhythm or mouse gestures where the analysis can be done continuously without interrupting or interfering with user activities. For a reliable verification process, the dynamic methods require the continuous registration of the chosen feature of behaviour. The following features are the subject of the analysis in the most common biometric methods:

- The profile of the voice. Every person possesses the unique voice which depends on sex, the building of body, the shape of the mouth itp. Verification is holds through the comparison the expressed, settled earlier phrase with pattern kept in the base of the data. The system of the identification of the voice takes into account about 20 the parameters of the voice, such as: depth, dynamics, the shape of the wave, etc.

- The written signature - this is one from the oldest methods of the person verification. Modern systems analyse the not only visual profile of the signature, but also the way how he became folded. It complies with the special feathers and graphic tablets which allow reading with the aim of the delivery fee of given to the analysis or even the change of the pressure exerted on tablet.

- The profile of writing on the keyboard. National Science Foundation and National Institute of Standards and Technology proved that the way people type using the keyboard was unique for every man. Methods of analysing the key strokes and profiles of writing using the keyboard allow monitoring entries on the keyboard reaching 1000 strokes a minute.

Other common biophysical features, which can be used as a source for pattern recognition, are:

- face recognition
- body (face) thermogram
- fingerprints

- eye-based features
- palm features
- motion analysis

2 Motion Capture Technique

Motion Capture (MoCap, MC) [2, 10, 12] is the process of recording alive figures movement, and then transforming him to the digital figure in the aim of analysis or reconstruction. Intercepted given the information can contain about the changes of the position in the space of specific points on the actor whole body or only information about the movements of face or chosen muscles. The techniques of the interception of the movement deliver the very precise data which processed suitably find the use in many fields, as:

- Computer Graphics – allow shortening the time for creation of realistic animations simultaneously thus considerably reducing the development costs.
- The medical application systems help doctors work they are able to detect the disorders of the equilibrium which can lead to the Parkinson's disease or other similar diseases. MoCap systems can also help in the process of rehabilitation, generating exact information about a patient's progress to the doctor.
- The sport system MoCap allows generating the detailed analysis of an athlete movements, thanks to what significant improvements can be made in sportsman's performance.
- The biometrics can be used for remote identification of various people.
- Actor's motion can be captured using various types of sensors and can be recorded in 3D virtual space.
- Data can be saved using predefined representation taking into consideration the skeleton's hierarchy
- Using MC technique many subtle details can be recorded, which normally can be very difficult to simulate using analytical motion model.

There are two most often used types of this technique. In the first case reflective markers are fixed on joints of a live actor and the motion of markers is tracked. In the latter case magnetic sensors are fixed on actor joints. These sensors are tracking disturbances of magnetic field during motion. In order to achieve realistic animation there is recorded motion of each human joint. This causes that it isnecessary to describe motion with a large set of data. Such data are hard to process in some fields of applications. This problem is especially visible in use of multimedia databases. Managing the tremendous amounts of data is often supported by clustering and classification methods. It is not easy to find such methods for motion sequences. In the proposed approach we try to solve this problem. In consecutive sections we describe problems and propose solutions that make up the method of motions clustering and classification.

2.1 The Possibility of Human Identification-Based Motion Capture 3D Technique

Using 3D systems Motion Capture for human identification is not a simple task. The process of the registration of the data about the movement has several very specific requirements:

The recording should be held the tudio in special, like by the means of the special costume.

The person identified should have well-fitting clothes on her in the measure.

In the reality it is rather hard to meet these requirements. What represents serious limitations for wider utilization of the Motion Capture technology for the identification of human figures is the fact that these systems to would have to be installed in the separate areas, where people could cross-walk in the definite spaces of the time. The registration of the human movement is using a single 2D camera. This offers possibility of using such a system in an open surface area. In many studies [24, 25, 26, 27] it was demonstrated that motion capture systems were able to identify people from the considerable distance using the painting registered by the single camera of the video effectively. Most of these systems were using the edge detection algorithms that first recognise the figure shape in the registered recording and then find characteristic points of the pattern related to the skeleton of the human body. The changes of their position are recorded and compared from given placed in the database.

2.2 The Principles of 3D Motion Capture Systems

In systems MoCap to the characteristic points of the actor's body special sensors called the markers are attached. To be also executed this be by the means of the special costume which already has built-in markers. The quantity of attached sensors depends on the level of the realism of the animation what we want to reach and from the computational power of the computer, used to the processing intercepted given. Using the digital cameras, what the definite individual of the time (called frame), it registers the changes of their position. These data are sent to the computer which processes them then one receives the figure three-dimensional model in the result. There are several types of Motion Capture systems. Among the most popular are:

- *Optical Systems.* In optical systems to the actor's body the small markers are attached which reflect visible or invisible for the man light generated through special devices.
- *Inertial systems.* Inertial systems make possible the registrations of the movements of the whole body in simple and profitable way.
- *Electromagnetic Systems.* Positions their change are recorded on the basis of the changes of the electromagnetic field which allows to the exact qualification of distance and the orientation of given sensor.

2.3 The Data Standards of in Motion Capture Systems

The development of the Motion Capture technology manifests itself in the formation of ever more advanced systems of the interception of the movement. This is due to applying various models of hierarchic skeletons and their dimensionality. The format created through the company Biovision is one of the most widespread formats - .bvh. The equally popular way of data recording was promoted by Acclaim. Intercepted data are written down in two files are given - .asf and .acm. First describes the applied model of the skeleton however information about the positions of ponds is placed in the second. The format of the Acclaim company has a more technical character [32]. The exchanged standards of the record of the data are the most widespread and the majority of the packets of graphic applications operate them. The BVH file format was originally developed by Biovision, a motion capture services company, as a way to provide motion capture data to their customers. The name BVH stands for Biovision hierarchical data. A BVH file has two parts, a header section which describes the hierarchy and initial pose of the skeleton; and a data section which contains the motion data. After this hierarchy is described it is permissible to define another hierarchy, this too would be denoted by the keyword "ROOT". In principle, a BVH file may contain any number of skeleton hierarchies.

3   Data Representations

- Motion capture is an attractive method for creating movements for computer animations.
- Application of quaternion based model leads to taking into consideration important motion features
- Proposed method of extended motion model gives a very interesting results
- Application of filtering for motion sequences (equivalent of Gaussian blur for a quaternion) results in significant improvement of clustering quality.

Actor's motion is captured using different kinds of sensors and recorded in 3D virtual space. In this technique many subtle details can be recorded, which are difficult to simulate using analytical motion model. Data are saved using proper representation taking into consideration skeleton hierarchy

3.1 Related Works Concerning Biometric Motion Analysis

In every biometric problem it is crucial to distinguish basic properties of the examined object. In the considered problem, motion is the object which is taken into account. The basic approach is to treat motion as a multi dimensional time-series of rotations (or positions) of each bone in the skeleton. It is also possible to compute accelerations or velocities of different parts of the body. The question is whether all parts have to be taken into account. Murray [27] suggested that the motion of hips and legs differs significantly among different people. In the paper [2] rotations of shanks, hips and the neck are considered. The approach has similar problems as the previous – those parts of the body are usually hidden under the clothes. On the contrary another part which could be analyzed are legs. The approach is presented in [3, 4, 5, 6, 7]. The main advantage is the possibility to determine the position of these parts of the body also on the image. Legs usually move in more regular way comparing to other parts of the body (like hands). The motion is more periodical, which is not interfered with other behaviors like carrying or gesticulations. Taking into consideration those conditions, we decided to examine the motion of legs, especially two bones: hips and shanks. We use motion capture data obtained from the commercial system to check the performance of the presented methods. In all experiments presented in this paper we use *LifeForms (LifeForms* software is produced by *Credo Interactive Inc. [21])* motion library, which contains different action sequences played by real actors and processed using *motion capture* technique1.

3.2 Data Representation

The following data representation techniques are mostly used in Motion Capture applications:

- Absolute translations
- Relative translations
- Euler's angles (transformation matrices)
- Unit quaternions
- Multidimensional time-series
- Space-time constraints

The natural way to represent data of human motion is to describe rotations of each bone. Rotations can be expressed as the vector of three variables:

$$\overline{m}_k(t) = \begin{vmatrix} x_k(t) \\ y_k(t) \\ z_k(t) \end{vmatrix} \tag{3.1}$$

where k denotes a bone in the skeleton (Fig. 1) and t denotes time moment (e.g. frame number) in the sequence. Complete rotation is described as a rotation around three axes independently.

This type of representation is called *Euler's rotations*. The main disadvantage of Euler's rotations is the necessity to preserve the same order of the axes. The simplification of the approach is to analyze motion which is performed into the perpendicular direction to the camera. This is the situation which takes place during video recording analysis (using the Hough transformation [35] it is fairly easy to distinguish moving legs of the person). The simplification was proposed in [6] and we base our experiments on that approach.

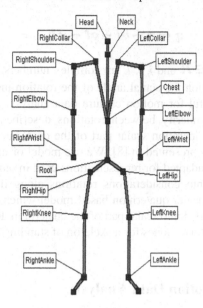

Fig. 1. Complex skeleton representation [31]

The novelty presented in our paper is using similar approach both for frequency as well as for time-series analysis of motion capture data. Moreover we propose to use also different data representation. We base our experiments on three dimensional motion capture data.

In *Absolute Translations* (AT) representation, each joint position is stored in each frame as translations which are absolute in regard of global coordinates. This approach has several limitations. All further transformations are very difficult and there is a risk of changing bones lengths. Moreover analysis of time-series data depends on the hierarchy. Some solutions to the limitations of both absolute translations and Euler's angles are obtained computing Timmer splines [36] out of raw data. Nevertheless, computed parameters are only an approximation of original data. To overcome those constraints, data can be saved in a relative way. It leads to *Relative Translations* (RT) representation. On the contrary to the AT, translations for each dimension concern a bone with reference to its local coordinates. Trajectory obtained in such a way is invariant with regard to transformations of other bones in skeleton.

3.3 Quaternions

One of the most convenient alternative representations for the rotation is a unit quaternion. Quaternions are generalization of complex numbers proposed by Hamilton [8]. The most important feature is that the unit quaternion (quaternion of the length 1) can represent rotations. More detailed application of quaternions for three-dimensional transformations can be found in [22].

$$q = a + bi + cj + dk \tag{3.2}$$

where: $a, b, c, d \in R$ and i, j, k are complex numbers. Using quaternions we can take into consideration the actual angle of the rotation instead of rotation along one axis. It can be useful for motion capture data analysis. We can also define a specific measure of similarity between rotations described by quaternions. The measure takes into consideration scalar part of the quaternion and has some interesting features as it was presented in [8]. We use model of motion of legs based on [10, 19, 20] which was adapted for representations used in our experiments. In some experiments we take into considerations rotations along different axes. For the comparison we also consider quaternion based model which uses adjusted information from all axes converted into scalar part of the quaternion. Rotations are measured against the initial position of bones for a skeleton of standing person.

4   Methods of Motion Data Analysis

One propose to use *Dynamic Time Warping (DTW)* [5, 6, 18] method to compare motion sequences. The method is based on dynamic programming and is widely used for different time-series comparison applications (like voice recognition [28]). The application for motion processing was presented in different papers [9, 11].

One propose to use *Dynamic Time Warping (DTW)* [5, 6, 18] method to compare motion sequences. The method is based on dynamic programming and is widely used for different time-series comparison applications (like voice recognition [28]). The application for motion processing was presented in different papers [9, 11]. The proposal of database structure based on DTW was proposed in [23]. Two experiments were prepared. In each we selected one motion capture sequence called *armout.bvh* as a template sequence (it is one of the most standard motions representing walking along a straight line). Each motion from the group of sequences was compared to the template. The warping cost represents a measure of similarity between sequences. The properties of comparison algorithm allow to process signals with different lengths. In the first experiment warping costs for rotations of a hip along three axes were computed. The results are presented in Fig. 2.

Warping costs for different rotations are placed on different axes. This method of data visualization is based on paper [7]. The novelty is to use it for *DTW* analysis of motion capture data. The closer points are placed on the graph, the more similar are analyzed sequences. Similar experiment was carried out using different parameters. This time we analyzed only *X* rotations, but taking into consideration different parts of the body: left hip, right hip and left shank. Because of the asymmetrical nature of such comparison, it is possible to detect any disorders in motion disturbing symmetrical character of normal gait.

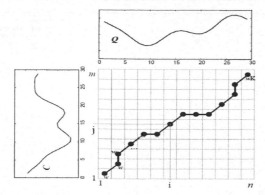

Fig. 2. Typical warping path; C – template signal, Q – test signal [12, 14]

4.1 Spectrum Analysis

There are some suggestions that spectrum analysis of the motion signal can lead to interesting results concerning person identification [3 – 7]. Spectrum of the signal can be computed using Fourier transform defined:

$$F_k = \sum_{k=-\infty}^{\infty} f_k e^{jk\omega_0 t} \tag{4.1}$$

where

$$\omega_0 = \frac{2\pi}{T} = 2$$

$$f_k = \frac{1}{T} \int_{t_0}^{t_0+T} f(t) e^{-jk\omega_0 t} dt, \quad k = 0, \pm 1, \pm 2, \dots$$

and

$$f_k = \sum_{k=-\infty}^{\infty} F_k e^{-jk\omega_0 t} \tag{4.2}$$

Discrete Fourier Transformation (DFT) is give by formula below:

$$X_k = \sum_{n=0}^{N-1} x_n e^{\frac{-2\pi i}{N}kn}, \quad k = 0, 1, \dots, N - 1 \tag{4.3}$$

Computed in this way spectrum can be afterwards analyzed and be the source of interesting biometric information. Kuan [17] showed that the spectrum parameters differ significantly for different persons. In [6, 7] the approach was extended with the application of phase-weighted magnitude. We use similar data visualization

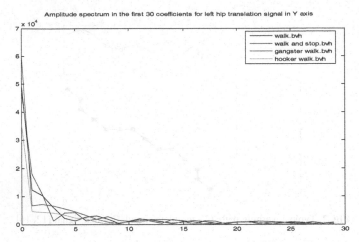

Fig. 3. A spectrum analysis for sequences SprintTo Walk and RunTo Walk [31]

in order to compare the results with other results The example of frequency analysis for two different sequences is presented in Figure 3. Both spectrums were computed for rotational signals along X axis of the right hip. As one can observe, the result of spectrum analysis differs significantly for two sequences, which differs only with some details (*Sprint* and *Walk* represent motions of the same type).

The application of technique Motion Capture in the systems of the movement identification is the difficult task, that is why should also analyse the possibility of use the 2D video cameras systems. The data was given over from four files BVH:

- walk.bvh,
- walk and stop.bvh,
- gangster walk.bvh,
- hooker walk.bvh

All of them describe the walking actor [21].The values of the deviations of the from settled planes figure chosen joint were marked then. It limited oneself only to bottom limbs, because they execute the periodical movement on which executed by studied actions have the small influence.

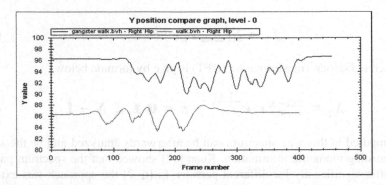

Fig. 4. Comparing right hip shifts from the surface 0 for a *gangster's walk.bvh, walk.bvh[31]*.

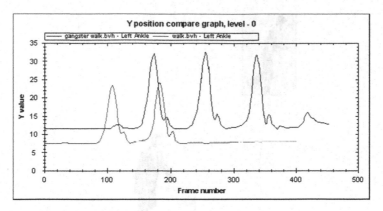

Fig. 5. Comparing left ankle shifts from the surface 0 for a *gangster's walk.bvh, walk.bvh[31]*

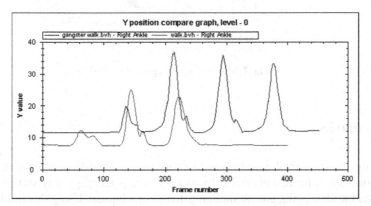

Fig. 6. Comparing right ankle shifts from the surface 0 for a *gangster's walk.bvh, walk.bvh[31]*.

The axis X represents the consecutive frames of the captured sequence and on the axis Y denotes the height of the given joint (Figs. 4-6) [31]

4.2 Experiment: Multidimensional Spectrum Analysis

One of the most difficult tasks is to distinguish noised group of motions. Some elements of other groups do not form a clearly defined cluster. It is possible to indirectly check the performance of the comparison method. The idea is to run clustering algorithm and then to check the quality of group clustering. It is feasible to apply this idea, because the origin of each element is known before clustering. Generated groups of amplitude coefficients can give much better results. The next step is to find a method which could deal with very similar motions like *Run-To Walk* and *JogTo Walk*. Those motions are so similar that even taking into account many amplitude coefficients would not give satisfactory recognition (Fig. 7). The idea is to use combined time-spectrum analysis. It could join the result obtained from *Dynamic Time Warping* comparison and spectrum analysis [4, 29, 30].

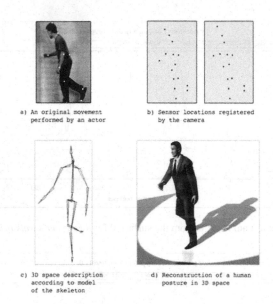

a) An original movement
performed by an actor

b) Sensor locations registered
by the camera

c) 3D space description
according to model
of the skeleton

d) Reconstruction of a human
posture in 3D space

Fig. 7. Motion Capture application [13]

5   Remarks

Human motion is a complex process and its various characteristics can be used in biometric analysis. Information carried by motion of legs and hips may serve as valuable basis for identifying unique motion patterns of a human. Using this information allows discovery and identification of captured human motion. Developing methods of such analysis will enable the future solutions for efficient identification of a person from a distance. This work presents possibilities created by applying various methods of motion analysis and suggests directions of advancing biometric identification. Motion capture systems have been under development for entertainment industry and medical applications. Desire to limit the cost of movie production stimulates the research that aim at improving methods of computer animation. Recordings gathered by motion capture systems make it easier to create animated creatures. Since human motion is a very complex process, it is necessary to distinguish most important features which would be useful for biometric analysis. The entire process of the identification is based on those features, so the choice of these features is crucial for the reliability of the whole system. One of the most common approaches is to consider human motion as a group of signals which are functions variable of time. Its proper representation is also very important and can affect on properties of other algorithms [1]. Murray [26] found that thigh and legs motions are very different for each person. In [3] rotations of thigh, shanks and hips vibrations are taken into account relatively to the position of the neck. The approach has similar limitation to the previous one: both thighs as well as neck are often very difficult to distinguish during motion recording. It is difficult to find their exact position because of the covering of clothes (in methods different than motion capture). An interesting modification of the inner measure of similarity was proposed in [12, 13]. In this chapter the most important approaches concerning motion analysis were presented. The aim was to check the possibility to use selected motion parameters to human identification. Motion capture is an attractive technique for creating movements for computer animations. Application of quaternion based model leads to taking into consideration important motion features. The proposed method of extended motion model as well as the application of filtering for motion sequences (equivalent of Gaussian blur for a quaternion) [11, 15] provides very interesting results and a significant improvement of clustering quality. Human motion can be an interesting source of information for a study of biometric posture recognition. The motion analysis of legs seems to be the most important for studying of biometric gate recognition. Both frequency based analysis (FFT) as well as time-series analysis (DTW) are powerful tools for biometric identification. Application of quaternion based model leads to taking into consideration important motion features. Application of filtering for motion sequences (equivalent of Gaussian blur for quaternions) provides a significant improvement of clustering quality while a combination of information from time-series and frequency analysis remains a challenging area of biometric motion identification research.

References

[1] Barr, A., Currin, B., Gabriel, S., Houghes, J.: Smooth Interpolation of Orientations with Angular Velocity Constraints. Computer Graphics 26 (1992)
[2] Bodenheimer, B., Rose, C., Rosenthal, S., Pella, J.: The Process of Motion Capture: Dealing with the Data (1997)
[3] Bruderlin, A., Williams, L.: Motion Signal Processing. Computer Graphics 29, 97–104 (1995)
[4] Keogh, E., Ratanamahatana, C.A.: Exact Indexing of Dynamic Time Warping, University of California–Riverside, USA China (2002)
[5] Chaczko, Z., Sinha, S.: Strategies of Teaching Software Analysis and Design Interactive Digital Television Games. In: ITHET 2006, Sydney, Australia (July 2006)
[6] Cunado, D., et al.: Automatic Extraction and Description of Human Gait Models for Recognition Purposes. Computer Vision and Image Understanding, 90(1) (2003)
[7] Cunado, D., Nixon, M.S., Carter, J.N.: Using Gait as a Biometric, via Phase-Weighted Magnitude Spectra. In: Proc. of 1st Int. Conf. on Audioand Video-based Biometric Person Authentication (2003)
[8] Eberly, D.: Quaternion Algebra and Calculus. Magic Software (2002),
http://www.magicsoftware.com/Documentation/
Quaternions.pdf
[9] Itakura, F.: Minimum Prediction Residual Principle Applied to Speech Recognition. IEEE Trans on Acoustics Speech and Signal Processing AS23.1, 67–72 (1975)
[10] Jablonski, B.: Porownywanie generatorow ruchu animowanych postaci ludzkich. In: Klempous, R. (ed.) Diversitas Cybernetica, Warsaw (2005)
[11] Jablonski, B., Klempous, R., Kulbacki, M.: PDE-based Filltering of Motion Sequences. Journal of Comp. and Applied Mathematics 189(1/2), 660–675 (2006)
[12] Jablonski, B., Klempous, R., Majchrzak, D.: Models and Methods for Biometric Motion Identification. Annales UMCS, vol. 4, Lublin (2006)

[13] Jablonski, B., Klempous, R., Majchrzak, D.: Feasibility Analysis of Human Motion Identification Using Motion Capture. In: Proc. of the Int. Conference on Modelling, Identification and Control. Acta Press (2006)

[14] Jablonski, B., Kulbacki, M., Klempous, R., Segen, J.: Methods for Comparison of Animated Motion Generators. In: Proc. of IEEE International Conference on Computational Cybernetics, Hungary, (2003)

[15] Jablonski, B.: Anisotropic Filtering of Multidimensional Rotational Trajectories as a Generalization of 2D Diffusion. In: Multidimensional Systems and Signal Processing. Springer, Netherlands (2008)

[16] Jablonski, B., Kulbacki, M., Klempous, R., Segen, J.: Methods for Comparison of Animated Motion Generators. In: Proc. ICCC 2003, Siófok, Hungary (2003)

[17] Kuan, E.L.: Investigating Gait as a Biometric, Technical Report, Departament of Electronics and Computer Science, University of Sothampton (1995)

[18] Kulbacki, M., Bak, A., Segen, J.: Unsupervised Learning Motion Models Using Dynamic Time Warping. In: Intelligent Information Systems 2002, Sopot, Poland, June 3-6, pp. 217–226 (2002)

[19] Kulbacki, M., Jablonski, B., Klempous, R., Segen: Learning from Examples and Comparing Models of Human Motion. Journal of Advanced Computational Intelligence and Intelligent Informatics 8(5), 477–481 (2004)

[20] Lee, J., Shin, S.Y.: Multiresolution Motion Analysis with Applications. In: International Workshop and Human Modeling and Animation, Seoul, pp. 131–143 (2000)

[21] LifeForms, Credo Interactive Inc., http://www.credo-interactive.com/

[22] Maillot, P.: Using Quaternions for Coding 3D Transformations. In: Graphics Gems I., pp. 498–515. Academic Press Inc., Boston (1990)

[23] Majchrzak, D.: Identy_kacja ruchu postaci. Analiza mo_zliwo_sci, metody, algorytmy", M.Sc.Eng., Thesis (in Polish) Wroclaw University of Technology, Faculty of Electronics (2000)

[24] Myers, C.S., Rabiner, L.R.: A Comparative Study of Several Dynamic Time-Warping Algorithms for Connected Word Recognition. The Bell System Technical Journal 60(7), 1389–1409 (1981)

[25] Mowbray, S.D., Nixon, M.S.: Automatic Gait Recognition via Fourier Descriptors of Deformable Objects. In: Proc. of Audio Visual BiometricPerson Authentication (2003)

[26] Murray, M.P.: Gait as a Total Pattern of Movement. Amer. J. Phys. Med. 46(1) (1967)

[27] Murray, M.P., Drought, A.B., Kory, R.C.: Walking Patterns of Normal Men. J. Bone Joint Surg. 46-A (2), 335–360 (1996)

[28] Rabiner, L.R., Juang, B.: Fundamentals of Speech Recognition. Prentice-Hall, Englewood Cliffs (1993)

[29] Yam, C.Y.: Automated Person Recognition by Walking and Running via Model-based Approaches. In: Pattern Recognition, vol. 37(5). Elsevier Science, Amsterdam (2003)

[30] Yam, C.Y., Nixon, M.S., Carter, J.N.: Extended Model-based Automatic Gait Recognition of Walking and Running. In: Bigun, J., Smeraldi, F. (eds.) AVBPA 2001. LNCS, vol. 2091, pp. 278–283. Springer, Heidelberg (2001)

[31] Grzegorek, T.: Motion Capture based identification methods using Life Forms database, Wroclaw University of Technology, Faculty of Electronics (2008)

[32] http://www.darwin3d.com/gamedev/articles/col0198.pdf

[33] http://www.claremontvc.com/images/focus1.pdf

[34] http://www.biom.cornell.edu/

[35] http://www.markschulze.net/java/hough/

[36] http://www.books.google.pl/books?isbn=3527406239

Through Wall Tracking of Moving Targets by M-Sequence UWB Radar

Dušan Kocur, Jana Rovňáková, and Mária Švecová

Dept. of Electronics and Multimedia Communications, Technical University of Košice
Letná 9, Košice, Slovakia
dusan.kocur@tuke.sk

Abstract. In this paper, the trace estimation method for through wall moving target tracking by M-sequence UWB radar is described as a complex procedure consists of such phases as raw radar data pre-processing, background subtraction, detection, time of arrival estimation, wall effect compensation, localization and tracking itself. The significance of the particular phases will be firstly given and then, a review of signal processing methods, which can be applied for the phase task solution, will be presented. The trace estimation method performance is demonstrated based on UWB radar signal processing obtained for scenario represented by through concrete wall tracking of single moving target. The obtained results confirm the excellent performance of the method. In the contribution, the extension of the trace estimation method for multiple target tracking is also outlined.

1 Introduction

Electromagnetic waves occupying a spectral band below a few GHz show reasonable penetration through most typical building material, such as bricks, wood, dry walls, concrete and reinforced concrete. This electromagnetic wave penetration property can be exploited with advantage by UWB radars operating in a lower GHz-range base-band (up to 5 GHz) for through wall detection and tracking of moving and breathing persons [2]. There are a number of practical applications where such radars can be very helpful, e.g. through wall tracking of moving people during security operations, through wall imaging during fire, through rubble localization of trapped people following an emergency (e.g. earthquake or explosion) or through snow detection of trapped people after an avalanche, etc.

There are two basic approaches of through wall tracking of moving target. The former approach is based on radar imaging techniques, when the target locations are not calculated analytically but targets are seen as radar blobs in gradually generated radar images [1]. For the radar image generation, different modifications of back-projection algorithm can be used [10], [4], [19]. With regard to fundamental idea of

I.J. Rudas et al. (Eds.): Towards Intelligent Engineering & Information Tech., SCI 243, pp. 349–364.
springerlink.com   © Springer-Verlag Berlin Heidelberg 2009

the method - the radar image generation based on raw radar data, the method is sometimes referred as imaging method. In order to detect, localize and track the moving target, the methods of image processing have to be applied in the case of imaging method.

The later approach of through wall tracking of moving target by using M-sequence UWB radar equipped with one transmitting and two receiving antennas has been originally introduced in [14]. Here, target coordinates as the function of time are evaluated by using time of arrival (TOA) corresponding to target to be tracked and electromagnetic wave propagation velocity along the line transmitting antenna-target-receiving antenna. According [14], moving target tracking, i.e. determining target coordinates as the continuous function of time, is the complex process that includes following phases-tasks of radar signal processing: raw radar data pre-processing, background subtraction, detection, TOA estimation, localization and tracking itself. Because TOA values taken on the corresponding observation time instances form so-called target trace [14], this method is referred as the trace estimation method.

The rough comparison of the imaging and trace estimation method performance has shown that the both methods can provide almost the same precision of through wall tracking of moving target. For target tracking, the imagining method uses 2D signal processing methods (image processing) mostly, whereas the trace estimation method is based on 1D signal processing. Therefore, the similar precision of both methods is reached at cost of the higher complexity of imagining method in comparison with that of trace estimation method.

With regard to these facts, we consider the estimation method of through wall target tracking to be new and very perspective. In [14], the basic principle of this method has been outlined only. Therefore, the key intention of this contribution is to provide a clear and fundamental description of trace estimation method as the promising approach for through wall tracking of moving target. Besides, method fundamental outlined in [14] will be extended by a wall effect compensation method and target tracking by using linear Kalman filters.

In order to fulfill this intention, our contribution will have the following structure. In the next section, a real through wall scenario of single target tracking will be described. As the radar device considered for the scenario, M-sequences UWB radar equipped with one transmitting and two receiving antennas will be used [2], [17]. In the Sections 3-9, the particular phases of trace estimation methods (i.e. raw radar data pre-processing, background subtraction, detection, TOA estimation, wall effect compensation, localization and tracking itself) will be described. In these sections, the significance of the particular phases will be given firstly and then, a review of signal processing methods, which can be applied for the phase task solution, will be presented. Finally, the outputs of the particular phases will be subsequently illustrated by processing of real raw radar data obtained according to the scenario outlined in the Section 2. Because of the limited range of the contribution, the detail descriptions of the signal processing algorithms used within Sections 3-9 will not be presented. Instead of them, a number of source references devoted to particular algorithms will be given. We believe, that this approach can help readers to see trace estimation inside very clearly and then by using the references, a reader can understand the proposed radar signal processing methods in details. Conclusions and final remarks concerning the next research in the field of through wall tracking of moving target by UWB radar will be drawn in Section 10.

2 Basic Scenario of through Wall Tracking of Moving Target

The basic scenario analyzed in this contribution is outlined in Fig. 1. The target has been represented by a person walking along perimeter of rectangular room with size 3.9 m x 2.6 m, from Pos. 1 through Pos. 2, Pos. 3 and Pos. 4 to Pos. 1. The walls of the room (Fig. 2b) were concrete with thickness of 0.5 m and 0.27 m, relative permittivity $\varepsilon_r = 5$ and relative permiability $\mu_r = 1$.

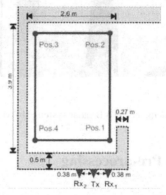

Fig. 1. Measurement scenario. A person was walking along perimeter of the room from Pos.1 through Pos. 2, Pos. 3 and Pos. 4 back to Pos. 1

The raw radar data analyzed in this contribution were acquired by means of M-sequence UWB radar with one transmitting and two receiving channels [2], [17], [18]. The system clock frequency for the radar device is about 4.5 GHz, which results in the operational bandwidth of about DC-2.25 GHz. The M-sequence order emitted by radar is 9, i.e. the impulse response covers 511 samples regularly spread over 114 ns. This corresponds to an observation window of 114 ns leading to an unambiguous range of about 16 m. 256 hardware averages of environment impulse responses are always computed within the radar head FPGA to provide a reasonable data throughput and to improve the SNR by 24 dB. The additional software averaging can be provided by basic software of radar device. In our measurement, the radar system was set in such a way as to provide approximately 10 impulse responses per second. The total power transmitted by radar was about 1 mW.

The radar has been equipped by three double-ridged horn antennas placed along line (Fig. 2a). Here, one transmitting antenna has been located in the middle between two receiving antennas. During measurement, all antennas were placed 1.25

m elevation above the floor and there was no separation between the antennas and the wall. The distance between adjacent antennas was set to 0.38 m (Fig. 1).

Raw radar data obtained by measurement according to above described scenario can be interpreted as a set of impulse responses of surrounding, through which the electromagnetic waves emitted by the radar were propagated. They are aligned to each other creating a 2D picture called radargram, where the vertical axis is related to the time propagation (t) of the impulse response and the horizontal axis is related to the observation time (τ).

(a)   (b)

Fig. 2. (a) Experimental M-sequence UWB radar system, (b) Measurement room interior

3   Raw Radar Data Pre-processing

The intention of the raw radar data pre-processing phase is to remove or at least to decrease the influence of the radar systems by itself to raw radar data. In our contribution, we will focus on time-zero setting.

In the case of M-sequence UWB radar, its transmitting antenna transmits M-sequences periodically around. The exact time instant at which the transmitting antenna starts emitting the first elementary impulse of M-sequence (so-called chip) is referred to as time-zero. It depends e.g. on the cable lengths between transmitting/receiving antennas and transmitting/receiving amplifiers of radar, total group delays of radar device electronic systems, etc., but especially on the chip position at which the M-sequence generator started to generate the first M-sequence. This position is randomly changed after every power supply reconnecting. To find time-zero means rotate all received impulse responses in such a way as their first chips correspond to the spatial position of the transmitting antenna. There are several techniques for finding the number of chips needed for such rotating of impulse responses. Most often used method is that of utilizing signal cross-talk [27]. The significance of the time-zero setting follows from the fact that targets could not be localized correctly without the correct time-zero setting.

The examples of radargrams obtained by the measurement according to scenario given in the Section 2.1 with correct time-zero setting utilizing signal cross-talk method are given for the first receiving channel (Rx_1) and the second receiving channel (Rx_2) in Figs. 3a and 3b, respectively.

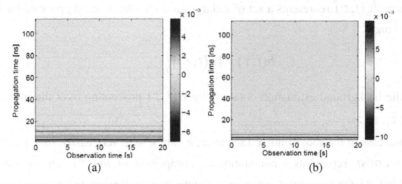

Fig. 3. Pre-proccessed radargram. (a) Receiving channel Rx_1, (b) Receiving channel Rx_2

4 Background Subtraction

It can be observed from Figs. 3a and 3b, that it is impossible to identify any target in the radargrams. The reason is the fact, that the components of the impulse responses due to target are much smaller than that of the reflections from the front wall and cross-talk between transmitting and receiving antennas or from other large or metal static object. In order to be able to detect, localize and track a target, the ratio of signal scattered by the target to noise has to be increased. For that purpose, background subtraction methods can be used. They help to reject especially the stationary and correlated clutter such as antenna coupling, impedance mismatch response and ambient static clutter, and allow the response of a moving object to be detected.

Let us denote the signal scattered by the target as $s(t,\tau)$ and all other waves and noises are denoted jointly as background $b(t,\tau)$. Let us assume also that there is no jamming at the radar performance and the radar system can be described as linear one. Then, the raw radar data can be simply modeled by following expression:

$$h(t,\tau) = s(t,\tau) + b(t,\tau), \tag{1}$$

As it is indicated by the name, the background subtraction methods are based on the idea of subtracting of background (clutter) estimation from pre-processed raw radar data. Then, the result of the background subtraction phase can be expressed as

$$h_b(t,\tau) = h(t,\tau) - \hat{b}(t,\tau) = s(t,\tau) + [b(t,\tau) - \hat{b}(t,\tau)], \qquad (2)$$

where $h_b(t,\tau)$ represents a set of radargram with subtracted suppressed background and

$$\hat{b}(t,\tau) = [h(t,\tau)]_{\tau_1}^{\tau_2} \qquad (3)$$

is the background estimation obtained by $h(t,\tau)$ processing over the interval $\tau \in <\tau_1, \tau_2>$.

In the case of the above outlined scenario, it can be seen very easily, that $s(t,\tau)$ for $t = const.$ represents a non-stationary component of $h(t,\tau)$. On the other hand, $b(t,\tau)$ for $t = const.$ represents a stationary and correlated component of $h(t,\tau)$. . Therefore, the methods based on estimation of stationary and correlated components of $h(t,\tau)$ can be applied for the background estimation.

Following this idea, the methods such as basic averaging (mean, median) [11], exponential averaging [28], adaptive exponential averaging [28], adaptive estimation of Gaussian background [26], Gaussian mixture method [20], moving target detection by FIR filtering [9], moving target detection by IIR filtering [10], prediction [25], principal component analysis [24], etc. can be used for background subtraction. These methods differ in relation to assumptions concerning clutter properties as well as to their computational complexity and suitability for online signal processing.

Because of simplicity of the scenario discussed in this contribution, a noticeable result can be achieved by using e.g. the simple exponential averaging method where the background estimation is given by

$$\hat{b}(t,\tau) = \alpha \hat{b}(t,\tau-1) + (1-\alpha)h(t,\tau) \qquad (4)$$

where $\alpha \in (0,1)$ is a constant exponential weighing factor controlling the effective length of window over which the mean value and background of $h(t,\tau)$ is estimated.

The results of background subtraction by using exponential averaging method applied for raw radar data processing given in Figs. 3a and 3b are presented in Figs. 4a and 4b. In these figures, high-level signal components representing signal scattered by moving target can be observed. In spite of that fact, there are still a number of impulse responses where it is difficult or impossible to identify signal components due to electromagnetic wave reflection by a moving target.

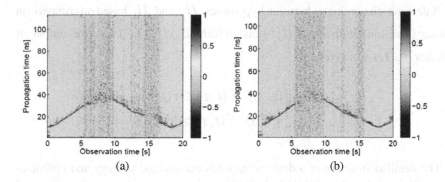

Fig. 4. Radargram with subtracted background. (a) Receiving channel Rx_1, (b) Receiving channel Rx_2

5 Detection

Detection is the next step in the radar signal processing which comes after background subtraction. It represents a class of methods that determine whether a target is absent or present in examined radar signals.

The solution of target detection task is based on statistical decision theory [8], [12], [23]. Detection methods analyze the radargram with subtracted background $h_b(t,\tau)$ along a certain interval of propagation time $t \in = t_1, t_1 + 1, \ldots, t_2$ and reach the decision whether a signal scattered from target $s(t,\tau)$ is absent (hypothesis H_0) or it is present (hypothesis H_1) in $h_b(t,\tau)$. The hypotheses can be mathematically described as follows:

$$H_0 \; : \; h_b(t,\tau) = n_{BS}(t,\tau) \tag{5}$$

$$H_1 \; : \; h_b(t,\tau) = s(t,\tau) + n_{BS}(t,\tau) \tag{6}$$

where n_{BS} represents residual noise obtained by $h(t,\tau)$ processing by a proper background subtraction method. Following expressions (1) and (2), $n_{BS}(t,\tau)$ can be expressed as follows

$$n_{BS}(t,\tau) = b(t,\tau) - \hat{b}(t,\tau) \tag{7}$$

A detector discriminates between hypotheses H_0 and H_1 based on comparison testing (decision) statistics $X(t,\tau)$ and threshold $\gamma(t,\tau)$. Then, the output of detector $h_d(t,\tau)$ is given by

$$h_d(t,\tau) = \begin{cases} 0, & if \ X(t,\tau) \le \gamma(t,\tau), \\ 1, & if \ X(t,\tau) > \gamma(t,\tau) \end{cases} \tag{8}$$

The detailed structure of a detector depends on selected strategy and optimization criteria of detection [8], [12], [23]. The selection of detection strategies and optimization criteria results in a testing statistic specification and threshold estimation methods.

The most important groups of detectors applied for radar signal processing are represented by sets of optimum or sub-optimum detectors. Optimum detectors can be obtained as a result of solution of an optimization task formulated usually by means of probabilities or likelihood functions describing detection process. Here, Bayes criterion, maximum likelihood criterion or Neymann-Pearson criterion are often used as the bases for detector design. However a structure of optimum detector could be extremely complex. Therefore, sub-optimum detectors are also applied very often [23].

For the purpose of target detection by using UWB radars, detectors with fixed threshold, (N,k) detectors, IPCP detectors [23] and constant false alarm rate detectors (CFAR) [3] have been proposed. Between detectors capable to provide good and robust performance for through wall detection of moving target by UWB radar, CFAR detectors can be especially assigned. They are based on Neymann-Person optimum criterion providing the maximum probability of detection for a given false alarm rate.

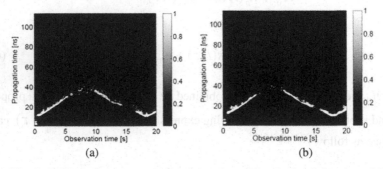

Fig. 5. CFAR detector output. (a) Receiving channel Rx_1, (b) Receiving channel Rx_2

There are a number of varieties of CFAR detectors. For example, the CFAR detector developed especially for UWB radar signal processing has been proposed in [3]. In spite of its simple structure and the assumption of Gaussian model of clutter, it has proved very good and robust performance for a lot of scenarios of through wall detection of single moving target. The illustration of its performance is given in Figs. 5a and 5b. Here, we can see the CFAR detector outputs obtained by signal processing represented by radargrams with subtracted background given in Figs. 4a and 4b.

6   TOA Estimation

If a target is represented by only one non-zero sample of the detector output for observation time instant $\tau = \tau_k$, then the target is referred as a simple target. However in the case of the scenario analyzed in this contribution, the radar range resolution is considerably higher than the physical dimensions of the target to be detected. It results in that the detector output for $\tau = \tau_k$ is not expressed by only one non-zero impulse at $\tau = TOA(\tau_k)$ expressing the target position by TOA for observation time τ_k, but the detector output is given by a complex binary sequence $h_d(t, \tau_k)$ (Figs. 5a–5b). The set of non-zero samples of $h_d(t, \tau_k)$ represent multiple-reflections of electromagnetic waves from the target or false alarms. The multiple-reflections due to the target are concentrated around the true target position at the detector outputs. In this case, the target is entitled as the *distributed target*. In the part of $h_d(t, \tau_k)$ where the target should be detected not only non-zero but also zero samples of $h_d(t, \tau_k)$ can be observed. This effect can be explained by a complex target radar cross-section due to the fact that the radar resolution is much higher than that of target size and taking into account different shape and properties of the target surface. The set of false alarms is due to especially weak signal processing under very strong clutter presence.

Because of the detector output for a distributed target is very complex, the task of distributed target localization is more complicated than for a simple target. For that purpose, an effective algorithm has been proposed in [14]. Here, the basic idea of distributed target localization consists in substitution of the distributed target with a proper simple target. Then a distributed target position can be determined by using the same approach as for a simple target.

Let us assume the scenario with one moving target. If the distributed target is substituted by one simple target, the target trace is defined as a sequence $h_T(\tau)$ where $h_T(\tau_k)$ expresses TOA of a simple target substituting a distributed target in such a way as the simple target is located at the propagation time instant $t = TOA(\tau_k)$ for the observation time instant $\tau = \tau_k$. It follows from this idea, that TOA corresponding to a simple target should be estimated based on detector outputs.

The TOA estimation algorithm applied in this radar signal processing phase has been originally introduced in [14]. Its punctual description can be found in [18]. Finally, a new version of TOA estimation algorithm capable to overcome the algorithm has been proposed in [16]. Its performance is illustrated in Figs. 6a and 6b, where the target traces for Rx_1 and Rx_2 are presented. It can be observed from theses figures, that target traces are represented by simple curves expressing TOA for particular observation time instants. However, it can be seen from these figures also there are some "missing" parts of the target traces. This imperfection of the trace estimation is due to high level of noise presented in raw radar data along corresponding intervals of observation time. This problem will be solved within target tracking phase by using prediction method.

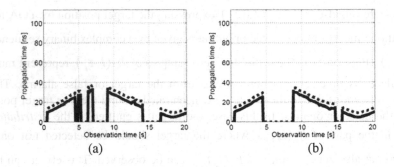

Fig. 6. Target traces. Dotted curve – estimated TOA without correction, Solid curve – TOA after wall effect compensation. (a) Rx_1, (b) Rx_2

7   Wall Effect Compensation

In the case of target localization by trace estimation method, target coordinates are evaluated by using TOA corresponding to target to be tracked as well as electromagnetic wave propagation velocity along the line transmitting antenna-target-receiving antenna. In many applications of target tracking, it can be assumed that the environment, through which the electromagnetic waves emitted by the radar are radiated, is homogenous (usually air). This is not true for through wall moving target localization because the wall is medium with different permittivity and permeability than that of the air and therefore, the electromagnetic wave propagation velocity in the air and wall are different. Besides mentioned quantities, wall thickness (d_w) has also strong influence on target location precision. This effect, which is sometime referred to as wall effect, displaces targets outside of their true positions, if the target localization is based on frequently used simplified assumption for through wall scenario that the electromagnetic wave propagation velocity is constant and equal to velocity of light. With regard to these facts, the precision of through wall target location can be improved if additional information such as permittivity, permeability and thickness of the wall (so-called wall parameters) are used for target position computation.

For that purpose, so-called wall effect compensation methods can be applied. These methods are based on estimation of time difference, referred to as delay time $(t_{delay}(\tau))$, which is applied for target trace correction by using expression [22], [15]:

$$h(\tau) = h_T(\tau) - t_{delay}(\tau) \qquad (9)$$

For trace estimation method, two methods of that kind referred to as target trace correction of the 1st and 2nd kind have been proposed in [15]. The target trace correction of the 1st kind is based on a simplified assumption that the electromagnetic waves emitted and received by radar propagate always in the perpendicular way with regard to wall plane. Then, the delay time can be determined by

$$t_{delay}(\tau) = \frac{d_w}{c}(\sqrt{\varepsilon_r \mu_r} - 1) \qquad (10)$$

for all possible position of the target and observation time instant. The target trace correction of the 2nd kind does not use this simplified assumption, i.e. for different position of the target different delay time can be evaluated. Here, delay time is obtained as a time difference between TOA obtained by UWB radar signal processing and TOA corresponding to the same target position under assumption that no wall is located between radar and target [15].

In order to illustrate the wall effect compensation significance, we have determined the correction of target traces given in Figs. 6a and 6b by solid lines. For that purpose, we have used the target trace correction of the 2nd kind including (9). The resulting traces of the target are given in Figs. 6a and 6b by dash lines. It is difficult to observe any impact of the wall compensation effect from these figures. However, its significance will be demonstrated very clearly by target trajectory estimation.

8 Localization

The aim of the localization task is to determine target coordinates in defined coordinate systems. Target positions estimated in consecutive time instants create target trajectory.

Let us assume, that TOA_k is the sample of the corrected trace of the target taken at the observation time instant τ for Rx_k and c is light propagation velocity. Then, the distance among Tx, target (T) and Rx_k is given by

$$d_k = cTOA_k \tag{11}$$

For arbitrarily placed transmitting antenna Tx=$[x_r, y_r]$ and receiving antennas Rx_k=$[x_k, y_k]$, the most straightforward way of estimation of target position, i.e. determining of its coordinates T=$[x,y]$, is to solve a set of equations created by using 11 taking into account the known coordinates of the transmitting and receiving antennas. Then, the following set of nonlinear equations can be built up based on measurements for the scenario with one transmitting and two receiving antennas [21], [5].

$$d_k = c.TOA_k = \sqrt{(x - x_t)^2 + (y - y_t)^2} + \sqrt{(x - x_k)^2 + (y - y_k)^2} \quad for \ k = 1,2 \tag{12}$$

Each range d_k and the pairs $Tx = [x_t, y_t]$ and $Rx_k = [x_k, y_k]$ $k = 1,2$ form two ellipses $k = 1,2$ with the foci $Tx = [x_t, y_t]$ and $Rx_k = [x_k, y_k]$ $k = 1,2$ and with the length of the main half-axis $a_k = \frac{d_k}{2}$ (Fig. 7).

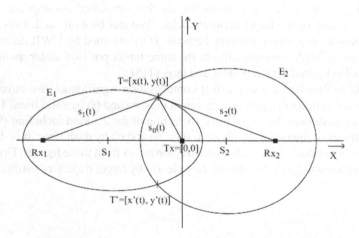

Fig. 7. Simple target localization for the scenario with one transmitting and two receiving antennas

It can be seen from this figure, that the target $Tx = [x, y]$ $(Tx' = [x', y'])$ lies on the intersection of these ellipses. Because of the ellipses are expressed by the polynomials of the second order, there are two solutions for their intersections. However, there is the only solution determining the desirable true coordinates of the target. Therefore, one of the obtained solutions has to be excluded for the scenario with one moving target. Usually, it can be done based on knowledge of a half-plane where the target is located. One of the solutions can be eliminated also if the solution (target coordinates) is beyond the monitored area or it has no physical interpretation (e.g. complex roots of (12)). For the solution of (12), so-called direct calculation method can be used. The detail description of direct calculation method is extensive and therefore beyond this contribution. Its punctual description can be found e.g. in [21] or [5]. In Fig. 8, the true target trajectory and target trajectory estimation by the described localization method for the scenario outlined in the Section 2 is presented. For the target trajectory estimation, the target traces given in Figs. 6a and 6b has been used. It can be observed from Fig. 8 that the target trajectory estimation by localization method is similar to true target trajectory, but large errors of target position estimation can be also found for many target positions. With regard to that fact, the target trajectory estimation should be significantly improved. For that purpose, target tracking and wall effect compensation methods can be used.

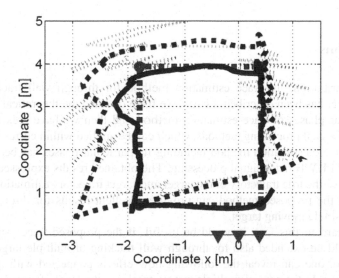

Fig. 8. Target trajectories. Dashdot curve − true target trajectory, thin dotted curve − estimated target trajectory after localization, thick dotted curve − estimated target trajectory after Kalman filtering, solid curve − estimated target trajectory after Kalman filtering with wall effect compensation

9   Tracking

Target tracking provides a new estimation of target location based on its foregoing positions. Usually, the target tracking will result in the target trajectory error decreasing including trajectory smoothing. The most of tracking systems utilize a number of basic and advanced modifications of Kalman filters as e.g. linear, nonlinear and extended Kalman filters and particle filters [13], [6]. Besides Kalman filter theory, further methods of tracking are available. They are usually based on smoothing of the target trajectory obtained by the target localization methods. Here, the linear least-square method is also widely used (e.g. [7]).

The significance of target tracking and wall effect compensation is illustrated in Fig. 8. For the moving target tracking, linear Kalman filter has been used. The missing samples of the target traces (Figs. 6a and 6b) have been completed through using of their predictions [6]. The final estimations of target trajectories by using Kalman filters are given in Fig. 8. The thick dotted curve presents the target trajectory estimation based on non-corrected target traces whereas the solid curve expresses the target trajectory estimation with application of target trace correction of the 2nd kind. The obtained results have shown that the application of Kalman filtering as tracking algorithm and the wall effect compensation can improve the target trajectory estimation in a significant way. The average localization error of the moving target obtained by that approach (Kalman filtering and the wall effect compensation method) for the scenario considered in this contribution is 0.16 m.

Conclusions

In this contribution, the trace estimation method for through wall tracking of moving target has been described. Firstly, we have outlined the theoretical base of the particular phases of trace estimation method and then we have presented an overview of signal processing methods, which can be applied within corresponding phase. The trace estimation method including its particular phases has been illustrated by real UWB radar signal processing. The obtained results expressed by the comparison of the true trajectory of the target and target trajectory estimations have shown, that the proposed method can provide excellent results for through wall tracking of single moving target.

With regard to this fact, it would be useful, if the proposed trace estimation method could be extended also for through wall tracking of multiple targets. For this scenario, one can reveal the following new effects connected with multiple target tracking. In the case of multiple targets tracking, the level of signal components scattered by the different targets will be usually different. For example, the target located close to radar antenna system is able to produce very strong reflection, but the second target located far from antenna system will reflect only very weak signals. In order to solve the effect of coincidental presence of targets reflected strong and weak signals, the advanced methods of background subtraction and target detection have to be developed and applied. Besides, the effect of mutual shadowing due to multiple targets can be presented. It can result in target disappearing from radargrams. Similar effects can be observed if the target is moved with stopping. For the purpose of temporary disappearing of target, more efficient methods of target trace estimation and multiple target trackers have to be used. In the case of direct calculation method for multiple target localization at the scenario with one transmitting and two receiving antennas, the so-called ghost effect can be reveal, too. It means e.g. that in the case of two targets tracking, four "potential targets" can be identified at the localization phase. They will be expressed by four intersections of two pairs of ellipses located in the same half-plane of scanned area. Two "potential targets" will correspond to true targets however the remainder "potential targets" will represent so-called ghosts. In order to separate true targets and ghosts, a suitable true target identification algorithm should be included into target tracking procedure. Preliminary analyse of the ghost identification problem has indicated that an analyses of the target trace properties could be very strong tool for true target identification. The above outlined effects and problems due to multiple target presence and possible bases for their solution has shown that the trace estimation method described in this contribution for single target tracking could be extended for multiple target tracking, too. The extension of the method will consist in application of advance signal processing methods for particular phases of target tracking procedure and in insertion of true target identification phase into discussed process of UWB radar signal processing. The solution of these tasks will be the subject of the next research of ours.

Acknowledgments. This work was supported by the Slovak Research and Development Agency under the contract No. LPP-0287-06, by European Commission under the contract COOPCT-2006-032744 and by Cultural and Educational Grant Agency of Slovak Republic under the contract No. 3/7523/09.

References

[1] Beeri, A., Daisy, R.: High-Resolution Through-Wall Imaging. In: Proc. of SPIE - Sensors, and Command, Control, Communications, and Intelligence (C3I) Technologies for Homeland Security and Homeland Defense V, 6201 (2006)

[2] Crabbe, S., et al.: Ultra Wideband Radar for Through Wall Detection from the RADIOTECT Project. In: Fraunhofer Symposium, Future Security, 3rd Security Research Conference Karlsruhe, Karlsruhe, Germany, p. 299 (2008)

[3] Dutta, P.K., Arora, A.K., Bibyk, S.B.: Towards Radar-Enabled Sensor Networks. In: The Fifth International Conference on Information Processing in Sensor Networks. Special track on Platform Tools and Design Methods for Network Embedded Sensors, pp. 467–474 (2006)

[4] Engin, E., Ciftcioglu, B., Ozcan, M., Tekin, I.: High Resolution Ultrawideband Wall Penetrating Radar. Microwave and Optical Technology Letters 49(2), 320–325 (2007)

[5] Fang, B.T.: Simple Solutions for Hyperbolic and Related Position Fixes. IEEE Transactions on Aerospace and Electronic Systems 26(5), 748–753 (1990)

[6] Grewal, M.S., Andrews, A.P.: Kalman Filtering: Theory and Practice. Prentice-Hall, Englewood Cliffs (2003)

[7] Hellebrandt, M., Mathar, R., Scheibenbogen, M.: Estimating Position and Velocity of Mobiles in a Cellular Radionetwork. IEEE Transactions on Vehicular Technology 46(1), 65–71 (1997)

[8] Minkler, G., Minkler, J.: CFAR. Magellan Book Company (1990)
[9] Nag, S., Barnes, M.: A Moving Target Detection Filter for an Ultra-Wideband Radar. In: Proceedings of the IEEE Radar Conference 2003, pp. 147–153 (2003)
[10] Nag, S., Fluhler, H., Barnes, H.: Preliminary Interferometric Images of Moving Targets Obtained Using a Time-modulated Ultra-Wide Band Through-Wall Penetration Radar. In: Proceedings of the IEEE - Radar Conference 2001, pp. 64–69 (2001)
[11] Piccardi, M.: Background Subtraction Techniques: a Review. In: Proceedings of IEEE - SMC International Conference on Systems, Man and Cybernetics, The Hague, The Netherlands (2004)
[12] Poor, H.V.: An Introduction to Signal Detection and Estimation. Springer, Heidelberg (1994)
[13] Ristic, B., Arulampalam, S., Gordon, N.: Beyond the Kalman Filter: Particle Filters for Tracking Applications. Artech House (2004)
[14] Rovňáková, J., Švecová, M., Kocur, D., Nguyen, T.T., Sachs, J.: Signal Processing for Through Wall Moving Target Tracking by M-sequence UWB Radar. In: The 18th International Conference Radioelektronika, Prague, Czech Republic, pp. 65–68 (2008)
[15] Rovňáková, J.: Compensation of Wall Effect for Through Wall Moving Target Localization by UWB Radar. In: The 8th Scientific Conference of Young Researchers, Košice (2008)
[16] Rovňáková, J.: Complete Signal Processing Procedure for Through Wall Target Tracking: Description and Evaluation on Real Radar Data. In: The 9th Scientific Conference of Young Researchers, Košice (accepted for publication) (2009)
[17] Sachs, J., et al.: Detection and Tracking of Moving or Trapped People Hidden by Obstacles using Ultra-Wideband Pseudo-Noise Radar. In: The 5th European Radar Conference (EuRAD), Amsterdam, Netherlands (2008)
[18] Sachs, J., Zaikov, E., Kocur, D., Rovňáková, J., Švecová, M.: Ultra Wideband Radio application for localisation of hidden people and detection of unauthorised objects. D13-Midterm report on person detection and localisation. Project RADIOTECT, COOP-CT-2006-032744 (2008)
[19] Sachs, J., Zetik, R., Peyerl, P., Friedrich, J.: Autonomous Orientation by Ultra Wideband Sounding. In: International Conference on Electromagnetics in Advanced Applications, Torino, Italy (2005)
[20] Stauffer, C., Grimson, W.: Learning Patterns of Activity Using Real-Time Tracking. IEEE Transactions on Pattern. Analyses and Machine Intelligence 22(8), 747–757 (2000)
[21] Švecová, M.: Node Localization in UWB Wireless Sensor Networks. Thesis to the dissertation examination. Fakulta elektrotechniky a informatiky Technickej univerzity v Košiciach, Košice (2007)
[22] Tanaka, R.: Report on SAR imaging. Internal Report, Technical University Delft, Netherlands (2003)
[23] Taylor, J.D. (ed.): Ultra-wideband Radar Technology. CRC Press, Boca Raton (2001)
[24] Tipping, M.E., Bishop, C.M.: Mixtures of Probabilistic Principal Component Analysers. Neural Computation 11(2), 443–482 (1999)
[25] Toyama, K., Krumm, J., Brumitt, B., Meyers, B.: Wallflower: Principles and Practice of Background Maintenance. In: International Conference on Computer Vision, pp. 255–261 (1999)
[26] Wren, C., Azarbayejani, A., Darrell, T., Pentland, A.: Pfinder: Real-Time Tracking of the Human Body. IEEE Transactions on Pattern. Analyses and Machine Intelligence 19(7), 780–785 (1997)
[27] Yelf, R.: Where is True Time Zero? In: Proceedings of the 10th International Conference on Ground Penetrating Radar, vol. 1, pp. 279–282 (2004)
[28] Zetik, R., Crabbe, S., Krajnak, J., Peyerl, P., Sachs, J., Thoma, R.: Detection and Localization of Persons Behind Obstacles Using m-Sequence Through-The-Wall Radar. In: Proceedings of SPIE - Sensors, and Command, Control, Communications, and Intelligence (C3I) Technologies for Homeland Security and Homeland Defense, 6201 (2006)

Advanced Industrial Communications

Petr Krist

Department of Applied Electronics and Telecommunications, Faculty of Electrical
Engineering, University of West Bohemia in Pilsen
Univerzitni 8, 306 14 Pilsen, Czech Republic
krist@kae.zcu.cz

Abstract. The paper deals with the new advanced industrial communication solutions concepts. It describes basic structures of industrial distributed control systems and physical and application layers of the most commonly used standards. The main differences between industrial real-time communications and information data network communications are classified and convergence trends of both of them are emphasized. The Ethernet standard is mentioned and the requirements for industrial Ethernet implementation declared. The progressive leading solution *ETHERNET Powerlink* is introduced. The contribution outlines the structure of an advanced distributed control node and declares the new conception of the virtual application communication bus with an example *CANopen - ETHERNET Powerlink*.

Keywords: bit-rate, bus, coding, communication, CAN, CANopen, Ethernet, Ethernet Powerlink, Fast Ethernet, frame, fieldbus, layer, master, network, NMT, node, PDO, protocol, RS-485, SDO, slave, TDMA.

1 Introduction

The rapid progress of the silicon and semiconductor technologies and the microelectronic systems has brought about huge deployment of the microcomputers and microcontrollers widely spread all over the world and implemented in many application areas. Formerly used local and complex stand-alone central control systems have been replaced by sophisticated structures of distributed control systems consisting of smart control application nodes communicating each other in order to exchange their local application data. Nowadays it is impossible to imagine an industrial control system without a communication capability and, therefore, the industrial communications play the key role in the design and implementation of any industrial control system.

I.J. Rudas et al. (Eds.): Towards Intelligent Engineering & Information Tech., SCI 243, pp. 365–376
© Springer-Verlag Berlin Heidelberg 2009

2   Current Communication Structures

Currently we recognize the two following industrial communication structures. The fist centralized *master-slave* topology (Fig. 1) represents command-based systems with application algorithm running on the master node.

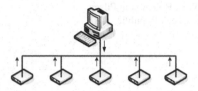

Fig. 1. Master-slave topology

The second one more advanced decentralized *multi-master* topology (Fig. 2) represents information-based systems with locally spread distributed application algorithm in every single node. The paper presents the advanced extension of this arrangement into the multiprotocol virtual application communication bus described further.

Fig. 2. Multi – master topology

3   Industrial Communication Standards

To guarantee mutual interoperability between communication devices of different manufacturers, the nodes of a control network communicate with each other according to the predefined set of rules – communication protocol. It defines a communication medium, an access method to the shared bus, the format of the messages being transmitted between devices, and the actions expected when one device sends a message to another. Typical fieldbus industrial communication standard operating in a real-time system often implements only reduced set of communication layers according to the seven-layer data communication reference model – only layers 1, 2 and 7 are usually implemented (Fig. 3).

Speaking of industrial communication protocols, it is desirable to mention the worldwide used proof standards RS-485 and CAN. They have in common their physical layer specification based on differential signal distribution method, convenient from the viewpoint of electromagnetic immunity in a harsh industrial environment.

7	**Application Layer**
6	**Presentation Layer**
5	**Session Layer**
4	**Transport Layer**
3	**Network Layer**
2	**Link Layer**
1	**Physical Layer**

Fig. 3. ISO/OSI data communication reference model

3.1 RS-485

The standard defines electrical transceiver characteristics related to the physical layer. It enables differential balanced bidirectional data transmission (Fig. 4) in two-wire half-duplex mode or in four-wire full-duplex master-slave mode.

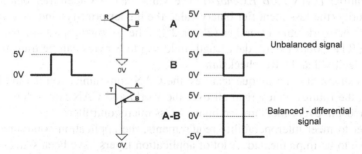

Fig. 4. RS-485 balanced differential transmission

In dependence on the transceiver type, bit rates up to 12 Mbit/s and distances up to 1200 meters can be achieved (of course not at the same time) and up to 32 unit load can be connected to the bus. The characteristic impedance of the twisted-pair cable is 120 Ω.

3.2 CAN

Controller Area Network specification was originally developed by Bosch Company for automotive purposes and due its features it quickly expanded into all industrial application areas. The physical layer of the high-speed version CAN ISO 11898 enables differential data transfer at bit rates up to 1 Mbit/s on a distance up to 40 meters with the characteristic impedance of the twisted-pair cable 120 Ω. In

addition to the physical layer, CAN standard specifies the link layer, implementing non-destructive CSMA/CR - *Carrier Sense Multiple Access/Collision Resolution* access method, based on multiple bit arbitration (Fig. 5).

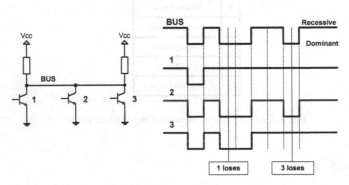

Fig. 5. CAN CSMA/CR access method

The CAN frame transmission starts by 11-bit identifier *(CAN 2.0A Standard)* or
29-bit identifier *(CAN 2.0B Extended)*. The value of CAN identifier determines
frame priority (the less identifier binary value the higher priority) and its value indicates the network object number (2^{11} or 2^{29}). The following parts of the CAN
frame are RTR control field, data length field (up to 8 bytes can be transmitted),
acknowledge field and CRC checksum.

Due its simplicity and unique features, the CAN communication is (and likely
will be in the future) widespread all over the world and CAN controllers are included almost in every medium and higher-class microcontroller.

In order to meet interoperability requirements, the application communication
layers have to be implemented. A lot of application layers have been standardized
and a lot of interested specialized groups have been established. Let's mention for
example general industrial application standard MODBUS which can be used together with RS-485 stated above. Nevertheless, to conform to the purposes of this
contribution, CAN application layer protocols are preferred in the following description. These are Device Net, SDS, CAN Kingdom, SAE 1939, and especially
CANopen - the most commonly used standard in the Europe region. The importance of the CANopen underlines the fact that this specification was adopted by
ETHERNET Powerlink standard described further.

3.3 CANopen

The CANopen specification defines the comfort application protocol layer for distributed industrial automation systems based on CAN. The heart of the CANopen
protocol is *Object Directory - OD* definition. It contains general configuration,
communication, and application data accessible via predefined service set *Service
Data Objects - SDO*. Real-time sensitive technological data can be mapped into

the *Process Data Objects - PDO* to manage fast real-time communication responses. The protocol defines by default 4 transmit TPDO and 4 receive RPDO objects of the 8 bytes length, identified by their unique ID. The protocol *Network Management - NMT* provides the services for node configuration, monitoring, and overhead. The CANopen communication profiles (CiA/DS 301 and CiA/DS 302) and device profiles (CiA 4xx) are defined by CiA group.

4   Data Network and Industrial Bus Comparison

Consistently with the goals of the paper, it is suitable to discuss the differences, identities, and relations between industrial communication bus and information data networks. A control network resembles in many ways a computer data network *LAN - Local Area Network*. The basics of the industrial communications and systems have been adopted from data networks concepts, where ISO/OSI data communication reference model completely describes structure and behavior of a communication device (Fig. 3). There are, however, some significant differences between data networks and industrial buses.

A **data network** is optimized for moving large amounts of data and the design of data network protocols assumes that occasional delays in data delivery and response are acceptable. The media access method used in data networks is mainly indeterminate. The complete protocol set of the ISO/OSI data communication reference model is implemented. Data networks operate usually in an office environment. In the most cases they are based on the Ethernet specification IEEE 802.3 - physical and link layer specification.

An **industrial bus** (often called fieldbus) on the other hand uses shorter data blocks to be transferred during short time periods in order to manage real-time responses of the system – the design of industrial control network protocols respects fast response requirements of control. The fieldbus media access method is mostly determinate. An industrial bus operates in a harsh environment, and increased electromagnetic immunity of the communication is required together with mechanical robustness. A fieldbus typically implements only physical, link, and application layers of the ISO/OSI data communication reference model (Fig. 3).

5   Data and Industrial Communications Convergence

In spite of the fact that the data network is not suitable well for the real-time processing and completely differs from the industrial communication bus, a lot of effort has been attended to utilize data networks for industrial communications and technological data transfer. The reasons are long distance internet network connectivity based on the standardized set of generally supported IP protocol specifications (TCP and UDP). Various gateways between an industrial bus and Ethernet network, industrial Web servers and terminal Ethernet devices are supplied by many vendors and companies. While a gateway between an industrial bus and Ethernet network enables connection between two different communication systems, industrial Web servers and terminal Ethernet devices view the Ethernet network as the fieldbus. This fact results in the increased requirements for implementation. In compliance with these requirements the conception of the **industrial Ethernet** has been defined. The necessary operating features are the increased electromagnetic immunity of the communication media together with mechanical robustness with regard to the harsh industrial environment. In dependency on the application, the real-time capability of the communication is required. This can be achieved by using of the Fast Ethernet operating at 100 Mbit/s bit rate (i.e. ten times faster than the standard Ethernet), by the network segmentation avoiding collisions, eventually by TDMA methods and global network time sharing. In the real-time control systems with the critical timing, the non deterministic media access method is not permissible and this case is solved using special Ethernet protocol specifications based on the deterministic media access method. A several industrial Ethernet protocol specifications have been developed – for example *ETHERNET Powerlink* (Bernecker & Rainer), *Profinet IRT* (PNO), *EtherNet/IP* (Rockwell), *SERCOS-III* (Sercos Interface e.V.), *Ethercat* (Ethercat Technology Group), *Modbus TCP* (IEC PAS 62030), *SynqNet* (Danaher Motion), *Vnet/IP* (Yokogawa). The most advanced progressive leading solution ETHERNET Powerlink is described further.

With regard to the importance of the Ethernet communication in industry, its key features are stated first.

6   Ethernet

The most commonly used standard of the data information network was developed in the beginning of the seventies by Xerox Company and later extended in cooperation with Digital and Intel companies. The standard was accepted by IEEE in 1995 under project specification IEEE 802.3 defining physical and link layers OSI-ISO. There are various versions of the standard available nowadays – base 10 Mbit/s bit rate version, most frequently used *Fast Ethernet* featuring 100 Mbit/s bit rate, *Gigabit Ethernet* version 1 Gbit/s, and optionally 10 Gbit/s bit rate version. The shared media access is executed due to *CSMA/CD (Carrier Sense Multiple Access/ Collision Detection)* access method managed by MAC link sub-layer. The MAC layer defines the Ethernet frame illustrated on the top of Fig. 6. Commonly used frame type at present is the *Ethernet II*.

The Ethernet specification defines various physical layers. Nowadays there are two best suitable and widespread physical layers for industrial applications. These use two twisted pairs of wires:

- *IEEE 802.3i*, referred to as *10Base-T* at the bit rate 10 Mbit/s, Manchester coding, two independent UTP or STP twisted pairs, category from 3 to 5, star network topology using active Ethernet hubs on maximal segment length of 100 m. The characteristic impedance of the twisted-pair cable is 100 Ω.

- *IEEE 802.3u*, Fast Ethernet referred to as *100Base-TX* at the bit rate 10 Mbit/s, 4B/5B, NRZI, and MLT-3 coding, two independent UTP or STP twisted pairs, category 5, star network topology using active Ethernet hubs on maximal segment length of 100 m. The characteristic impedance of the twisted-pair cable is 100 Ω.

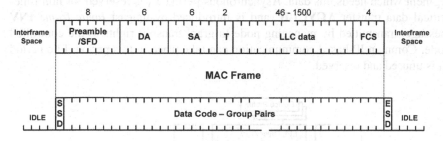

Fig. 6. Fast Ethernet MAC frame

7 ETHERNET Powerlink

The ETHERNET Powerlink, often called the **fieldbus of the second generation** provides the protocol based on the Fast Ethernet standard but, at the same time, it meets the real-time requirements and allows deterministic data transfer with implemented cycle times as low as 200 µs and ultra-precise timing better than 1 µs. It is the only protocol with this performance not violating any Ethernet standards. It can be implemented with any standard Ethernet chips and processor architecture and there are no dependencies on customized chips and vendors. ETHERNET Powerlink works according to TDMA - Time Division Multiple Access and polling principles (Fig. 7).

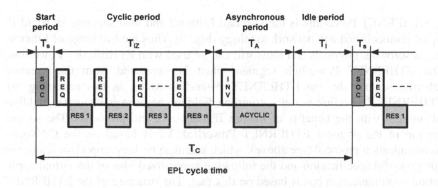

Fig. 7. ETHERNET Powerlink cycle

The basic ETHERNET Powerlink cycle T_C (EPL cycle time) consists of the following four periods – start period T_S, cyclic period T_{IZ}, asynchronous period T_A, and idle period T_I. Start period T_S includes *Start of cyclic frame* **SOC** broadcast message, transmitted by the managing node in order to synchronize all controlled nodes in the EPL segment (see Fig. 9). Cyclic isochronous data exchange interval T_{IZ} consists of the row of double frames – unicast *Poll-request frame* **REQ**, sent by managing node to the unique controlled node, followed by its multicast *Poll-response frame* **RESi** which can be received by any node in the EPL segment which needs this data. Asynchronous period T_A is reserved for non-time-critical data transfer **ACYCLIC** and is introduced by unicast *Invite frame* **INV** message, transmitted by managing node to grant transmit rights to the controlled node. Common IP-based communication can take place in this period. Idle period T_I is unused and reserved.

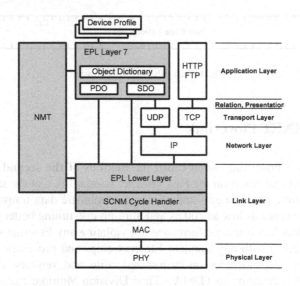

Fig. 8. ETHERNET Powerlink V2 protocol structure

ETHERNET Powerlink is the standard Ethernet and also supports standard IP based protocols and any network topology (Fig. 8). Thus the full range of Ethernet based software, protocols and tools still can be used with ETHERNET Powerlink. The ETHERNET Powerlink segment must be separated from the standard Ethernet segment by the ETHERNET Powerlink router in this case (Fig. 9). ETHERNET Powerlink is not a common fieldbus because it integrates fieldbus advantages with the benefits of mature IP networks and protocols. The second version of the protocol ETHERNET Powerlink V2 is based on the CANopen communication protocol (see above), which seems to be very important feature of this powerful specification and the following generalized idea of the virtual application communication bus is based on this fact. The structure of the ETHERNET Powerlink protocol is shown on Fig. 8 which outlines the basic layout of the protocol In addition to the standard IP based protocols (UDP, TCP, FTP, HTTP and so on) the communication profile *CANopen* is implemented in the application layer with its main components *Object Dictionary, Process Data Objects - PDO* and *Service Data Objects - SDO*. As is illustrated, the SDO protocol is implemented via UDP/IP layer also and therefore using standard IP messages. This enables direct access to the object dictionaries of EPL devices by the devices and applications outside the EPL system via EPL routers.

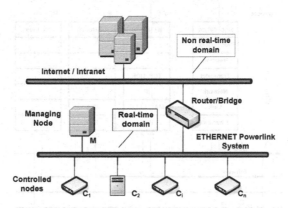

Fig. 9. ETHERNET Powerlink communication system structure

8   Virtual Application Communication Bus

Having all necessary important facts classified, it is possible to define the requirements for the features of an **advanced application distribution control node** as the fundament of the communication arrangement based on **virtual application communication bus**.

The heart of the advanced application distribution control node should be a powerful microcontroller featuring sufficient computing performance (preferably 32-bit), memory system resources (FLASH, RAM, EEPROM) and rich peripheral set to satisfy the application requirements which means appropriate GPIOs, PWMs, Capture modules, RTC, A/D and D/A converters, DMA channels, SPI, I^2C, advanced interrupting system etc. But the most important feature from the point of communications is a multiple communication interface, including several USART controllers, CAN controller, and especially MAC Ethernet controller. An additional USB interface can be useful but not necessary, because it is not an industrial communication interface. Corresponding interface circuits to meet interface specifications requirements are supposed, as well as all other input and output signals to be connected with the controlled technological system.

Let's have such a distributed control node structure defined above and equipped by multiple communication interface CAN, RS-485 and Ethernet, for example. Then we can define the new conception of the virtual application communication bus. Fig. 10 shows the principle of its arrangement.

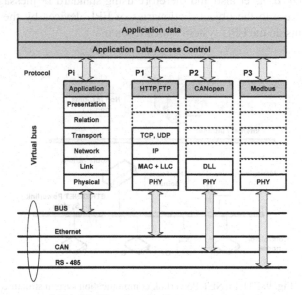

Fig. 10. Virtual application communication bus

Every physical communication interface (generally labeled as **BUS** in Fig. 10) has got the corresponding protocol driver associated (**Pi** label in Fig. 10). This single protocol driver Pi implements either full or partial ISO/OSI protocol layer set according to the particular communication standard specification and accesses to the **Application data**. The application data structures are unique in the range of the node and all the protocol drivers share them - it means that there is the only instance of the application data. From the point of application data consistency it is obvious that concurrent accesses of multiple protocol drivers must be resolved. Therefore arbitration **Application Data Access Control** sub-layer must be inserted between application data and protocol drivers to conform data consistency and application algorithms integrity. It is clear that the same common unified access interface to the application data structure must be targeted by every single protocol driver Pi.

It is not relevant which of the protocol drivers Pi serves to the communication from the viewpoint of the application. It depends on the connectivity accessible at the moment. It is possible to exploit just one or two communication interfaces only or all implemented communication interfaces. This way we get generally purposed versatile communication solution made up of all implemented protocol drivers accessing common application data. This application – communication structure is possible to view as the only flexible **virtual application communication bus** with the effect of continuous data flow between different communication buses. This can resemble gateway configuration but it is necessary to emphasize that the communication structure stated above due to its attachments to the application algorithm does not act as a simple gateway, but features the possibilities beyond the gateway behavior, where only the data transfer between two or more interfaces is provided. Nevertheless, in the case of absence of any application algorithm, the structure converts to the gateway role.

There are some examples of typical protocol drivers implemented in the Fig. 10 – CAN with CANopen, RS-485 with MODBUS and Ethernet with HTTP or FTP protocol implementation. Communication protocol specifications used in the structure differs in the various different features – for example bit rate, data frame length, data frame periods generation, data disturbance check and correction, error confinement, data acknowledgement, type of services, etc. Taking into account this fact it is obvious that the **careful accurate analysis of the protocol and application algorithm timing and behavior is necessary and desirable prior to the implementation** in order to achieve fluent data flow and acquisition. Generally, the slowest item of the structure determines execution, computing speed of the distributed control node, and data throughput of the whole system.

The interesting example of the virtual application communication bus generally described above is the **CANopen – ETHERNET Powerlink** implementation. The features the both of the protocol specifications were mentioned in the previous paragraphs. Fig. 11 shows its configuration. There are the following communication interfaces – *CAN ISO 11898 2.0A* with CANopen application layer, and *Fast Ethernet IEEE 802.3u* as the low layers of the ETHERNET Powerlink protocol standard.

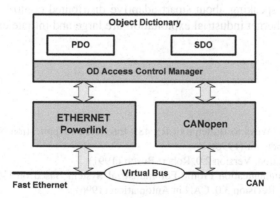

Fig. 11. Virtual application communication bus CANopen – ETHERNET Powerlink

The good advantage of this arrangement is the fact that the ETHERNET Powerlink application layer is built directly on the CANopen application layer and contains its all important elements and properties (see Fig. 8).

The idea of the virtual application communication bus defined above and especially the example structure of CANopen - ETHERNET Powerlink meets the principles and requirements for information data network and industrial communication bus convergence stated in paragraph 5. This trend seems to be the future of the industrial communications together with industrial Ethernet utilization.

Conclusion

The expanding branch of industrial communications plays the key role in the future automation world. The proof typical industrial contemporary communication standards face the new progressive trends. There are communication solutions completely migrating to the new standards on one hand and, on the other hand, there are hybrid systems coexisting together. The main problems of such systems are interoperability, openness and standardization.

The new advanced trend of the industrial communication systems – industrial Ethernet exploitation in automation – is more and more up-to-date.

This contribution reflects these facts and outlines the new approaches in the communication structure design. It summarizes the current state of the perspective proof industrial communication standards, examines the Ethernet standards features and defines completely new conception of the virtual application communication bus. In conjunction with that the new industrial Ethernet standard is introduced – ETHERNET Powerlink.

Due to its features the Ethernet specification is the future of industrial communications. It provides the long distance connectivity, the proof TCP/IP based protocol set, security, distant diagnostics and system reconfiguration. The remote long-distance application algorithm debugging of distributed control systems is possible too, not speaking about smart adaptive distributed control systems. The chances of the Ethernet industrial exploitation are large and initiate extensive field of applications.

References

[1] Goldie, J.: Ten Ways to Bulletproof RS 485 Interfaces - Application Note 1057. National Semiconductor (1996)
[2] CAN Specification, Version 2.0. Robert Bosch (1991)
[3] CANopen Communication Profile For Industrial System Based on CAL, CiA Draft Standard 301 - Revision 3.0. CAN in Automation (1996)
[4] Pfeiffer, O., et al.: Embedded Networking with CAN and CANopen. RTC Books (2003) ISBN 0-929392-78-7
[5] Cisco Corp.: Internetworking Technologies Handbook – 1-58705-001-3
[6] IEEE 802.3 Standard – Part 3 – Carrier Sense Multiple Access with Collision Detection (CSMA/CD) Access Method and Physical Layer Specifications
[7] EPSG brochure 2005 – ETHERNET Powerlink – An Open Standard for Safety Real-Time Communication. EPSG (2005) MM-E00506.243
[8] ETHERNET Powerlink V2.0 Communication Profile Specification, Draft Standard, Version 1.0.0. ETHERNET Powerlink Standardization Group (2006)

Logical Consequences in Partial Knowledge Bases

Mirko Maleković, Mirko Čubrilo, and Kornelije Rabuzin

Faculty of Organization and Informatics, University of Zagreb, Croatia
mmalekovic@inet.hr

Abstract. In this paper, we consider logical consequences for reasoning about functional, multivalued, and join dependencies in partial knowledge bases. The standard consequence, strong consequence, and the weak consequence are characterized. We prove that reasoning based on the standard logical consequence is equivalent to reasoning based on the strong logical consequence. In addition, we prove that reasoning based on the standard logical consequence and reasoning based on the weak logical consequence are not equivalent. We also define a formal system FSED for reasoning about existence dependencies. We prove that the formal system FSED is sound and complete. We state the connection between existence and functional dependencies: reasoning about existence dependencies based on the corresponding logical consequence and reasoning about functional dependencies based on the standard logical consequence are equivalent.

Keywords: partial knowledge bases, dependencies, logical consequences, formal systems, inference rules, soundness, completeness.

1   Introduction

Information is often missing in the real world. Therefore, we need some way of dealing with such missing information in our knowledge bases. The theory of missing information can be found in [1] and [2]. A critic of the 3VL (three-valued logic) and 4VL (four-valued logic) approach can be found in [3], [4], and [5]. A very interesting metadata approach to missing information is described in [9]. A careful consideration of the partial knowledge bases shows that the integrity-constraint component of partial knowledge bases, especially reasoning about dependencies in partial knowledge bases, needs much more attention of database researchers. We could say that what is missing about the missing information is an exploration of the types of logical consequences and their relationships. This paper is an extention and improvement of our paper [8], where some properties of reasoning about functional and existence dependencies are characterized. Here we consider a natural framework, namely logical consequences and their respective

I.J. Rudas et al. (Eds.): Towards Intelligent Engineering & Information Tech., SCI 243, pp. 377–388.
springerlink.com   © Springer-Verlag Berlin Heidelberg 2009

formal systems, for reasonig about dependencies in partial knowledge bases (functional, multivalued, join, and existence dependencies are included). We show that there are strong connections between the introduced logical consequence types. We also state some contraintuitive rules of reasoning about dependencies. This can be taken, in some way, as a support of the critic of 3VL approach mentioned above.

The paper comprises five sections and an Appendix containing the proofs of some of the propositions.

In Section 2, we describe the basic notions of partial knowledge bases. The definitions of functional dependencies, multivalued dependencies, join dependencies, and existence dependencies are given too. In Section 3 we consider standard, strong and weak logical consequences for functional, multivalued and join dependencies. The main result states that reasoning based on the standard logical consequence and reasoning based on the strong logical consequence are equivalent, whereas reasoning based on the standard logical consequence and reasoning based on the weak logical consequence are not equivalent. In Section 4 we define a formal system with only two inference rules, FSED, for reasoning about existence dependencies. We prove there that the formal system FSED is sound and complete. We define a logical consequence for existence dependencies and prove that reasoning about existence dependencies based on the corresponding logical consequence and reasoning about functional dependencies based on the standard logical consequence are equivalent. Conclusions are given in Section 5. Two open problems are also given there.

2   Basic Notions

In this Section we describe partial knowledge bases, PKB, and a set of dependencies in PKB. The set of dependencies includes: functional dependencies, multivalued dependencies, join dependencies, and existence dependencies.

2.1 Partial Knowledge Bases

A partial knowledge base is a triple PKB = (EDB, IDB, IC), where EDB is a partial extent ional (relational) database, IDB is an intentional database, and IC is a set of integrity constraints. A partial extentional database, EDB, is a data base: $r_1(R_1)$, ..., $r_m(R_1)$ over a database scheme DBS: R_1, .., R_m, where $r_1(R_1)$, ..., $r_m(R_m)$ are partial relations. An intentional database, IDB, is the set of rules for generating new relations from relations in EDB and previously defined relations. The set of integrity constraints, IC, can include, in addition to some other constraints (primary key, entity integrity, referential integrity), the set of dependencies in relational databases (functional, multivalued, join, and existence dependencies). Our main goal in this paper is to consider some properties of logical consequences which are a foundation for reasoning about dependencies mentioned above. A partial relation r over a set of attributes $R = \{A_1, .., A_m\}$, denoted $r(R)$, is a finite set of partial tuples over the set R. We say that R is a relational scheme. Ler $R = \{A_1, .., A_m\}$ be a relational scheme; $Dom(A_i) = D_i$, $i = 1,.., m$;

$D = D_1 \cup .. \cup D_m$; $D? = D \cup \{?\}$.

A partial tuple t over R is a function t: $R \rightarrow D?$ such that

    (1) $(\forall A_j \in R)[t(A_j) \in D_j$ or $t(A_j) = ?]$ and

    (2) $(\exists A_k \in R)[t(A_k) \in D_k]$.

Notation $t(A_j) = ?$ means that the value of the tuple t for the attribute A_j, $t(A_j)$, is not known.

A partial tuple t over R is total iff $(\forall A_i \in R)[t(A_i) \in D_i]$.

This definition says that a total tuple does not have the presence of the symbol ?.

Let r be a partial relation over R. If we change all the presences of symbol ? with appropriate domain elements, then we obtain a completion of the relation r, denoted $co(r)$. It is easy to see that $co(r)$ consists of only total tuples, that is, $co(r)$ is a total relation. Let $Co(r)$ be the set of all completions of relation r.

2.2 Dependencies in PKB

Functional Dependency

Let R be a relational scheme, $X, Y \subseteq R$. In addition, let $Trel(R)$ be the set of all total relations over R.

The expression $X \rightarrow Y$ is a functional dependency over R. The set of all functional dependencies over R is denoted by $FD(R)$.

Let $X \rightarrow Y \in FD(R)$, $r \in Trel(R)$. We say that $X \rightarrow Y$ holds in r, denoted

$r \Vdash X \rightarrow Y$, iff $(\forall t_1, t_2 \in r)(t_1[X] = t_2[X] \Rightarrow t_1[Y] = t_2[Y])$.

Join Dependency

Let $d(R) = \{ R_{1,...,} R_k \}$ be a decomposition of a relational scheme R. The expression

$\bowtie(R_{1,...,} R_k)$ is a join dependency over R. The set of all join dependencies over R is denoted by $JD(R)$. Let $\bowtie(R_{1,...,} R_k) \in JD(R)$, $r \in Trel(R)$. We say that

$\bowtie(R_{1,...,} R_k)$ holds in r, denoted $r \Vdash \bowtie(R_{1,...,} R_k)$ iff $r = \sqcap[R_1](r) \bowtie .. \bowtie \sqcap[R_k](r)$.

Here, $\sqcap$ is the projection operator.

Let R be a relational scheme, $X, Y \subseteq R$. The expression $X \twoheadrightarrow Y$ is a multivalued dependency over R. It is a special case of a join dependency. Namely, $X \twoheadrightarrow Y$ is defined by $X \twoheadrightarrow Y = \bowtie(X \cup Y, X \cup (R \setminus (X \cup Y)))$.

Existence Dependency

Let R be a relational scheme, X, Y $\subseteq$ R. In addition, let Prel(R) be the set of all partial relations over R. The expression X $\Rightarrow$ Y, read X requires Y, is an existence dependency over R. We denote the set of all existence dependencies over R by ED(R). In order to describe existence dependency semantics, we introduce a unary predicate T, where T(t) means that a tuple t is total. Let X $\Rightarrow$ Y $\in$ ED(R), r $\in$ Prel(R). We say that X $\Rightarrow$ Y holds in r, denoted r $\Vdash$ X $\Rightarrow$ Y, iff ($\forall$t $\in$ r)[T(t[X]) $\Rightarrow$ T(t[Y])].

This definition says that X $\Rightarrow$ Y holds in a partial relation r in and only if for all tuples in r the totality of t[X] implies the totality of t[Y].

Example 1

Let PKB = (EDB, IDB, IC) be a partial knowledge base, where

EDB: r_1(A B C) r_2(B C D) r_3(A D)

```
=======    ========    ======
 0  2 2      0  0 1      2  2
 1  2 3      2  ? 1      3  ?
             1  ? ?
```

IDB: p_1: s_1(x, y) $\leftarrow$ r_1(x, x_1, y)

      p_2: s_2(x, y, z) $\leftarrow$ r_1(x, x_1, y), r_2(x_1, y_1, z)

IC: ic_1: $1 \leq A \leq 5$ for r_3(A, D)

    ic_2: A $\rightarrow$ B for r_1(A, B, C)

    ic_3: C $\rightarrow$ B for r_1(A, B, C)

    ic_4: $\bowtie$(AC, BC) for r_1(A, B, C)

    ic_5: C $\Rightarrow$ D for r_2(B, C, D)

We have that r_1 in EDB is total, whereas r_2 and r_3 are not total. The following relation, r_4,

r_4(A D)

```
======
 2  2
 3  2
```

is a completion of the relation r_3, that is, $r_4 \in$ Co(r_3). IDB consists of two rules p_1 and p_2 that generate relations s_1 and s_2. IC comprises five integrity constraints: ic_1 (says that the values of the attribute A must be between 1 and 5 for all instances of r_3); ic_2 (states that the functional dependency $A \to B$ must hold in each instance of r_1); ic_3 (says that the multivalued dependency $C \twoheadrightarrow B$ must hold for all instances of r_1); ic_4 (constrains that the join dependency $\bowtie(AC, BC)$ must hold in each instance of r_1); and ic_5 (says that the existence dependency $C \Rightarrow D$ must hold in each instance of r_2).

3 Logical Consequences for Functional and Join Dependencies

3.1 Standard Logical Consequence

Let $FJD(R)$ be the set of all functional and join dependencies over R, that is, $FJD(R) = FD(R) \cup JD(R)$. Now let $f_1, .., f_m, f \in FJD(R)$. We say that f is a standard logical consequence (implication) of $f_1, .., f_m$, denoted $f_1, .., f_m \models f$, iff $(\forall r \in Trel(R))((r \Vdash f_1 \wedge .. \wedge r \Vdash f_m) \Rightarrow r \Vdash f)$.

3.2 Strong Logical Consequence

Let $f \in FJD(R)$ and $r \in Prel(R)$. We say that f strongly holds in r, denote $r \Vdash [s] f$, iff $(\forall p \in Co(r))(p \Vdash f)$. This definition states that a dependency f strongly holds in a partial relation r iff the dependency f holds in each completion p of the relation r. A strong logical consequence is characterized as follows. Let $f_1, .., f_m, f \in FJD(R)$. f is a strong logical consequence of $f_1, .., f_m$, denoted $f_1, .., f_m \models [s] f$, iff $(\forall r \in Prel(R))((r \Vdash [s]f_1 \wedge .. \wedge r \Vdash [s]f_m) \Rightarrow r \Vdash [s]f)$.

3.3 Weak Logical Consequence

Let $f \in FJD(R)$ and $r \in Prel(R)$.
We say that f weakly holds in r, denote $r \Vdash [w] f$, iff $(\exists p, q \in Co(r))(p \Vdash f \wedge q \nVdash f)$.

Accordingly, a dependency f weakly holds in a partial relation r iff we have some completions p, q of r such that f holds in p and does not hold in q. Let $f_1, .., f_m, f \in FJD(R)$. f is a weak logical consequence of $f_1, .., f_m$, denoted

$f_1, .., f_m \models[w] f$, iff $(\forall r \in \text{Prel}(R))((r \Vdash[w] f_1 \wedge .. \wedge r \Vdash[w] f_m) \Rightarrow r \Vdash[w] f))$. In the next proposition, we characterize the connection between the standard and strong logical consequences.

3.4 Proposition (standard, strong)

(for all $f_1, .., f_m, f \in \text{FJD}(R))[(f1, .., fm \models f) \Leftrightarrow (f1, .., fm \models[s] f)]$

Proof (Appendix) Proposition (standard, strong) states that reasoning based on the standard logical consequence is equivalent to reasoning based on the strong logical consequence. Therefore, we say that 'the standard world' and 'the strong world' are equivalent with regard to reasoning about logical consequences. Consequently, if we have an implication problem in 'the strong world', it can be solved by solving the respective implication problem in 'the standard world', where there are well known algorithms for solving implication problems, [1], [6]. Regarding the connection between the standard and the weak consequences, we have the following proposition.

3.5 Proposition (standard, weak)

(for some $f_1, .., f_m, f \in \text{FJD}(R))[(f1, .., fm \models f) \wedge (f1, .., fm \not\models[w] f)]$

Proof (counterexample) We know that $A \rightarrow B, B \rightarrow C \models A \rightarrow C$. On the other hand,

$A \rightarrow B, B \rightarrow C \models[w] A \rightarrow C$ does not hold as the following example shows.

```
r(A  B  C)
=======
 1  2  2
 1  ?  3
```

Here we have $r \Vdash[w] A \rightarrow B$, $r \Vdash[w] B \rightarrow C$, and $r \not\Vdash[w] A \rightarrow C$

Therefore, 'the standard world' and 'the weak world' are not equivalent with regard to reasoning about logical consequences. Briefly, f1, .., fm $\models$f) $\Leftrightarrow$ f1, .., fm $\models[w]$ f. This result is in some way counterintuitive. Namely, our intuition once developed in 'the standard world' does not function in 'the weak world' and the interaction between the naïve users (including some professionals too) and partial knowledge bases can be confused and unexpected. We would say that working with something we don't know could make us put against the wall.

As a direct consequence of these propositions, we have that 'the strong world' and 'the weak world' are not equivalent, either.

It is well known that Armstrong's formal system, denoted AFS, is sound and complete for functional dependencies. AFS consists of the following rules:

as_1: $\vdash X \to Y$ if $Y \subseteq X$ (triviality)

as_2: $X \to Y \vdash XZ \to YZ$ (augmentation)

as_3: $X \to Y, Y \to Z \vdash X \to Z$ (transitivity)

The soundness of AFS means that its rules are correct, that is, the respective (standard) logical consequences

slc_1: $\models X \to Y$ if $Y \subseteq X$ (triviality)

slc_2: $X \to Y \models XZ \to YZ$ (augmentation)

slc_3: $X \to Y, Y \to Z \models X \to Z$ (transitivity)
hold, too.

It is interesting that none of AFS rules holds in 'the weak world'. We could say that 'the strong world' intuition is expired in 'the weak world'.

3.6 Proposition (AFS, weak)

None of FAS-rules holds in 'the weak world', that is,

wlc_1: $\not\models[w]$ $X \to Y$ if $Y \subseteq X$

wlc_2: $X \to Y \not\models[w] XZ \to YZ$

wlc_3: $X \to Y, Y \to Z \not\models[w] X \to Z$

Proof (Appendix) Accordingly, the 'weak world' is really something very different from 'the standard world' and 'the strong world'

4 Formal Systems for Existence Dependencies

Let $X \Rightarrow Y \in ED(R)$, $r \in Prel(R)$. We defined in Section 2 that $X \Rightarrow Y$ holds in r, denoted $r \Vdash X \Rightarrow Y$, iff $(\forall t \in r)[T(t[X]) \Rightarrow T(t[Y])]$.

Consequently, for all tuples in r we have that the totality of t[X] implies the totality of t[Y].

Example 2

r(A B C)
=======
 1 2 1
 2 ? 3

We can conclude that r ⊩ A ⇒ C holds and

r ⊮A ⇒ B does not hold, that is, r ⊮A ⇒B holds.

Now we introduce a formal system for existence dependencies. The formal system, called FSED-system, consists of two inference rules as follows. Let X, Y, Z, W ⊆ R be arbitrary subsets of a relation scheme R.

ed_1: ⊢X ⇒Y if Y⊆X   (triviality)

ed_2: X ⇒Y, YZ ⇒W ⊢XZ ⇒ W (generalized transitivity)

4.1 Proposition (FSED soundness, completeness)

FSED-system is sound and complete.

Proof
soundness
Firstly, we need to define a logical consequence for existence dependencies. Let f_1, .., f_m, g ∈ ED(R) be existence dependencies. We say that g is an existence logical consequence of f_1, .., f_m , denoted f_1, .., f_m ⊨[e] f, iff (∀r ∈ Prel(R))((r ⊩f1 ∧ .. ∧ r ⊩fm) ⇒ r ⊩f).

The soundness of ED-system means that its rules ed_1 and ed_2 are correct, that is, the respective existence logical consequences hold.

ed_1: ⊨[e]X ⇒Y if Y ⊆ X (triviality)

ed_2: X ⇒Y, YZ ⇒W ⊨[e] XZ ⇒ W   (generalized transitivity)

Proof (ed1): Let r ∈ Prel(R) be an arbitrary partial relation. We have to show that r ⊩X ⇒Y if Y ⊆X. Let t be an arbitrary tuple in r. Since T(t[X]) ⇒ T(t[Y]) holds if Y ⊆ X, we have that r ⊩X ⇒Y holds.

Proof (ed2): Let r ∈ Prel(R) be an arbitrary partial relation. Assume r ⊩X ⇒Y and r ⊩YZ ⇒W. We would like to prove r ⊩XZ ⇒W. Let t ∈ r be an arbitrary tuple such that T(t[XZ]). It follows T(t[X]) and T(t[Z]). Because of r ⊩ X ⇒Y, we obtain T(t[Y]). From T(t[Y]) and T(t[Z]) it follows T(t[YZ]). Finally, from T(t[YZ]) and r ⊩YZ ⇒ W, we have T(t[W]), as desired.

completeness
We have to prove that the following implication

i_1: (f_1, .., f_m ⊨[e] f) ⇒ (f_1,.., f_m ⊢[FSED] f) holds, where f_1,.., f_m ⊢[FSED]f indicates that dependency f is derivable from dependencies f_1, .., f_m by using FSED-system.

The implication i1 is equivalent to the implication

i_2: $(f_1,.., f_m \nvdash [FSED] f) \Rightarrow (f_1, .., f_m \nvDash[e] f)$.

Now we prove i_2. Let $f = X \Rightarrow Y$, $F = \{ f_1,.., f_m \} \subseteq ED(R)$.

Assume that A1: $(f_1,.., f_m \nvdash [FSED] X \Rightarrow Y)$ holds. We need to show that A2: $(f_1,.., f_m \nvDash[e] X \Rightarrow Y)$ holds. We define the set $Ecl(X, F) = \{A \in R: F \vdash[FSED] X \Rightarrow A\}$. $Ecl(X, F)$ is the existence closure of the set of attribute X with respect to the set of existence dependencies F. It is easy to show that the equivalence

eq: $(F \vdash[FSED] X \Rightarrow Y) \Leftrightarrow (Y \subseteq Ecl(X, F))$ holds.

Let $r \in Prel(R)$ be a partial relation that consists of only one tuple t such that $(\forall A \in Ecl(X, F))(T(t[A]))$ and $(\forall B \in (R - Ecl(X, F)))(t[B] = ?)$. From A1 and eq we obtain that $Ecl(X, F) \neq R$. Now, it is not difficult to prove that $r \Vdash F$ and $r \nVdash X \Rightarrow Y$, as needed.

Example 3

We show that $X \Rightarrow Y \vdash[FSED] XZ \Rightarrow YZ$.

1 $X \Rightarrow Y$ hypothesis

2 $YZ \Rightarrow YZ$ ed_1

3 $XZ \Rightarrow YZ$ 1., 2., and ed_2

As a final result in this Section 4, we characterize the connection between 'the standard world' and 'the existence world'. The connection is given in the following proposition.

4.2 Proposition (standard, existence)

(for all $X_1, Y_1,.., X_m, Y_m, X, Y \subseteq R$)$[(X_1 \rightarrow Y_1,.., X_m \rightarrow Y_m \vDash X \rightarrow Y)$ iff $(X_1 \Rightarrow Y_1,.., X_m \Rightarrow Y_m \vDash[e] X \Rightarrow Y)]$

Proof: A sound and complete formal system for reasoning about functional dependencies, FS2, given in [7], consists of only two rules: reflexivity and generalized transitivity. Accordingly, the formal systems FS2 for functional dependencies and the formal system FSED for existence dependencies have the same rules (reflexivity and generalized transitivity). Therefore,

$(X_1 \rightarrow Y_1,.., X_m \rightarrow Y_m \vDash X \rightarrow Y)$ iff $(X_1 \rightarrow Y_1,.., X_m \rightarrow Y_m \vdash[FS2] X \rightarrow Y)$ iff $(X_1 \Rightarrow Y_1,.., X_m \Rightarrow Y_m \vdash[FSED] X \Rightarrow Y)$ iff $(X_1 \Rightarrow Y_1,.., X_m \Rightarrow Y_m \vDash[e] X \Rightarrow Y)$, as wanted.

This proposition says that reasoning based on the standard logical consequence for functional dependencies is equivalent to reasoning based on the existence logical consequence for existence dependencies. Therefore, we can say that 'the standard world' restricted to functional dependencies and 'the existence world' are equivalent with regard to the respective logical consequences. It also means that each implication problem in 'the existence world' can be solved by solving the respective implication problem in 'the standard world', where, as we have already mentioned, there are very efficient algorithms for solving implication problems.

Conclusions

We have considered standard, strong and weak logical consequences for functional, multivalued and join dependencies in partial knowledge bases. We have showed that reasoning based on the standard logical consequence is equivalent to reasoning based on the strong logical consequence, whereas reasoning based on the standard logical consequence and reasoning based on the strong logical consequence are not equivalent to reasoning based on the weak logical consequence. We have introduced a formal system FSED (with only two inference rules) for reasoning about existence dependencies. We have proved that the formal system FSED is sound and complete. We have defined a logical consequence for existence dependencies, called existence logical consequence, and proved that reasoning about existence dependencies based on the existence logical consequence and reasoning about functional dependencies based on the standard logical consequence are equivalent. We also have proved that none of the inference rules of Armstrong's formal system, AFS, holds for the weak logical consequence. In addition to formally stated propositions (about) logical consequences, we have described the semantics of the propositions more metaphorically by using 'the standard world', 'the strong world', 'the weak world', and 'the existence world' syntagms. Namely, we have intentionally done it having in mind a possibility of applying the properties of reasoning from these worlds in characterizing reasoning about knowledge in multi-agent systems. We are going to explore this possibility in some of forthcoming papers.

We also state here the following open problems.

P1: Define a sound and complete formal system for reasoning about functional dependencies based on the weak logical consequence.

P2: Is there a sound and complete formal system for existence dependencies with only one inference rule?

Appendix

Proof (Proposition (standard, strong))

We want to prove

$(f_1, .., f_m \models f) \Leftrightarrow (f_1, .., f_m \models [s] f)$.

$(\Rightarrow)$

Assume (S1) $f_1, .., f_m \models f$. We need to show (S2) $f_1, .., f_m \models [s] f$. Let $r \in Prel(R)$ be an arbitrary partial relation r. Suppose (S3) $r \Vdash [s]f_1, .., r \Vdash [s]f_m$. We would like to prove (S4) $r \Vdash [s]f$. Let $p \in Co(r)$ be an arbitrary completion of r. We need to show (S5) $p \Vdash f$. Because of (S3), we obtain

$p \Vdash f_1, .., p \Vdash f_m$. It follows by assumption (S1) that (S5), that is, (S4) holds, as wanted.

$(\Leftarrow)$

The proof of this part is similar to the proof of $(\Rightarrow)$.

Proof (Proposition (AFS, weak))

We need to prove

wlc_1: $\not\models[w]$ $X \to Y$ if $Y \subseteq X$

wlc_2: $X \to Y$ $\not\models[w]$ $XZ \to YZ$

wlc_3: $X \to Y, Y \to Z$ $\not\models[w]$ $X \to Z$

We have already proved wlc_3 (Proposition (standard, weak)). Now we firstly prove wlc_1 (triviality does not hold).

Let r be a partial relation over $R = AB$.

```
r(A  B)
=====
 1  ?
```

We have that $(\forall p \in Co(r))(p \Vdash AB \to B)$. Therefore, $r \not\Vdash [w]$ $AB \to B$, as wanted.

In order to prove, wlc_2, (augmentation does not hold), we construct a partial relation r over $R = ABC$ as follows.

```
r(A  B  C)
========
 1  2  3
 1  ?  4
```

We have $r \Vdash [w]$ $A \to B$ and $r \not\Vdash [w]$ $AC \to BC$, as needed.

References

[1] Abiteboul, S., Hull, R., Vianu, V.: Foundation of Databases. Addison-Wesley, Reading (1995)

[2] Cood, E.F.: The Relational Model for Database Management. Addison-Wesley, Reading (1990)

[3] Date, C.J.: Database in Depth: Relational Theory for Practitioners. O'Reilly, Sebastopol (2005)

[4] Date, C.J.: Why Three- and Four-Valued Logic Don't Work, pp. 2000–2006. Apress (2006)

[5] Date, C.J.: Logic and Databases: The Roots of Relational Theory. Trafford Publishing (2007)

[6] Maleković, M.: A Combined Algorithm for Testing Implications of Functional and Multivalued Dependencies. Informatica, An International Journal of Computing and Informatics 17, 277–283 (1993)

[7] Maleković, M.: A Sound and Complete Axiomatization of Functional Dependencies: A Formal System with Only Two Inference Rules. Informatica, An International Journal of Computing and Informatics 19, 407–408 (1995)

[8] Maleković, M.: Reasoning about Dependencies in Partial Knowledge Bases. In: Proceedings of 9th IEEE International Conference on Intelligent Engineering Systems, INES 2005, pp. 135–139 (2005)

[9] Pascal, F.: Practical Issues in Database Management. Addison-Wesley, Reading (2000)

Distributed Detection System of Security Intrusions Based on Partially Ordered Events and Patterns

Liberios Vokorokos, Anton Baláž, and Martin Chovanec

Abstract. The proposed system architecture of intrusion detection [1][2] uses a two-layer hybrid model for detecting intrusions. The system operates on the basis of partial network flows in real communication operation and provides processing of these data in real time. First layer consists of detection sensors which provide basic processing of input data on behalf of statistical methods with a direct connection to countermeasure modules. Performance and accuracy of the modeling system is ensured by using central distributed processing, in which the detection of generalized description of partial ordered events is used, preventing the intrusion itself. By doing so the attack variability issues of the same type are provided.

Keywords: distributed system, intrusion detection system, architecture, partially ordered events, patterns, sensor.

1 Introduction

Communication security is the basic aspect in every computer system. Technologies covering network connections, resource sharing and global security profiles create inclination on system attacks. To ensure prevention of these situations there is a possibility to implement intrusion detection systems in communication networks. These systems provide ability to detect intrusions and prevention against attacks on security systems. Currently it makes a great deal of difficulty to create secure system and maintain the security level during system functioning. In some cases it is impossible to apply global security politics covering the variety of attacks in time periods. Detection of intrusion from variety of sources represents a complex task. Searching for intrusion attributes has a non-deterministic character, where the same

[1] Supported by VEGA 1/4071/07

[2] This work was supported by the Slovak Research and Development Agency under the contract No. APVV-0073-07

I.J. Rudas et al. (Eds.): Towards Intelligent Engineering & Information Tech., SCI 243, pp. 389–403.
springerlink.com
© Springer-Verlag Berlin Heidelberg 2009

intrusion can be realized by different permutations of the same events. The aim of this article is to introduce a new architecture of hybrid-distributed system of detecting intrusion based on statistical methods partially ordered events realized by means of Petri nets. This research is realized in cooperation with the Department of Computers and Informatics and the Institute of Computer Technology of the Technical University in Košice, supported by VEGA 1/4071/07 and APVV-0073-07.

An intrusion detection system (IDS) inspects all inbound and outbound network activity and identifies suspicious patterns that may indicate a network or system attack from someone attempting to break into or compromise a system. This system are called network based IDS. In case, that the sensors perform some basic classification and data processing, systems can be called distributed IDS. On the base of collected information's, IDS can react on intersection attempts and reconfigure system parts to avoid potential security breach[1].

Standard structure of IDS[5]:

- events generator,
- analytical module,
- storage mechanism,
- countermeasure module.

Distributed architecture is enhancement of classic IDS, which adding follow elements into IDS structure[9]:

- distributed sensor,
- fusion manager,
- detection module,
- intrusion prevention module,
- centralized database,
- user front-end.

2 Designed Architecture of Intrusion Detection System

The proposed model presumes multilayer processing and distributing of the detection into remote topology (Fig. 1).

Elements of designed system include[6]:

- modified sensor system,
- system of distributed processing,
- basis of descriptive rules,
- modular system of countermeasures.

System working by this principle supposes the existence of cooperating sensor systems, to achieve secure high speed data flow. Supposing the use of sensors as minimal systems of vast amount there is a supposition of lower computing power hardware platforms, and so in some cases lower rate of detection. This fact in the research minimized by the use of correct and suitable method in the independent

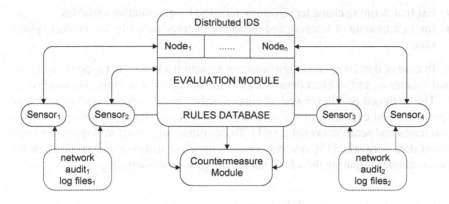

Fig. 1. Architecture of proposed hybrid DIDS

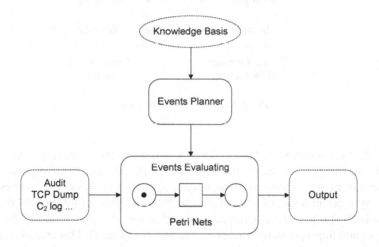

Fig. 2. Core of evaluation module of proposed DIDS

systems and appended by the exact distribution detection in the core of the entire system (Fig. 2) by means of intrusion accounts - realized as partially ordered events in the form of Petri nets. [8].

3 Design of Distributed Intrusion Sensor

Sensor as the main element of the first layer in the intrusion detection system, is created by the principle of individual detector with the use of statistical methods of detection[12]. The core of the system is created by a statistical characteristic of the spotted behavioral system stored in database element, which supports methods of system behavioral discretion[7]:

- Explicit defining character of communication by quantitative variables.
- Implicit module of learning and obtaining characteristics by the correct system state.

In case of detection the design of sensor system itself does not suppose its maximal exactness, and so exact detection and classifying the continuing occurrence.

The designed multilayer system supposes the use of superior system to ensure sensitivity and exactness of the detection. For this case the designed sensor communicates and sends collected data[1]. To minimize infrastructure load of the high-speed data networks [15] systems contains filter for merging and minimizing the event samples requiring the additional handling or processing.(Fig. 3).

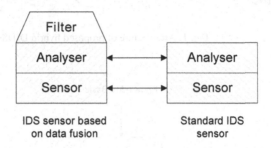

Fig. 3. Structure of sensor subsystem

In case of the sensor part of the system, there is a need to effectively define areas of intrusion detection and their responsive actions. Based on practical experiments a lower sensitivity was confirmed on minor changes for such a system, which resulted that the system did not detect the attack and failed to prevent this attack. For this purpose in the project, there was a need to define the dispersal boundary between the proper and improper state in the protected system. (Fig. 4). This area of research is defined as inexplicit or uncertain area and its scope was derived from the experimental environment on the basis of Poisson methods of distribution with the use of mathematical statistics.

Fig. 4. States of designed system

Based on these distributions areas behavioral profiles and steps for an automatic response were assigned to the individual areas. Defined areas satisfying those actions:

- *Normal State*: no action, update of time-dependent profile,
- *Undefined State*: alarm and generating data samples for further analysis,

- *Abnormal State*: alarm, the creation, application of countermeasures in the last step to generate data sample.

On this bases of distribution it is possible to specify the difference between uncertain and anomalous state just by applying the proactive policy, where in the first case there is no data flow constrained, there is only a waiting for the classification of the central system. In the second case the probability of the existence of this event is high, therefore data flow is blocked and the central system in this case has only the function of the possible correction module.

4 Distributed Intrusion Detection System

The main part of a distributed system presents a summary of computing nodes for analysis and identification of events by the means of described rules. This method provides high accuracy of detection, which is directly dependent on the signature of time-dependent event in the system. In terms of processing, this method requires a comparison of large quantities of rules which may cause time delay in system detection[4]. In order to eliminate the quantity description of intrusions, the system uses partially ordered events, which significantly reduce the number of evaluation by reducing the variants of the same attack[8].

Despite the cluster of intrusion descriptions by the partially ordered events, during the experimental verification the system contained a large number of descriptive rules to describe the representation of major events. The real set on was conditioned by the large number of conditions in the shortest possible time - reaching the real time processing. The isolated systems did not keep processing the system load in the mentioned limit, and so distributed processing was implemented into the system.

The research is considered on the distribution of data base in terms of:

- source of vectors - data obtained from the network audit,
- descriptive vectors - rules describing adverse events in the system.

Mathematical expressions of load balancing is represented as follows:

$$Z_{q,i} = \frac{t_z(i)}{\sum_{x=1}^{n} t_z(x)}; DZ_i = Z_{q,i} \times sizeof(dataVector) = \frac{t_z(i)}{\sum_{x=1}^{n} t_z(x) \times (dataVector)}$$
(1)

In terms of testing both variants and implementation, testing distributed data of both models has been chosen.

5 The Core Detection of Intrusions in Partially Ordered States

One of the main problems concerning the detection of intrusions in systems lies in the variability of possible attacks. The same attack can be implemented in several ways. The proposed DIDS (Fig. 1) architecture makes use of the analyses of partially-ordered states (Fig. 2) in the core, unlike the classical analysis of the transitions between states of the monitored system. The attacks are represented as a sequence of transient states[9] in the classical analysis scheme. The states in the system correspond with the states in the scheme of the attack, these are Boolean arguments associated with these conditions. These expressions must be net in order to transition to the next state. Individual conditions are linked by oriented edges, which represent events or conditions in the change of the state. Such a diagram represents the current state of the monitoring system. The change of the state is considered as the intrusion or penetration by the means of events carried out by an attacker, which begin in the initial state and end in the compromised state. The initial state represents the state of the systems before the starting the actual penetration. A final state represents a compromised state a compromised state of the system resulting in the termination of the penetration. Between the initial and the final state there are crossing states, which the violator must accomplish to achieve the final result of the system penetration. The figure 5 is an example[13] of an attack consisting of four states of an attack.

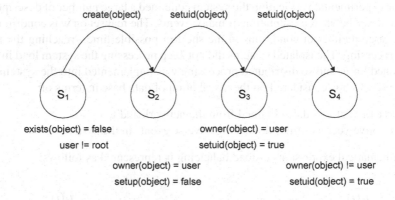

Fig. 5. States transition diagram

The conventional method strictly analyzes the features of the transitions states as an ordered sequence of states[10], without the possibility of overlapping the event sequence. The proposed DIDS architecture increases the flexibility of the state sequence by using partially arranged or ordered events. Partially ordered events specify the ability when the events are ordered one to another, whilst others are without this feature. Analysis of the partially ordered states allows multiple sequence of events in the states diagram. Using the partially ordered states in comparison to complete ordered states gives us only one description of the permutation of the same attack.

In the proposed architecture the transitions of partially ordered states are generated by means of partially ordered planning. Representation using partially ordered plan is essentially a more complete expression with a respect to fully organized representation. It allows the planner to postpone or completely ignore the unnecessary arrangement withdrawals. In the analysis of the transition states, the number of complete orders grows exponentially with the growing number of states. This attribute of complexity in the complete order is eliminated in the case of partially ordered planning. Using partially ordered notation and its degradation properties it is possible to deal with complex domain penetrations without the need to use exponential complexity.

Flexibility gained through the representation of partially ordered plans allows planer to process wide range of domain problems as well as detection of system intrusion. Partially order scheme provides a better representation of intrusion characters as a fully ordered representation - since only necessary dependencies between events are considered. Figure 6 shows only one dependence between operation *touch* and *chmod*.

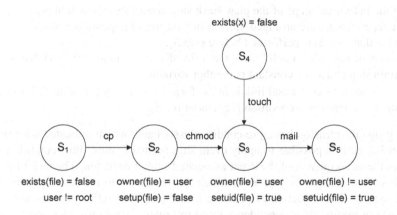

Fig. 6. Partially ordered states of penetration

It is not clear what relations are necessary between individual states - as shown on the figure 5. Figure 6 shows clearly which events precede which. Compromised state on the figure 6 is possible to represent using formulation of first order logic:

$$\exists \, /var/spool/mail/root \; x$$
$$/var/spool/mail/root \in x \wedge$$
$$owner(/var/spool/mail/root) = root \wedge$$
$$setuid(/var/spool/mail/root) = enable$$
$$\Rightarrow compromised(x) = true$$

The proposed intrusion analyzes approaches the need to identify at least minimal set of intrusion a necessary dependencies between these characters. Minimal set of characters assumes the elimination of irrelevant characters of the intrusion itself.

6 Planning Sequence of the Intrusion

Intrusion can be defined as a set o events with the target to compromise the integrity, confidentiality, availability of resources. The proposed IDS architecture includes a part for planning construction sequence event plans in the core, which consists of the intrusion. Planning consists of objectives, states and events. Depending on what needs to be done in the final plans, planning combines current state of the environment with information based on the final outcome of the events.

The analysis of transitions states is characterized by penetration of the performed sequences of events, which result form the initial system state to the compromised state.

Planning is defined as follows[11]:

1. Set the individual steps of the plan. Each step constitutes the action plan.
2. Set dependencies are arranged. Each is in the form of dependence $S_i \prec S_j$, which means that step S_i is performed before step S_j.
3. Ties set of variables. Each link in the form of $v = x$, where v the variable is in a certain step and x is a constant or another variable.
4. Set of causal links. Causal link is in the form $S_i \xrightarrow{c} S_j$. Out of state S_i is using c state S_j, where c pre-condition is essential for S_j.

Any sign of intrusion has a pre-condition associated, which indicated what must be met before it is possible to apply event sing related feature. Post-condition expresses the consequence of the event associated with a intrusion. The task of planning is to find a sequence of events, which has resulted into the actual penetration to the system. The objective of planning in the proposed IDS system is to find a sequence of events, their dependencies and to create a final character order of the intrusion.

Algorithm of partially ordered planning starts with a minimum part plan (subplan) and in each step this plan is expanded by the means of achievable pre-condition step.

6.1 Planning

The aim of this section is representation of algorithm planning in the proposed IDS architecture. The plan is to represent penetration by a triplet $\langle A, O, L \rangle$, where A is a set of events, O is a set of ordered dependencies over set A, and L is a set of causal connections. The planner starts its operation with an empty plan, and gradually refines this plan and has to meet the conditions of consistency defined in the set O. Key step in this activity is to store states of past decisions and also conditions for these decisions. To ensure consistency between the various events storing relations between events by means of causal connections is done. Causal connection is structure consisting of two connections to the events of the plan (producer A_p and consumer A_c) and expression Q, which is a result of A_p and pre-condition A_c. Entry is represented as $A_p \xrightarrow{Q} A_c$ and the connections are stored in the set L. Causal connections are used to detect the interference between new and past decisions. Classed as a threat. In other words, if $\langle A, O, L \rangle$ represents a plan and $A_p \xrightarrow{Q} A_c$ is the L. Let A_t is another event in the A, A_t threatening $A_p \xrightarrow{Q} A_c$ if:

- $O \cup \{A_p < A_t < A_p\}$ and
- A_t has $\neg Q$ as result

If the plan includes the threat of this plan may not satisfy the specified conditions defined on $\langle A, O, L \rangle$. The planner must take into account threats when creating the final plan. The algorithm can add additional ordered dependencies by insuring the implementation of A_t before A_p. This method of removing threats represents so called - dismantling plan.

Planning $(\langle A, O, L \rangle, agenda, \Lambda)$

1. **Completion:** If the agenda is empty, return $\langle A, O, L \rangle$.
2. **Target selection:** $\langle Q, A_{need} \rangle$ is a pair in the agenda (according $A_{port} \in A$ and Q is conjunction of preconditions A_{need}).
3. **Event selection:** $A_{need} = event\ selection$ that adds to Q one of the new events from Λ, or the event already in A, possible to be ordered according to A_{need}. If there does not exist any of the mentioned events, return error. Let $L' = L \cup \{A_{add} \xrightarrow{Q} A_{need}\}$, and $O' = O \cup \{A_{add}\}$ and $O' \cup O' \{A_0 < A_{add} < A_\infty\}$ otherwise $A' = A$).
4. **Update set of events:** Let $agenda' = agenda - \{\langle Q, A_{need} \rangle\}$. If A_{need} is a new instance, than for each conjunction Q_i of its precondition add $\langle Q_i, A_{add} \rangle$ to $agenda'$.
5. **Protection of casual connections:** For each operation A_t, that can threaten the casual connections $A_p \xrightarrow{R} A_c \in L'$ select consistent ordered dependencies:
   - *Factorization :* Add $A_t < A_p$ to O'
6. **Recursive calling:**
   Planning $(\langle A', O', L' \rangle, agenda', \Lambda)$

The core of planning is represented by planning algorithm above, which searches the state space plans. The algorithm starts with an empty plan, step by step executes the non-deterministic event sequence choice until all pre-conditions are met by means of causal connections. Partially arranged dependencies of the final plan represents only a partial ordered plan, which solves the chosen problem of planning. The arguments of the algorithm are planning structure and plan agenda. Each agenda item is a pair of $\langle Q,A \rangle$, where Q is conjunction of pre-conditions A_i.

7 Implementation and Experimental Validation of Proposed Model

Implementation of designed model was realized based on object oriented coding, in term of direct access to systems source and to devices of distributed processing, was chosen language C++, which in term of:

1. Sensors system promoted:

   - cases
   - reindexed field
   - data structure
   - communication and sampling bookcases
   - high performance and speed of processing

2. central distributed systems :

   - distributed environment - MPI
   - supply of data structure

Experimental architecture attest was realized in two environments, within the first testing of portability and architecture function was used laboratory conditions. After that, for efficiency revision of parameters was system tested in real conditions for securing asked data flow and amount of watched datagram.

Verification of sensor subsystem

Isolated system of sensors was attached to a protected computer network, where its function was verified in the actual operation (Fig. 7).

Graphs reflecting the abilities and limits to the applicability of proposed sensor system where created on the basis of performance measurements. The first parameter was the difficulty of the system resources - in terms of CPU. Analysis of flow in real time required more architectures because of their spatial and pricing complexity(Fig. 8).

From the measured values, it is clear that the tops of the sensor system were stored in the profile during the transitional intervals. During time interval of one

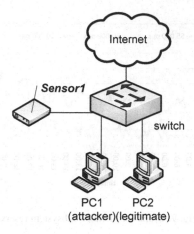

Fig. 7. Experimental topology of sensor subsystem

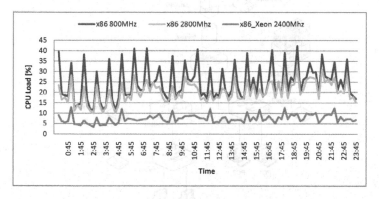

Fig. 8. Graph of computing capacity requirements

hour the tops did not exceed the value of 42% of system load configuration. Memory intensity was another observed parameter, where it is necessary to verify the occupancy of memory elements during the operation of sensor system.(Fig. 9).

The use of RAM (operating memory) in real time during the detection did not exceed the value of 100 megabytes not even by using 55 000 pps which provides the possibility to use the sensor in high-speed networks or in networks with great load during their operation.

Verification of distributed system

Experimental verification of this system was done in a more complex topology, where it was necessary to process data from several sensor subsystems. The process has been implemented in the MPI GRID system, realized in a mutual workplace of FEI and SAV(Fig. 10).

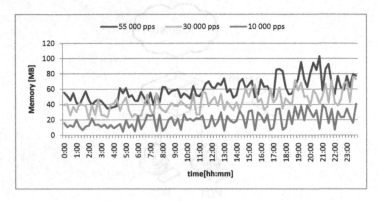

Fig. 9. Graph for memory subsystem reservation

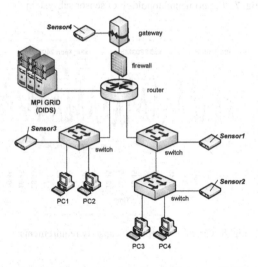

Fig. 10. Experimental topology of proposed DIDS

Testing showed that the use of production nodes in the GRID system is efficient in the terms of load distribution and increase of the processing speed by means of descriptive rules. Based on the existence assumption of large quantities of descriptive rules distributed processing has been chosen. Testing comparison of this analysis was observed on thirteen nodes of the tested system. The results suggested the possibility of processing large number of rules(Fig. 11).

The sum characteristic of these rules showed that processing the number of 700 000 rules (Fig. 12) in the distributed system is possible in six seconds by using the current hardware configuration.

Based on the experimental verification it was possible to derive parameters of evaluation efficiency, acceleration and overbalance:

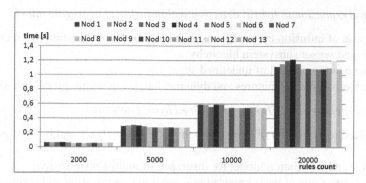

Fig. 11. Graph of rule distribution in GRID

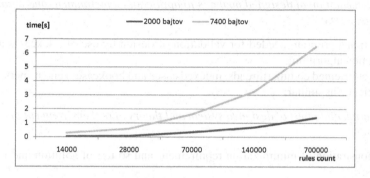

Fig. 12. Graph of time dependency for rule execution in DS

$$Z_n(p) = \frac{T_n(1)}{T_n(p)} = \frac{T_{\frac{7400}{700000}}(1)}{T_{\frac{7400}{700000}}(13)} = \frac{57134567[\mu s]}{5088769[\mu s]} = 11.22934 \tag{2}$$

$$E_n(p) = \frac{Z_n(p)}{p} = \frac{Z_{\frac{7400}{700000}}(13)}{13} = \frac{11.22934}{13} = 0.8637 \tag{3}$$

$$V_n(p) = \frac{T_n(1)_{max} - T_n(1)_{min}}{T_n(1)_{min}} = \frac{T_{\frac{7400}{20000}}(1)_{max} - T_{\frac{7400}{20000}}(1)_{min}}{T_{\frac{7400}{20000}}(1)_{min}} = \frac{4688769 - 4016918\mu s}{4016918\mu s} = 0.16725 \tag{4}$$

8   Goals of Designed IDS Architecture

Starting from initiated study, it was possible to derive next contributions and advances in listed domain:

Method expansion in hybrid detection within fusion of aggregate attack:

- detection of intrusion by modified statistic method with assistance of reconfiguration and sensor subsystem hierarchy;
- extract handling model of undefined state;
- detection based on signatures and dynamic balancing within processing DS;

Design of architecture DIDS for high-speed hybrid networks:

- architecture expansion of sensor system by event analyzer and by configuration data;
- arrange and association model for insurance of multi-sensor fusion;
- model of revision of basis knowledge's, amount minimization and verification of rules;

Implementation of designed methods to application environment and creating of prototype DIDS:

- functional program needed for validation of design by use of bookcases and innovation algorithm;
- object-oriented coding (threads, indexed cases and bookcases) and its perspective efficiency in similar;

Experimental attest of designed system and theory in real environment of extreme character:

- demonstration of minimization requirement and so use of solution and its methods;
- detection exactness of controlled system, results in detection of DOS and DDOS attacks;
- important results in processes of balancing and dividing of load applicable in several DS;

Benefits of this work are an important contribute to VEGA 1/4071/07 projects (Security architecture of heterogeneous distributed and parallel computing system and dynamical computing system resistant against attacks) and APVV 0073-07 (Identification methods and analysis of safety threats in architecture of distributed computer systems and dynamical networks). Extension of this work is feasible in the application of creation and automated generation of description rules within system detection on behalf of sign intrusion. Efficient decoding of datagrams to a higher layer of network communication to ensure the detection to the application level (deep packet inspection).

9 Conclusion

Data security is one of the priorities in the communication technology. This paper deals with the methods of intrusion detection in high-speed networks. The paper points out analysis methods of data input flow, proposing its expansion and the specification on the basis of trends in current systems. Description of the methods on a partially ordered events provide not only a time-dependent description of events in the system, but also an effective way of reporting events preceding the existence of distortion and minimizing the required number of attack descriptions of the same type. Hybrid detection associated with detection has a principle of monitoring the signature, decreases the reaction time of the system and brings up methods for real time monitoring. Distributed methods of detection can be accelerated by using distributed description rules. Experimental verification results in practical implementation results, performance measurement and efficiency of proposed methods. Work focuses on the structure and classification, distributed approach and its implementation in IDS. Emphasis is placed on communication security and a proposed model of distributed intrusion detection.

References

1. Vokorokos, L., Chovanec, M., Látka, O.: Security of Distributed Intrusion Detection System Based on Multisensor Fusion. In: 6th International Symposium on Applied Machine Intelligence and Informatics SAMI, pp. 19–24 (2008) ISBN: 978-1-4244-2106-0
2. Vokorokos, L., Chovanec, M.: Improvement of security protocols based on asymmetric key mechanism. In: ECI 2006, Herľany - Slovakia (2006) ISBN 963 7154
3. Vokorokos, L.: Principles of Data Flow Controlled Computer Architectures, Copycenter Košice (2002) ISBN 80-7099-824-5
4. Vokorokos, L., Baláž, A., Chovanec, M.: Intrusion Detection System Using Self Organizing Map. Acta Electrotechnica et Informatica 6(1), 81–86 (2006)
5. Axelsson, Š.: The base-rate fallacy and the difficulty of intrusion detection. Information and System Security 3(3), 186–205 (2000)
6. Bass, T.: Intrusion detection systems and multisensor data fusion. Communications of the ACM 43(4), 99–105 (2000)
7. Chovanec, M., Látka, O.: Communication security of Sensors in Distributed Intrusion Detection Systems. In: 7th Scientific Conference TU Košice, Košice, Slovakia, 23.5, 2007 Elfa, s.r.o, I., 7 (2007) ISBN 978-80-8073-803-7
8. Baláž, A.: Architecture of the Intrusion Detection System Based on Partially Ordered Events and Patterns, Dissertation thesis, Košice (2008)
9. Bace, R.G.: Intrusion Detection. Macmillan Publishing Co., Inc. (2000)
10. Zhouwei, L., Amitabha, D., Jianying, Z.: Theoretical Basis for Intrusion Detection (2005)
11. Russel, S., Norvig, P.: Artificial Intelligence: A Modern Approach. Prentice-Hall, Inc., Englewood Cliffs (1995)
12. Anderson, Lunt, Javits, Tamaru, Valdes: Detecting Unusual Program Behavior Using the Statistical Components of NIDES (1995)
13. Curry, D.A.: Improving the Security of your UNIX system (1990)
14. Fan, W., Miller, M., Stolfo, S., Lee, W., Chan, P.: Using artificial anomalies to detect unknown and known network intrusions, pp. 507–527 (2004)
15. Kruegel, C., Valeur, F., Vigna, G., Kemmerer, R.: Stateful intrusion detection for high-speed networks (2002)

Wireless Sensors Networks – Theory and Practice

István Matijevics

Department of Informatics, University of Szeged, Árpád tér 2, H-6720 Szeged, Hungary
mistvan@inf.u-szeged.hu

Abstract. Additional advantages of WSNs are the possibility of control and monitoring these applications from remote places and having a system that can provide large amounts of data about those applications for longer periods of time. This large amount of data availability usually allows for new discoveries and further improvements. [3]

Keywords: Wireless sensor networking, Greenhouse, Embedded systems, Sun SPOT.

1 Sensors

A **sensor** is a device that measures a physical quantity of signals and converts it into a voltage or current, analog or digital signal which can be read by an observer, instrument or by an computer (microcontroller) based instrument. Sensors are used in everyday objects. For accuracy, all sensors need to be calibrated against known standards. [16]

The following list of sensors is sorted by sensor types [16]:

- Acoustic, sound, vibration
- Automotive, transportation
- Chemical
- Electric current, electric potential, magnetic, radio
- Environment, weather
- Flow
- Ionising radiation, subatomic particles
- Navigation instruments
- Position, angle, displacement, distance, speed, acceleration
- Optical, light, imaging
- Pressure, force, density, level
- Thermal, heat, temperature
- Proximity, presence

I.J. Rudas et al. (Eds.): Towards Intelligent Engineering & Information Tech., SCI 243, pp. 405–417.
springerlink.com
© Springer-Verlag Berlin Heidelberg 2009

2   Wireless Sensors Network

Nowadays the modern technologies and scientific developments in sensor techniques and wireless communication technologies have made available a new type of communication network. These equipments are battery-powered, integrated wireless sensor devices. Wireless Sensor Networks (WSNs), are self-configured and are without infrastructures. WSN collects data from the environment and sends it to a destination site where the data can be observed, memorized and analyzed. Wireless sensor devices responds to a "control site" on specific requests, or can be equipped with actuators to realize commands. We can divide these networks in two groups, Wireless Sensor and Actuator Networks.

At present time, due to economic and technological reasons, most available wireless sensor devices are very constrained in terms of computational, memory, power, and communication capabilities. The research on WSNs is concentrated on design of improving the energy- and computationally effectiveness. The main goal is to find new algorithms and protocols. The application field has been restricted to simple data-oriented monitoring and reporting applications. All this is changing very rapidly, as WSNs are capable of performing more advanced functions and handling multimedia data are being introduced. New network architectures with heterogeneous devices and expected advances in technology are eliminating current limitations and expanding the spectrum of possible applications for WSNs considerably.

Wireless Sensor Device has 6 main parts (Figure 1):

- sensor,
- I/O interface,
- memory,
- processor,
- radio and
- battery.

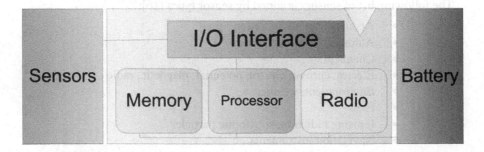

Fig. 1. General architecture of a wireless sensor device

3   The Architecture of Wireless Sensor Network

A **wireless sensor network** (WSN) is a wireless network consisting of spatially distributed autonomous devices using sensors to cooperatively monitor physical or environmental conditions, The development of wireless sensor networks was originally motivated by military applications such as battlefield surveillance. However, wireless sensor networks are now used in many industrial and civilian application areas, including industrial process monitoring and control, machine health monitoring, environment and habitat monitoring, healthcare applications, home automation, and traffic control. In addition to one or more sensors, each node in a sensor network is typically equipped with a radio transceiver or other wireless communications device, a small microcontroller, and an energy source, usually a battery.

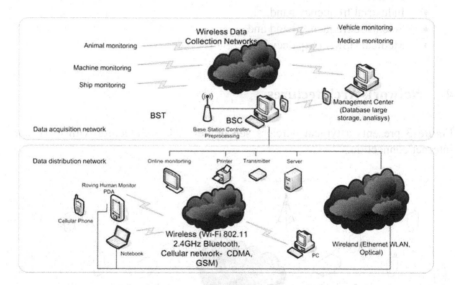

Fig. 2. Wireless sensor network

The envisaged size of a single sensor node can vary from shoebox-sized nodes down to devices the size of grain of dust, although functioning 'motes' of genuine microscopic dimensions have yet to be created. Normally a sensor network is a wireless ad-hoc network, each sensor supports a multi-hop routing algorithm. A typical Wireless Sensor Network is shown in Figure 2.

All Wireless sensors are equipped with low-power 16 or 32 bit RISC microcontrollers running at low frequencies, their computationally capacities are lowpitched. The microcontrollers can be set to sleep mode.

The Random Access Memory (RAM) is restricted to a few hundreds of kilobytes to store application related data. The Read-Only Memory (ROM) is to store program code, include a few hundreds of Electrically Erasable Read-Only Memory (EEPROM).

The radio transceiver is a complete half-duplex radio link, contains all circuits to send and receive data over the wireless media:

- modulator,
- demodulator,
- D/A digital to analog converter,
- A/D analog to digital converter,
- amplifiers,
- filters,
- mixers and
- antenna.

The used frequencies in WSN technology are:

- 400 MHz,
- 800–900 MHz,
- 2.4 GHz,
- Industrial frequency band,
- Scientific frequency band and
- Medical (ISM) frequency band.

4   Network Architectures

Figure 3 presents a typical wireless sensor network architecture with radio and internet connection.

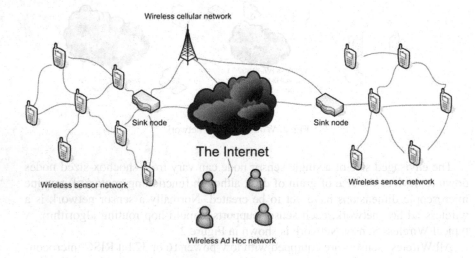

Fig. 3. General wireless sensor network architectures

The presented architecture has two advantages:

- It is possible use very different equipments and
- The modification of system is always possible with new elements.

5 Type of Wireless Technologies

Some representation of wireless transmission technology are shown in Table 1. with characteristic parameters [7].

Table 1. Comparison of four wireless network technologies

	Wi-Fi	Bluetooth	ZigBee
Standard	IEEE 802.11b,g	IEEE 802.15.1	IEEE 802.15.4
Frequency band	2.4 GHz	2.4 GHz	868/915 MHz, 2.4 GHz
Data rate	Up to 22 Mb/s	1 Mb/s	20/40/250 kb/s
Network topology	32 active nodes	8 active nodes	255 active nodes
Range	100 m	10 /100 m	10/100 m
Battery life	h	Days	Years
Cost	Relatively high	Relatively low	Lowest
Applications	Wireless internet access	Data and voice access, Ad-hoc networking	Remote monitoring and control

6 Small Programmable Object Technology (Sun SPOT) Sun™

The current part describes the implementation and configuration of the wireless sensor network using the Sun SPOT platform. Sun SPOT is a small electronic device made by Sun Microsystems. They have a wide variety of sensors attached to it. Sun SPOTs are programmed in a Java programming language, with the Java VM run on the hardware itself. It has a quite powerful main processor running the Java VM "Squawk" and which serves as an IEEE 802.15.4 wireless network node. The SPOT has flexible power management and can draw from rechargeable battery, USB host or be externally powered. The Sun SPOT is designed to be a flexible development platform, capable of hosting widely differing application modules.

6.1 Wireless Sensor Networks Using Sun SPOT Devices

Wireless sensor networks with Sun SPOTs consist of tiny devices that usually have several resource constraints in terms of energy, processing power and memory. In order to work efficiently within the constrained memory, many operating systems for such devices are based on an event-driven model rather than on multi-threading. Continuous advancements in wireless technology and miniaturization have made the deployment of sensor networks to monitor various aspects of the environment increasingly possibilities. Wireless Sensor Networks have recently received a lot of attention within the research community since they demand for new solutions in distributed networking. A common scenario associated with these networks is that tiny nodes, equipped with several sensors and hardware for wireless communication, are deployed randomly and in large numbers within a certain area. In order to report the data they gather in their proximity to an interested application or user, nodes connect to their neighbors and send valuable information on a multi-hop path to its destination.

The concept of wireless sensor networks is based on a simple equation:

$$Sensing + CPU + Radio = Thousands\ of\ potential\ applications$$

As soon as people understand the capabilities of a wireless sensor network, hundreds of applications spring to mind. It seems like a straightforward combination of modern technology. However, actually combining sensors, radios, and CPU's into an effective wireless sensor network requires a detailed understanding of the both capabilities and limitations of each of the underlying hardware components, as well as a detailed understanding of modern networking technologies and distributed systems theory.

Wireless sensor networking is one of the most exciting technologies to emerge in recent years. Advances in miniaturization and MEMS-based sensing technologies offer increases by orders of magnitude in the integration of electronic networks into everyday applications. Traditional microcontroller design strategies have not reached the best possible power consumption, especially for the specialized application set of sensing networks. Power efficiency is a prime concern in wireless sensors, whether powered by a battery or an energy-scavenging module. Trends in miniaturization suggest that the size of wireless sensors will continue to drop, however there has not been a corresponding drop in battery sizes.

6.2 Sun SPOT

One of the most poplar technologies in the WSN area is Sun SPOT (Small Programmable Object Technology). It contains 32-bit ARM9 CPU, 512 K memory, 2 Mb flash storage and wireless networking is based on ChipCon CC2420 following the 802.15.4 standard with integrated antenna and operates in the 2.4 GHz to 2.4835 GHz ISM unlicensed bands. The IC contains a 2.4 GHz RF transmitter/receiver with digital direct sequence spread spectrum (DSSS) baseband modem with MAC support. Figure 4 presents the Sun SPOTs processor board.

Fig. 4. Sun SPOT processor board

The sensor board integrates multiple sensors, monitoring LED and interactive switches into one board. All the facilities of this board are programmable in Java.

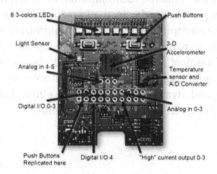

Fig. 5. Sun SPOT sensor board

The facilities of the sensor board are:
- One 2G/6G 3-axis accelerometer
- One temperature sensor
- One light sensor
- Two 8-bit tri-color LEDs
- 6 analog inputs
- Two momentary switches
- 5 general purpose I/O pins

The internal battery is a 3.7 V rechargeable lithium-ion prismatic cell. The battery has an internal protection circuit to guard against over discharge, under voltage and overcharge conditions. The battery can be charged from either the USB type mini-B device connector or from an external source with a 5 V power supply.

6.3 Java

The Java programming language is a general-purpose concurrent class-based object-oriented programming language, specifically designed to have as few implementation dependencies as possible. It allows application developers to write a program once and then be able to run it everywhere on the Internet. The most popular developing environment for Java is Netbeans IDE.

A major benefit of using bytecode is porting. However, the overhead of interpretation means that interpreted programs almost always run more slowly than programs compiled to native executables would, and Java suffered a reputation for poor performance. This gap has been narrowed by a number of optimization techniques introduced in the more recent JVM implementations.

One such technique, known as just-in-time (JIT) compilation, translates Java bytecode into native code the first time that code is executed, then caches it. This results in a program that starts and executes faster than pure interpreted code can, at the cost of introducing occasional compilation overhead during execution. More sophisticated VMs also use dynamic recompilation, in which the VM analyzes the behavior of the running program and selectively recompiles and optimizes parts of the program. Dynamic recompilation can achieve optimizations superior to static compilation because the dynamic compiler can base optimizations on knowledge about the runtime environment and the set of loaded classes, and can identify hot spots - parts of the program, often inner loops, that take up the most execution time. JIT compilation and dynamic recompilation allow Java programs to approach the speed of native code without losing portability.

6.4 Examples

The sensor board includes a 3D accelerometer, a temperature sensor, a light sensor, eight LEDs, two switches, five general-purpose I/O pins and four high current output pins. Because of its Java implementation, programming the Sun SPOT is surprisingly easy. The task of developing and deploying applications on a Sun SPOT using the Sun SPOT Software Development Kit (the Sun SPOT SDK) can be demonstrated with a few examples.

6.5 Reading Data from 3-Axis Accelerometer

The three-axis accelerometers measure acceleration in three dimensions. Accelerometers are very handy for measuring the orientation of an object relative to the earth, because gravity causes all objects to accelerate towards the earth. The range of value for each axis is 4 to 929. The default value without having any acceleration is 461 to 463 for X, Y and Z axis. The example below shows the proper usage.

```
import com.sun.spot.sensorboard.EDemoBoard;
import com.sun.spot.sensorboard.peripheral.IAccelerometer3D;
private IAccelerometer3D accel = EDemoBoard.getInstance().getAccelerometer();
int tiltX = (int)Math.toDegrees(accel.getTiltX()); // returns [-90, +90]
double GForceX = accel.getAccelX();
```

6.6 Reading Data from Temperature Sensor

This sensor is capable of detecting the environmental temperature. The temperature value which is read from the sensor is a raw value that represents a temperature number without the standard of Celsius or Fahrenheit. The model of the temperature sensor is ADT7411 with a 10-bit temperature to digital converter which is capable of detecting -40 to +125 Celsius. The range of the raw value is 0 to 1023. The following mathematical algorithm can be applied to convert the raw reading into Celsius or Fahrenheit.

```
import com.sun.spot.sensorboard.EDemoBoard;
import com.sun.spot.sensorboard.io.ITemperatureInput;
private ITemperatureInput tempSensor = EDemoBoard.getInstance().getADCTemperature();
double tempF = tempSensor.getFahrenheit(); // The value converted to Farenheight
double tempC = tempSensor.getCelsius(); // The value converted to Celcius
```

6.7 Reading Data from Light Sensor

The light sensor measures the range from darkness to lightness and converts it to a raw value. The range for the raw light value is 0 to 1023. The value represents the intensity of detected light.

```
import com.sun.spot.sensorboard.EDemoBoard;
import com.sun.spot.sensorboard.peripheral.ILightSensor;
private ILightSensor lightSensor = EDemoBoard.getInstance().getLightSensor();
int lightLevel = lightSensor.getValue();
```

6.8 Configuring the 8 Tri-Color LEDs

There are eight tri-color LEDs located on the sensor board. These LEDs are able to flash with a range of colors. The typical use of these two LEDs is to show the status when an application is in a particular state. Therefore, it might be easier for a developer to see the current progress of an application. The color of the LEDs is shown with red, green and blue. The range of each color for intensity is 0 to 255.

```
import com.sun.spot.sensorboard.EDemoBoard;
import com.sun.spot.sensorboard.peripheral.ITriColorLED;
import com.sun.spot.sensorboard.peripheral.LEDColor;
private ITriColorLED [] leds = EDemoBoard.getInstance().getLEDs();
leds[i].setOff(); // turn off LED
leds[i].setRGB(int red, int green, int blue ); // set color
leds[i].setColor(LEDColor.MAGENTA);
leds[i].setOn(); // turn on LED
```

7   Distant Monitoring and Control for Greenhouse Systems Using Sun SPOT

This Web based distant greenhouse monitoring system uses a PIC microcontroller, that will allow you to monitor and control the greenhouse laboratory from any Internet connection. The described system is highly customizable and easily adaptable to particular deployment requirements. The main idea is to develop the greenhouse laboratory which can be used by students of our faculty from home. The hardware part consists of several modules. Some of them such as the developer board and monitoring console are essential and the rest can be adjusted according user requirements. This study presents the development of a web-based remote control laboratory using a greenhouse scale model for teaching greenhouse climate control techniques using different hardware and software platforms including the Sun SPOT devices.

7.1 Introduction

The glass house is a construction where plants are grown. The glass house is constructed of glass; it heats up because the sun's electromagnetic rays heat the plants, soil and other things inside the glass house. The heated air remains inside the walls and roof of the glass house.

In the remote greenhouse laboratory, students, as future professional engineers, develop practical skills that enhance their qualities of develop and control the model of the real system. Climate control is vitally important to the operation of greenhouses. New, specialized, electronic climate control systems have become available on the market. From an energy efficiency point of view, significant savings in heating, cooling, watering costs are possible with these systems. For example, strawberries are a plant that you can grow easily in the greenhouse. The difference in growing strawberries to that of tomatoes in the greenhouse is following: on the tomato plant only the leaves that are on the top of the plant will get sunlight. The top leaves will shade the lower leaves. The strawberry plant grows outward, so all the leaves on this plant will receive light, making the strawberry plant an easy choice for beginner gardeners. Because the strawberries do not get light from other sources, the amount of sunlight they receive will have to be regulated. A good rule of thumb is ten to twelve hours of light, sunlight, or indoor light for good production.

All types of strawberry plants will require cooler temperatures in order to produce strawberries. The weather outdoors also has to be taken into consideration. If there were three or four weeks of cooler temperatures in the spring months, or possibly six to seven weeks; this is what has to be to replicated in the greenhouse.

7.2 Solution

Assembling the web server happened in several steps, as there was a new task after the realization of each step. The web surface is simple and transparent, the various hardware components can be easily controlled. The image of the web cam can be seen on the internet, thus the user has a visual image of the virtual laboratory. The source code is simple, understandable and therefore, applicable for further development. The web server's basic function is the web surface, through which the specific components can be controlled. The Sun SPOT device is used for temperature monitoring and lightness measuring.

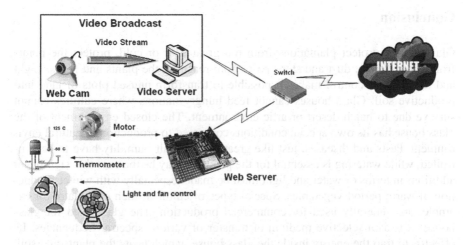

Fig. 6. The whole remote control system realized so far

Through this the various elements of the laboratory which are connected to the
server can be controlled from any point of the earth using only the web browser,
furthermore, feedback from the controlled objects will also be sent.

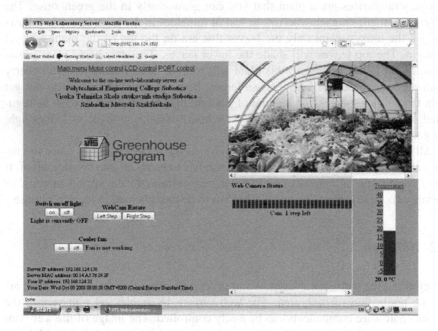

Fig. 7. Screenshot from the system

For the display of the real time video streaming the web browser is also of vital
importance, as it can be only accessed via the internet.

Conclusion

Glass houses protect plantations from too great heat or cold, protect the plants
from snow storms, dust and stops pests from reaching the plants and fruits. Light
and temperature control makes it possible to transform unused plots of land into
productive soil. Glass houses cannot feed hungry nations where cultures cannot
survive due to harsh desert or artic environment. The closed environment of the
glass house has its own special conditions compared to production in natural environment. Pests and diseases, just like great heat and air humidity have to be controlled, while watering is essential for the plants. It may be the case that significant
additions in terms of water and light must be made, especially with winter production of warm period vegetables. Special types of cultures, such as tomato, for example, are generally used for commercial production. The glass used for glass
houses acts as a selective medium of transfer of various spectral frequencies. Its
effect is to trap the energy inside the glass house, which heats the plants and soil.
This also warms up the air near the soil, which cannot rise and escape.

Acknowledgments. The paper was supported by TÁMOP-4.2.2/08/01 project titled "Sensor network based data collection and information processing".

References

[1] Römer, K., Mattern, F.: The Design Space of Wireless Sensor Networks. IEEE Wireless Communications 11(6), 54–61,
`http://www.vs.inf.ethz.ch/publ/papers/`
`wsn-designspace.pdf` doi:10.1109/MWC.2004.1368897

[2] Lewis, F.L.: Wireless Sensor Network

[3] Labrador, M.A., Wightman, P.M.: Topology Control in Wireless Sensor Networks: with a Consort Model Agency for Doctrine and Research. Springer, Heidelberg (2009)

[4] Otto, C., Milenkovic, A., Sanders, C., Jovanov, E.: System Architecture of a Wireless Body Area Sensor Network for Ubiquitous Health Monitoring. Journal of Mobile Multimedia 1(4), 307–326 (2006)

[5] Stoianov, I., Nachman, L., Madden, S., Tokmouline, T.: Information Processing in Sensor Networks. In: Proceedings of 6th International Conference on Information Processing in Sensor Networks, pp. 264–273 (2007) ISBN:978-1-59593-638-X

[6] Gascón, D.: 802.15.4 vs ZigBee (November 17, 2008), `http://www.sensor-networks.org/index.php?page=0823123150`

[7] Xuesong, S., Wu, C., Ming, L.: Wireless Sensor Networks for Resources Tracking at Building Construction Sites. Tsinghua Science And Technology 13(S1), 78–83 (2008)

[8] Sun™ Small Programmable Object Technology (Sun SPOT) Owner's Manual Release 3.0, Sun Microsystems, Inc. (2007)

[9] Sun Spot Developer's Guide. Sun Microsystems (2005)

[10] Demo Sensor Board Library. Sun Microsystems (2005)

[11] Scaglia, S.: The Embedded Internet. Addison-Wesley, Reading (2008)

[12] Gosling, J.: The Java™ Language Specification, 3rd edn. Addison-Wesley, Reading (2005)

[13] István, M., János, S.: Advantages of Remote Greenhouse Laboratory for Distant Monitoring. In: Proceedings of the Conference ICoSTAF 2008, Szeged, Hungary, pp. 1–5 (2008)

[14] János, S., István, M.: Distant Monitoring and Control for Greenhouse Systems via Internet. In: Zbornik radova konferencije Yuinfo 2009, Kopaonik, Srbija, pp. 1–3 (2009)

[15] István, M., János, S.: Comparison of Various Wireless Sensor Networks and their Implementation. In: Proceedings of Conference SIP 2009, Pécs, Hungary, pp. 1–3 (2009)

[16] Wikipedia: Sensor, `http://wikipedia.org`

Image Processing Using Polylinear Functions on HOSVD Basis

András Rövid and László Szeidl

John von Neumann Faculty of Informatics, Budapest Tech
Bécsi út 96/B, H-1034 Budapest, Hungary
rovid.andras@nik.bmf.hu; szeidl@bmf.hu

Abstract. The main aim of the paper is to introduce a new representation domain advantageously usable also in the field of image processing. In this case the image is represented by polylinear functions on HOSVD basis. The paper gives an overview, how these functions can be reconstructed and how they can be applied in case of image processing. Comparing to other domains, using the proposed domain the image can be expressed by a reduced number of components (polylinear functions) by maintaining its quality. The efficiency of this representation form is demonstrated by performing an image resolution enhancement trough this new domain by maintaining its quality significantly.

1 Introduction

Digital image processing plays an important role in most of the nowadays applications. Numerous applications of numerous fields are strongly related to reconstruction of surfaces from a given set of discrete points [4] [5].

Numerical reconstruction or recovering of a continuous intensity surface from discrete image data samples is considered for example when the image is resized or remapped from one pixel grid to another one. In case of image enlargement the colors of such missing pixels should be estimated. One common way for estimating the color values of missing points is interpolating the discrete source image. There are several issues which affect the perceived quality of the interpolated images: sharpness of edges, freedom from artifacts and reconstruction of high frequency details [10].

There are numerous methods approximating the image intensity function based on the color and location of known image points, such as the bilinear, bicubic or spline interpolation all working in the spatial domain of the input image. Depending on their complexity, these use anywhere from 0 to 256 (or more) adjacent pixels when interpolating. The more adjacent pixels they include, the more accurate they can become, but this comes at the expense of much longer processing time [12].

I.J. Rudas et al. (Eds.): Towards Intelligent Engineering & Information Tech., SCI 243, pp. 419–434.
springerlink.com   © Springer-Verlag Berlin Heidelberg 2009

Bilinear Interpolation determines the value of a new pixel based on a weighted average of the 4 pixels in the nearest 2 x 2 neighborhood of the pixel in the original image. The averaging has an anti-aliasing effect and therefore produces relatively smooth edges with hardly any jaggies [12].

Bicubic goes one step beyond bilinear by considering the closest 4x4 neighborhood of known pixels. Since these are at various distances from the unknown pixel, closer pixels are given a higher weighting in the calculation. Bicubic produces noticeably sharper images than the bilinear one, and is perhaps the ideal combination of processing time and output quality. This is the method most commonly used by image editing software [12]. Because of the spatial domain the image in these cases is represented as a set of discrete color values without any useful predefined properties.

There are also image zooming methods based on partial differential equations [8], multiscale geometric representations [9], some are merging a set of low-resolution images into a high-resolution image [7], some methods are extending the above basic interpolation methods by considering also the image local features [11].

It is well known, that many kind of image enhancement procedures can be performed much more efficiently when working in other domain. Representing an image in other domain can give us new possibilities by the processing. These representations are related to expressing the image intensity function as a linear combination of simpler functions (components) having useful predefined properties.

In case of Fourier series these functions are trigonometric and the domain connected to this representation is called the frequency domain. Many procedures like detecting edges, smoothing, image compression, etc. can be performed much more effectively in this domain then in spatial one. On the other hand to represent the image in frequency domain without meaningful quality decline, relatively large number of trigonometric components is needed.

In this paper we propose a new representation form for digital images using polylinear functions on Higher Order Singular Value Decomposition (HOSVD) basis. The main goal of the paper is to numerically reconstruct these polylinear functions using the pixels of the original image. As demonstration of the effectiveness of this concept these functions will be used for resolution enhancement of an image. In case of HOSVD-based representation the number of polylinear functions (components) expressing the image without disturbing its quality will be much more less then the number of trigonometric components in case of Fourier based representation of the same image.

The paper is organized as follows: Section theory gives a closer view how to express a multidimensional function using polylinear functions on HOSVD basis, and how to reconstruct these polylinear functions, Section 3 describes how this representation can be applied in case of digital images, and how the enlargement of an image in this case works. Section 4 shows the experimental results and finally conclusions are reported.

2   Theoretical Background

The approximation methods of mathematics are widely used in theory and practice for several problems. If we consider an n -variable smooth function

$$f(x), \ x = (x_1,...,x_N)^T, \ x_n \in [a_n, b_n], \ 1 \le n \le N,$$

then we can approximate the function $f(x)$ with a series

$$f(x) = \sum_{k_1=1}^{I_1}...\sum_{k_N=1}^{I_N} \alpha_{k_1,...,k_N} \, p_{1,k_1}(x_1) \cdot ... \cdot p_{N,k_N}(x_N). \tag{1}$$

where the system of orthonormal functions $p_{n,k_n}(x_n)$ can be choosen in classical way by orthonormal poloynomials or trigonometric functions in separate variable and the numbers of functions I_n playing role in (1) large enough. With the help of Higher Order Singular Value Decomposition (HOSVD) a new approximation method was developed in [2] and [3], [4], [5] in which a specially determined system of orthonormal functions can be used depending on function $f(x)$, instead of some system of orthonormal polynomials or trigonometric functions.

Assume that the function $f(x)$ can be given with some functions $\tilde{w}_{n,i}(x_n), x_n \in [a_n, b_n]$ in the form

$$f(x) = \sum_{k_1=1}^{I_1}...\sum_{k_N=1}^{I_N} \alpha_{k_1,...,k_N} \, \tilde{w}_{1,k_1}(x_1) \cdot ... \cdot \tilde{w}_{N,k_N}(x_N). \tag{2}$$

Denote by $\mathcal{A} \in \mathbb{R}^{I_1 \times ... \times I_N}$ the N -dimensional tensor determined by the elements $\alpha_{i_1,...,i_N}$, $1 \le i_n \le I_n$, $1 \le n \le N$ and let us use the following notations (see: [1]).

- $\mathcal{A} \boxtimes_n \mathbf{U}$: the n -mode tensor-matrix product,
- $\mathcal{A} \boxtimes_{n=1}^{N} \mathbf{U}_n$: multiple product as $\mathcal{A} \boxtimes_1 \mathbf{U}_1 \boxtimes_2 \mathbf{U}_2 ... \boxtimes_N \mathbf{U}_N$.

The n -mode tensor-matrix product is defined by the following way. Let $\mathbf{U}$ be an $K_n \times M_n$ -matrix, then $\mathcal{A} \boxtimes_n \mathbf{U}$ is an $M_1 \times ... \times M_{n-1} \times K_n \times M_{n+1} \times ... \times M_N$ -tensor for which the relation

$$(\mathcal{A} \boxtimes_n \mathbf{U})_{m_1,...,m_{n-1},k_n,m_{n+1},...,m_N} \overset{def}{=}$$

$$\sum_{1 \le m_n \le M_n} a_{m_1,\dots,m_n,\dots,m_N} U_{k_n,m_n}$$

holds. Detailed discussion of tensor notations and operations is given in [1]. We also note that we use the sign $\boxtimes_n$ instead the sign $\times_n$ given in [1] Using this definition the function (2) can be rewritten as a tensor product form

$$f(x) = A \boxtimes_{n=1}^{N} \tilde{w}_n(x_n), \tag{3}$$

where $\tilde{w}_n(x_n) = (\tilde{w}_{n,1}(x_n),\dots,\tilde{w}_{n,I_n}(x_n))^T$, $1 \le n \le N$. Based on HOSVD it was proved in [2] and [3] that under milde conditions the (3) can be represented in the form

$$f(x) = \mathcal{D} \boxtimes_{n=1}^{N} w_n(x_n), \tag{4}$$

where

- $\mathcal{D} \in \mathbb{R}^{r_1 \times \dots \times r_N}$ is a special (so-called core) tensor with the properties:

  1 $r_n = rank_n(A)$ is the n-mode rank of the tensor A, i.e. rank of the linear space spanned by the n-mode vectors of A:

  $$\{(a_{i_1,\dots,i_{n-1},1,i_{n+1},\dots,i_N},\dots,a_{i_1,\dots,i_{n-1},I_n,i_{n+1},\dots,i_N})^T :$$

  $$1 \le i_j \le I_n, \ 1 \le j \le N\},$$

  2 all-orthogonality of tensor $\mathcal{D}$: two subtensors $\mathcal{D}_{i_n=\alpha}$ and $\mathcal{D}_{i_n=\beta}$ (the nth indices $i_n = \alpha$ and $i_n = \beta$ of the elements of the tensor $\mathcal{D}$ keeping fix) orthogonal for all possible values of n, α and β : $\langle \mathcal{D}_{i_n=\alpha}, \mathcal{D}_{i_n=\beta} \rangle = 0$ when $\alpha \ne \beta$. Here the scalar product $\langle \mathcal{D}_{i_n=\alpha}, \mathcal{D}_{i_n=\beta} \rangle$ denotes the sum of products of the appropriate elements of subtensors $\mathcal{D}_{i_n=\alpha}$ and $\mathcal{D}_{i_n=\beta}$,

  3 ordering: $\|\mathcal{D}_{i_n=1}\| \ge \|\mathcal{D}_{i_n=2}\| \ge \cdots \ge \|\mathcal{D}_{i_n=r_n}\| > 0$ for all possible values of n ($\|\mathcal{D}_{i_n=\alpha}\| = \langle \mathcal{D}_{i_n=\alpha}, \mathcal{D}_{i_n=\alpha} \rangle$ denotes the Kronecker-norm of the tensor $\mathcal{D}_{i_n=\alpha}$).

- Components $w_{n,i}(x_n)$ of the vector valued functions

$$w_n(x_n) = (w_{n,1}(x_n),\dots,w_{n,r_n}(x_n))^T, \ 1 \le n \le N,$$

are orthonormal in L_2-sense on the interval $[a_n,b_n]$, i.e.

$$\forall n : \int_{a_n}^{b_n} w_{n,i_n}(x_n) w_{n,j_n}(x_n) dx = \delta_{i_n,j_n},$$

$$1 \le i_n, j_n \le r_n,$$

where $\delta_{i,j}$ is a Kronecker-function ($\delta_{i,j} = 1$, if $i = j$ and $\delta_{i,j} = 0$, if $i \ne j$)

The form (4) was called in [2] and [3] HOSVD canonical form of the function (2).

Let us decompose the intervals $[a_n, b_n]$, $n = 1..N$ into M_n number of disjunct subintervals $\triangle_{n,m_n}$, $1 \le m_n \le M_n$ as follows:

$$\xi_{n,0} = a_n < \xi_{n,1} < \dots < \xi_{n,M_n} = b_n,$$

$$\triangle_{n,m_n} = [\xi_{n,m_n}, \xi_{n,m_n-1}).$$

Assume that the functions $w_{n,k_n}(x_n)$, $x_n \in [a_n, b_n]$, $1 \le n \le N$ in the equation (2) are piece-wise continuously differentiable and assume also that we can observe the values of the function $f(x)$ in the points

$$y_{i_1,\dots,i_N} = (x_{1,i_1}, \dots, x_{N,i_N}), \ 1 \le i_n \le M_n. \tag{5}$$

where

$$x_{n,m_n} \in \Delta_{n,m_n}, \ 1 \le m_n \le M_n, \ 1 \le n \le N$$

Based on the HOSVD a new method was developed in [2] and [3] for numerical reconstruction of the canonical form of the function $f(x)$ using the values $f(y_{i_1,\dots,i_N})$, $1 \le i_n \le M_n$, $1 \le i_n \le N$. We discretize function $f(x)$ for all grid points as:

$$b_{m_1,\dots,m_N} = f(\mathbf{y}_{m_1,\dots,m_N}).$$

Then we construct N dimensional tensor $\mathcal{B} = (b_{m_1,\dots,m_N})$ from the values $b_{m_1,\dots,m_N}$. Obviously the size of this tensor is $M_1 \times \dots \times M_N$. Further, discretize vector valued functions $\mathbf{w}_n(x_n)$ over the discretization points x_{n,m_n} and construct matrices $\mathbf{W}_n$ from the discretized values as:

$$\mathbf{W}_n = \begin{pmatrix} w_{n,1}(x_{n,1}) & w_{n,2}(x_{n,1}) & \cdots & w_{n,r_n}(x_{n,1}) \\ w_{n,1}(x_{n,2}) & w_{n,2}(x_{n,2}) & \cdots & w_{n,r_n}(x_{n,2}) \\ \vdots & & \ddots & \vdots \\ w_{n,1}(x_{n,M_n}) & w_{n,2}(x_{n,M_n}) & \cdots & w_{n,r_n}(x_{n,M_n}) \end{pmatrix} \tag{6}$$

Then tensor $\mathcal{B}$ can simply be given by (4) and (6) as

$$\mathcal{B} = \mathcal{D} \boxtimes_{n=1}^{N} \mathbf{W}_n. \tag{7}$$

Consider the HOSVD decomposition of the discretization tensor

$$\mathcal{B} = \mathcal{D}^d \boxtimes_{n=1}^{N} \mathbf{U}^{(n)} \tag{8}$$

where $\mathcal{D}^d$ is the so-called core tensor, and $\mathbf{U}^{(n)} = \left(U_1^{(n)} \ U_2^{(n)} \ \cdots \ U_{M_n}^{(n)} \right)$ is an $M_n \times M_n$-size orthogonal matrix $(1 \leq n \leq N)$.

Let us introduce the notation: $\tilde{r}_n^d = rank_n\mathcal{B}, \ 1 \leq n \leq N$ and consider the $\tilde{r}_1^d \times \ldots \times \tilde{r}_N^d$ -size reduced version $\tilde{\mathcal{D}}^d = (\mathcal{D}_{m_1,\ldots,m_N}^d, 1 \leq m_n \leq r_n, 1 \leq n \leq N)$ of the $M_1 \times \ldots \times M_N$-size tensor $\mathcal{D}^d$. The following theorems were proved in [2] and [3]. Denote

$$\Delta = \max_{1 \leq n \leq N} \max_{1 \leq i_n \leq M_n} (\xi_{n,m_n} - \xi_{n,m_n-1}) \quad \text{and}$$

$$\rho = \prod_{n=1}^{N} \rho_n, \ \rho_n = (b_n - a_n)/M_n.$$

Theorem 1. If Δ is sufficiently small, then $\tilde{r}_n^d = r_n, 1 \leq n \leq N$ and the convergence $\sqrt{\rho}\tilde{\mathbf{D}}^d \to D, \ \Delta \to 0$ is true.

Let us denote the elements of matrix $\mathbf{U}^{(n)}$ by $U_{i,k}^{(n)}$ and introduce the step functions $u_{n,i}(x), 1 \leq i \leq r_n$ in the following way

$$u_{n,i}(x) = \frac{1}{\sqrt{\rho_n}} U_{i,k}^{(n)} I(x \in \Delta_{n,k}), 1 \leq k \leq M_n$$

Theorem 2. If $\Delta \to 0$ then

$$\int_{a_n}^{b_n} \left(w_{n,i}(x) - u_{n,i}(x) \right)^2 dx \to 0, \quad 1 \leq i \leq r_n, 1 \leq n \leq N$$

The proposed method is based on the above approach and on the properties of singular values of the HOSVD.

3 Color Image Resolution Enhancement on HOSVD Basis

Let $f(x), x = (x_1, x_2, x_3)^T$ represent the image function, where x_1 and x_2 correspond to the vertical and horizontal coordinates of the pixel respectively. Variable x_3 is related to color components of the pixel, i.e. the red, green and blue color components in case of RGB images. Function $f(x)$ can be approximated (based on notes discussed in the previous section) in the following way:

$$f(x) = \sum_{k_1=1}^{I_1}\sum_{k_2=1}^{I_2}\sum_{k_3=1}^{I_3} \alpha_{k_1,k_2,k_3} \tilde{w}_{1,k_1}(x_1) \cdot \tilde{w}_{2,k_2}(x_2) \cdot \tilde{w}_{3,k_3}(x_3). \qquad (9)$$

We can store each color component of each pixel in a $m \times n \times 3$ tensor, where n and m correspond to the width and height of the image respectively. Let $\mathcal{B}$ denote this thesor. The first step is to reconstruct the functions $\tilde{w}_{n,k_n}, 1 \leq n \leq 3, 1 \leq k_n \leq I_n$ by decomposing the tensor $\mathcal{B}$ using the HOSVD (see Fig. 1) as follows:

$$\mathcal{B} = \mathcal{D}^d \boxtimes_{n=1}^{3} \mathbf{U}^{(n)} \qquad (10)$$

where $\mathcal{D}^d$ is the so called core tensor. Vectors corresponding to the columns of matrices $\mathbf{U}^{(n)}, 1 \leq n \leq 3$ as described in the previous section are representing the discretized form of functions $\tilde{w}_{n,k_n}(x_n)$ corresponding to the appropriate dimension n, $1 \leq n \leq 3$.

Our goal is to demonstrate the efficiency of image processing when the image is represented in this HOSVD-based form. For this purpose we will increase the resolution of an image, i.e. enlarge it by maintaining its quality as high as possible.

Let s denote the zooming factor. First let consider the first column $U_1^{(1)}$ of matrix $\mathbf{U}^{(1)}$ the function $\tilde{w}_{1,1}$ is corresponding to. The function value $\tilde{w}_{1,1}(1)$ corresponds to the 1st element of $U_1^{(1)}$, $\tilde{w}_{1,1}(2)$ to the 2nd element ... $\tilde{w}_{1,1}(M_n)$ to the $M_n{}^{th}$ element of $U_1^{(1)}$. If the image is enlarged by a factor s the assigment of elements of $U_1^{(1)}$ to the function values should be performed as follows: The corresponding value to $\tilde{w}_{1,1}(1)$ will remain the same as above, i.e. it will be the vector element $U_1^{(1)}(1)$, to $\tilde{w}_{1,1}(1+s)$ the element $U_1^{(1)}(2)$ will correspond, ...,

$\tilde{w}_{1,1}(1 + M_n s)$ will have the value of $U_1^{(1)}(M_n)$. In this case the missing function values can be determined by interpolation. We used cubic spline interpolation
in order to obtain the missing $\tilde{w}_{1,1}(x_1)$ values, considering the known
$\tilde{w}_{1,1}(1 + is), 1 \le i \le M_n$.

The next step is to perform the same procedure regarding the remaining func-
tions $\tilde{w}_{1,2}$, $\tilde{w}_{1,3}$,..., $\tilde{w}_{1,I_n}$ as well by considering the corresponding $U_2^{(1)}$,
$U_3^{(1)}$,..., $U_{I_n}^{(1)}$ vectors.

The above assigments have to be applied similarly for the 2nd dimension, as well,
i.e. for the variable x_2.

After every function $\tilde{w}_{n,k_n}, 1 \le n \le 2, 1 \le k_n \le I_n$ has been determined the
enlarged image can be created using the equation (9).

4   Examples

The pictures are illustrating the effectiveness of the proposed method by comparing
it to the results obtained by bilinear and bicubic image interpolation methods. Image
2 illustrates the original low resolution image with indicated rectangular areas. The
10x zoomed version of these areas can be followed in images 3-10. The 5x zoomed
version of areas indicated in Fig. 11 can be followed in Figs. 12-14. By the processing bilinear, bicubic interpolation and the proposed method have been used. The
experiment was performed on color images, i.e. their below grayscale equivalents
reflect the efficiency of the proposed method partly.

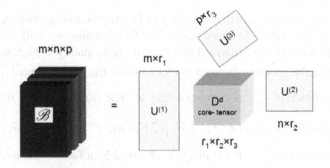

Fig. 1. Illustrating the HOSVD of an RGB image

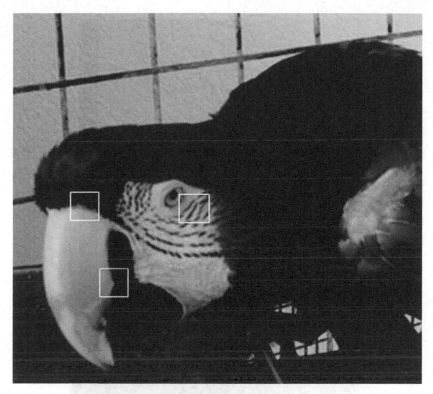

Fig. 2. The original image with indicated areas of interest

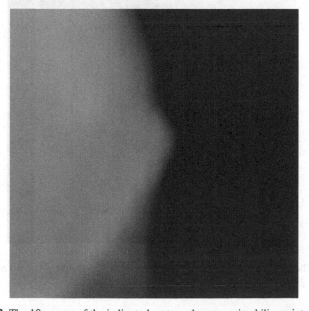

Fig. 3. The 10x zoom of the indicated rectangular area using bilinear interpolation

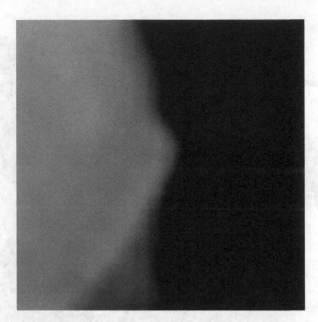

Fig. 4. The 10x zoom of the indicated rectangular area using bicubic interpolation

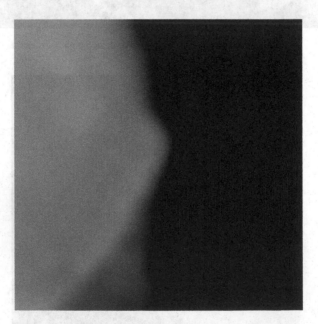

Fig. 5. The 10x zoom of the indicated rectangular area using the proposed method

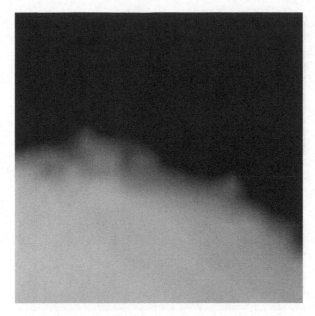

Fig. 6. The 10x zoom of the indicated rectangular area using bilinear interpolation

Fig. 7. The 10x zoom of the indicated rectangular area using bicubic interpolation

A. Rövid and L. Szeidl

Fig. 8. The 10x zoom of the indicated rectangular area using the proposed method

Fig. 9. The 10x zoom of the indicated rectangular area using bilinear interpolation

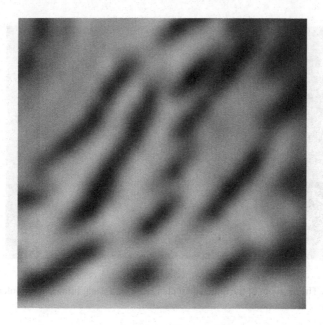

Fig. 10. The 10x zoom of the indicated rectangular area using the proposed method

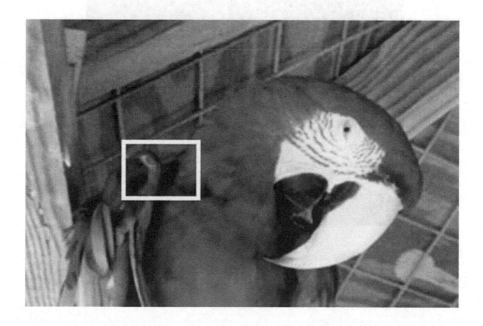

Fig. 11. The original image with indicated areas of interest

Fig. 12. The 5x zoom of the indicated rectangular area using bilinear interpolation

Fig. 13. The 5x zoom of the indicated rectangular area using bicubic interpolation

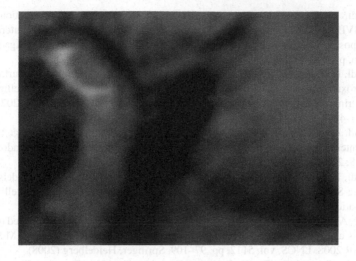

Fig. 14. The 5x zoom of the indicated rectangular area using the proposed method

Conclusions

In the paper a new image representation form was proposed for supporting the processing of digital images. The representation is based on polylinear functions on HOSVD basis. The effectiveness of the representation was tested on images which were enlarged based on the reconstruction of the mentioned polylinear functions. By this technique the resulted image maintains the edges more clearly then the other well known image interpolation methods, e.g. bilinear, bicubic. The proposed form of image representation can advantageously be used for many other purposes like image compression, filtering, etc. [4].

Acknowledgments. This paper was supported by the János Bolyai Research Scholarship of the Hungarian Academy of Sciences and by the Hungarian Scientific Research Fund OTKA T048756.

References

[1] De Lathauwer, L., De Moor, B., Vandewalle, J.: A Multilinear Singular Value Decomposition. SIAM Journal on Matrix Analysis and Applications 21(4), 1253–1278 (2000)

[2] Baranyi, P., Szeidl, L., Várlaki, P.: Numerical Reconstruction of the HOSVD-based Canonical Form of Polytopic Dynamic Models. In: Proc. of IEEE 10th Int. Conf. on Intelligent Engineering Systems, London, United Kingdom, June 26-28, 2006, pp. 196–201 (2006)

[3] Szeidl, L., Baranyi, P., Petres, Z., Várlaki, P.: Numerical Reconstruction of the HOSVD-based Canonical Form of Polytopic Dynamic Models. In: 3rd International Symposium on Computational Intelligence and Intelligent Informatics, Agadir, Morocco, pp. 111–116 (2007)

[4] Szeidl, L., Rudas, I., Rövid, A., Várlaki, P.: HOSVD-based Method for Surface Data Approximation and Compression. In: 12th International Conference on Intelligent Engineering Systems, Miami, Florida, February 25-29, 2008, pp. 197–202 (2008) 978-1-4244-2083-4

[5] Rövid, A., Várlaki, P.: Method for Merging Multiple Exposure Color Image Data. In: 13th International Conference on Intelligent Engineering Systems, Barbados, April 16-18, 2009, pp. 27–31 (2009)

[6] Szeidl, L., Várlaki, P.: HOSVD-based Canonical Form for Polytopic Models of Dynamic Systems. Journal of Advanced Computational Intelligence and Intelligent Informatics 13(1), 52–60 (2009)

[7] Horé, A., Deschênes, F., Ziou, D.: A New Super-Resolution Algorithm Based on Areas Pixels and the Sampling Theorem of Papoulis. In: Campilho, A., Kamel, M.S. (eds.) ICIAR 2008. LNCS, vol. 5112, pp. 97–109. Springer, Heidelberg (2008)

[8] Gao, R., Song, J.-P., Tai, X.-C.: Image Zooming Algorithm Based on Partial Differential Equations Technique. International Journal of Numerical Analysis and Modeling 6(2), 284–292 (2009)

[9] Nickolaus, M., Yue, L., Do Minh, N.: Image Interpolation Using Mulitscale Geometric Representations. In: Proceedings of the SPIE, vol. 6498, pp. 1–11 (2007)

[10] Su, D., Willis, P.: Image Interpolation by Pixel-Level Data-Dependent Triangulation. Computer Graphics Forum 23(2), 189–201 (2004)

[11] Yuan, S., Abe, M., Taguchi, A., Kawamata, M.: High Accuracy Bicubic Interpolation Using Image Local Features. IEICE Transactions on Fundamentals of Electronics, Communications and Computer Sciences E90-A(8), 1611–1615 (2007)

[12] Bockaert, V.: Digital Photography Review - Interpolation (Cited, April 2009)

Intelligent Short Text Assessment in eMax

Dezső Sima, Balázs Schmuck, Sándor Szöllősi, and Árpád Miklós

Budapest Tech
Bécsi út 96/B, H-1034 Budapest, Hungary
sima@bmf.hu; schmuck.balazs@nik.bmf.hu; szollosi.sandor@nik.bmf.hu;
miklos.arpad@nik.bmf.hu

Abstract. Rapidly increasing student numbers and spreading distance learning systems strengthen the urgent need for using Knowledge Assessment Systems (KAS's). Recent KAS's have however, the deficiency of not providing intelligent assessment modules for e.g. the evaluation of freely formulated short answers including a few sentences or partially solved mathematical problems. The eMax KAS, developed at the Intelligent Knowledge Management Innovation Center of IBM Hungary and the John von Neumann Faculty of Informatics at Budapest Tech, aims to provide these capabilities. Our paper gives an introduction to the intelligent assessment of short texts component of eMax by presenting the approach used for the formal description of the "answer space" defined as well as the methods chosen for the syntactic analysis, semantic analysis and scoring. Finally, we present the standard course of examination using eMax, then summarize test results and our conclusions.

Keywords: Answer Space, Intelligent Assessment Systems, Knowledge Assessment Systems, Semi-automatic assessment, Short Text Evaluation.

1 Introduction

1.1 Motivation

In the last decade student enrollment to our faculty has steadily grown, nearly five-fold in ten years. The annual intake to our Engineer Informatics program approaches already 400 direct and 100 part time students. With such a large number of students, the assessment of midterm and final exams require a huge effort from our faculty members. In order to substantially ease the burden of assessing large numbers of exam questions while avoiding the constraints of passive answer types, such as multiple-choice questions, an internal project was launched five years ago with the goal of developing an intelligent knowledge assessment system (KAS). In the first phase of the project, a wide range literature research was conducted

I.J. Rudas et al. (Eds.): Towards Intelligent Engineering & Information Tech., SCI 243, pp. 435–445.
springerlink.com
© Springer-Verlag Berlin Heidelberg 2009

[1], [2] and a test system called Evita [3] was developed to check the feasibility of our ideas concerning the assessment of freely formulated short texts. Based on positive results obtained, an Intelligent Knowledge Management Innovation Center was established in 2005 jointly by IBM Hungary and the John von Neumann Faculty of Informatics at Budapest Tech to develop an appropriate intelligent KAS. The research and development work performed has resulted in the beta version of the eMax system, now under testing. Beyond the usual passive question types, eMax is able to semi-automatically assess freely formulated short student answers and partially solved mathematical problems as well. Unlike Automated Essay Grading (AEG) published widely [4], [5] the short text module of our eMax is constrained only to a few (say 2-3) sentences and is based on a formal semantic description rather than on methods used in AEG, such as statistical or Natural Language Processing (NLP) techniques. The remainder of this paper is organized as follows: Following an introduction in Section 1, Section 2 gives a brief introduction to the eMax KAS. In Section 3 we present the formal description of the "answer space" that serves as the basis for the evaluation of short student answers as well as describe the approach to the way short texts are assessed in eMax. Next in Section 4 we describe the standard process of electronic examination using eMax. Finally, Sections 5 and 6 summarize the test results obtained as well as our conclusions.

2   The eMax System

The eMax system is designed to support various examination scenarios, ranging from simple ad-hoc small class tests to midterm exams to final exams built upon various passive and active question types. The implementation of eMax places it in the following major design space aspect categories [1], [2]: a) it is an open network protocol based system, providing access to the server via either a local or global network, so it can be made available via the Internet; b) temporal access to the system is currently restricted in order to avoid an easy depletion of the initial question bank; c) its servers are a platform-dependent set of .NET executables and libraries (note, however, that the underlying framework is being made available for several platforms, which will help eMax automatically enhance its platform coverage over time); d) for security and user experience considerations, it uses a fat client architecture, although the design has the potential for an easy to implement thin client version in the future; and e) clients are based on the same architecture as the servers, that is, the points made in paragraphs c) and d) apply here as well.

In summary, eMax is an open network based system with semi-platform-dependent servers and semi-platform-dependent fat clients that have time-restricted access to the servers. A few options were deliberately left open in terms of positioning the system within the design space discussed above in order to accommodate future development requirements.

3 The Intelligent Text Evaluation Subsystem of eMax

3.1 Aim and Expected Features of the Intelligent Text Evaluation Subsystem

The eMax module is aimed at the semi-automatic assessment of short, freely formulated answers produced in student exams. In the alpha version of eMax, short answers are restricted to single sentences, while the beta version, completed by the beginning of 2008, is able to assess student answers consisting of two or three sentences as well. Although this limitation might seem too restrictive at first, practically most exam questions of the short answer type may easily be broken down to a few sub-questions such that each sub-question may be responded to in a short answer not exceeding two or three sentences.

As far as the targeted "level of intelligence" of the short text evaluation module is concerned, the beta version is expected to assess at least 90% of student answers automatically. With this level of intelligence less than 10% of the student answers will be rejected by eMax and should be manually evaluated by the lecturer or instructor. Assuming e.g. an enrollment of 300 students, having to assess 300 handwritten student answers or only 30 typed ones per exam question clearly makes a significant difference.

Concerning the scoring of student answers, our target in the beta version is to have a no higher than ±7.5% deviation compared to the scenario of having the same short answers checked manually. As eMax is conceived to be easily enhanced by extending or improving the formal description of the answer space and/or extending the database containing synonyms or antonyms, we expect noticeably improved performance parameters in the beta version.

3.2 Basic Approach to Assessing Freely Formulated Short Student Answers Used in eMax

1) Entering Short Text Type Exam Questions by Lecturers: If an exam question is of the short text type, the lecturer or instructor enters the question to be asked and the expected correct answer in eMax. Obviously, the term "correctness" here does not relate to the formulation (i.e. the syntax) of the answer, but only to its semantic content. Furthermore, in the freely formulated correct answer the lecturer needs to mark relevant semantic elements that are expected to be included in student answers. In addition, they check the synonyms offered by eMax for each of the relevant semantic elements in the correct answer and either accept or reject the items offered, while missing items can be added as needed. Finally, the lecturer needs to provide the scoring scheme for student answers by allocating weights to the semantic elements expected to occur in student answers in the following form:

$$R = \begin{cases} \sum_i w_i * s_i \ if & R \geq L \\ 0 & otherwise \end{cases}$$

where   R is the score given to the student's answer,
                w_i is weight of the associated semantic elements,
                $s_i = 1$ if the semantic element s_i is included in the student's
                answer at the right place,
                otherwise $s_i = 0$,
                L is the minimum score required to accept the student's
                answer at all.

Exam questions are held in a question bank in eMax and can be selected during the preparation of a concrete exam.

2) Main Steps of Assessing Student Answers: During an exam the questions asked are presented to the students, and their answers are collected in the answer bank belonging to that particular exam. Short texts are assessed in the following three main steps:

- syntactic analysis,
- semantic analysis, and
- scoring.

The syntactic analysis is based on a formal description of the answer space, discussed in the next section.

3) The Answer Space (AS): This is an appropriate subset of the natural communication language used that is more easily manageable in a formal way than the natural language. The formal description of the answer space is accomplished by a formal grammar, designated as Answer Space Grammar (ASG). The ASG is described by a 3-tuple, as follows:

$$ASG = (A_i, B_{i,j}, C_{i,k})$$

where A_i: is the set of answer types defined,
          $B_{i,j}$: are the sets of basic syntactic structures belonging to the answer types Ai such that for each answer type A_i the basic syntactic structure $B_{i,1}$ is considered to be the normal form and all other structures $B_{i,j}$ ($j \neq 1$) are alternative syntactic structures; and
          $C_{i,k}$: are the sets of possible grammatical structure constructors (key expressions) occurring in the basic syntactic structures such that in their normal form $B_{i,1}$ they occur in the increasing order of their middle indices (e.g. $C_{i,1}, C_{i,2}, C_{i,3}$) whereas in the alternative syntactic structures their order is defined by the particular syntactic structure.

Based on the evaluation of a large number of exams we defined in the beta version of eMax the following five answer types that include appropriate constructors in Hungarian and English languages:

- two-term conditional answers ("if-then" form),
- three-term conditional answers ("if-then-else" form),
- cause-definition answers
- purpose-definition answers
- comparisons.

Also, a sixth answer type exists, comprising answers that does not include constructors. Obviously, no syntactic analysis is needed for this answer type.

The basic syntactic structures are set up of Constructors ($C_{i,k}$) and Statements (S_k), as illustrated in Fig. 1.

$C_{1,1}$	S_1	$C_{1,2}$	S_2	$C_{1,3}$	S_3

Fig. 1. Example of a basic syntactic structure. (The normal form of conditional three-term answers.)

These Statements are built up of Subjects and Predicates, such that one or more Predicates may be allocated to each Subject and vice versa, one or more Subjects may own the same Predicate or Predicates. The beta version of eMax only allows Statements including a single Subject and a single Predicate, where either the Subject or the Predicate may be omitted. Subjects and Predicates are considered to be syntactic elements (s_i)

Beyond Constructors, Subjects and Predicates, eMax considers all other terms occurring in student's or teacher's answers as don't care terms.

The formal description of AS requires for each answer type Ai the specification of possible basic syntactic structures $B_{i,j}$ along with the sets of possible constructors $C_{i,k}$ occurring in the considered basic syntactic structures.

As an example, Fig. 2 shows the basic syntactic structures and the associated constructor sets for the three-term conditional answer type.

$B_{I,1}$: $C_{I,1}$ S_1 $C_{I,2}$ S_2 $C_{I,3}$ S_3.   where: $C_{I,1}$ = {if, when, provided, assuming, given, granted, in case }

$B_{I,1}$: S_2 $C_{I,1}$ S_1 $C_{I,3}$ S_3.

$C_{I,2}$ = {then, in that case }

$B_{I,1}$: $C_{I,2}$ S_3 $C_{I,1}$ ¬S_1 $C_{I,3}$ S_2. $C_{I,3}$ = {else, otherwise, in other cases }

$B_{I,1}$: $C_{I,1}$ S_1 S_2 $C_{I,3}$ S_3.

Fig. 2. Basic syntactic structures and associated constructor sets for three-term conditional answers (the if-then-else form.)

Here we note that the described method for providing a formal description of the AS can presumably be applied to most character based languages. The presented grammar (ASG) generates all possible formulations ($F_{i,r}$) to each answer type A_i.

Based on its generator elements, the AS can be represented graphically by a tree where the nodes belonging to subsequent levels of the tree are defined by the sets A_i, $B_{i,j}$, $C_{i,k}$, as shown in Fig. 3.

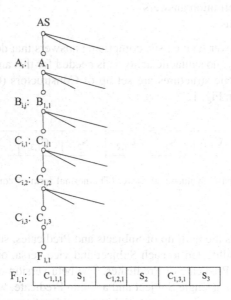

Fig. 3. Representation of the AS as a tree

The leaves of the tree represent the set of all possible formulations $F_{i,r}$ that can be identified and processed in the AS. If a student chooses to use a formulation not included in $F_{i,r}$, eMax rejects the evaluation of this answer and asks for a manual assessment. Nevertheless, with a continuous enhancement applied to the formal description of the AS during usage (by enriching the sets of A_i, $B_{i,j}$ and $C_{i,k}$), more and more possible formulations $F_{i,r}$ will be identified and the rejection rate of the syntactic analysis will decrease.

As to the expressiveness of our answer space grammar, the number of available formulations $F_{i,r}$ processed by eMax for the answer type A_i is given by the sum of all combinations of $C_{i,k}$ associated to each basic form $B_{i,j}$ belonging to that particular answer type A_i.

For instance, in the recent implementation of eMax the number of alternative formulations that can be identified e.g. for three-term conditional answers amounts to:

$$(7*2*3) + (7*3) + (2*7*3) + (7*3) = 126$$

4) Syntactic Analysis: In the syntactic analysis phase of short texts eMax first scans the teacher's answer and then all students' answers for constructors. The number and actual values of the constructors found in a particular answer identify the formulation $F_{t,u}$ used by the teacher or individual students. The syntactic analysis may have three distinct results as follows:

- the number and the actual values of the constructors found do not match to any of the formulations $F_{i,r}$ included in the A_S. In this case eMax marks this answer as not recognized and transfers it to the list of answers to be manually assessed.
- Typically, the number and the actual values of the constructors found matches a single formulation $F_{s,v}$. Then $F_{s,v}$ uniquely determines both the answer type A_s and the basic syntactic structure B_s,w used in the answer. Here we assume that the teacher's answer uniquely matches to a single formulation $F_{t,u}$, and thus both the answer type A_t and the basic syntactic form $B_{t,z}$ used by the teacher becomes known.
- Seldom, the number and the actual values of the constructors found match more than one formulation, say $F_{a,x}$ and $F_{b,y}$. Then the semantic analysis, discussed below, covers both possible answer types and basic syntactic structures. In the following we assume an appropriate processing of this exceptional case and do not point out the tasks related to it.

5) Semantic Analysis: The semantic analysis phase of the evaluation basically consists of three tasks:

- Transformation of the teacher's answer to the normal form $B_{t,1}$ belonging to the answer type used (A_t) and omitting all don't care terms from this answer, as illustrated in Fig. 4.

Fig. 4. The normal form of teacher's answer, assuming a three-term conditional answer type, after omitting "don't care" terms.

- Transformation of all recognized student answers to the normal form belonging to the answer type identified, as shown in Fig. 5.

Fig. 5. Normal form of a particular student's answer assuming a three-term conditional answer type. Individual words in the answer are designated by W.

Searching for the semantically relevant statements of the teacher's answer in the matching sections of each student's answer. This task is greatly simplified by the preceding syntactic analysis and the transformation to the normal form. Also, view to the basic approach chosen for the formal semantic description of the Answer Space, issues like semantic ambiguity do not play in eMax such a sensitive role as in AEG. Nevertheless, it remains a quite complex task that can only be roughly described in this paper due to length limitations. So points, like handling negations, multi-word syntactic elements or intentional cheating are not discussed here.

During the search, each matching section of the student's answer is scanned for any specified synonyms of the semantically relevant terms included in the related statement within the teacher's answer.

The semantic analysis yields a set of return values specifying whether or not the related semantically relevant term of the statement in concern could be found in the student's answer at the right position, as indicated in Fig. 6.

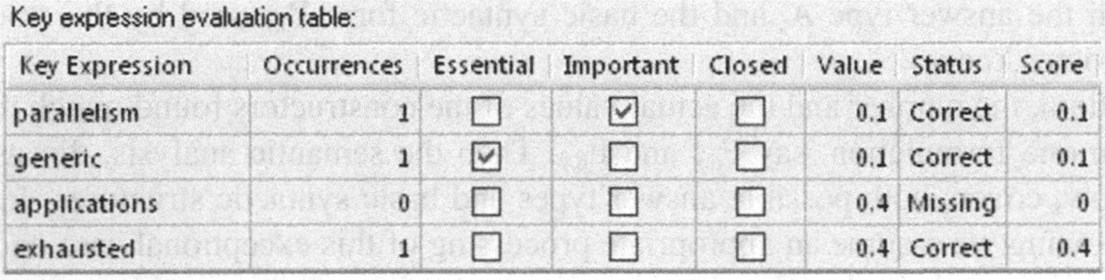

Key expression evaluation table:

Key Expression	Occurrences	Essential	Important	Closed	Value	Status	Score
parallelism	1	☐	☑	☐	0.1	Correct	0.1
generic	1	☑	☐	☐	0.1	Correct	0.1
applications	0	☐	☐	☐	0.4	Missing	0
exhausted	1	☐	☐	☐	0.4	Correct	0.4

Fig. 6. Return values and scoring of an evaluated student answer with relevant semantic elements

6) Scoring of Students' Answers: Scoring is performed according to the scheme specified by the lecturer or instructor with the return values delivered by the semantic analysis. As an example, Fig. 6 shows the composition of the final score of an individual student's answer (60%), assuming the following scoring scheme:

$$R = w_1{}^*s_1 + w_2{}^*s_2 + w_3{}^*s_3 + w_4{}^*s_4$$

with $s_1 = 1$, $s_2 = 1$, $s_3 = 0$, $s_4 = 1$,

which yields $R = 0.1 * 1 + 0.1 * 1 + 0.4 * 0 + 0.4 * 1 = 0.6$.

4   Standard Course of Examination

The basic model of operation implemented in the beta version, release in January 2008, is illustrated by the following flow chart:

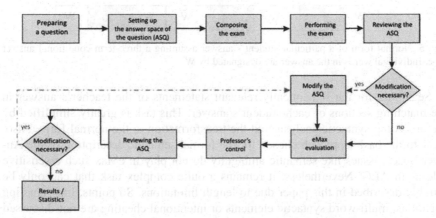

Fig. 7. High level overview of the examination process

First, professors enter questions and define the initial professor answer space of the question (ASQ), containing fully correct, expected answers (the system support multiple correct answers for a single question). The ASQ is made up of expected knowledge elements, their logical and semantic relationships including synonyms, antonyms and known wrong expressions within the defined context as well as a marking/scoring strategy for each correct answer. This is followed by setting up exam occasions by entering relevant parameters (e.g. length, score range limits for grading etc.). Next comes the examination itself, using the eMax system. Once students have completed an exam, an evaluation preparation stage is worth considering, comprising the review of a sample of 10 to 20 student answers for each question and expanding the initial professor answer space if necessary (e.g. by adding new synonyms etc.). This ASQ review improves the accuracy and automation level of evaluation, and this feedback path is indicated by the dotted red line in Fig. 7. Evaluation preparation is followed by automatic evaluation itself.

A new feature added in the beta version is that eMax assesses the "reliability" of each automatically obtained score based on a set of review rules, and submits those either not guaranteed to be accurate or not automatically evaluable for professor review. This process greatly enhances evaluation accuracy.

When reviewing student answers marked by the system as needing professor review, the answer space can again be enhanced and modified as needed. This additional ASQ review, indicated by the dotted green line in Fig. 7, enhances automatic evaluation accuracy and coverage even further. Obviously, this stage needs to be followed by another automatic evaluation session.

5   Test Results

Live testing for the beta version, released in January 2008, was performed during the Spring and Fall 2008 terms for a single course (Computer Architectures). While the testing process in the Spring term included automatic as well as manual evaluation of every answer in parallel, followed by a detailed comparison of every score obtained using these two different evaluation methods, testing in the Fall term was based on random sampling and comparison of evaluation results. In 2009, several professors within the faculty — having seen the successful test results — decided to introduce electronic examination in their courses. A total of 6 to 7 courses have been using eMax in the Spring 2009 semester for electronic examination, with test results underway.

Table 1 and the pie charts in Figures 8 and 9, to be presented below, contain our Spring 2008 test results. While the table contains exact numerical summaries, the pie charts demonstrate percentages of the appropriate cases.

Table 1. Summary of live test results

Exam No.	Number of student answers	Number of cases manual and automatic evaluation results differ	Number of answers submitted by the system for professor review	Number of erroneously evaluated answers submitted for professor review	Number of erroneously evaluated answers NOT submitted for professor review
1.	105	22	46	16	6
2.	176	42	53	27	15
3.	330	47	70	23	21
Total:	611	111	169	66	42
		18%	28%	11%	7%

The above data are illustrated in the following two pie charts:

■ Number of answers submitted by the system for professor review

■ Number of automatic evaluated answer

Fig. 8. Results of initial evaluation session: Automation level of evaluation

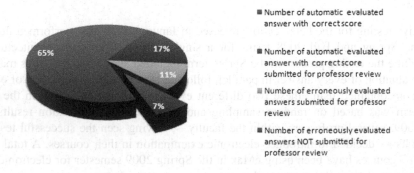

■ Number of automatic evaluated answer with correct score

■ Number of automatic evaluated answer with correct score submitted for professor review

■ Number of erroneously evaluated answers submitted for professor review

■ Number of erroneously evaluated answers NOT submitted for professor review

Fig. 9. Results after professor review: Accuracy of evaluation

Table 1 shows that after the initial evaluation, the system (using the built-in configurable set of review rules) submitted 169 of a total 611 student answers for professor review, indicated in percentage format in Fig. 8.

Fig. 9 demonstrates the fact that the professor review serves to vastly enhance the accuracy of evaluation. Professor review of student answers submitted for review according to Table 1 yielded the following results: out of a total 169 answers submitted for review, professors had to correct the automatically obtained score in 66 cases (11%), while the automatically obtained score was the same as the professor's score in 103 cases (17%) — that is, automatic evaluation by the system was actually correct. Finally, we found a total of only 42 student answers that had a different score for automatic and manual evaluation and that were not submitted by the system for professor review, which represents 7% of all student answers.

Conclusions

eMax is an innovative prototype knowledge assessment system that is able to semiautomatically evaluate short free text responses as well as partially solved mathematical problems.

This publication has presented the formal description methodology underlying automatic evaluation, and demonstrated key evaluation steps. Test results obtained using the beta version released in January 2008 are very close to the original goals set forth for the system, as it is capable of automatically evaluating 80 to 90% of answers, and the evaluation results — also considering the evaluation of answers after the additional professor review stage — has a less than 7.5% deviation from the results of fully manual assessment.

In the Spring 2009 semester, a total of 6 to 7 courses have already been using eMax for electronic examination, and further proliferation is expected in the near future.

References

[1] Szöllősi, S., Sima, D., Schmuck, B.: The Design Space of the Services of Knowledge Assessment Systems. In: Proc. 7th International Conference on Information Technology Based Higher Education & Training (ITHET), Sydney, pp. 392–399 (2006)

[2] Schmuck, B., Sima, D., Szöllősi, S.: The Design Space of the Implementation of Knowledge Assessment Systems. In: Proc. 7th International Conference on Information Technology Based Higher Education & Training (ITHET), Sydney, pp. 408–414 (2006)

[3] Csink, L., György, A., Raincsák, Z., Schmuck, B., Sima, D., Sziklai, Z., Szöllősi, S.: Intelligent Assessment Systems for e-Learning. In: Proc. 4th European Conference on E-Activities, E-Comm-Line 2003, Bucharest, pp. 224–229 (2003)

[4] Valenti, S., Neri, F., Cucchiarelli, A.: An Overview of Current Research on Automated Essay Grading. Journal of Information Technology Education 2, 319–330 (2003), http://jite.org/documents/Vol2/v2p319-330-30.pdf

[5] Williams, R., Dreher, H.: Formative Assessment Visual Feedback in Computer Graded Essays. In: Proc. Informing Science and Information Technology Education Joint Conference 2005, Flagstaff, Arizona, USA, pp. 23–32 (2005)

Determination the Basic Network Algorithms with Gains

András Bakó[1,2], Péter Földesi[2], and István Szűts[1]

[1] Budapest Tech, Hungary
 bako@bmf.hu, szuts@bmf.hu
[2] Széchenyi István University, Hungary

Abstract. Several optimization algorithms have been proposed for the solution of the basic network flow algorithms, such as the minimal and the multiterminal minimal path of a network having cost (distance) function, maximal flow of a capacitated network.

In this paper we present these algorithms in a special network in which on the edges a gain function is given. On the edge (x,y) of the network a t(x,y) transportation cost is defined. In the course of the transportation on the edge (x,y) the goods loose a part of there weight. If one unit of goods is transported from point x to point y then k(x,y) unite of goods arrive at point y, where 0<k(x,y)<1. The above mentioned 3 algorithms are presented in the network having gains.

These problems arc interested not only theoretically, but in the practice too. For example the maximal flow problem could be used in the case of electrical network, where we are interested to know the amount of electricity on the edges of the network.

Keywords: Gain function, shortest path with gain, maximal flow, multiterminal minimal path.

1 Introduction

In this paper 3 algorithms are presented: the shortest path, the multiterminal shortest path and the maximal flow. On the edges of the network a gain function is defined.

Let N be a set containing a finite number of points denoted by small Latin letters. The arc connecting the points x, $y \in N$, $x =l= y$ will be referred to as the edge of the network and the edge from x to y will be denoted by (x, y). The set of edges is denoted by E.

Let $s \in N$ and $z \in N$ be two fixed points of the network, s be called source and z be called sink. The transportation cost of one unit of goods from point x to point y, $(x,y) \in E$ be denoted by $t(x,y)$. In the course of the transportation along the arc (x, y) the goods lose a part of their weight. If one unit of goods is transported from

I.J. Rudas et al. (Eds.): Towards Intelligent Engineering & Information Tech., SCI 243, pp. 447–456.
springerlink.com   © Springer-Verlag Berlin Heidelberg 2009

point x, then k (x, y) units of goods arrive at point y. Thus the weight loss of one unit of goods en route from x to y is $1 - k(x,y)$.

The functions t (x, y) and k (x, y) have to satisfy the following conditions

$$t(x, y) \geq 0 \text{ for all } x, y \in N, (x,y) \in E$$

$$t(x, y) = \infty \quad \text{if no transport is possible on the arc } (x,y), \tag{1}$$

$$0 \leq k(x,y) \leq 1 \text{ for all } x, y \in N, (x,y) \in E.$$

The user has at his disposal an unlimited quantity of goods at point s and his purpose is to transfer one unit of goods to point z at minimum cost, taking into consideration the weight loss on the arcs.

Let $P = (s = x_0, x_1, ..., x_k = z)$ be one path from s to z. Let us denote by $l(x_0, x_i)$ the transportation cost of one unit of goods along the route P to the point x_i and by $m(x_0, x_i)$ the weight of the goods, starting from s with one unit weight, along the route P at the point x_i. Obviously, these functions can be given by the following recursion formulae

$$l(x_0, x_0) = 0$$

$$l(x_0, x_i) = l(x_0, x_{i-1}) + m(x_0, x_{i-1}) d(x_{i-1}, x_i)$$

and $\tag{2}$

$$m(x_0, x_0) = 1$$

$$m(x_0, x_i) = m(x_0, x_{i-1}) k(x_{i-1}, x_i).$$

2   Minimal Path Problem

The problem is to find the route P for which

$$\frac{l(x_0, z)}{m(x_0, z)} = \text{minimum} \tag{3}$$

i.e. the path on which the cost of transportation of one unit of goods arriving to the points z is minimal. In order to solve the problem, termed the *primal*, we attach a dual problem to the previous, so-called primal one.

The dual problem can be obtained by the following economical consideration.

The transportation can be made by a transporter who gives the price of one unit of goods at each point $x \in N$ of the network. Let this price be $d(x)$. The user is willing to buy the goods at a point y only if the price is smaller than his cost as if he transported the goods himself from an other point x. This constraint for the arc (x,y) can be formulated as follows:

$$d(y) \le \frac{d(x) + t(x, y)}{k(x, y)}.$$

The aim of the transporter is to maximize his profit i.e. $d(z)$.

The dual problem on the basis of the foregoing paragraphs can be formulated as follows. Let us determine the prices $d(x) \ge 0$, $x \in N$ which satisfy the conditions

$$k(x, y) d(y) - d(x) \le t(x, y) \tag{4}$$

and maximize

$$d(z). \tag{5}$$

A set of values $d(x)$, $x \in N$ satisfying the conditions (4) is called a *feasible solution* of the dual problem. The dual problem has a feasible solution, i.e. $d(x) = 0$, $x \in N$ satisfies the conditions (4).

The proof of the following theorem is constructive thus it gives an algorithm to solve the problem (3).

Theorem: If the functions l, m are given by the relations (2), then

$$\min_{P} \frac{l(x_0, z)}{m(x_0, z)} = \max d(z),$$

where the minimum refers to the path from s to z and the maximum refers to the prices satisfying the conditions (4).

The following lemma, which also shows the connections of the feasible solutions of the primal and dual problems, will be used in the proof of the theorem.

Lemma:

$$\frac{l(x_0, z)}{m(x_0, z)} \ge d(z)$$

Proof: The theorem is proved by induction for i showing that for each point x_i of a route P the relation

$$\frac{l(x_0, x_i)}{m(x_0, x_i)} \ge d(x_i) \tag{6}$$

is valid.

a) For $i = 0$, $X_0 = s$ and from the definitions of l, m and d follows that

$$\frac{l(x_0, s)}{m(x_0, s)} = d(s).$$

Let us assume that the inequality (6) is valid for $i - 1$, i.e.

$$\frac{l(x_0, x_{i-1})}{m(x_0, x_{i-1})} \ge d(x_{i-1}).$$

Using the assumption (4) we obtain the inequality

$$\frac{l(x_0, x_{i-1})}{m(x_0, x_{i-1})} \geq k(x_{i-1}, x_i) \cdot d(x_i) - t(x_{i-1}, x_i);$$

rearranging and using the recursion formulas (2) we get

$$l(x_0, x_i) \geq m(x_0, x_i) d(x_i).$$

Thus the lemma is proved.

If we find a path, and prices satisfying (4) and in the case of $x_i \in P$ the equality is valid in (6), then they will necessarily supply the minimum of the primal and the maximum of the dual problem.

Proof of the theorem. Let us assume that the function $d(x_i)$ is chosen so that d (z) is maximal. Let us divide the points of the set N in two sets S and S' according to the following rules:

a) $s \in S$,

b) $x \in S$ if there is such an $y \in S$ that for (x, y) the relation (4) becomes equality.

Let be $S' = N - S$.

If $z \in S$, then the path P, along which the relation (6) becomes equality, is determined.

If $z \in S$, then

$$k(x, y) . d(y) - d(x) < t(x,y) \text{ for all } x \in S, y \in S'.$$

Let be:

$$\varepsilon = \min_{\substack{x \in S \\ x \in S'}} (\frac{t(x, y) + d(x)}{k(x, y)} - d(y)) > 0.$$

We construct a new "price" function by ε from the function $d(x)$

$$d(x) = \begin{cases} d(x), & if \ x \in S \\ d(x)+\epsilon, & if \ x \in S' \end{cases}$$

It will be shown that the new price function satisfies the conditions (4). Obviously

$$\overline{d} \ (s) \, d \, (s) = O.$$

The condition $k(x, y) . d(y) - d(x) \leq t(x, y)$ has to be investigated in the following 4 cases:

a)   If $x \in S, y \in S$, then $d(x) = \overline{d} \ (x)$ thus condition (4) is satisfied.

b)

$$If \ x \notin S, y \notin S', then \ k(x, y).d(y) - d(x) = k(x, y)[d(y)+\varepsilon]d(x) \leq$$

$$k(x, y)\left[d(y) + \frac{t(x, y) + d(x)}{k(x, y)} - d(y)\right] - d(x) = t(x, y).$$

c) If x ∈ S', y ∈ S, then

$$k(x, y) . \bar{d}(y) - \bar{d}(x) = k$$
$$(x, y) . d(y) - d(x) - \varepsilon \le$$
$$k(x, y). d(y) - d(x) \le t(x, y).$$

d) If $x ∈ S'$, $y ∈ S'$, then

$$k(x, y) . d(y) - d(x) = k(x, y) . d(y) - d(x) -$$
$$\varepsilon [1 - k(x, y)] \le$$
$$k(x, y) . d(y) - d(x) \le t(x, y).$$

So we get $\bar{d}(z) > d(z)$ because $\varepsilon > 0$ in contradiction with the supposition.

3 Multiterminal Minimal Path

The multiterminal minimal path problem is to determine the cheapest (shortest) between every pair of nodes.

We can distinguish between two kinds of methods for the solution of the multiterminal minimal path problem. There are methods which use dynamic approach [2], [5], [6] and there are methods based on the triangle inequality valid for the shortest path [6], [8], [9], [12], [13]. Methods of the first kind use the following recurrence relations

$$t_{ij}^{(0)} = t(x_i, x_j)$$

$$t_{ij}^{(k)} = \min_{1 \le l \le n}(t_{il}^{(k-1)} + t_{lj}^{(k-1)}), \qquad k = 1, 2, ..., n.$$

Among methods using the triangle inequality Warshall's method is considered to be the best one. It gives the optimal solution in the following form

$$t_{ij}^{(0)} = t(x_i, x_j)$$

$$t_{ij}^{(k)} = \min(t_{ij}^{(k-1)}, t_{ik}^{(k-1)} + t_{kj}^{(k-1)}), \quad k = 1, 2, ..., n$$

Our algorithm works along the line of Warshall's algorithm.

The algorithm consists of subsequent steps. To every step we determine matrices $L^{(k)} = (l_{ij}^{(k)})$, $M^{(k)} = (m_{ij}^{(k)})$. Here k runs from 0 to n, where n is the number of nodes. The initial matrices belong to $k = 0$.

The initial matrix $L^{(0)} = (l_{ij}^{(0)})$ has the following entries

$$l_{ij}^0 = \begin{cases} d(x_i, x_j), & if (x_i, x_j) \in E \\ 0 & if \ i = j \\ K & otherwise, \end{cases} \tag{7}$$

where K is greater than the maximum value of $d(x_i, x_j)$ for all $(x_i, x_j) \in E$.
We also give the initial matrix $M^{(0)} = (m_{ij}^{(0)})$ by

$$m_{ij}^{(0)} = \begin{cases} k(x_i, x_j), & if (x_i, x_j) \in E \\ c & otherwise \end{cases} \tag{8}$$

where $c \langle \prod_{(x_i, x_j) \in E} k(x_i, x_j).$

The recurrence relations at the kth step are the following: we compute the values m_{ij} and l_{ij}

$$m_{ij} = m_{ik}^{(k-1)} \cdot m_{kj}^{(k-1)}, \ i \neq k, j \neq k, i \neq j, \tag{9}$$

$$l_{ij} = l_{ik}^{(k-1)} + m_{ik}^{(k-1)} l_{kj}^{(k-1)};$$

and determine the entries of $L^{(k)}$ and $M^{(k)}$ as follows

$$l_{ij}^{(k)} = \begin{cases} l_{ij} & if \ \dfrac{l_{ij}}{m_{ij}} < \dfrac{l_{ij}^{(k-1)}}{m_{ij}^{(k-1)}} \\ l_{ij}^{(k-1)} & \end{cases} , otherwise, \tag{10}$$

$$m_{ij}^{(k)} = \begin{cases} m_{ij} & if \ \dfrac{l_{ij}}{m_{ij}} < \dfrac{l_{ij}^{(k-1)}}{m_{ij}^{(k-1)}} \\ m_{ij}^{(k-1)} & \end{cases} otherwise. \tag{11}$$

$L^{(n)}$ and $M^{(n)}$ solve the problem. The proof of this fact is presented in the next section.

Now we validate the algorithm. First we prove two lemmas. Let $P_1 = (x_1, x_2, ..., x_j)$ and $P_2 = (x_j, x_{j+1}... , x_r)$ be two paths and $P = P_1 \cup P_2$.

Lemma 1. For the path P we have

$m(x_1, x_r) = m(x_1, x_j) m(x_j, x_r)$.

Proof. The definition of the function m implies

$$m(x_1, x_r) = \prod_{l=2}^{r} k(x_{l-1}, x_l) =$$

$$= \prod_{l=2}^{j} k(x_{l-1}, x_l) \prod_{l=j+1}^{r} k(x_{l-1}, x_l) = m(x_1, x_j) m(x_j, x_r),$$

which is the required equality.
The following lemma can be proved similarly.

Lemma 2. For the path P we have

$l(x_1, xr) = l(x_1, xj) + m(x_1, xj)l(xj, xr)$.

The validity of our algorithm is expressed by the following theorem.

Theorem. The elements of the matrices $L^{\{n\}}$ and $M^{\{n\}}$ belong to the minimal path leading from x_1 to xj.

Proof. In order to prove this theorem it is sufficient to show that in k^{th} step the value $\dfrac{l_{ij}^{(k)}}{m_{ij}^{(k)}}$ belonging to the matrices $L^{(k)}$, $M^{(k)}$ is mij minimal if the optimal path leading from x_i to xj goes through only a subset of points $x_1 x_2 , ... , x_k$.

For $k = I$ the theorem is obvious because $\dfrac{l_{ij}^{(1)}}{m_{ij}^{(1)}} \leq \dfrac{l_{ij}^{(0)}}{m_{ij}^{(0)}}$ for every i, j. Strict inequality holds if and only if x_1 enter the path.

Let us assume that the statement of the theorem is valid for $k - 1$. At the step there are two cases:

(a)   The optimal path going through the subset of the points $x_1, x_2. , xk$ does not contain the point xk. In this case $l_{ij}^{(k)} = l_{ij}^{(k-1)}$, $m_{ij}^{(k)} = m_{ij}^{(k-1)}$. Thus (10) and (11) imply the statement for k.

(b)   The point xk *belongs* to the path P leading from xi to xr. *Then* xk joins two paths both of which go through certain subsets of $x_1, x_2, ..., x_{k-1}$. Because of the assumption these paths are optimal relative to the points $x_1, x_2, ..., x_{k-1}$. Considering the Lemmas (1) and (2) we can use the recurrence formulas for these two paths. The quotient belonging to the union of these two paths is less than the former quotient because of the formulas (10), (11).   Q.E.D.

This algorithm needs $n(n - 1)^2$ additions, divisions and comparisons and $2n3$ multiplications. If we solved this problem by the simple application of the algorithm given in [1] so that we find the minimal path between each pair of points, than we should have to compute more because in this case the number of additions is $\dfrac{2}{3}n^3(n-1)$ and the number of comparisons, divisions and subtractions is

$\dfrac{1}{3}n^3(n-1)$.

If we want to determine the path itself we need another matrix S. This is the so-called label matrix.

At the beginning the entries of the label matrix are $s_{ij}^{(0)} = j$

At step k the entries are the following

$$
s_{ij}^{(k)} = \begin{cases} (k-1) \, if \, \dfrac{l_{ij}}{m_{ij}} < \dfrac{l_{ij}^{(k-1)}}{m_{ij}^{(k-1)}}, \\ s_{ik} \\ s_{ij}^{(k-1)} \end{cases} \qquad \text{otherwise.}
$$

Having determined subsequently $S^{(1)}, ..., S^{(n)}$, from the last matrix we can find the optimal path itself in the usual way (Dreyfus [5], Ford -Fulkerson [6], Hu [8]).

4   Maximal Flow Problem

The maximal flow problem is to find a feasible flow through a single-source, single-sink flow network that is maximum.

On each edge (x, y), $(x, y) \in E$ is given a nonnegative capacity $c(x, y) \geq 0$. We distinguish two types of nodes of the network: a source x_s, and a sink x_t. For the convenience we assume that every vertex lies on some path from the source to sink, that is for each point of $x \in P$ there is a path from x_s to x and from x to x_t. The graph is therefore connected.

The flow in graph (N, E) is a real valued function, which satisfies the following properties:

Capacity constrains   x, y $\in$P, (x, y) $\in$ E $f(x, y) \leq c(x, y)$

Skew symmetry for all x, y $\in$P (x, y) $\in$ E , $f(x, y) = -f(y, x)$ (12)

Flow conservation   each x $\in$P, x $\neq$ x_s, x_t we require $\sum\limits_{y\varepsilon P} f(x, y) = 0$.

The value of $f(x, y)$ which can be positive, zero or negative is called the flow from point x to point y. The value of a flow from source s to sink t is defined by

$$
F = \sum_{x \varepsilon P} f(s, x) = \sum_{x \varepsilon P} f(x, t) \qquad (13)
$$

The maximal flow problem is to find such a $f(x, y)$ values which fulfill the given conditions (12) and maximize (13).

In the case of network having gains other formalization is needed.

The meaning of the gain function $k(x, y)$, $(x,y) \in E$ is the same as earlier: if one unite of flow is transported from point x than $k(x, y)$ will arrived to the point y $(0 \le k(x, y) \le 1)$. Now we formulate flow of this network. A function $f(x,y)$ is called a flow from source x_s to sink x_t when it fulfills the following conditions:

$$\sum_{(x_i,x_j) \in E} f(x_i,x_j) - \sum_{(x_j,x_i) \in E} f(x_j,x_i)k(x_j,x_i) = 0 \qquad x_i \in N, \; x_i \ne x_s, \; x_i \ne x_t$$

$$\tag{14}$$

$$\sum_{(x_s,x_j) \in E} f(x_s,x_j) - \sum_{(x_j,x_s) \in E} f(x_j,x_s)k(x_j,x_s) < 0 \tag{15}$$

$$\sum_{(x_t,x_j) \in E} f(x_t,x_j) - \sum_{(x_j,x_t) \in E} f(x_j,x_t)k(x_j,x_t) < 0 \tag{16}$$

and

$$0 \le f(x_i,x_j) \le b(x_i,x_j) \qquad (x_i,x_j) \in E \tag{17}$$

The condition (14) means that the flow value in each point- except the source and sink is equal to zero. The condition (15) and (16) mean that the flow value going out of the source is positive similarly the flow value going into the sink also positive. The maximal flow problem in this network: we have to determine such a flow which fulfills the conditions (14)-(17) and maximize the value (16). The problem will be solved by simplex method.

In order to present the structure of the LP we define the model. Let us denote the set of point by N and the set of edges by E:

$$E = (e_1, e_2, \quad ,e_n) \quad N=(x_1, x_2, ...,x_m)$$

Let us denote the flow value on e_1, e_2, $\quad$,e_n by f_1, f_2, ...,f_n , the capacity on that edges by b_1, b_2, ...,b_n. The flow value on the source is denoted by f and on the sink by f. The loss function will be denoted by k_1, k_2, ...,k_n.

Let us denote by e_1, e_2, ...,e_r the edges going into x_i, $x_i \in N$, $x_i \ne s$, $x_i \ne t$ and by e_{r+1}, e_{r+2}, ..., e_v the edges going out of the point x_i.

Using this denotation the following equations are valid

$$\sum_{i=r+1}^{v} f_i - \sum_{j=1}^{r} f_j k_j = 0 \tag{18}$$

These are m linear equations. The upper bounds

$$0 \le f_i \le b_i \tag{19}$$

give also n conditions for the function values $f_1, f_2, ...,f_n$.

Two additional columns are added to the matrix. The column related to the source function value f_s contains zero except the row belongs to source s where the value is 1. The column related to the sink value f_t consist of zeros except the row belongs to t, where the value is -1.

The right hand side vector b is the following

$$b=(0,0,\dots,0, k_1, k_2, \dots, k_n)$$

where the number of zeros equals to the number of points.

The objective vector values

$$C= (0, 0, \dots\dots,0, 1)$$

where the number of zeros equals to the number of arcs.

Using this denotation we formulate the maximal flow problem. The maximal flow problem in a network having gain is to determine the flow vector $F=(f_1, f_2, \dots,f_n, f_s, f_t)$ which fulfills the conditions ****and maximizes the objective function

$$cb= (0, 0, \dots 0, 1) (f_1, f_2, \dots, f_n, f_s, f_t)$$

The matrix of the above defined LP is similar to the matrix of the transportation problem because each column contains two nonzero value namely -1, $-k_{ij}$ (except the upper bound). The number of upper bounds quite large and equal to the number of arcs. In order to reduce the size of the matrix and the computation time it is advisable to apply a special upper bound algorithm.

Acknowledgments. The preliminary version of this paper appeared in the Proceedings of the IEEE 13th International Conference on Engineering Systems, INES 2009 (Barbados, April 16-18, 2009), CD publishing.

References

[1] Bakó, A.: On the Determination of the Shortest Path in a Network Having Gains. Operationsforschung und Statistik 4, 63–68 (1973)
[2] Bellman, R.: On a Routing Problem. Quarterly of Appl. Math. 16, 87–90 (1958)
[3] Charnes, A., Raike, W.M.: One-Pass Algorithms for Some Generalized Network Prblems. Operations Research 14, 914–923 (1966)
[4] Cormen, T.H., Leiserson, E.C., Rivert, R.L.: Introduction to Algorithms, p. 1062. MIT Press, Cambridge (2001)
[5] Dantzig, G.B.: All Shortest Routes in a Graph, Operations Research House, Stanford University, TR 66-3 (1966)
[6] Dreyfus, S.: An Appraisal of Some Shortest Path Algorithms. Operations Research 17, 395–412 (1969)
[7] Ford Jr., L.R., Fulkerson, D.R.: Flows in Networks. Princeton University Press, Princeton (1962)
[8] Glover, F., Klingman, D., Napier, A.: A Note on Finding All Shortest Path, Management Science Report Series, Univ. of Colorado, Bulder, Colorado (1972)
[9] Hu, T.C.: A Decomposition Algorithm for Shortest Path in a Network. Operations Research 16, 91–102 (1968)
[10] Jarvis, J.J., Jezior, A.M.: Maximal Flow with Gains through a Special Network. Perations Researc. 19, 678–688 (1971)
[11] Jewell, W.S.: Optimal Flow through Network with Gains. Operations Research 10, 476–499 (1962)
[12] Warshall, S.: A Theorem on Boolean Matrices. J. of ACM 9, 11–12 (1962)
[13] Yen, J.Y.: On Hu's Decomposition Algorithm for Shortest Path in a Network. Operations Research 19, 983–985 (1971)

The Role of RFID in Development of Intelligent Human Environment

Márta Seebauer

Regional Education and Innovation Center, Budapest Tech
Budai út 45, H-8000 Székesfehérvár, Hungary
seebauer.marta@roik.bmf.hu

Abstract. Ambient systems are a new and perspective field of informatics. The scope of Intelligent Home project in Budapest Tech is the implementation of ambient research methods in all day's life of people. The information flow between a family and the information society were attached through World Wide Network, and mobile communication lines. The problem is the synchronization of the information flow to material flow by the incoming retail article's recognition at the Intelligent Home. This problem was solved by using radio frequency identification techniques.

Keywords: intelligent systems, ambient systems, radio frequency identification, RFID.

1 Introduction

The European Union in the Work Program of the 7th Framework Program (FP7) in field of Information and Communication Technologies (ICT) described a new vision of information society [1]. One of priorities for 2009-10 years is *Ambient Assisted Living* (AAL) Joint National Programme where actually 14 projects were founded. The motivation of AAL program is the problem of growing age of the population in Europe. The accepted projects focus on the visual context recognition, on energy, emergency, and security monitoring, on authentication, on interactive design process, pervasive computing, real-time context-aware, and integration the physical with the digital world [2].

The common aim each of ambient projects is the realization of intelligent systems for the human environment using wireless communication, plug and play technology, natural feeling human interface adaptive to user requirements, advanced encryption techniques, and applying such technologies as for example mobile intelligent agents and fuzzy systems [3].

I.J. Rudas et al. (Eds.): Towards Intelligent Engineering & Information Tech., SCI 243, pp. 457–465.
springerlink.com
© Springer-Verlag Berlin Heidelberg 2009

2   The Intelligent Home Project

The Intelligent Home project in Budapest Tech was started in year 2006. The aim
of *Intelligent Home project* was the optimization of family household processes
through the integration of them in the processes of information society. The distinctive feature of this project from the above mentioned projects is the direct sat-
isfaction the people's all day's needs.

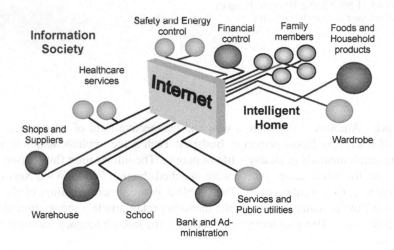

Fig. 2.1. Intelligent Home in the Information Society

A family is the smallest unit of a society. Through the growing penetration of
networked computers and mobile phones the families became to the part of virtual
world that means the information flows between the family members and the information society utilities via Internet, and mobile communication services are established and works in growing extent. The people use widely Web-shops, home
banking, electronic school reports, railway, and airport information systems, the
services of administration, healthcare, and public utilities. A distance learning or
distance working can be carried on in this way (Fig. 2.1).

The people apply computers and network most of all for storing data and communication but not for supporting and optimization of own all day's processes.
The automatic checking, planning, and control of needs and consumption can optimize the household processes and the flow of incoming products sparing time,
energy and costs.

The Intelligent Home project which consists of four activities covers the main
processes of a household

- Intelligent Fridge (started 2006)
- Intelligent Washing Machine (started 2007)
- Intelligent Lighting System (started 2006)
- Intelligent Security System (started 2007)

By the Intelligent Fridge, and Intelligent Washing Machine project cropped up the problem of recognition incoming retail articles.

3   Automatic Identification Technologies

For the synchronization of material and data flow at the inputs of information systems each piece of products most be individually identified. The *automatic identification* technologies became to a standard in the production, trade, transport, logistics, and personal identification in every cases where correct and fast collection and entry of date was needed.

3.1 Bar Code

The *bar code technology of identification* looks more than 50 years ago. In this period the bar codes founded a wide range of applications in sale, manufacturing, and services while this technology is very simple, and cheap. There are a lot of standards for type of data, encoding, and printed formats but the common feature of this techniques is that the represented data most be optically readable. This is the most grievous disadvantage of this technology. The symbols have been printed on the surface area of objects what can follow to the damaging of code.

The other main disadvantage of bare code technology is the limited quantity of represented data however the need of more information will be supplied by growing use of 2D bare codes.

Fig. 3.1. Bar Code Applied in Retail

The packing of retail article will be filled increasingly with compulsory and mandatory product information whereupon they are printed already by unreadable small letter all thought this holds very important information about the ingredients, store conditions, sell-by date, use or misuse, manufacturer, origin etc. of product.

3.2 Radio Frequency Identification

The technology of *radio frequency identification* (RFID) opened a new horizon for automatic and mobile identification [4]. RFID technology support the recognition of each peace of object by radio waves therefore can be widely used for identification of persons, or several types of products.

The main advantage of RFID technology is that unlike bar code identification, this technology doesn't require direct or optical contact. RFID data can be read through cover, wrapper, cloths, animals' or human body, too. Any restrictions of RFID use can cause the metallic and liquid substances. In depends of type the data capacity of RFID tags mounts from couple of bytes up to more Kbytes.

A typical RFID system consists of three components: transponder (RF tag), reader, and host computer (Fig. 3.2).

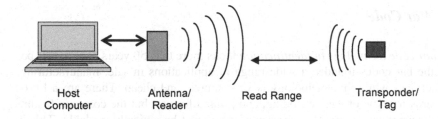

Fig. 3.2. A Typical RFID System

The transponder usually contains two parts, one of them is an integrated circuit for data storing and processing, modulating and demodulating radio frequency signals, and other functions, for example security coding. The other part of a tag is an antenna for receiving and transmitting signals. RFID tags exist in wide range of size and shape from the smallest microchip for implants, placed under the animals' skin until the credit card formats used for the access applications. The largest sizes of transponders – up to 10-15 cm – are used in cargo systems for container tracking.

RFID tags are either active or passive. The active tags use internal battery and the stored data can be modified or rewritten. Active RFID tags in an application can control processes giving any other objects instructions, furthermore receiving and collecting information about the process state. Usually the active tags have not only the most memory size - up to 1 MB, but they have almost the most read range, and in consequence they have the highest costs and the shortest operational life.

The passive RFID tags operate without internal battery, the power supply will generated by reader. The passive tags have less read range, less memory capacity – usually 4-8 bytes, theoretical unlimited lifetime, they are smaller and cheaper as active tags. In the applications using passive tags, the preprogrammed data loaded from the tags play only the role of key information and the interactive data related to the object will be stored in the database of application.

The reader emits radio waves in the given frequency defining the operating distance

- **Low frequency** (LF) system operating in range 125-135 kHz has short read distance (up to 10 cm), low costs and highest level of security.
- **High frequency** (HF) system operating in range 13.56 MHz has middle read range (about 1 m) and are tolerant of metal or fluids.
- **Ultra high frequency** (UHF) system operating in 850-950 MHz range has a long read ranges (about 30 m) and high reading speed, more system costs.
- **Microwave frequency** system operates by 2.4-2.5 GHz very short reading time but the reader is very expensive.

The reader decodes the data encoded in the microchip of tags and sends the data to the host computer passing a standard serial port of them. The host computer will used for data storage and processing.

Since 2005 the Wall-Mart Stores Inc. in United States of America requires that the top suppliers apply RFID tags to all of products improving the supply chain management. This requirement promoted the RFID production. The vendors use special *Electronic Product Code* (EPC) in form of bar code on the surface of label with integrated RFID tag in it.

The authors of the last RFID reports [5] give a full analysis of RFID market and suggest the stabile growing number of application all of the type of RFID tags and labels in the entire world but first of all in the USA and then in East Asia what results a boom of RFID in apparel.

4 Application of RFID in Intelligent Home Project

RFID are widely used in ambient systems, all of them for the personal or position recognition [5, 6]. In Intelligent Home Project was presupposed that retail products are supplied with RFID tags which store all of the important information about them. Until widespread applying RFID tags on retail products this parameters are stored in database of information system. For the product identification were passive tags used.

4.1 The Intelligent Fridge

Intelligent fridges exist on the market. The user can send and receive e-mails, browse in the World Wide Web, look videos, listen to music, and nothing more.

The aim of *Intelligent Fridge* project was the detection and optimization of family needs. The hardware model of intelligent fridge consists on a pocket personal computer with touch screen, RFID reader and experimental tags visible in the front of Fig. 4.1, a GSM module for the mobile communication on the right side and a weighing machine (not illustrated), as well.

Fig. 4.1. Hardware Model of Intelligent Fridge

The software application consists on database, shopping list, calendar, statistics, message board, and entertainment modules. All of important parameters of articles are stored in database (Fig. 4.2). Input of article's data can occurs by RFID entry or manual if RFID tags are not available.

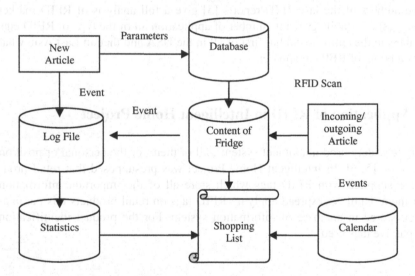

Fig. 4.2. Data Flow of Intelligent Fridge

Incoming food products will be checked by weight and putting in the fridge recorded in the content table of fridge. Each event is registered in log file used for the statistics. The database contents the basic parameters of program, and the main parameters of family members.

The following date of purchase and for the purchase responsible person is always fixed in calendar module. If nobody is fixed than the default person will receive the shopping list that will be sent by SMS or E-mail depending on setup. Each family member can enter up several events in the calendar module. If this event has a history in statistics module then this entry will automatically modify the needs. Actual shopping list will compile using the information from the statistics and calendar module, and actual content of fridge, but can be revised manually. The application regularly checks the shelf-life of foods.

The intelligent fridge program compares the shopping list with the incoming products. If a product will ignored for a given period, it will become inactive but not deleted from the database. The database can be extended from the fridge foods to all of household products in the future.

In the perspective are two functional extensions. One of them is an expert system of family's best recipes taking proposals on required foods for given dishes by backward chaining inference, or in opposite conclude the supposed dishes from given in fridge foods using forward chaining reasoning. The other perspective is the realization of an on-line information and logistic system for the direct food supply from the local shops to the fridge.

4.2 The Intelligent Washing Machine

The *Intelligent Washing Machine* project is implemented using the same hardware model as the intelligent fridge. The system is extended only by laundry baskets and a virtual washing machine provided with RFID reader each of them.

The purpose of this project is the optimized handle of family's wardrobe. By the data entry is supposed that each peace of clothes has a RFID label and can automatically recognized (Fig. 4.3) failing which the clothes features must be entered manual (Fig. 4.4).

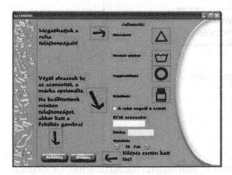 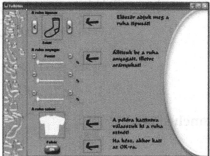

Fig. 4.3. Input of RFID Code   **Fig. 4.4.** Input of New Cloth's Date

The basic dates of clothes are stored in the database (Fig. 4.5).The pieces of clothes throwing into one of the laundry baskets will recorded in the laundry basket table.

The sorting program regularly checks this table and respecting the washing parameters of each piece try to complete and optimize the washing doses. The selected dose will moved to the washing table which can be revised by manual loading pieces from the laundry basket or put it back.

On the base on all of given parameters the program using fuzzy logic calculates the optimal washing program, washing time, water's temperature, quantity of water, washing-powder, and softener, the time and phase of spin-drier. The reader in the washing machine checks the material process and compare with the actual content of the washing table.

The results of washing process will be recorded into the log file. Based on log file the program can handle any exceptions, and parameter's update in the database. The junked pieces of clothes will be removed from database. In the perspective is the hardware implementation of fuzzy control.

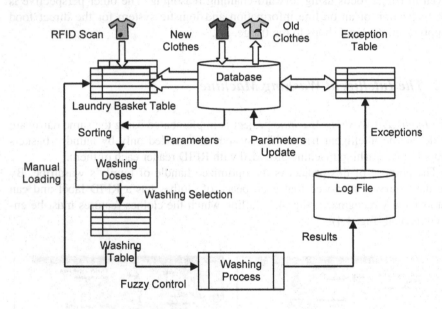

Fig. 4.5. Data Flow of Intelligent Washing System

Conclusions

For building an Intelligent Home system the automatic identification of objects was needed. In the retail widely used bar code holds not enough information about the products, and not allows the individually identification each peace of article. Looking for the widespread implementation of RFID technology two projects for intelligent home supporting the human needs were completed. By the RFID implementations were short read range passive RFID tags used throughout the data entry in the database system is correct and simply.

Acknowledgments. The author would like to acknowledge the studious work of young researcher team of Intelligent Home project Nézics A, Takács G, Gotthard Z, Hoffmann Zs, Mikó T, and Csapó M and the financial support of the project on beside of IBM® Hungary.

References

[1] European Commission, ICT – Information and Communication Technologies. Work Programme 2009-10 (2007),
http://cordis.europa.eu/fp7/ict/docs/ict-wp-2009-10_en.pdf
(Accessed, 24 April 2009)

[2] IST Advisory Group, Scenarios for Ambient Intelligence in 2010 (2001), ftp://ftp.cordis.europa.eu/pub/ist/docs/istagscenarios2010.pdf (Accessed, 24 April 2009)

[3] Augusto, J.C.: Ambient Intelligence: the Confluence of Ubiquitous/Pervasive Computing and Artificial Intelligence. In: Schuster, A. (ed.) Intelligent Computing Everywhere. Springer, Heidelberg (2007)

[4] Want, R.: An Introduction to RFID technology. Pervasive Computing 5(1), 25–33 (2006)

[5] Das, R., Harrop, P.: RFID Forecast, Players and Opportunities 2009-2019 (2009),
http://www.idtechex.com/research/reports//
rfid_forecasts_players_and_opportunities_2009_2019_000226.
asp (Accessed, 24 April 2009)

[6] Berger, M., Fuchs, F., Pirker, M.: Ambient Intelligence – From Personal Assistance to Intelligent Megacities. In: Augusto, J.C., Shapiro, D. (eds.) Advanced in Ambient Intelligence, pp. 21–35. IOS Press, Amsterdam (2007)

[7] Berényi, Z., Charaf, H.: Retrieving Frequent Walks from Tracking Data in RFID-Equipped Warehouses. Library of Congress: 2007905110 (2008) ISBN: 1-4244-1543-8

Part V
Machines, Materials and Manufacturing

History and Challenges of Mechanism and Machine Science within IFToMM Community

Marco Ceccarelli

LARM: Laboratory of Robotics and Mechanics, DIMSAT, University of Cassino
43 Via G. Di Biasio, 03043 Cassino, Italy
ceccarelli@unicas.it
http://webuser.unicas.it/weblarm/larmindex.htm

Abstract. Mechanism and Machine Science (MMS) has been the core of mechanical engineering and indeed of industrial engineering since the beginning of engineering practice and particularly in modern times. A historical short survey is presented to outline the main characteristics of mechanisms and their evolution also with the aim to identify challenges and role of MMS in future developments of Technology for the benefit of the Society. The significance of IFToMM, the International Federation for the Promotion of MMS is also stressed as the worldwide community that in the last forty years has contributed to aggregate common views and developments and can have a important role for future improvements yet.

1 Introduction

Mechanics of systems has been a main concern for system developments and mechanisms have been used since the beginning of a growth of Society and. Over the time changes of needs and task requirements in Society and Technology have required continuous evolution of mechanisms and their uses, with or without a rational technical consciousness. In past evolution, technical knowledge has made possible to propose more and more solutions enhancing mechanisms and their uses in order to satisfy demands from Technology and Society. Today it seems that we have reached a saturation and a so high level of knowledge on mechanisms that several professionals and even researchers from other fields think that there is nothing else to be discovered or conceived in MMS.

Since the past historical background and developments have been outlined even from several technical viewpoints (beside in circuits of History of Science) in several works with the aim to track historical evolution of Technology and Engineering, and to recognize the paternity of machine achievements, like for example in (Chasles 1837 and 1886; Reuleaux 1875; De Groot 1970; Crossley 1988; De Jong 1943; Dimarogonas 1993; Ferguson 1962; Hartenberg and Denavit 1956; Nolle 1974; Roth 2000) just to cite few relevant sources with fairly simple availability.

I.J. Rudas et al. (Eds.): Towards Intelligent Engineering & Information Tech., SCI 243, pp. 469–488.
springerlink.com   © Springer-Verlag Berlin Heidelberg 2009

Recently a specific conference forum has been established within IFToMM (The International Federation for the Promotion of MMS) as HMM Symposium in which several views and studies are discussed, (Ceccarelli Ed., 2000 and 2004; Yan and Ceccarelli, 2008). Some more specific emphasis has been addressed on historical trends on recent research activity in papers like (Bottema and Freudenstein 1966; Hunt 1984; Roth 1983; Shah 1966). Even the author has attempted to outline historical developments with the aim to track the past to identify directions for future work in (Ceccarelli, 2001 a and b; 2004 a and b; 2007).

In this survey a vision is outlined with a historical perspective in order to show that, although many new issues in Mechanism Design can be based on basic concepts that have been developed in the past, we have still several challenging issues to approach for giving proper solutions to new and updated problems in the evolving Technology and Society. New systems and updated performance are asked for mechanism applications that deserve attention starting from the theoretical bases before to update or conceive algorithms for design and/or operation with optimal characteristics.

Two main considerations can be observed in order to claim that MMS is still a discipline with necessary strong activity in teaching, research, and practice. Namely they are:

Human beings operate and interact with environment and systems on the basis of actions of mechanical nature; therefore mechanisms will be always an essential part of systems that help or substitute human beings in their operations and other manipulations.

There is a continuous need to update problems and solutions in Technology since Society continuously evolves with new and updated needs and requirements; thus, even mechanisms are asked for new and updated problems that require a continuous evolution and update of knowledge and means for proper applications of mechanisms.

2   A Short Account of History of Mechanism Design

Mechanisms and Machines have addressed attention since the beginning of Engineering Technology and they have been studied and designed with successful activity and specific results. But TMM (Theory of Machines and Mechanisms) have reached a maturity as independent discipline only in the 19th Century.

The historical developments of Mechanisms and Machines can be divided into periods with specific technical developments that, according to my personal opinion, can be identified and characterized by referring to significant starting events such as:

- Utensils in Prehistory
- Antiquity: 5-th cent. B.C. (Mechanos in Theater plays)
- Middle Ages: 275 (sack of the School of Alexandria and destroy of Library and Academy)

- Early design of machines: 1420 (Zibaldone by Filippo Brunelleschi)
- Early discipline of mechanisms: 1577 (Mechanicorum Liber by Guidobaldo Del Monte)
- Early Kinematics of mechanisms: 1706 (Traitè des Roulettes by Philippe De La Hire)
- Beginning of TMM: 1794 (Foundation of Ecole Polytechnique)
- Golden Age of TMM: 1841 (Principles of Mechanism by Robert Willis)
- World War Period: 1917 (Getriebelehre by Martin Grubler)
- Modern TMM: 1959 (Synthesis of Mechanisms by means of a Programmable Digital Computer by Ferdinand Freudenstein and Gabor N. Sandor)
- MMS Age: 2000 (re-denomination of TMM by IFToMM)

A preliminary version of a brief account on History of Mechanism Design has been presented by the author at 2004 IFToMM World Congress (Ceccarelli 2004). However, many other authors have outlined historical evolution of the field in the past and even recently in order to emphasize new research lines, as for example those who have been mentioned in the introduction.

Most of the historical views and achievements are usually referred to western countries as related to the development of Industrial Revolution and Modern Technology that occurred mainly in western countries. Thus, historical technical studies often ignore what happened in the rest of the world mainly because the achievements that occurred there, did not affect directly the technical evolution during the Industrial Revolution. However, those technical achievements in historical developments in other countries and civilizations can be and are considered of great interest also to understand the worldwide acceptance of Industrial Revolution and cultural evolution of Mechanical Engineering at large. But they are often underestimated and they should be better considered in more clear studies and presentations by engineers as oriented to engineers.

The historical evolution to the current MMS can be outlined by looking at developments that occurred since the Renaissance. Mechanisms and machines were used and designed as means to achieve and improve solutions in other fields. Specific fields of mechanisms grew in results and awareness and first personalities were recognized as brilliant experts, like for example Francesco Di Giorgio Martini and Leonardo Da Vinci among many others, as emphasized in (Ceccarelli 2008). At the end of Renaissance Mechanics of Machinery addressed a great attention also in Academic world, starting from the first classes given by Galileo Galilei in 1593-98. The designer figure evolved to a professional status with strong theoretical bases finalizing a process in 18th Century that in the Renaissance saw the activity of closed small communities of pupils/co-workers after 'mastros' and 'maestros'. Lot of academic activity increased basic knowledge for rational design and operation of mechanisms. First mathematizations were attempted and fundamentals on mechanism kinematics were proposed by first investigators who were specifically dedicated to mechanism issues, like for example Philippe De la Hire among many others. The successful practice of mechanisms was fundamental for relevant developments in the Industrial Revolution during which many practitioners and researchers implemented the evolving theoretical knowledge in practical applications and new powered machines. The 19th Century can be considered the Golden Age of MMS since relevant novelties were proposed both in theoretical an practical fields. Mechanism were the core of any machinery and any technological advance. A community of professionals was identified and specific academic formation was established. TMM gained and important role in the development of Technology and Society and several personalities expressed the fecundity of the field with their activity, like for example Franz Reuleaux among many others. The first half of 20th Century saw the prominence of TMM in mechanical (industrial) engineering but with more and more integration with other technologies. A great evolution was experienced when with the advent of Electronics it was possible to handle contemporaneously several motors in multi-d.o.f. applications of mechanisms and to operate 3D tasks with spatial mechanisms. The increase of performance (not only in terms of speed and accuracy) required more sophisticated and accurate calculations that have been possible with the advent of Informatics means (computers and programming strategies).

Today, a modern machine is a combination of systems of different natures and this integration has brought to the Mechatronics concept. Most of the recent advances in machinery are thought in fields other than MMS. But Mechanism Design can be still recognized as a fundamental discipline for developing successful systems that operate in the mechanical world of human beings. Tasks and systems for human beings must be of mechanical nature and a careful Mechanism Design is yet fundamental to obtain systems that help or substitute human beings in their operations. Most of those tasks are already obtained with mechanism solutions that can be seen as traditional successful ones that nevertheless could need further update or re-consideration because of new operation strategies and/or new material and components (scaled designs). Therefore, Mechanism Design can be considered still a discipline for current research interests. But: What are open problems and challenges for today Mechanism Design? Can they be considered new issues or should they be rediscovered from past ideas?

3   IFToMM Activity and Its Role

The identity of a person and even a Community can be indicated by a name giving a synthetic description of the personality and main capability or characteristics. The names of IFToMM (the International Federation for the Promotion of MMS), TMM (Theory of Machines and Mechanisms), and MMS (Machine and Mechanism Science) identify IFToMM Community who refers to MMS at large. The names of IFToMM, TMM, and MMS are related to fields of Mechanical Engineering concerning with Mechanisms in broad sense.

TMM is often misunderstood even in the IFToMM Community, although it is recognized as the specific discipline of Mechanical Engineering related with mechanisms and machines. The meaning of TMM, now MMS, can be clarified by looking at terminology.

IFToMM terminology (IFToMM 1991 and 2003) gives:
- Machine: mechanical system that performs a specific task, such as the forming of material, and the transference and transformation of motion and force.
- Mechanism: system of bodies designed to convert motions of, and forces on, one or several bodies into constrained motions of, and forces on, other bodies.

The meaning for the word "Theory" needs further explanation. The Greek word for Theory comes from the corresponding verb, whose main semantic meaning is related both with examination and observation of existing phenomena. But, even in the Classic language the word theory includes practical aspects of observation as experiencing the reality of the phenomena, so that theory means also practice of analysis results. In fact, this last aspect is what was included in the discipline of modern TMM when Gaspard Monge (1746-1818) established it in the Ecole Polytechnique at the beginning of 19th Century, (Chasles 1886), (see for example the book by Lanz and Betancourt (1808), whose text include early synthesis procedures and hints for practical applications). Later (se for example Masi 1888) and up today (see for example Uicker et al. 2003) many textbooks have been entitled Theory of Mechanisms since they describe both fundamentals and applications of mechanisms in machinery.

The term MMS has been adopted within the IFToMM Community since the year 2000 (IFToMM 2004) after a long discussion (see (Ceccarelli 1999) in the IFToMM Bulletin), with the aim to give a better identification of the modern enlarged technical content and broader view of knowledge and practice with mechanisms. Indeed, the use of the term MMS has also stimulated an in-depth revision in the IFToMM terminology since the definition of MMS has been gives as, (IFToMM 2003):
- Mechanism and Machine Science: Branch of science, which deals with the theory and practice of the geometry, motion, dynamics, and control of machines, mechanisms, and elements and systems thereof, together with their application in industry and other contexts, e.g. in Biomechanics and the environment. Related processes, such as the conversion and transfer of energy and information, also pertain to this field.

Summarizing, since the modern assessment, TMM has been considered as a discipline, which treats analysis, design and practice of mechanisms and machines. This will be also in the future for the area today named as MMS, since modern and futures systems will still include mechanisms and machines with mechanical designs and operations as related with life and working of human beings. These mechanical devices need to be designed and enhanced with approaches from mechanical engineering because of the mechanical reality of the environment where the human beings will always live, although new technologies will substitute some components or facilitate the operation of mechanical devices.

Technically, MMS can be seen as an evolution of TMM as having a broad content and view of a Science, including new disciplines. Historically, TMM has included as main disciplines: History of TMM; Mechanism Analysis and Synthesis; Theoretical Kinematics; Mechanics of Rigid Bodies; Mechanics of Machinery; Machine Design; Experimental Mechanics; Teaching of TMM; Mechanical Systems for Automation; Control and Regulation of Mechanical Systems; RotorDy-namics; Human-Machine Interfaces; BioMechanics. The modernity of MMS has augmented TMM with new vision and means but also with many new disciplines, whose the most significant can be recognized in: Robotics; Mechatronics; Compu-tational Kinematics; Computer Graphics; Computer Simulation; CAD/CAM for TMM; Tribology; Multibody Dynamics.

The evolution of the name from TMM to MMS, that has brought also a change in the denomination of the IFToMM Federation from "IFToMM International Federation of TMM" to "IFToMM, the International Federation for the Promotion of MMS", can be considered as due both to an enlargement of technical fields to an Engineering Science but even to a great success in research and practice of TMM with an increase of engineer community worldwide.

The developments in TMM have stimulated cooperation all around the world at any level. One of the most relevant results has been the foundation of IFToMM in 1969, Figs. 1 and 2. IFToMM was founded as a Federation but as based on the ac-tivity of individuals within a family frame with the aim to facilitate co-operation and exchange of opinions and research results in all the fields of TMM. Many in-dividuals have contributed and still contribute to the success of IFToMM and re-lated activity, (see IFToMM webpage: www.iftomm.org) under a vision coordina-tion of IFToMM Presidents over time, Fig. 3.

The modernity and relevance of IFToMM activity can be recognized in the common frame of views and results on MMS, in the many different technical fields. Thus, the role of IFToMM can be still recognized, like stated in its constitu-tion, as instrumental in stimulating enhancements and giving common frames and views for the evolution of MMS both with technical aims and benefits for the So-ciety.

Fig. 1. A historical moment of the foundation of IFToMM, the International Federation for the Theory of Machines and Mechanisms, in Zakopane (Poland) on 27 September 1969, (Courtesy of IFToMM Archive) in which one can recognize: 1) Prof. Ivan Ivanovic Artobolevskii (USSR); 2) Prof. Adam Morecki (Poland); 5) Prof. Nicolae I. Manolescu (Romania); 6) Prof. Erskine F. Crossley (USA); 7) Prof. Giovanni Bianchi (Italy); 8) Prof. Aron E. Kobrinskii (USSR); 9) Prof. Werner Thomas (Germany); 10) Prof. Jan Oderfeld (Poland)

We, the undersigned chief delegates at the Inaugural
Assembly of the International Federation for the Theory
of Machines and Mechanisms (IFTOMM) here at Zakopane Poland on 27th September 1969, declare that we have founded the above-mentioned Federation and that we have adopted its Constitution which is attached hereto and decided
to the following categories (see Article 8.4 of the Constitution).

Territory	Chief delegate	Proposed Category	Signature
Australia	JACK PHILLIPS		
Bulgaria	George Rusanov		
German Democratic Republic	Wolfgang Rössner		
German Federal Republic	Werner Thomas		
Hungary	Lulu TERPLAN		
India	J. S. RAO		
Italy			
Poland	Adam Morecki		
Rumania	Nicolae I. Manolescu		
United Kingdom			
U.S.A.	Douglas Muster		
U.S.S.R.			
Yougoslavia			

Fig. 2. IFToMM foundation act (Courtesy of IFToMM Archive)

Fig. 3. IFToMM Presidents at the Symposium HMM2000 in Cassino on May 2000 (from left to right): Giovanni Bianchi (1984-'87 and 1988-'91), Arcady Bessonov in substitution of Ivan I. Artobolevsky (1969-71 and 1972-'75), Bernard Roth (1980-'83), Jorge Angeles (1996-2000), Kenneth J. Waldron (2000-'03 and 2004-'07), Leonard Maunder (1976-'79), Adam Morecki (1992-'95), and Marco Ceccarelli (2008-2011; at the time of the photo Chairman of HMM2000 IFToMM Symposium on History of Machines and Mechanisms) (years in the brackets indicate the President term)

4   Old and New Solutions in MMS towards Future Challenges

The main current interests for research in MMS can be summarized in the follow-
ing topics:
- 3D Kinematics and its application in practical new systems and methodologies
- Modeling and its mathematization
- Multi-d.o.f. multibody systems
- Spatial mechanisms and manipulators
- Unconventional (compliant, underactuated, overconstrained) mechanisms
- Scaled mechanisms
- Creative design
- Mechatronic designs
- Reconsideration and reformulation of theories and mechanism solutions

Those topics are also motivated by needs for formation and activity of professionals who will be able to conceive and transmit innovation both into production
and service frames.

Teaching in MMS requires attention on modern methodologies that can use efficiently computer and software means, which are still evolving rapidly. Thus,
there is a need to update also the teaching means that makes use of simulations
and computer oriented formulation. In addition, mechatronic lay out of modern
machinery suggest to teach mechanisms as integrated with other components like
actuators and sensors since the beginning of the formation curricula.

Activity by professionals asks for novel applications and high performance designs of machines as they are continuously needed in evolving/updating systems
and engineering tasks. In addition, there is a need to make understandable new
methodologies to professionals for practical implementation both of their use and
results. New solutions and innovation is continuously asked not only for technical
needs but even for political/strategic goals of company success.

Are the above-mentioned topics really new arguments in the History of MMS?
What challenges are still facing the MMS community?

It is quite clear that modern developments in Technology and Science have
stimulated and required developments and novelties in machines and mechanisms
too, since the growth of new and updated needs, but also because of current practical possibilities for mechanism solutions that were utopist in the past.

Past solutions and efforts can be helpful to understand situations and developments that are still needed to reach successful achievements fulfilling today and
even future requirements. In the following, few examples are briefly discussed to
illustrate what has been developed in the past that can be considered as new today
and therefore of inspiration for today research in MMS together with new challenges.

Lot of today activity is addressed to new models and new mathematizations for
mechanisms. This attention is not new, but procedures and means can be considered new since they are related to goals of using modern means of calculus both in
term of numerical algorithms and computer facilities. Today a high level of

abstraction is used in treating kinematic chains and motion characteristics of mechanisms. However, even in the past abstraction was used to study properly a general but specific motion of a rigid body, like the early studies of point trajectories in the form of curve formulation since the late renaissance and up the beginning of Industrial Revolution. Even today, this study does not refer directly to any practical application and seems to be a pure academic exercise, but one can easily recognize that results of such a theoretical analysis can be implemented to improve successfully machinery operations both in terms of theoretical capabilities and mechanical design characteristics.

Today dualistic viewpoints are proposed as based on graphical and mathematical approaches yet. Thus, for examples graphs and algebraic groups are used to describe kinematic chains of mechanisms and their functionalities. Similarly, Graph Theory with a matrix algebra and Group Theory are introduced to formulate the operation of mechanisms and then new design algorithms. Lot of work is addressed to a suitable mathematization of mechanism aspects that can be efficiently treated through computer means. This issue, indeed, is an old need, i.e. designers and researchers worked on engineering views as a function of computational issues with means of the time and therefore, theories and algorithms are today needed to be formulated or re-formulated accordingly to the capacity of those new available means. Computer means can make available mathematical means for practical engineering purposes with great computational accuracy, when they are properly adjusted to each other. An example from the past is the Algebra of matrices that today is commonly used in Robotics. It was not used until computer calculation since the end of 1950s' made it feasible and more efficient than traditional graphical procedures. New attention is addressed today to other algebras and even using quaternions and biquaternions in a considerable growing literature. Today computational issues seem to motivate also basic research and re-formulation regarding problems that were considered as definitively solved even in a recent past. This is because a new mathematization and related computational algorithms can give further insight both in solutions and design algorithms. Emblematic is the attention to the kinematics of four-bar linkages whose fecundity (as Hartenberg pointed out in the 1950s') still gives new insights and procedures for designing also more complex systems or new solutions for practical applications (see for example compact actuation mechanisms for backdoors of cars). Issues on mathematization are still of current interest non only for computational purposes, but even for further investigation fields as for example for 3D kinematics. The great attention on 3D kinematics has been motivated from engineering viewpoint since when there has been the possibility to operate and regulate spatial mechanisms in practical applications. This has happened when it has been possible to control and to sensor spatial motion by using electronic components. Indeed, even the increase of spatial tasks for manipulations and industrial processes have made of practical interest mechanisms that were studied since the second half of 19th Century, but mainly for pure academic interests. Screw systems are today extensively modeled and formulated to design and operate manipulators and spatial mechanisms in practical applications in several fields other than industry, like for example in medical engineering. Schemes from Screw Theory are used and although they are presented in new algorithms, basic concepts were conceived in the past. An example is shown in Fig. 4 in which the so-called Screw Triangle is outlined by Bricard in 1926 for a design algorithm that today still addresses great attention mainly for suitable computer-oriented formulation in a very rich growing literature. Thus, the concepts of the Screw Theory have been outlined since the 18th Century (through the works of Euler, D'Alembert, Bernoulli, Frisi, and Mozzi (see Ceccarelli 2000) and then it has been clearly formulated in 19th Century, by starting with the work of Ball. But, a useful mathematization and consequent practical implementation of Screw Theory have been developed in modern terms and are still under development as function of mathematical and informatics means through several approaches.

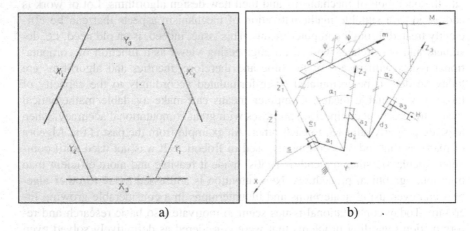

a)   b)

Fig. 4. A scheme for Screw Triangle of a 3R manipulator: a) by Bricard in 1926,) b) as drawn today

Spatial mechanisms are thought to be conceived in the last five decades. But the possibility to use them is not new. One of those last systems can be considered cable parallel manipulators that have addressed attention of researchers and designers only recently. But in Fig. 5 Filippo Brunelleschi seems to have used such a system in a crane to increase payload and mechanical versatility, already in 1420.

Problems that are related to manipulation mechanisms such as for grasping, are today of fundamental relevance, and considerable attention is addressed to the variety of grasps and to general models and formulation of the mechanics of grasp. Those needs have been determined by a large variety of tasks and objects that can be grasped in today applications both in industry and diary life. In Fig. 6 a study by Mariano Di Jacopo (il Taccola) in the 14th Century is reported in which different grippers are examined in terms of fingers and locking systems, likewise in today investigations. Today, we have very powerful technology for sensors and force control, but still the mechanics of the grasp is fundamental and even more influential on the grasp success is the mechanism design and its functionality.

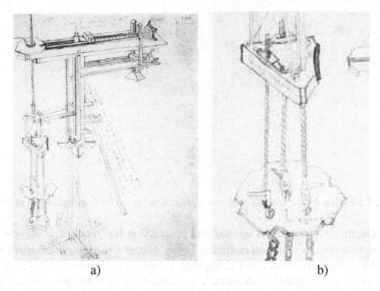

a)   b)

Fig. 5. An early cable parallel manipulator in a crane by Filippo Brunelleschi in 1420: a) the crane system; b) zoomed view of an early cable parallel manipulator

In Fig. 7 schemes are shown as used by Francesco Masi in 1897 to study the stability of a grasp with multiple contacts, likewise today approaches for in multi-fingered robotic hands. Indeed the sketches recall very much the today schemes of force closure for investigating on the stability of multi-contact grasps. The grasp versatility of human hand with its compact design is still a challenge both for robotic applications and prosthesis implementation.

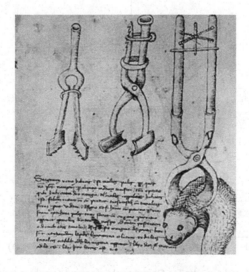

Fig. 6. A variety of two-finger grippers with different locking systems for fish grasping by Mariano Di Jacopo in 14th Century

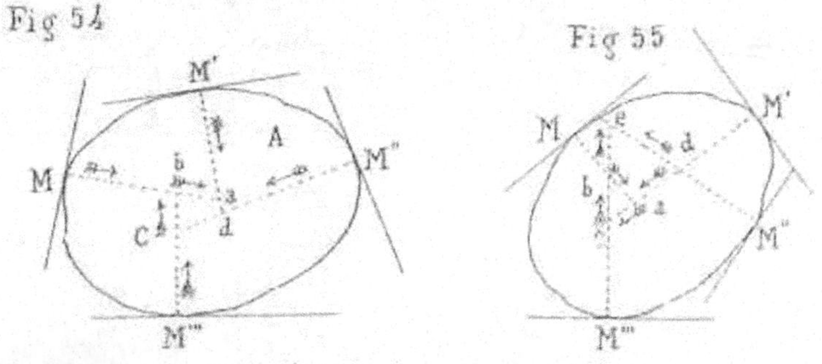

Fig 54

M

M'

A

b

a

3

C

d

M''

M'''

Fig 55

M

e

d

M'

b

a

M''

M'''

Fig. 7. Sketches for the analysis of planar multi-contact grasp by Francesco Masi in 1897

3D kinematics has been deepened and abstraction has reached good results so that is it used also in many other fields. Today one of the most successful field of novel applications is considered Computer Vision and Graphics, since when vision systems are available with suitable advanced capabilities. But already in the last decade of 19th Century there was a successful activity in applying Kinematics to Graphics with perspective to Vision, but for practical applications in technical Drawing. Emblematic is the example dated 1880 in Fig. 8 in which computations results from kinematics in Fig. 8a) are used for shadowing a complex object in Fig. 8b). The topic was specifically addressed to enhance technical drawing towards its standardization. This is again the case today as related to user-oriented CAD techniques that can give results useful also for Vision applications.

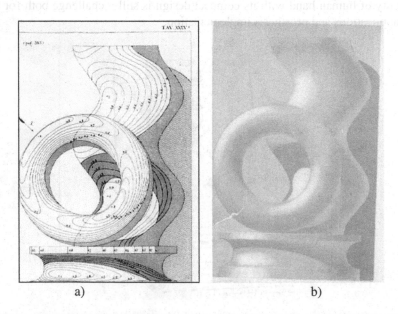

a)　　　　　　　　　　　　　　　　　　　b)

Fig. 8. An example of early results for Computer Vision and Graphics by using Kinematics by Domenico Tessari in 1880: a) computed results for isotonic curves of a 3D object; b) corresponding pictorial reconstruction of the illuminated object

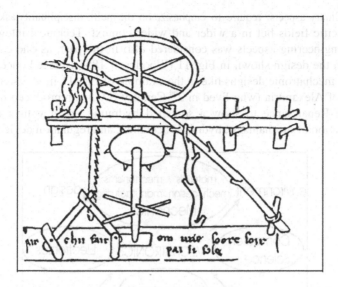

Fig. 9. An automatic wood sawing machine by Villard de Honnecourt in the 13th Century

Nowadays automatic machinery requires multi-d.o.f. mechanisms that have been possible only with the advent of modern control engineering and advanced electric motors. But solutions were designed and used also in the past by using mechanical devices and/or ingenious designs not only in the form of automtons. Figure 9 shows the design of an automatic wood sawing machine by Villard de Honnecourt in 13th Century. In the machine one can identify the axle of the water turbine transmitting motion both to the saw linkage mechanism and wood feeding slide. In addition, the saw linkage mechanism can be interpreted as two surprising operations, namely the saw is guided by using a coupler point (usually James Watt is said to have been the first using coupler curves for body guiding, but in 1742) and/or the mechanism can be understood even as a five-bar linkage by looking at the drawn bars as movable links when the second input can be used to adjust the saw operation yet.

Mechatronics is usually considered a last achievement of modern engineering by which modern systems are designed and operated because of integration of several components of different natures with a multi-disciplinary engineering approach, Fig. 10. The significant role of mechanisms in Mechatronics can be understood since it is fundamental a mechanical system to interact with the environment or to perform the task. However, for a mechanical system, but even for a mechatronic one within mechanics aspects mechanism can be considered together with material, mechanical design, and manufacturing as main aspects to be considered for a proper design and functionality. Although engineer formation was and is still achieved by teaching separately courses on specific disciplinary subjects, nevertheless machines have been always treated by looking at the integration of different aspects. Of course, nowadays the multitude and sophistication of those multidisciplinary aspects require to emphasize on the multi-disciplinarity asking expertise in specific fields but in a wider and wider context. Technical integration of different engineering aspects was considered also in the past, as one can see for example in the design shown in Fig. 11 (Woodcroft 1851). Indeed once can find even early mechatronic designs like in the example of Fig. 11 in which a machine by Heron of Alexandria (who lived in 2nd Century B.C.) is reproduced in a drawing during Renaissance to show a so-called hydraulic organ with a combination/integration of mechanisms, hydraulic actuators, and regulation devices.

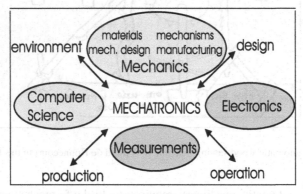

Fig. 10. A scheme for definition of Mechatronics

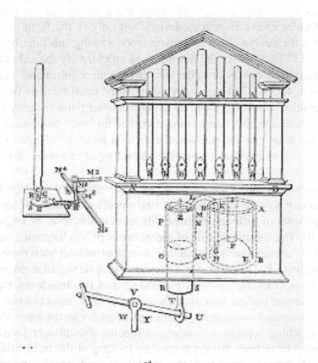

Fig. 11. A hydraulic organ designed in the 2nd Century B.C. by Heron of Alexandria as redrawn in 15th Century

Another modern issue is related with machinery materials and related tribological problems. Today, but even in the past, main limits of life duration of machines (or at least of their efficiency) are considered as related to tribological issues due to friction and wear. This critical issue was understood since the beginning of using moving connected bodies and several solutions were studied and attempted both in terms of materials and manufacturing/operation techniques in order to have a control and/or an estimation of the effects. In Fig. 12 an example of such an attention and adopted consequent action is reported from the time of antique Egyptians. A man at the foot of the statue is specifically devoted to spray a liquid (perhaps vene not pure water) as lubricant on the sliding surface with the to reduce the friction for moving the heavy load of a statue. There, we have the principle of attaching a tribological problem by using intermediate lubricant material, suitable surface design, and even controlled operation, likewise today, although today we have a larger variety of materials and technologies.

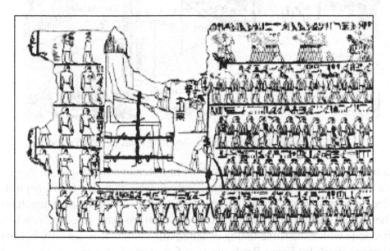

Fig. 12. A tribological solution with lubricant application for transportation of a statue at the time of Antique Egypt

Another challenging trend in MMS can be identified in scaled mechanisms/machines and their applications both to very small and very large sizes. Miniaturization has been experienced in the last decades and today micromechanisms are almost usual machines so that research is now directed even to molecular scales. On the other hand the increase of power needs has required larger and larger size of mechanical devices. Even spatial exploration and exploitation have stimulated a great consideration of very large mechanism with deployable mechanism structures. While miniaturization at today levels is indeed a novelty in the history of engineering, the scaling need is not new and engineering approaches were developed to adapt the mechanism size to the application requirements even by considering peculiarities of the scaling in terms of actions and relative significance, as in the example in Fig. 13 (Besson 1578), for a mechanism to lift large boats at the time of late Renaissance. In particular, the history of

Mechanism Design is mainly related to an increase of tasks by enlarging mechanisms in terms of power, motion, and productivity. Thus once more, the new scaling of design and operation has required and still require to re-consider the basic principles of mechanisms. Similarly, the challenge of micro-mechanisms and even more nano-mechanisms can be considered in adapting and therefore in redesigning known mechanisms by considering the different new situations and environments at the very small scale applications.

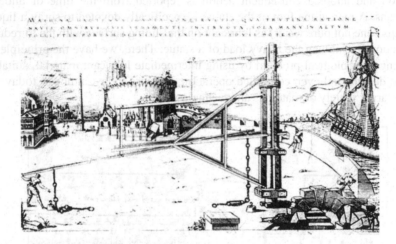

Fig. 13. A scaled large mechanism for lifting boats in 16th Century

One of the greatest attention for novelties in Mechanism Design can be considered the conception of new mechanisms and systems that can fulfill (optimally) identified new problems. Today is very difficult to conceive new mechanisms from theoretical viewpoint, after that in the past, efforts have obtained mechanism classifications with exhaustive listing of kinematic chains. It is not only because of the milestone work by Franz Reuleaux that for example Francesco Masi in Italy and many others up to in other countries have completed by listing many types of mechanisms (up to millions of different architectures; (relevant is the recent encyclopedic work of Artobelevski in 1975), but even in a very past, the variety of mechanisms was considered in unifying approaches that helped in conceiving new mechanisms. Figure 14 is a brilliant example of such an early activity for mechanism classification that Francesco di Giorgio Martini discussed for pumping systems to derive an early concept of kinematic inversion of mechanism as in the example of the second drawing of the second column of Fig. 14.

However, new architectures of mechanisms have been attempted even in the past by using different concepts from traditional mechanism design like for example with compliance or underactuated and overconstrained mobility. Indeed they are extensively used only in modern times, even if in the past we can find pioneering solutions, like for example the case in Fig. 15 in which elastic bodies are used for the functionality of a lock mechanisms. Indeed elastic/compliant operation of links were also used in Chinese locks since the Antiquity yet.

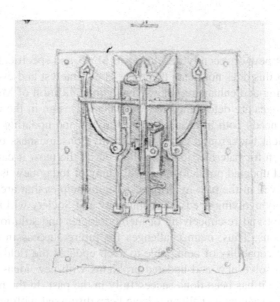

Fig. 15. An early compliant mechanism for door lock by Leonardo da Vinci

In the last decade, mechanism design has been revitalized as central role for machine design by a development of so-called Creative Design in which the creative skill of a designer is considered together with technical methodologies and computational algorithms in a well defined design process by which all the possibilities are explored for a given task. Indeed, Creative Design is still understood in a wider concept but always considering as central the intellectual creativity, which can be also independent of a strong technical background. The challenge can be recognized in the possibility to be free from the well defined schemes of the well established technical frames for mechanism design with the aim to discover both new systems or simple efficient solutions even from the large existing variety. Indeed, most of the inventions of the past have been based on the ingenuity of inventors, who first attempted the construction and operation of their inventions and then they or even other studied and systematized a related theoretical background for further designs and enhancements.

Summarizing this brief account that has been supported with few significant illustration examples from a huge literature and engineering history, in MMS there is much of new in what is old but still novelties can be conceived not only by looking at the past.

A fundamental aspect, which is in common to new ideas from the past and challenges for real today novelties, can be identified in issues that are related to mathematization and computation of theories and mechanism solutions. Challenging problems, like in the past, are related to derive methodologies (i.e. knowledge) that can be used in practical engineering at current modern levels of efficiency, and are feasible for conceiving and updating tasks and requirements.

Conclusions

Not everything is new or recently developed in MMS and specifically in Mechanism Design. But this does not mean that there is not interest and even no need to work on developing and enhancing knowledge and application of Mechanism Design. New challenges are determined for Mechanism Design in the new Technology and Society needs both in term of new solutions and updating past systems. An aware historical background can give not only consciousness of past efforts and solutions, even for paternity identification, but at the most it can help to find ideas for new and updated problems to solve. Many of today new issues in MMS have been conceived in the past in terms of basic principles that are often forgotten. But the rapidly evolving needs of Technology and Society will require a continuous re-thinking and re-conceiving of methodologies and solutions in suitable updated applications. Thus, main challenges for future success in MMS can be recognized in the capability of being able to keep updated the field and therefore in being ready to solve new and updated problems with new ideas or refreshing past solutions, like it has been done successfully in the past. In the paper the matters for historical survey and challenges have been discussed with general considerations by using few emblematic examples.

References

The reference list is limited for space limits to main works for further reading and to author's main experiences.

[1] Artobolevsky, I.I.: Mechanisms in Modern Engineering. Mir Publ. Moscow 5 (1975-1980)
[2] Besson, J.: Théatre des instruments Mathématiques et Mécaniques, Lyon (1578)
[3] Ceccarelli, M.: Mechanism Schemes in Teaching: A Historical Overview. ASME Journal of Mechanical Design 120, 533–541 (1998)
[4] Ceccarelli, M.: On the meaning of TMM over time. Bulletin IFToMM Newsletter 8(1) (1999), http://www.iftomm.org
[5] Ceccarelli, M.: Preliminary Studies to Screw Theory in XVIIth Century. In: Ball Conference, Cambridge, CD Proceedings
[6] Ceccarelli, M. (ed.): International Symposium on History of Machines and Mechanisms - Proceedings of HMM 2000. Kluwer, Dordrecht (2000)
[7] Ceccarelli, M.: From TMM to MMS: a Vision of IFToMM., Bulletin IFToMM Newsletter 10(1) (2001), http://www.iftomm.org
[8] Ceccarelli, M.: The Challenges for Machine and Mechanism Design at the Beginning of the Third Millennium as Viewed from the Past. In: Proceedings of Brazilian Congress on Mechanical Engineering COBEM2001, Uberlandia, 2001. Invited Lectures, vol. 20, pp. 132–151 (2001)
[9] Ceccarelli, M.: A Historical Perspective of Robotics Toward the Future. Fuji International Journal of Robotics and Mechatronics 13(3), 299–313 (2001)
[10] Ceccarelli, M. (ed.): International Symposium on History of Machines and Mechanisms - Proceedings of HMM2004. Kluwer, Dordrecht (2004)
[11] Ceccarelli, M.: IFToMM Activity and Its Visibility., Bulletin IFToMM Newsletter 13(1) (2004), http://www.iftomm.org

[12] Ceccarelli, M.: Classifications of Mechanisms Over Time. In: Proceedings of International Symposium on History of Machines and Mechanisms HM M2004, pp. 285–302. Kluwer Academic Publishers, Dordrecht (2004)

[13] Ceccarelli, M.: Evolution of TMM (Theory of Machines and Mechanisms) to MMS (Machine and Mechanism Science): An Illustration Survey. In: Keynote Lecture, 11th IFToMM World Congress in Mechanism and Machine Science, 2004, Tianjin, vol. 1, pp. 13–24 (2004)

[14] Ceccarelli, M.: Early TMM in Le Mecaniche by Galileo Galilei in 1593. Mechanisms and Machine Theory 41(12), 1401–1406 (2006)

[15] Ceccarelli, M.: Renaissance of Machines in Italy: from Brunelleschi to Galilei through Francesco di Giorgio and Leonardo. Mechanism and Machine Theory 43, 1530–1552 (2008)

[16] Chasles, M.: Apercu historique sur l'origin et le développement des méthodes en géométrie., 2nd edn., Mémoires couronnés par l'Académie de Bruxelles, Paris, vol. 11 (1983)

[17] Chasles, M.: Exposé historique concernant le cours de machines dans l'enseignement de l'Ecole Polytechinique, Gauthier-Villars, Paris (1886)

[18] Crossley, E.F.R.: Recollections from Forty Years of Teaching Mechanisms. ASME Jnl of Mechanisms, Transmissions and Automation in Design 110, 232–242 (1988)

[19] De Groot, J.: Bibliography on Kinematics. Eindhoven University, Eindhoven (1970)

[20] De Jonge, A.E.R.: A Brief Account of Modern Kinematics. Transactions of the ASME, 663–683 (August 1943)

[21] Dimarogonas, A.D.: The Origins of the Theory of Machines and Mechanisms. In: Erdman, A.G. (ed.) Modern Kinematics – Developments in the Last Forty Years, pp. 3–18. Wiley, New York (1993)

[22] Ferguson, E.S.: Kinematics of Mechanisms from the Time of Watt. In: Contributions from the Museum of History and Technology, Washington, paper 27, pp. 186–230 (1962)

[23] Koetsier, T.: Mechanism and Machine Science: its History and its Identity. In: International Symposium on History of Machines and Mechanisms HMM2000, pp. 5–24. Kluwer, Dordrecht (2000)

[24] Hain, K.: Applied Kinematics. McGraw-Hill, New York (1967)

[25] Hartenberg, R.S., Denavit, J.: Men and Machines ...an Informal History. Machine Design, 75–82 (May 3, 1956); 101–109 (June 14, 1956); 84–93 (July12, 1956)

[26] FToMM 1991, IFToMM Commission A. Standard for Terminology. Mechanism and Machine Theory 26(5) (1991)

[27] IFToMM 2003, special issue Standardization and Terminology, Mechanism and Machine Theory 38(7-10) (2003)

[28] IFToMM, IFToMM Constitution and By-Laws 2007 (2007), http://www.iftomm.org

[29] IFToMM webpage (2009), http://www.iftomm.org

[30] Lanz, J.M., Betancourt, A.: Essai sur la composition des machines, Paris (1808)

[31] Masi, F.: Teoria dei meccanismi, Bologna (1897)

[32] Nolle, H.: Linkage Coupler Curve Synthesis: A Historical Review –I and II. IFToMM Journal Mechanism and Machine Theory 9(2), 147–168, 325-348 (1974)

[33] Reuleaux, F.: Theoretische Kinematic, Braunschweig, ch. 1 (1875)

[34] Roth, B.: The Search for the Fundamental Principles of Mechanism Design. In: International Symposium on History of Machines and Mechanisms - Proceedings of HMM2000, pp. 187–195. Kluwer, Dordrecht (2000)

[35] Shah, J.J. (ed.): Research Opportunities in Engineering Design – Final Report to NSF. In: NSF Strategic Planning Workshop, ASME DETC, Irvine (1996)

[36] Tessari, D.: La Teoria delle Ombre e del Chiaro-Scuro, Torino (1880)

[37] Uicker, J.J., Pennock, G.R., Shigley, J.E.: Theory of Machines and Mechanisms. Oxford University Press, New York (2003)
[38] Woodcroft, B.: The Pneumatics of Hero of Alexandria, London (1851)
[39] Yan, H.S., Ceccarelli, M. (eds.): International Symposium on History of Machines and Mechanisms - Proceedings of HMM 2008. Springer, Dordrecht (2008)

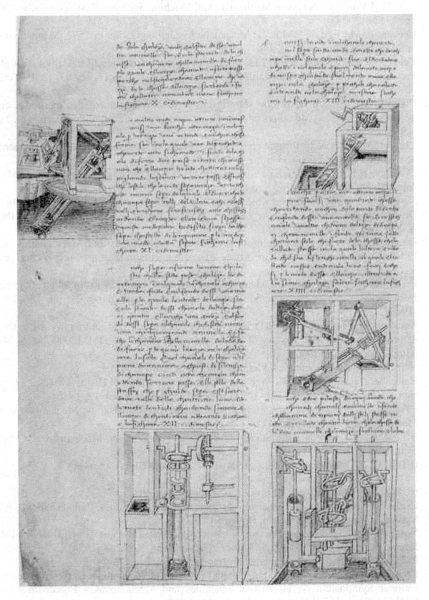

Fig. 14. An early classification of mechanism for pumping systems by Francesco Di Giorgio Martini in 15th Century

A Service-Orientated Arhitecture for Holonic Manufacturing Control

Theodor Borangiu

Dept. of Automation and Applied Informatics, University Politehnica of Bucharest,
RO-060032 Bucharest, Romania
borangiu@cimr.pub.ro

Abstract. The paper describes a solution and implementing framework for the management of changes which may occur in a holonic manufacturing system. This solution is part of the semi-heterarchical control architecture developed for agile job shop assembly with intelligent robots-vision workstations. Two categories of changes in the manufacturing system are considered: (i) changes occurring in resource status at process level: breakdown, failure of (vision-based) in-line inspection operation, and depletion of local robot storages; (ii) changes in production orders at business (ERP) level: rush orders. All these situations trigger production plan update and rescheduling (they redefine the list of Order Holons) by pipelining CNP-type resource bidding at shop-floor horizon with global product scheduling at aggregate batch horizon. Failure- and recovery management are developed as generic scenarios embedding the CNP mechanism into production self-rescheduling. Implementing solutions and experimental results are reported for a 6-station robot-vision assembly cell with twin-track closed-loop pallet transportation system, Cartesian pallet feeding station, dual assembly component feeder with robot-vision tending and product tracking RD/WR devices. Future developments will consider manufacturing integration at enterprise level.

Keywords: holonic manufacturing, distributed control, reconfigurable systems, robotics, applied AI.

1 Introduction

Some of the problems that discrete, repetitive manufacturing industry faces are: resource availability (unexpected failure or recovery of a resource and insertion or removal of resources from the production process) and treatment of "rush orders".

I.J. Rudas et al. (Eds.): Towards Intelligent Engineering & Information Tech., SCI 243, pp. 489–503.
springerlink.com   © Springer-Verlag Berlin Heidelberg 2009

To cope with these problems three concepts have been developed in past years: (i) *Flexible Manufacturing Systems* – FMS (Groover 1987; Upton 1992), (ii) *Multi-Agent* – MAS and *Holonic Manufacturing Systems* – HMS (Van Brussel et al. 1998; Leitao 2006) and (iii) *Product-Driven Control for Manufacturing* – PDCM (Petin et al. 2006; Gouyon et al. 2007). The first, FMS, deals with the physical composition of a manufacturing cell which has a minimal degree of flexibility allowing easy reconfiguration and also facing disturbances like resource breakdowns. The second concept, HMS, deals with the control part of a manufacturing cell, structuring it into basic building blocks characterized by autonomy and cooperation. Manufacturing tasks are solved by cooperation between these entities and to the exterior the system is seen as a single entity, making it easier to integrate such structures with the upper levels (ERP) of an enterprise. The last concept, of "intelligent product", assumes that a local intelligence is provided to the product (moving on a pallets) integrated via RFID devices in an Enhanced Information Management System (IMS-RFID) which is used to retrieve process-, resource- and cell- data for product routing.

The need of methods and tools to manage the process of change addresses both the level of business reengineering (including information technology infrastructures) and shop floor reengineering (production processes are executing). A particularly critical element in the shop floor reengineering process is the control system. Current control / supervision systems are not agile because any shop floor change requires programming modifications, implying the need for qualified programmers, usually not available in manufacturing SMEs. Even small changes (e.g. rush orders) might affect the global system architecture, which inevitably increases the programming effort and the potential for side-effect errors.

The methodology used for shop-floor reengineering, proposed in this paper, compensates for the deficiencies of both hierarchical and heterarchical enterprise control systems, and is based on new concepts for the design and implementing of manufacturing control systems in the frame of Holonic Manufacturing Execution Systems. Such concepts attempt to model a manufacturing system based on some analogies with other existing theoretical, natural or social organization systems (Babiceanu et al. 2004; Barata 2000; Van Brussel et al. 1998). The *agent-based* and *holonic* paradigms symbolize these new approaches; they deal with the reconfigurability in discrete, repetitive manufacturing by introducing an adaptive production control system that evolves dynamically between a more hierarchical (providing global efficiency / optimality) and a more heterarchical (self-adapting, fault-tolerant, agile) control architecture, based in self-organization and learning capabilities embedded in individual holons – information counterparts of resources, processes and products (Bellifemine et al. 2001).

A generic distributed enterprise control architectures for shop floor reengineering aims at accommodating the requirements:

• *Modularity*: production systems should be created by composing modularized manufacturing components, which become basic building blocks (developed on the basis of the processes they are to cater for).

- *Configuring rather than programming*: the addition or removal of any building block is done smoothly, with minimal programming effort. The system composition and its behaviour are established by configuring the links among modules, using contractual mechanisms.
- *High reusability*: the building blocks should be reused for as long as possible, and easily updated.
- *Legacy systems migration*: legacy and heterogeneous controllers are accepted in the global architecture.

The proposed multi-agent control architecture supports the reengineering process of shop floor control. This generic MAS architecture uses contracts to govern the relationships between coalition members (production agents), including the reengineering process within the life cycle. The control system architecture considers that manufacturing components can be reused and plugged/unplugged with reduced programming effort, supporting the *plug& produce* metaphor.

The Service Oriented Architecture (SOA) concept is used to face the interoperability problems in the autonomous, re-configurable architecture implemented as a HMES. Each device controller encapsulates functions and services that its associated physical device can perform). These services, that can be modified, added or removed (e.g. a new product can be handled by a robot after the aggregation of a new gripper), are then exposed to be invoked by other device controllers.

The SOA for production management and control integrates four areas: (1) Offer Request Management; (2) Management of Client Orders; (3) Order- & Supply-Holon (OH, SH) Management; (4) OH Execution & Tracking) – Fig. 1. The first area is responsible for generating offers in response to requests, based on: product knowledge (embedded in Product Holons - PH), resource capabilities (from Resource Holon - RH data), supply constraints and activities planning (CAPP). Once received customer orders, they are interpreted, validated and mapped into aggregate production orders (APO) at ERP level. APO is the input to the Global Production Scheduler (GPS) which, generates the lists of Supply- (SH) and Order Holons (OH).

A solution for implementing a SOA system is offered by IBM. The IBM SOA foundation is an integrated, open set of software, best practices, and patterns. The SOA foundation provides full support for the SOA lifecycle through an integrated set of tools and runtime components that allow leveraging skills and investments across the common runtime, tooling, and management infrastructure. The IBM SOA Foundation includes the following lifecycle phases: **Model, Assemble, Deploy** and **Manage** (Fig. 2).

There are a couple of key points to consider about the SOA lifecycle: (i) The SOA lifecycle phases apply to all SOA projects; (ii) Activities in any part of SOA lifecycle can vary in scale and level of tooling used depending on its adoption step.

The HMES implementing the MAS reference architecture assumes that there is a similarity between the proposed reengineering process and the formation of consortia regulated by contracts in networked enterprise units.

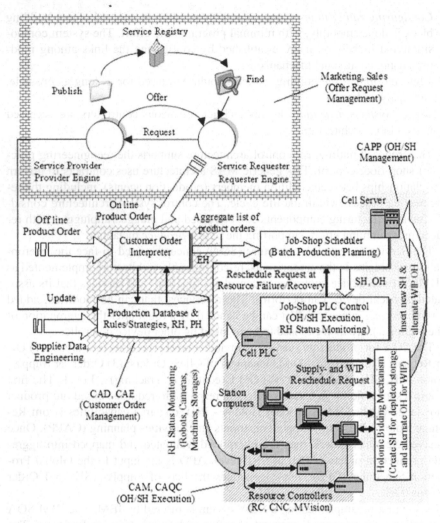

Fig. 1. SOA integrates job-shop, team-based manufacturing with holonic robot control

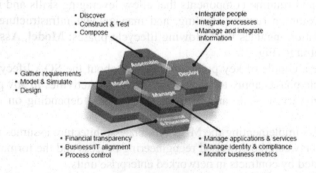

Fig. 2. IBM SOA Foundation Cycle

The adoption of web services in the HMES holarchy will satisfy the requirements (Jia and Fuchs 2002):

- Resources can be encapsulated with a service provider that acts like a bridge between the internal structure and the exposed interface.
- Some services can be composed by other services, creating a levelled structure of services (e.g. task-and product-oriented learning of virtual cameras).
- Interoperability in the MAS is addressed by using common communication semantics based on the use of open protocols or web technologies (services).
- Fault-tolerant attribute is provided (anomalies that may occur during the production processes, and identification possible disturbances are handled).

The scientific contribution of the presented research is the definition and design of an implementing frame for a holonic control architecture for agile job shop assembly with networked intelligent robots, based on the dynamic simulation of material processing and transportation. The holarchy is defined considering the PROSA reference architecture relative to which in-line vision-based quality control was added by help of feature-based descriptions of materials.

The paper describes in detail the methodology used for the management of changes – reallocating already scheduled production orders (OHs) in a perturbed environment. The control architecture is distributed, of semi-heterarchical type, in which the organizational control is arranged on two levels, referred to as *global* and *local*.

The global level assumes the responsibility for planning and coordination of shop-floor level activities and the resolution of conflicts between local objectives; the local level has autonomy over the planning and control of internal activities (e.g. the robot assembly team).

There is an entity placed on a superior decisional level – the Global Production Scheduler (GPS) – which sends aggregate product orders, optimally scheduled – Order Holons (OH), to entities on inferior levels – Device (e.g. Robot, Machine) Controller, cooperating to accomplish the orders. The schedules delivered by the GPS are not imposed to any of the individual resources; instead, they are only recommended to the decision-making entities – the Order Holons. These recommendations are followed as long as failures or changes do not occur in the system (*hierarchical operating mode*); they will be ignored at failure/change and recovery moments, being replaced by alternate schedules created from resource (robot) offers mutually agreed by cooperation mechanism (*heterarchical operating mode*). The holonic manufacturing control automatically switches between these two modes.

The holonic control strategy follows the key features of the PROSA reference architecture (Van Brussel et al. 1998; Valkaenars 1994), implemented as an extended HMES:

- Automatic switching between *hierarchical* (efficient / optimal use of resources) and *heterarchical* (agility to order changes, e.g. rush orders, and fault tolerance to resource breakdowns) production control modes.

- Automatic planning and execution via Supply Holons (SH) of part supply; automatic generation of self-supply tasks upon detecting local storage depletion.
- In-line vision-based part qualification and inspection of products in user-definable execution stages.
- Robotized processing (e.g. assembling, machine tending, fastening, assembling) under visual guidance

2   System Architecture

As suggested by the PROSA abstract, the manufacturing system was broken down into three basic holons:

1 **Resource Holons** (RH): they hold information about cell resources. Any resource may have a number of sub-resources, which are also seen as holons.
2 **Product Holons** (PH): they hold information about a product type. The product information is more than a theoretical description of the physical counterpart but not directly associated with one individual physical item, unlike the resource holon (Leitao and Restivo, 2006).
3 **Order Holons** (OH): represent all information necessary to produce one item of a product type. This holon is directly associated with the emerging item, it holds information about its status. OHs are created by the GPS from an Aggregate List of Product Orders generated at ERP level. Alternate OH are created in response to changes in product batches (*rush orders*) and failures (resource breakdown, storage depletion).

A holon designs a class containing data fields and functionalities. Beside the information part, holons possess a physical part too, like the *product_on_pallet* for OH.

The way in which different types of holons communicate and the type of information they exchange depends on the functionalities imposed to the manufacturing cell. Fig. 3 shows the interaction diagram of the basic holon classes as they were implemented into software to solve scheduling and failure management problems. A **HolonManager** hosts all holons and coordinates the data exchange.

The HolonManager entity is responsible with the planning (by help of **Expertise Holons** – EH) and management of OH as Staff Holons in the PROSA architecture do; in addition, it externally interfaces the application (maps the OH list in standard PLC files and tracks OH execution).

A **basic process plan** is generated initially, upon receiving from the ERP level an APO or raw orders, based on: (i) Knowledge-based scheduling (KBS, inspired by Kusiak 1990) or (ii) Resolved Scheduling Rate Planner (RSRP, Borangiu 2008). This basic process plan is computed at the global horizon of P products of the aggregate batches, and consists from a list of Supply Holons responsible for feeding the local robot storages and a list of Order Holons driving product execution. The OH list is mapped into PLC files for batch execution.

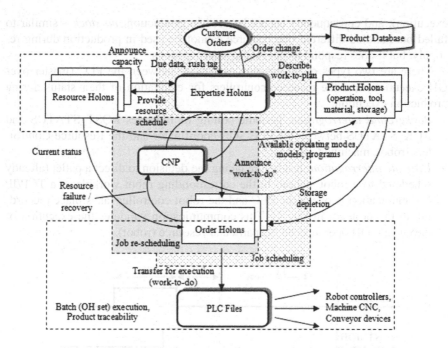

Fig. 3. Basic holon cooperation and communication structure in the semi-heterarchical control architecture

Alternative process plans, triggered by resource failure, local storage depletion or occurrence of rush orders, are pipelined automatically: (a) at the horizon of p_E products in course of execution in the system, based on heterarchical contract negotiation schemes (CNP-type) between valid resources; (b) at the global horizon of $P - p_T - p_E$ remaining products, p_T = number of terminated products, based on hierarchical GSP. Two categories of changes are considered:

1 Changes occurring in the <u>resource status</u> at shop floor level: (i) breakdown of one resource (e.g. robot, machine tool); (ii) failure of one inspection operation (e.g. visual measurement of a component/assembly); (iii) depletion of one robot workstation storage.
2 Changes occurring in <u>production orders,</u> i.e. the system receives a *rush order* as a new batch request (APO).

All these situations trigger a fail-safe mechanism which manages the changes, providing respectively fault-tolerance at critical events in the first category, and agility in reacting (via ERP) to high-priority batch orders. A *FailureManager* was created for managing changes in resource status. A virtually identical counterpart, the *RecoveryManager*, takes care of the complementary event (resource recovery).

Upon monitoring the processing resources (robots), their status may be at run time: *available* – the resource can process products; *failed* – the resource doesn't respond to the interrogation of the PLC (the entity responsible for Order Holon execution), and consequently cannot be used in production; *no stock* – similar to failed but handled different (the resource cannot be used in production during re-supply, but it does respond to PLC status interrogations.

There are two types of information exchanges between the PLC (master over OH execution) and the resource controllers for estimation of their status during production execution:

- *Background interrogation*: periodic polling of I/O lines RQST_STATUS and ACK_STATUS between the PLC –OH coordinator and the Resource Controllers (robot, machine tool, ASRS).
- *Ultimate interrogation*: just before taking the decision to direct a pallet (already scheduled to a robot station) to the corresponding robot workplace, a TCP/IP communication between the PLC and the robot controller takes place (according to the protocol in Fig. 4). This communication validates the execution of the current OH operation on the particular resource (robot).

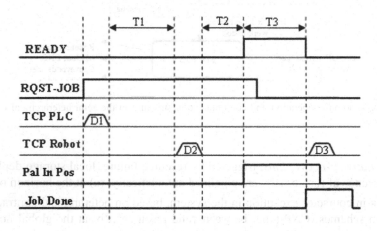

Fig. 4. Communication protocol between the PLC and a Robot Controller authorizing an OH operation execution

In this protocol, READY is a signal generated by the Robot Controller indicating the *idle* or *busy* state of the resource (robot). The PLC requests through its digital output line RQST-JOB to use the robot for an assigned OH operation upon the product placed on the pallet waiting to enter the robot workstation. D1 details the scheduled job via the TCP PLC transmission line from the PLC to the Robot Controller. The Robot Controller indicates in D2 job acceptance or denial via the TCP Robot transmission line.

When the job is accepted, the pallet is directed towards the robot's workplace, where its arrival is signalled to the Robot Controller by the Pal In Pos digital output signal of the PLC. Job Done is a signal indicating job termination (D3 details the way the job terminated: success, failure). T1 is the decision time on job acceptance (storage evaluation etc), T2 is the transport time to move the pallet from the main conveyor loop to the robot workplace, and T3 is the time for job execution.

Upon periodic interrogation, the entity coordinating OH execution – the PLC – checks the status of all resources, which acknowledge being *available* or *failed*. The ultimate interrogation checks only the state of one resource – the one for which a current operation of an OH was scheduled; during this exchange of information, the PLC is informed whether the resource is *available*, *failed* or valid yet unable to execute the requested OH operation upon the product due to components missing in its storage (*no stock* status).

3 Managing Resource Breakdown / Recovery

When the **failure** status of a resource is detected, the *FailureManager* is called, executing a number of actions according to the procedure given below (Fig. 4):

1 Stop immediately the transitions of executing OH, i.e. the circulation of *products_on_pallets* in the cell; production continues at the remaining valid resources (robots).
2 Update the resource holons with the new states of all robots.
3 Read Order Holons currently in execution in the cell.
4 Evaluate all products if they can still be finished, by checking the status of each planned OH:

   – if the OH was in the failing robot station, mark it as failed and evacuate its *product_on_pallet*;
   – if the OH is in the system, but cannot be completed anymore because the failed resource was critical for this product, mark it as failed and evacuate its *product_on_pallet*;
   – if the OH is not yet in the system, but cannot be completed due to the failure of the resource which is critical for that product, mark it as failed (n_e .is the total number of such OH).

5 For the remaining $n'_{wip} = n_{wip} - n_{fail}$ schedulable OH in the system, locate their *products_on_pallets* and initialize the transport simulation associated to the current operational configuration of the system. Authorise the n'_{wip} OH to launch Contract Net Protocol-based negotiations (HBM) with the remaining available Resource Holons for re-scheduling of their associated operations. n_{wip} are the OH currently introduced in the system (in the present implementation, $n_{wip} \leq 5$), and n_{fail} is the total number of OH currently in the system, which cannot be finished because they need the failed resource at some moment during their execution

6 Run the GPS algorithm for the $N - n_{fin} - n_{wip} - n_e$ OH not yet introduced in the system, a number of N OH being scheduled and n_{fin} OH were finished.

7 Delete the orders stored and transfer the updated orders to the system.
8 Resume *product_on_pallet* transfer within the transport system (allow OH transitions in the system).

It might happen that a failed robot is repaired before the current manufacturing
cycle is finished. In this **recovery** case, the cell regained the ability to run at full
capacity but the lined up orders do not make use of this fact, as they are managed
by the system in a degraded mode.

The procedure of rescheduling back the Order Holons is virtually identical to
the one used in case of failure; the main deference is that none of *the prod-
ucts_on_pallets* being currently processed need to be evacuated since there is no
reason to assume they could not be completed. Any orders that were marked as
failed due to temporary resource unavailability are now untagged and included in
the APO list for scheduling at the horizon of the rest of batch, as they may be
manufactured again due to resource recovery (Lastra and Delamerm 2006).

4 Management of Rush Orders

The system is agile to changes occurring in production orders too, i.e. manages
rush orders received as *new batch requests* from the ERP level while executing an
already scheduled batch production (a sequence of OH).

Because of the similarity between a task run on a processor and a batch of orders executed in a manufacturing cell (both are pre-emptive, independent of other
tasks or batches, have a release, a delivery date and an fixed or limited interval in
which they are processed), the Earliest Deadline First (EDF) procedure was used
to schedule new batches (rush orders) for the robotized assembly cell.

EDF is a dynamic scheduling algorithm generally used in real-time operating
systems for scheduling periodic tasks on resources, e.g. processors (Sha et al.
2004). It works by assigning a unique priority to each task, the priority being inversely proportional to its absolute deadline and then placing the task in an ordered queue. Whenever a scheduling event occurs the queue is searched for the
task closest to its deadline term (Borangiu. 2008a; Barata 2005; Rahimifard 2004).

A *feasibility test* for the analysis of EDF scheduling was done (Liu and Layland
1973); it shows that if: (1) all tasks are periodic, independent, fully pre-emptive;
(2) all tasks are released at the beginning of the period and have deadlines equal to
their period; (3) all tasks have a fixed time or a fixed upper bound which is less or
equal to their period; (4) no task can voluntarily stop itself; (5) all overheads are
assumed to be 0; (6) there is only one processor, then a set of n periodic tasks can

be scheduled if $\sum_{i=1}^{n} \frac{C_i}{T_i} \leq 1$, n = number of tasks, C_i = execution time, T_i = cycle time or,

in other words, if the utilization of the processor (resource) is less than 100%.

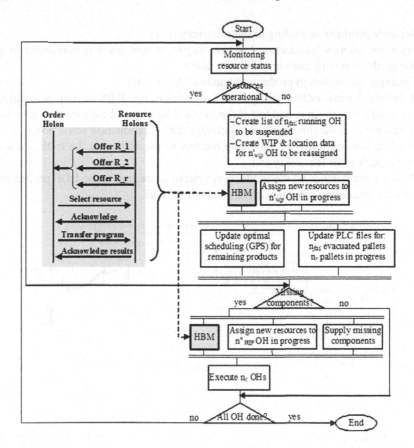

Fig. 5. Dynamic OH rescheduling at resource failure/storage depletion with embedded CNP job negotiation (monex)

A *batch* or Aggregate Product Order list (APO) is composed of raw orders (list of products to be manufactured); this is why two different batches are independent. Nevertheless, there is a difference between a task and a batch of products: a task is periodic while a batch is generally a periodic, i.e. instead of testing the feasibility of assigning batches to the production system with the equation above, the test can be used: "for an ordered queue (based on delivery date) of n batches with computed makespan, if $\sum_{j=1}^{i} \text{makespan}_j \leq \text{delivery\_date}_i, i = \overline{1, n}$, the batches can be assigned to the production cell with EDF without depassing the delivery dates".

This EDF approach is used to insert **rush orders** in a production already scheduled by the GPS; the steps below are carried out for inserting a new production batch during the execution of a previously created sequence of OH (Fig. 6):

0 Compute remaining time for finishing the rest of current batch (if necessary).
1 Insert new production data: product types, quantities, delivery dates.

2 Separate products according to their delivery date.
3 Form the entities "production batches" (a *production batch* is composed of all
   the products having the same delivery date).
4 Generate raw orders in production batches (APO lists).
5 Schedule the raw orders (using a GPS algorithm, e.g. KBS or Step Scheduler),
   compute the makespan and test if the inserted batch can be done (the makespan
   is smaller than the time interval to delivery date if production starts now).
6 Analyse the possibility of allocating batches to the cell using the EDF and second equation for feasibility test.
7 Allocate the batches on the production system according to the EDF procedure.
8 Resume execution process with new scheduled OH.

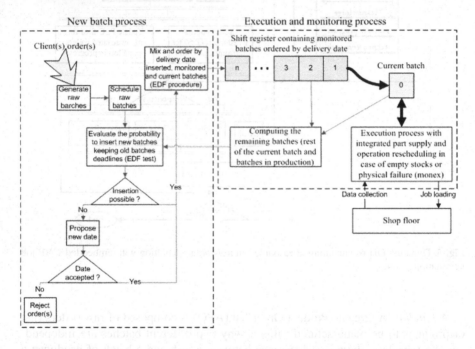

Fig. 6. *Add rush order* diagram. Integration with dynamic job re-scheduling and CNP (monex)

In this mechanism for managing the changes in production orders, an *inserted
batch* is a batch that arrives while another one is in execution. A *monitored batch*
is one whose orders are scheduled and assigned to the cell (it has a priority and is
waiting to enter execution). A *current batch* is that executing.

The capability of adding rush orders to production needs a new entity, the
batch. Thus, job scheduling is done at batch level (all orders with the same delivery date are scheduled together) and then batches are assigned to the cell according to their delivery date, using the EDF procedure (Table 1).

Table 1. The minimal structure of a *batch holon*

Type	Name	Description
string	batch_name	Name or index of the batch
Date	delivery_date	Delivery date of the orders
Product[ ]	requested_products	Vector containing the products to be executed
Resource[ ]	used_resources	Vector containing the configuration used for current batch planning
Order[ ]	orders_to_execute	Vector containing the entities OH already scheduled using a specified cell structure (defined by the variable used_resources)
int	makespan	Time interval needed for the current batch to be executed if started now and not interrupted (it is a result of scheduling)

Because batch execution is interruptible (pre-emptive system), new batches (rush orders) can be introduced exactly at the moment of their arrival. The insertion process is triggered by the arrival of a "new order" event; a real-time acceptance response can be provided (via the ERP level) to the customer if the rush order can be executed at requested delivery date.

Conclusions

The distributed control solution was implemented, tested and validated on a real manufacturing structure with six industrial assembly robots from Adept Technology (one Cartezian, three SCARA and two vertical articulated) and two 4-axis CNC milling machines, using the holonic approach. This holonic platform was finalized during 2008 in the Laboratory of Robotics and Artificial Intelligence of the Department of Automation and Industrial Informatics within the University Politehnica of Bucharest (Fig. 7).

The control is structure is fully operational, both in the normal hierarchical mode and upon switching automatically to the heterarchical one in response to discussed changes.

Production scheduling at batch level was implemented and tested using the EDF method; Fig. 8 shows the results obtained when two new batch orders $T_24 = (4, 17)$ and $T_25 = (1, 3)$ are received at time $T = 2$ after the execution of three planned batches: $T_11 = (2, 18)$, $T_12 = (3, 20)$, $T_13 = (7, 11)$ started. Here $T_j = (m, dd)$ signifies the number (j) of the batch for which execution was requested at date i; the batch has the makespan m and due delivery date, dd (both expressed in time units).

Fig. 7. Layout of the manufacturing cell with holonic control

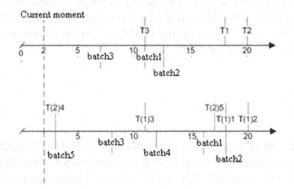

Fig. 8. Inserting new batches among executing ones with the EDF algorithm.

Acknowledgments. This work was partially supported from the scientific grant 146 / 2007 "Autonomous, intelligent robot-vision platforms for product qualifying, sorting / processing / packaging and quality inspection with Service-Oriented, Feature-based HolonIc Control aRchitecture – SOFHICOR" of the National Agency of Scientific Research (ANCS).

References

[1] Babiceanu, R.F., et al.: Framework for Control of Automated Material-Handling Systems Using Holonic Manufacturing Approach. Int. J. Prod. Res. 42(17), 3551–3564 (2004)

[2] Barata, J., Camarinha-Matos, L.M.: Shop Floor re Engineering to Support Agility in Virtual Enterprise Environments. In: E-Business and Virtual Enterprises, pp. 287–291. Kluwer Academic Publishers, London (2000)

[3] Barata, J.: Coalition-based Approach for Shop Floor Agility. Orion edn., Amadora-Lisbon (2005)

[4] Bellifemine, F., Poggi, A., Rimassa, G.: Developing Multi-Agent Systems with FIPA-Compliant Agent Framework. Software Practice and Experience 31(2), 103–128 (2001)

[5] Borangiu, T., Gilbert, G., Ivanescu, N.: A Rosu Holonic Robot Control for Job Shop Assembly by Dynamic Simulation. In: Proc. of the 16th Mediterranean Conference on Control and Automation – MED 2008, Ajaccio (June 2008)

[6] Borangiu, T., Ivanescu, N., Raileanu, S., Rosu, A.: Vision-guided Part Feeding in a Holonic Manufacturing System. In: Proc. of 16th Workshop on Robotics in Alpe-Adria-Danube Region RAAD 2009, Ancona, pp. 51–56 (2008a)

[7] Cheng, F.-T., Chang, C.F., Wu, S.L.: Development of Holonic Manufacturing Execution Systems Industrial Robotics: Theory, Modelling and Control, and Control, Advanced Robotics Systems, Vienna, Pro Literatur Verlag Robert Mayer-Scholz (2006)

[8] Gouyon, D., Pétin, J.F., Morel, G.: A Product-driven Reconfigurable Control for Shop Floor Systems. Studies in Informatics and Control, vol. 16 (2007)

[9] Groover, M.: Automation, Production Systems and CIM. Prentice-Hall, Englewood Cliffs (1987)

[10] Higuera, A.G., Montalvo, A.C.: RFID-enhanced Multi-Agent-based Control for a Machining System. Int. J. on Flexible Manufacturing Systems 19, 41–61 (2007)

[11] Jia, H.Z., Fuh, H.: Web-based Multi-Functional Scheduling System for a Distributed Manufacturing System. Concurrent Engineering 10(1), 27–39 (2002)

[12] Kusiak, A.: Intelligent Manufacturing Systems. Prentice Hall, Englewood Cliffs (1990)

[13] Lastra, J., Delamerm, I.: Semantic Web Services in Factory Automation: Fundamental Insights and Research Roadmap. IEEE Trans. on Industrial Informatics 2, 1–11 (2006)

[14] Leitao, P., Restivo, F.: ADACOR: A Holonic Architecture for Agile and Adaptive Manufacturing Control. Computers in Industry 57, 121–130 (2006)

[15] Liu, C.L., Layland, J.W.: Scheduling Algorithms for Multiprogramming in a Hard real-Time Environment. Journal of ACM 20, 46–61 (1973)

[16] Lusch, R.F., Vargo, S.L., Wessels, G.: Towards a Conceptual Foundation for Service Science: Contributions from Service-Dominant Logic. IBM Systems Journal 47(1) (2008)

[17] Marchand, H., Bournai, P., Le Borgne, M., Le Guernic, P.: Synthesis of Discrete-Event Controllers Based on the Signal Environment. Discrete Event Dynamic Systems: Theory and Applications 10, 325–346 (2000)

[18] Pétin, J.F., Gouyon, D., Morel, G.: Supervisory Synthesis for Product-Driven Automation and its Application to a Flexible Manufacturing Cell. Control Engineering Practice 15, 595–614 (2007)

[19] Rahimifard, S.: Semi-Heterarchical Product Planning Structures in the Support of Team-based Manufacturing. Int. J. Prod. Res. 42, 3369–3382 (2004)

[20] Sha, L., et al.: Real Time Scheduling Theory: A Historical Perspective. Real-Time Systems 28, 101–155 (2004)

[21] Upton, D.: Flexible Structure for Computer Controlled Manufacturing System. Manufacturing Review 5, 58–74 (1992)

[22] Valckenaers, P., Van Brussel, H., Bongaerts, L., Wyns, J.: Results of the Holonic Control System Benchmark at the KULeuven. In: Proceedings of the CIMAT Conference (CIM and Automation Technology), pp. 128–133. Rensselaer Polytechnic Institute, Troy (1994)

[23] Van Brussel, H., Wyns, J., Valckenaers, P., Bongaerts, L., Peeters, P.: Reference Architecture for Holonic Manufacturing Systems: PROSA. Computers in Industry, Special Issue on Intelligent Manufacturing Systems 37(3), 255–276 (1998)

Sensitivity of Power Spectral Density (PSD) Analysis for Measuring Conditions

Árpád Czifra

Bánki Donát Faculty of Mechanical and Safety Technique Engeenering, Budapest Tech
Népszínház u. 8, H-1081 Budapest, Hungary
czifra.arpad@bgk.bmf.hu

Abstract. Nowadays power spectral density (PSD) analysis is one of the leading characterisation techniques of surface topography. Fractal dimension obtained from PSD plays a significant role in recent friction, adhesion and wear models.

Knowledge of the sensitivity of PSD for the analysis and adjustment and measurement of parameters is relevant in extensive and reliable applications. In this study, a three-dimensional (3D) PSD analysis of engineering surfaces is carried out, and its sensitivity for frequency sampling, for line fitting – required to calculate the fractal dimension – and for sampling the distance of measurement are investigated.

Based the on results, it can be established that 3D PSD analysis provides stable results when using only few frequencies. Line fitting has a considerable impact on fractal dimension results, so the application of PSD in the calculation of fractal dimension needs circumspection. A huge amount of fractal information is in the height frequency range, so the sampling distance greatly influences the fractal dimension of surface. Thus, measuring conditions are required to be specified in order to get proper information about surface self-affinity.

1 Introduction

The operation, reliability, and lifetime of parts produced in different ways greatly depend on the quality of machined surfaces as well. Higher quality criteria require adequate accuracy of manufacturing as well as a deeper analysis of surface micro-topography.

Traditionally and in accordance with Hungarian and international standards, the microgeometry of operating surfaces has been characterized by two dimensions; however its information content is limited and depends on filtering – see Thomas [1]. Demand for 3D processing was presented as early as the second half of the

I.J. Rudas et al. (Eds.): Towards Intelligent Engineering & Information Tech., SCI 243, pp. 505–517.
springerlink.com   © Springer-Verlag Berlin Heidelberg 2009

80s. In the first half of the 90s, till then missing conditions, such as computers of adequate speed of operation and processing softwares became increasingly available, making it possible to realize 3D processing.

Nowadays 3D power spectral density (PSD) analysis is one of the techniques need one of the height processing capacity. Information obtained from the micro and nano topographies of operating surfaces appears as input in today's friction and wear models. Persson [2] takes surface topography into account when calculating the hysteresis component of the coefficient of friction, using the relation of the power spectral density (PSD) curve and the real contact area. In Schargott's [3] models, the PSD curve plays a part in the numerical simulation of adhesion and abrasion wear processes. [4] wear model is entirely based on surface microtopography. The auto correlation function and the power spectral density curve are used for describing surfaces. The reliability of tribological models is in close connection with the effectiveness of topographical analyses. Tribological processes can only be understood and accurately modeled through an in-depth knowledge of surface topography.

The aim of this study is to overview the mathematical background of 3D PSD analysis and – using the developed software – to investigate the effect of measuring and characterization parameters with analysis of engineering surfaces.

2 Mathematical Background of PSD

To characterize the measured topographies an algorithm was developed and interpreted as PSD analysis software. The theoretical base of 3D PSD analysis was [2] and [5].

Discrete Fourier transformation (DFT) of 3D topography can be written as follows:

$$F(q_x, q_y) = \Delta y \cdot \Delta x \sum_{d=1}^{N} \sum_{c=1}^{M} z(x_c, y_d) e^{-i2\pi(x_c q_x + y_d q_y)} \tag{1}$$

where: q_x, q_y frequencies in x and y directions,

$z(x_c, y_d)$ height coordinate located in x_c, y_d,   M

number of points in profile,

N   number of profiles,

$\Delta x, \Delta y$ sampling distances.

DFT gives complex results, so PSD „amplitude" is calculated:

$$A_{PSD} = \frac{\text{Re}^2 F + \text{Im}^2 F}{MN\Delta x \Delta y} \tag{2}$$

PSD topography can be reduced to Persson's PSD curve using (3). It means 2D representation, which can be easy handled, but contains 3D information about topography.

$$q = \sqrt{q_x^2 + q_y^2} \tag{3}$$

There are two possibilities of showing results. One is to represent the amplitude of PSD in the function of wavelength. The other prevalent method is logarithmic scale frequency-PSD amplitude visualization. The practical gain of the first method is that dominant wavelength components appear as a maximum point of the PSD curve. In the second method the height frequency range of the curve can be approximated by a line. The slope of the line is in correlation with the fractal dimension of surface. In the latter case, wavelengths smaller than the highest dominant wavelength play a considerable role. PSD amplitude becomes constant – the self-affinity character of the surface disappears – in a lower wavelength range. The slope of fitted line (s) to Persson-curve has correlation with fractal dimension of surface according to (4).

$$Df = 4 + \frac{s}{2} \tag{4}$$

3   Methodology

The primary result of the algorithm developed was fractal dimension. Investigation focussed on this parameter in case of five topographic measurements from three different points of view: first, the frequency sampling of PSD, secondly, the effect of line fitting, and thirdly, the effect of sampling distance is investigated.

Three topographies were machined "finely", while the other two were "rough" surfaces. "Fine" surfaces were produced by milling, grinding and lapping technologies at an average roughness (Ra) of 0.4 μm. "Rough" surfaces were of Ra=3.2 μm and were produced by turning and milling. Measuring conditions were similar in all cases: the measuring area was 1 by 1 mm, and the sampling distance was 2 by 2 μm. Fig. 1 shows "fine" topographies and Fig. 2 shows the "rough" ones. The following signs are used to identify the surfaces: "fine" milling = Mil04, grinding = Gri04, lapping = Lap04, turning = Tur32, "rough" milling = Mil32.

In investigating the effect of frequency sampling of PSD when analysing the above topographies, five different settings were used. In all cases, PSD surfaces containing 25x25, 50x50, 75x57, 100x100 and 125x125 points were calculated

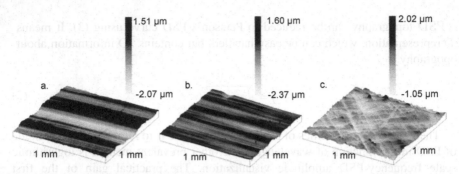

Fig. 1. Topographies of a) milled (Mil04), b) grinded (Gri04), c) lapped (Lap04) surfaces

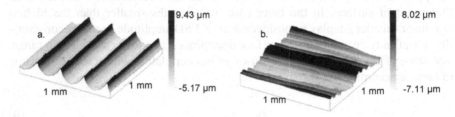

Fig. 2. Topographies of a) turned (Tur32), and b) milled (Mil32) surfaces

analysing the topographies containing 500x500 points. Partitions of frequencies in the minimal and maximal frequency range were automatically generated using a logarithmic division. The minimal value of frequency is the inverse number of measuring length: $q_{min}=1/(M*\Delta x)$, while the maximal value of frequency is the inverse number of double of the sampling distance: $q_{max}=1/(2*\Delta x)$. Discrete values of frequencies were calculated according to (5).

$$\Delta q = \sqrt[n-1]{\frac{q_{max}}{q_{min}}} \qquad (5)$$

where n is the number of frequencies.

The Persson-curves obtained from the PSD surface contain 625, 2500, 5625, 10000 and 16525 points, respectively. A line is fitted to the Persson-curve in certain frequency ranges to calculate the fractal dimension of the topography. To investigate the effect of sampling distance, 100x100 frequency was used in all cases and the original sampling distance of topography (2 μm) was enlarged to 4, 6, 8 and 10 μm. This enlargement was achieved by thinning the measuring points: measuring points 2, 3, 4 and 5; were left standing.

4   Results

4.1   Effect of Frequency Sampling of PSD

Figs. 3 and 4 show the PSD surface of Mil04 and Lap04 surfaces using 25x25 and 125x125 frequencies.

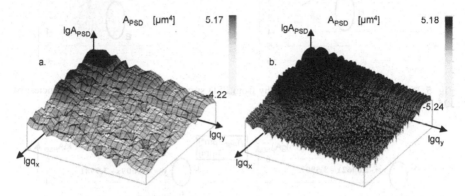

Fig. 3. PSD surface of Mil04 topography a) 25x25 frequency; b) 125x125 frequency

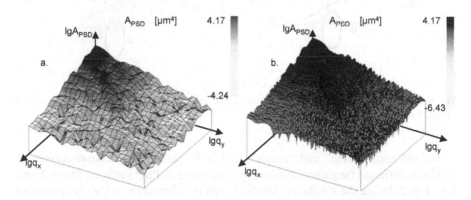

Fig. 4. PSD surface of Lap04 topography a) 25x25 frequencies; b) 125x125 frequencies

Visualisation of PSD results does not show significant differences in case of 25x25 and 125x125 frequencies. PSD surfaces in the cases presented have good correlation with the original topographies: on milled surfaces – with an orientation in direction "y"– the PSD amplitudes are higher in this direction; on lapped surfaces – with about 45° orientation – the PSD surface has similar orientation.

Figs. 5 and 6 show the Persson-curve – obtained from PSD surfaces containing 25x25 and 125x125 frequencies – of Gri04 and Tur32 topographies.

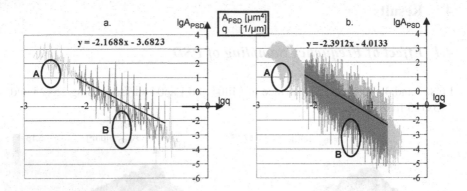

Fig. 5. Persson-curve of Gri04 topography from PSD surface contains a) 25x25 frequencies; b) 125x125 frequencies

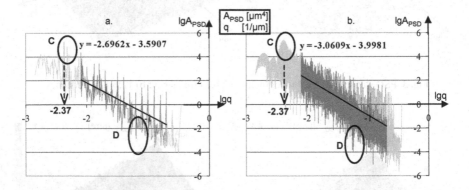

Fig. 6. Persson-curve of Tur32 topography from PSD surface contains a) 25x25 frequencies; b) 125x125 frequencies

In both cases ("fine" and "rough"), it can be stated that Persson-curves are similar in different frequency-resolutions – the same as PSD surfaces. Signs A and B of Fig. 5 denote the similarity, identical with the characters of the curves signed by C and D on Fig. 6 appearing on both curves; what is more, the break points of curves are at identical frequency values ($-2.37 = \lg 0.00427$). So the physical character of the surface is observable. Feeding marks – as shown in Fig. 2 – represent the largest dominant wavelength: about 250 µm, that is 0.004 1/µm frequency, similarly to the break point of curves.

Table 1 summarizes the effect of the number of frequencies used in PSD analysis. In all cases, fast convergation can be observed independently of the machining technology and of the average roughness of the investigated surface and it can be stated that the results obtained from only 75x75 point curves are very similar to the results calculated from Persson-curves containing 100x100 or 125x125 points. So the number of frequencies are not required to be increased because of the exponential increase of CPU time. Further analysis uses 100x100 frequencies to

Table 1. Fractal dimensions of topographies in function of number of frequencies

Number of frequencies	Mil04	Lap04	Gri04	Tur32	Mil32
25x25	2.85	2.46	2.92	2.65	2.85
50x50	2.74	2.46	2.77	2.51	2.46
75x75	2.76	2.46	2.82	2.50	2.51
100x100	2.76	2.46	2.78	2.48	2.49
125x125	2.76	2.46	2.80	2.47	2.50

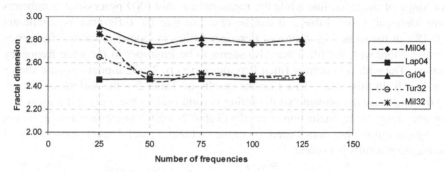

Fig. 7. Effect of frequency sampling of PSD to fractal dimension

calculate PSD surfaces. In case of topographies containing 500x500 points and using a "normal" PC it means 15 min CPU time.

4.2 Line Fitting of Persson-Curve

For calculating fractal dimensions, some characterisation parameters have to be fixed in order to make results comparable. One of these parameters is the frequency range of the Persson-curve where the line is fitted. Line fitting was accomplished by a Microsoft Excel line fitting module.

Topographies without any modification were analysed using 100x100 frequencies in all cases. The PSD surface was reduced to Persson-curve. Then line was fitted to Persson-curve in two different frequency range. Only the maximal frequency was modified, because the following step of examinations focuses on changes of sampling distance, which modifies the maximal frequency.

Table 2 shows that fractal dimension results may alter when the frequency range of line fitting changes. The difference from "original" results is 1-9%, where this latter is extremely high taking into consideration that fractal dimension must be in the range between 2 and 3. The highest difference can be found in case of lapped surface (2.46-2.24). This is 22% of the entire range of fractal dimension.

Table 2. Effect of frequency range of line fitting to fractal dimension

	Mil04	Lap04	Gri04	Tur32	Mil32
Wide freq. range: q [10^{-3} 1/μm]	5.62 – 199.5	10 – 199.5	7.10 – 199.5	7.95 – 199.5	6.31 – 199.5
Fractal dim. [-]	2.76	2.46	2.78	2.48	2.49
Narrow freq. range: q [10^{-3} 1/μm]	5.62 – 44.66	10 – 44.66	7.10 – 44.66	7.95 – 44.66	6.31 – 44.66
Fractal dim. [-]	2.94	2.24	2.75	2.41	2.64

Fig. 8 shows the impact of the width of the frequency range on the modification of slope of the fitted line while the measurement and PSD processing parameters are identical. From Table 2 it can be read out that the difference is significant (0.18); at the same Fig. 8 proves improper fitting in case of a narrow frequency range. Moreover, the fitted line also seems to be improper in the wide frequency range. The wide and narrow fitted Persson-curves of Tur32 topography are shown in Fig. 9. In this case – based on the results of Table 2 – the difference is only 0.07, and it can be proved that the fitting is right both in the wide and narrow frequency ranges. He fluctuation of results (Table 2) relates to the fact that this is not a regular failure, but draws attention to the fact that a careful application of fractal dimension results is needed.

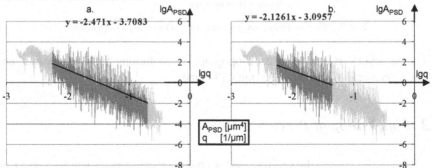

Fig. 8. Persson-curves of Mil04 fitted in a) wide and b) narrow frequency range

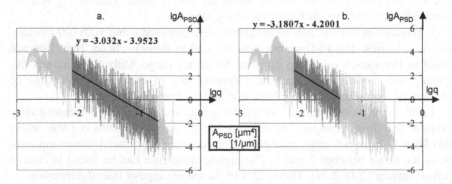

Fig. 9. Persson-curves of Tur32 fitted in a) wide and b) narrow frequency range

4.3 Effect of Sampling Distance

The third part of the investigation focuses on the correspondence of the sampling distance and the calculated fractal dimension. The PSD analysis used 100x100 frequency in all cases, as mentioned before. Line fitting was performed to the narrowest frequency range that can be calculated from the 10 μm sampling distance because sampling distance influences the maximal frequency of analysis (see above).

Table 3. Effect of sampling distance to fractal dimension

Sampling [μm]	Mil04	Lap04	Gri04	Tur32	Mil32
2	2.94	2.24	2.75	2.41	2.64
4	2.93	2.23	2.76	2.44	2.69
6	3.00	2.30	2.81	2.46	2.70
8	3.00	2.47	2.95	2.61	2.72
10	3.00	2.48	2.92	2.61	2.84

Taking all the above into consideration – and in the knowledge of the inaccuracy of line fitting – the results are summarised in Table 3. The visualisation of results (see Fig. 10) shows that the increase of the sampling distance increased the value of fractal dimension. The rising was about 0.2 in all cases, but it appeared differently: in some cases (e.g. Lap04, Mil04) only a small increase of sampling distance was enough to excite significant changes, while in case of Tur32 8 μm sampling distance caused this change.

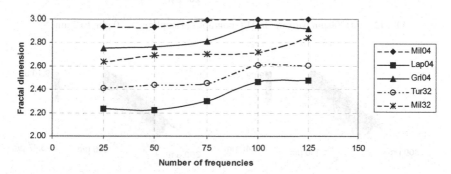

Fig. 10. Effect of sampling distance to fractal dimension

Fig. 11 represents the Persson-curves of ground surface (Gri04) with 2 and 10 μm sampling distance. Although Table 2 – in correlation Fig. 11 – shows that line fitting is right, the same as in case of topography Tur32 (see Fig. 9), modification of fractal dimension by sampling distance also exists. The question arises: Is the cause of this phenomenon in the mathematical background of PSD or is it conveyed by the physical "changes" of topographies?

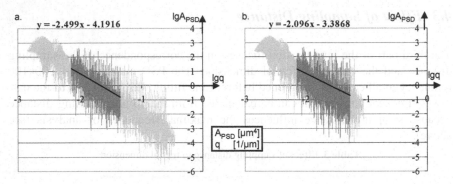

Fig. 11. Effect of sampling distance to fractal dimension of Gri04

200 by 200 μm and 300 by 300 μm parts of Gri04 and Tur32 topographies are shown in Figs. 12 and 13 in case of 2 μm and 10 μm sampling distance. The changes of fine ground surface are evident. When using 10 μm sampling many details disappear, but the global character remains. Topographic elements of the turned surface are relatively high so the modification of sampling distance does not change the visual character of the surface.

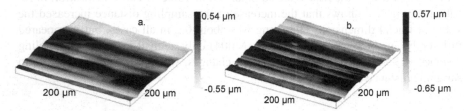

Fig. 12. 200x200 μm parts of Gri04 topography; sampling: a) 10 μm, b) 2 μm

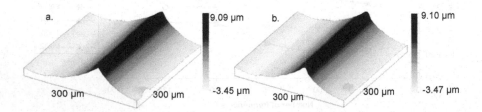

Fig. 13. 300x300 μm parts of Tur32 topography; sampling: a) 10 μm, b) 2 μm

2D PSD investigations were carried out to perform a deeper analysis, to follow the changes of details towards finding the real effect of sampling. Fig. 14 shows a profile of turned surface produced by 50 μm/revolution. Fig. 14a contains 500 points (2 μm sampling distance), while b. contains only 250 points (4 μm sampling distance). Profiles do not show any significant difference.

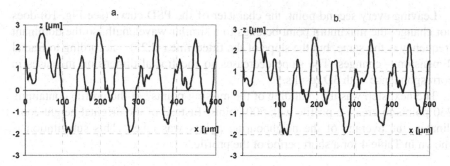

Fig. 14. Turned profile containing a) 500 b) 250 points

For the analysis of physical changes, not only the logarithmic scale of PSD was used, but also the linear one. Persson's approximation is not needed because the results of 2D PSD can be visualised in 2D diagram directly. Fig. 15a – linear scale PSD – represents the dominant wavelength (50 μm) of the profile. This maximum point can also be seen on a logarithmic scale (Fig. 15b, sign A).

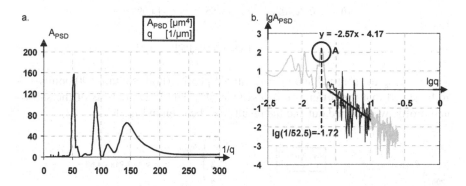

Fig. 15. PSD results of turned profile (500 point) a) linear scale b) logarithmic scale

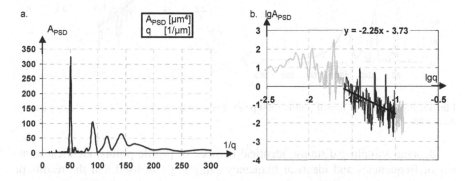

Fig. 16. PSD results of turned profile (250 point) a) linear scale b) logarithmic scale

Leaving every second point, the character of the PSD curve (see Fig. 16) does not change: the maximum point belongs to a similar wavelength, so the dominant frequency is the same, but the slope of the fitted line (-2.25) – according to fractal dimension – changes. The slope decreases so the fractal dimension increases in correspondence with 3D results.

The next step was the "building of the missing points" of the profile containing 250 points to create a profile with 500 points. Instead of the original height coordinates, the average of the neighbouring points was taken. This substitution is shown in Table 4 for a short period of the profile.

Table 4. Coordinates and its difference of original and substituted profile

x [μm]	z [μm], orig.	z [μm], subst.	Δz [%]
0	1.428619	1.428619	0
2	0.9023355	0.98341075	8.985045
4	0.5382025	0.5382025	0
6	0.4887393	0.6109306	25.00132
8	0.6836587	0.6836587	0
10	0.7840201	0.6734755	-14.0997
...	...	...	
		average diff. [%]	5.44

Comparing the original and the substituted profile (both contain 500 points!), it can be stated that the difference is less than 5%. The PSD analysis of the substituted profile was carried out and the results can be seen in Fig. 17.

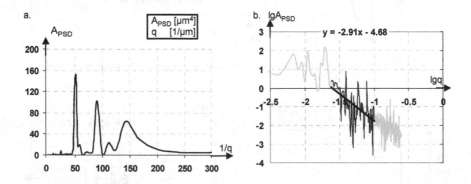

Fig. 17. PSD results of turned profile (500 point, linear substitution) a) linear scale b) logarithmic scale

Identical sampling distance, identical number of profile points, identical number of frequencies and identical frequency sampling was used, but the results of fractal dimension are different, meanwhile the character of the PSD curve is similar to those in Figs. 15 and 16. The differences of the fractal dimension of substituted profile from the original one and from the profile containing 250 points explain the changes of fractal dimension of topographies with different sampling distance. The increase of sampling distance results in a loss of information greatly influencing the fractal character of the surface. Namely, it is not that PSD is sensitive to sampling, but the physical content of the topography is different.

Several consequences can be drawn from this. The "absolute" usage of fractal dimension calculated from PSD analysis can be challenged because it may depend on the sampling of the measurement and can generate significant failures in tribological models.

PSD analysis is required to be used for comparing different surfaces to ascertain measuring and characterisation parameters.

Conclusions

The following conclusions can be drawn:

3D PSD analysis can be performed using only a few frequencies. Dominant wavelength, orientation of the surface and fractal dimension can be calculated from PSD topography containing a minimum of 75x75 frequency.

Fitting a line to the Persson-curve, the frequency range of fitting may have a great influence (20%) on the results of fractal dimension depending on the character of the topography. Visualisation can help identify accuracy.

The sampling distance of measurement is a dominant parameter in the analysis of the fractal character of the surface. The low spacing elements of surfaces contain a huge amount of fractal information, thus the omission of these elements reduces the surface and increases fractal dimension.

References

[1] Thomas, T.R., Rosén, B.G.: Determination of the Sampling Interval for Rough Contact Mechanics. J. Trib. Int. 33, 601–610 (2000)
[2] Persson, B.N.J., Albohr, O., Trataglino, U., Volokitin, A.I., Tosatti, E.: On the Nature of Surface Roughness with Application to Contact Mechanics, Sealing, Rubber Friction and Adhesion. J. Phys, Condens. Matter 17, R1–R62 (2005)
[3] Schargott, M., Popov, V.: Diffusion as a Model of Formation and Development of Surface Topography. J. Trib. Int. 39, 431–436 (2006)
[4] Ao, Y., Wang, Q.J., Chen, P.: Simulating the Worn Surface in a Wear Process. Wear 252, 37–44 (2002)
[5] Stout, K.J., Sullivan, P.J., Dong, W.P., Mainsah, E., Luo, N., Mathia, T., Zahouni, H.: The Development of Methods for Characterisation of Roughness in Three Dimensions. Printing Section University of Birmingham Edgbuston, Birmingham (1993)

Numerical Prediction of Friction, Wear, Heat Generation and Lubrication in Case of Sliding Rubber Components

Tibor J. Goda

Department of Machine and Product Design, Budapest University of Technology and
Economics, Műegyetem rkp. 3, H-1111 Budapest, Hungary
goda.tibor@gt3.bme.hu

Abstract. Modeling strategies and algorithms have been presented for the numerical prediction of hysteretic friction, wear, frictional heat generation and lubrication state of rubber components subjected to sliding friction. All the algorithms presented base on different numerical techniques - such as finite element and finite difference method-, and mathematical methods (e.g. discrete Fourier transformation, numerical integration, etc.). This numerical approach allows the integration of the algorithms and, as a direct consequence, the development of design tools. As an example for the design tool, in the second part of this contribution, a very recently published numerical model has been adopted an applied to a widely used, standardized hydraulic O-ring. The model takes into consideration the effect of surface roughness, deformation of seal, pressure dependency of viscosity and cavitation, respectively. As an asperity type contact model is incorporated into the model it can be used not only for full film but also for mixed lubrication. By using the design tool developed both the pressure distribution within the lubricating film, and the amount of fluid flow transport during outstroke and instroke (their difference defines the amount of leakage) have been predicted.

Keywords: rubber, sliding contact, hysteresis, wear, heat generation, lubrication, numerical modeling.

1   Introduction

As sliding rubber components are widely used in different industrial applications it is very important to predict the wear rate, the friction resistance (friction force) and the amount of leakage which can be defined as the difference between fluid transport during outstroke and instroke. In dry case, due to the high friction force the frictional heat generation may be very intensive thus its effect on the tribological behaviour can not be neglected. As rubber has non-linear, time- and temperature-dependent material behaviour the application of effective and accurate

I.J. Rudas et al. (Eds.): Towards Intelligent Engineering & Information Tech., SCI 243, pp. 519–530.
springerlink.com   © Springer-Verlag Berlin Heidelberg 2009

numerical models and techniques can be especially advantages. At the same time the surface roughness and the long time interval that must be simulated may result in huge CPU time.

In the absence of lubricant when contacting surfaces are dry and clean the rubber friction is mainly due to adhesion (especially at low sliding speeds) and hysteresis [1-3]. The hysteretic friction comes into being when the rubber is subjected to cyclic deformation by the macro and/or micro geometry (surface roughness) of the hard, rough substrate [2, 3]. When a rubber component slides on a hard, rough substrate the surface asperities of the substrate exert oscillating forces on the rubber surface leading to energy "dissipation" via the internal friction of the rubber. In the most engineering applications, rubber/metal sliding pairs are lubricated in order to decrease the friction force arising in dry case. In presence of lubricant, rubber friction is due to hysteretic losses in the rubber, boundary lubrication and fluid friction. The lubrication decreases the contribution of nano-roughness to hysteretic friction because lubricant fills out the nano-valleys that makes the penetration of the rubber impossible into these regions. In the case of fluid friction, friction force comes from the shearing of a continuous, relatively thick fluid film. At the same time, in regions where a very thin lubricating film separates contacting surfaces, the friction force comes primarily from the shearing of this thin boundary layer.

The main objective of this contribution is to present numerical models and algorithms for the investigation of hysteretic friction, wear, frictional heat generation and lubrication. The models to be presented can be considered as the basic components of an integrated design tool that can be very useful during the design of sliding rubber components. By using virtual models the tribological behavior of a newly design component can be investigated and predicted without producing and experimental testing of prototypes.

2   Numerical Modeling Strategies

2.1 Numerical Modeling of Hysteretic Friction

Basically, both the macro and the micro deformations can cause hysteretic friction. In case when a smooth steel ball slides or rolls over a rubber plate [4] the hysteretic friction is due to macro deformations caused by the ball. During sliding or rolling, work is done as the ball moves forward. This work is required to deform the rubber in front of the ball. Simultaneous with this elastic energy is recovered from the rear. Since rubber has viscoelastic material behavior it shows hysteresis and thus one portion of the work done is lost which can be considered as the work of the hysteretic friction force. At micro-level, where the hysteretic friction is due to the real micro-topography (surface roughness) the contact problem hast to be modeled at asperity level [5, 6]. By using FE technique, in addition to the material model of the rubber a surface model substituting the real, measured rough surface is also needed (see Fig. 1). As a possibility the micro-topography measured by AFM, stylus instrument or laser technique can be substituted e.g. by spline surface fitted to the measurement results. As an alternative, the measured micro-topography can be decomposed into harmonic components by using the discrete Fourier transformation. This approach allows engineers to predict the contribution of each component to the hysteretic friction. In order to take into consideration all the components of the micro-topography (from micro- to nano-level) in a single FE model a very fine mesh is required that may result in enormous CPU time.

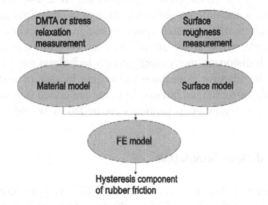

Fig. 1. Flowchart of the FE based hysteretic friction prediction

Considering the material behavior of the rubber the generalized Maxwell-model as a material model can be a reasonable choice because it is available in the most, commercial FE software. At the same time, it must be mentioned that its applicability depends strongly on the number of Maxwell elements. Even in case of 15-term Maxwell model, there is strong fluctuation in the loss factor-frequency curve that may cause inaccurate modeling results especially when the excitation frequency varies in a broad frequency range. Based on preliminary studies, in can be concluded that a 40-term Maxwell model is able to describe with high accuracy the time dependent material behavior of the rubber in a wide frequency range.

2.2 Numerical Modeling of Wear

Based on the literature one can conclude that there is no standardized method for including wear within finite element analysis. The most important characteristics of FE modeling strategies for wear simulation are as follows:

1 Simple wear models are used instead of the direct modeling of important wear
   mechanisms. These wear models are implemented within an FE based wear
   simulation tool.
2 Contacting surfaces are modeled as perfectly smooth ones.
3 The size of an individual finite element located within the contact region is
   much greater than the largest wavelength of the surface roughness. Thus a single element consists of a lot of asperities.
4 Parameters of the wear models are determined from a fit to the experimental
   measurements. The effect of surface topography, lubricant and temperature is
   taken into consideration through these parameters.

In a numerical simulation a steady change of geometry can not be realized because of the incremental solution strategy (the sliding motion is divided into small
periods). The geometry is modified according to the wear law at the end of these
periods only. At the same time no geometry change is taken into account during a
single period. Based on the computed volume change the thickness of the abraded
material layer is determined at the contact nodes. To eliminate deformation and
wear induced mesh distortion mesh modification technique must be used such as
the rezoning and the adaptive remeshing technique. In case of rezoning, only the
coordinates of nodes are changed but no elements are added or subdivided. At the
adaptive remeshing technique, new elements are added to the old mesh.

2.2.1   Simplified Wear Simulation

No transfer of state variables (stress and strain states) is considered at the transition between two simulation steps. Only a change of geometry is performed. The
simulation is restarted after every change with a „virgin" model, which does not
carry any initial stresses or temperatures. At the beginning of each run, the rubber
part is in its undeformed configuration. The history-dependent material behavior
of rubber can not be taken into account in this case. The wear simulation tool
works in a loop and performs a series of static FE simulations with updated surface geometries.

2.2.2   Wear Simulation with the Transfer of State Variables

In this case, there is a need to transfer state variables (stress and strain states) from
an old to a new, modified mesh before the restart of the simulation. Regarding the
existing capabilities of commercial FE programs such a restart is not trivial. The
history-dependent material behavior of rubber can be taken into account in this
case.

2.3 Numerical Modeling of Frictional Heat Generation

Heat generation can also be occurred due to the sliding friction between the reciprocating rod and rubber seal as well as the material friction (hysteresis) of rubber seal. As a consequence of the material friction one part of the mechanical work converts into heat while the rubber seal deforms owing to, for example, surface asperities. Generally, a measured coefficient of friction is used in the modeling of frictional heat generation which includes the effect not only of the adhesion but also of the hysteresis and other physical processes producing energy dissipation. On the other hand, the frictional heat generates a temperature distribution in the contacting bodies, which contributes to the deformations through thermal strains and influences both the mechanical and the thermal properties. The treatment of the complex interaction of the above mentioned phenomena requires a thermo-mechanical coupled analysis (see Fig. 2). In the thermo-mechanical coupled problems, two analyses are performed in each time increment: a thermal analysis (thermal pass) and a stress analysis (stress pass). Iterations can also be carried out within each increment to improve the convergence of the coupled thermo-mechanical solution. Firstly, a heat transfer analysis is performed, then a stress analysis. This coupled approach allows to solve a contact problem taking into consideration thermal strains, time- and temperature-dependent mechanical properties as well as temperature-dependent thermal properties (thermal conductivity, specific heat).

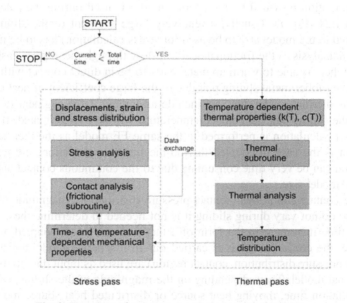

Fig. 2. Flowchart of a thermo-mechanical coupled analysis

Since the relative sliding velocity of contacting surfaces, the contact pressure distribution as well as the pressure- and sliding velocity-dependent coefficient of friction can vary during the simulation the frictional heat generated will be not constant with time. Therefore a transient heat transfer analysis is needed within the coupled analysis.

In order to define a pressure, sliding velocity and temperature dependent coefficient of friction frictional user subroutines can be used. The calculated contact pressure distribution is used in the thermal pass of the coupled analysis (Fig. 2) to determine the momentary magnitude of the heat source. The nodal friction force, relative sliding velocity and the node number of the given contact node are provided by the frictional subroutine. These data will be used within the thermal subroutine to calculate the frictional heat flux at a contact node.

By knowing the contact pressure distribution (p), the magnitude and variation of the sliding speed (v), as well as the coefficient of friction (μ) the frictional heat generated over a unit surface per unit time (q) can be calculated as

$$q = \mu \cdot v \cdot p. \tag{1}$$

Two different techniques can be used for FE thermal calculations. In the first case, the contact and the thermal calculations are performed in the same FE model. Therefore, a thermo-mechanical coupled analysis has to be applied, which is very time consuming. In the second case, the contact and thermal calculations are separated from each other i.e. there are two different FE models, one for the contact and one for the thermal calculation. In the thermal model, the contact pressure distribution provided by the contact model is used during the calculation of the heat flux (Eq. 1). Usually, a relatively large segment of the sliding system (compared to the model size to be used for stress calculation) has to be modeled in a thermal analysis as the frictional heat can flow through metal parts easily i.e. the frictional heat is able to warm up metal parts to be in direct contact with the metal side of the rubber/metal sliding pair due to the large coefficient of heat conductivity. If the thermal model does not include an enough large segment of the metal side of the sliding pair the contact temperature will be overestimated. If the stress or contact calculation is performed in the same FE model as the thermal calculation (that is the case in coupled thermo-mechanical simulation) the temperature prediction can be very time consuming due to the continuous contact analysis and the large model size.

If the contact conditions (contact pressure distribution, magnitude of the contact area) do not vary during sliding it is not needed to determine these quantities at each time increment in the form of a thermo-mechanical coupled analysis. In this case, the contact analysis is carried out in a separate model and the results (contact pressure distribution, contact region) are used as input data in the separate FE thermal model. Then, depending on the magnitude of the sliding velocity and the simulation time, moving heat source or distributed heat source model can be used to estimate the temperature distribution.

In case of moving heat source, the position of the heat source changes as a function of time. If the sliding velocity is high enough and the temperature distribution has to be calculated not in the initial phase then the moving heat source can be substituted by a distributed heat source at which the heat generated during a single cycle is distributed uniformly on the surface scrubbed by the rubber seal. In FE thermal simulations, the continuous movement of the heat source is discretized as follows. The heat source is staying in a given position for a short time then it is moving in other position. After a short time, the heat source is moving in a new position again. In a transient, sliding friction related FE thermal simulation with moving heat source these steps are repeated until the desired simulation time is reached. As there is an enormous difference in thermal conductivity of rubber and steel, as a first step, the heat generation can be estimated by a simple model in which the metal side of the sliding pair is included only. In such a model, the total frictional heat acts on the metal rod i.e. the heat partition is 1. The main advantage of this technique is that we do not make an effort to determine the heat partition between contacting bodies. Heat sources are in a given position for a time of dt=dx/v (v is the sliding speed), then they are moved forward by dx. These steps are repeated during the simulation. If the sliding component performs reciprocating motion and the desired simulation time is many times greater than the period (measured in seconds) the moving heat source can be substituted by a distributed heat source. In case of distributed heat source model, the frictional heat generated during a single cycle is distributed over the surface scrubbed by the rubber component in the same time period. Based on this condition one can calculate the heat flux as a distributed (over the total length) heat source. Using this technique, the heat generation can be studied in a wider time interval than in the case of moving heat source approach because the CPU time of a single cycle is significantly less.

2.4 Modeling of Lubrication

To predict the lubrication performance of a reciprocating seal engineers need effective and accurate models and algorithms. In the last few years, an intensive development was observable in the numerical modelling of lubrication of reciprocating seals.

One of the latest numerical models – proposed by Salant et al [7] – is able to take into consideration the deformation of seal, the effect of surface roughness and cavitation simultaneously. Cavitation usually occurs in the regions of diverging contact gap where the gap expands. Hydrodynamic pressure is achieved until the gap between contacting bodies expands to a level where the lubricant does not occupy the complete gap and cavitates. In the cavitated region, a gas-lubricant mixture fills out the gap between contacting surfaces. The cavitation model in [7] assumes the preservation of the lubricant mass continuity and a constant contact pressure (cavitation pressure) i.e. a zero pressure gradient in the cavitated region. As an asperity type contact model, namely the Greenwood-Williamson model, is also incorporated in the proposed model it is possible to study both the full film and the mixed lubrication in reciprocating seals. In the case of full film lubrication, there is a continuous lubricating film with load carrying capacity between surfaces which does not make it possible to come into direct, asperity type contacts. In this case, the total normal load is carried by the lubricating film. Contrary to this, in mixed lubrication, the lubricating film is not able to separate surfaces from each other completely i.e. local solid contacts can be formed among asperities. As a consequence, one part of the normal load is carried by the lubricating film while the rest is carried by asperity contacts. The one-dimensional form of the average Reynolds equation valid for incompressible lubricant with pressure dependent viscosity [8] is

$$\frac{d}{dx}\left(\phi_x \cdot h^3 \cdot e^{-\alpha \cdot \overline{p}} \frac{d\overline{p}}{dx}\right) = 6 \cdot \eta_0 \cdot U\left(\frac{d\overline{h}_T}{dx} + \sigma \frac{d\phi_{scx}}{dx}\right) \tag{2}$$

where x is the coordinate in axial direction, U is the sliding speed, η_0 is the dynamic viscosity at ambient pressure (the pressure dependency of viscosity is given by the Barus formula $\eta(p) = \eta_0 \cdot e^{\alpha \cdot p}$), α is the pressure-viscosity coefficient, ϕ_x is the pressure flow factor, $\overline{p}$ is the mean pressure, $\overline{h}_T$ is the average gap, σ is the RMS roughness, h is the nominal film thickness defined as the distance between the mean levels of rough surfaces and ϕ_{scx} is the shear flow factor. Salant and his co-workers [7] have incorporated the effect of cavitation in the universal, dimensionless form of Eq. 2:

$$\frac{d}{d\hat{x}}\left(\phi_x \cdot \hat{h}^3 \cdot e^{-\hat{\alpha} \cdot F \cdot \phi} \frac{d(F \cdot \phi)}{d\hat{x}}\right) = 6\frac{\eta_0 \cdot U \cdot L}{p_a \cdot \sigma^2}\left(\frac{d}{d\hat{x}}\left\{\left[1 + (1 - F)\phi\right] \cdot H_T\right\} + F \cdot \frac{d\phi_{scx}}{d\hat{x}}\right) \tag{3}$$

where $\hat{x}$ is the dimensionless axial coordinate ($\hat{x} = \dfrac{x}{L}$, where L is the length of the solution domain in x- or axial direction), $\hat{h}$ is the dimensionless nominal film thickness ($\hat{h} = \dfrac{h}{\sigma}$), $\hat{\alpha}$ is the dimensionless pressure-viscosity coefficient ($\hat{\alpha} = \alpha \cdot p_a$), p_a is the ambient pressure, H_T is the dimensionless average gap ($H_T = \dfrac{\overline{h}_T}{\sigma}$), F is the cavitation index and ϕ is a universal variable.

Eq. 3 is discretized using a control volume finite difference scheme. The discretized form of Eq. 3 is solved by iteration for ϕ and F. At the beginning of the iteration it is assumed that there is no cavitation inside the solution domain and the pressure is equal to the atmospheric pressure at these locations. The numerical model developed can be used for real reciprocating seals because, as it is able to take into account the effect of elastic deformation of the seal on the gap form. The reason why iteration is needed is that the pressure field induces elastic deformations in the seal that directly modify the gap form. At the same time, this modifies the Reynolds equation used for the computation of a new pressure field. Additionally, two inner iterations are needed to calculate the locations of the cavitated regions and to handle non-linearity caused by pressure-dependent viscosity. In the latter case the cause of iteration is that one wants to calculate pressure distribution at a given film thickness on the basis of the Reynolds equation which consists of a pressure-dependent viscosity.

2.4.1   Application of the Numerical Model to an O-ring

The main objective of this section was to examine the leakage of an O-ring by the numerical model presented above. By using the theory presented in [7] a one-dimensional, steady-state numerical model has been developed [9] for the modeling of lubrication in reciprocating hydraulic rod seals under the assumption that the lubricant is considered as incompressible. Typical outputs of the numerical simulation are the pressure distribution within the lubricating film (fluid pressure), the pressure distribution due to asperity contacts (contact pressure), and the amount of fluid flow transport during outstroke and instroke (their difference defines the amount of leakage). The numerical model developed was verified by using literature results for hydrodynamic lubrication [9]. The hydraulic O-ring studied has a size of 8x1.5 mm.

As a first step the static contact pressure distribution of the mounted in, pressurized O-ring was calculated by FE technique. For this purpose an axisymmetric FE model has been developed. On the left side of the O-ring the sealed pressure was p_{sealed}=6 bar. On the right side of the O-ring there was atmospheric pressure (p_a=1 bar). The static contact pressure distribution can be seen in Fig. 3b. The length of the contact area calculated by FE technique was 0.7986 mm. As a next step the static film thickness distribution was calculated by using the Greenwood-Williamson contact model. At this point, it was assumed that the contact pressure is the static contact pressure and the distance between the mean planes of contacting rough surfaces was calculated (RMS roughness=0.4 micron, asperity radius=1 micron, asperity density=10^{13} $1/m^2$, elastic modulus of the O-ring=10 MPa, Poisson ratio=0.48). In order to compute the radial deformation of the O-ring under a given pressure distribution the influence coefficient method was used. The influence coefficient method takes into consideration the effects of the forces at all nodes of the contact area on the deformation at a given node. In the present case, the number of nodes located within the contact area is 41. The distance between two neighboring nodes is 19.966 micron. The elements of the influence coefficient matrix (I_{ik}) were calculated by FE technique. I_{ik} represents the deformation at node i produced by a unit pressure at node k.

2.4.1.1 Results

The leakage simulation was performed under the following conditions: (a) sliding speed during in- and outstroke is U=50 mm/s, (b) dynamic viscosity of the fluid at atmospheric pressure is η_0=0.02 Pas, (c) pressure-viscosity coefficient is α=20x10^{-9} 1/Pa, (d) Both the inlet and the outlet of the solution domain are flooded that is the cavitation index is F=1 at these locations, (e) the cavitation pressure is assumed to be 0.1 MPa that is equal to ambient or atmospheric pressure. Fig. 3 shows the calculated film thickness, contact and fluid pressure distributions for outstroke. At the edges of the contact area the film thickness is larger than in the middle of it.

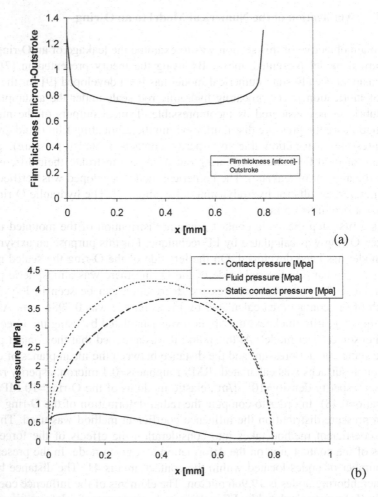

(a)

(b)

Fig. 3. Outstroke: (a) film thickness distribution, (b) fluid, contact and static contact pressure distribution (the fluid pressure at x=0 and x=0.7986 mm equals to the sealed pressure and the atmospheric pressure, respectively)

In addition, there is a cavitated region within the contact area where the fluid pressure equals to the atmospheric pressure. At 0.12 mm, the fluid pressure reaches its maximum value of 0.76 MPa. The computed fluid flow transport (volumetric flow rate per unit circumferential length) during outstroke was 0.0124 mm^2/s. As the diameter of the rod was 8 mm this fluid flow value corresponds to a volumetric flow rate of 0.312 mm^3/s.

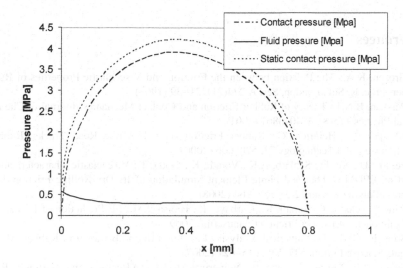

Fig. 4. Fluid, contact and static contact pressure distribution during instroke

In case of instroke, (see Fig. 4), there is no cavitation within the contact area and the maximum fluid pressure is equal to the sealed pressure (0.6 MPa). The calculated fluid flow transport during instroke was 0.0107 mm^2/s that corresponds to a volumetric flow rate of 0.269 mm^3/s. As the leakage is interpreted as the difference between the fluid flow transport during out- and instroke it has a value of 0.043 mm^3/s for the O-ring investigated.

Conclusions

FE based numerical models and modeling strategies have been presented for the prediction of hysteretic friction, wear, frictional heat generation and lubrication. Each of these numerical models can be considered as a design tool with which the investigation of a new component can be performed virtually. The integration of these models makes it possible to obtain accurate and reliable numerical predictions.

In the second part of this study, a one-dimensional numerical model taking into consideration cavitation, surface roughness, elastic deformations and pressure-dependent viscosity has been developed for the investigation of a standardized hydraulic O-ring. In addition to the film thickness distribution, the fluid, contact and static contact pressure distributions have also been predicted for both instroke and outstroke. Finally, the leakage rate has been evaluated. All the predictions seem realistic and are as anticipated.

Acknowledgments. This work was performed in the framework of an integrated project of EU (KRISTAL; Contract Nr.: NMP3-CT-2005-515837; www.kristal-project.org).

References

[1] Grosch, K.A.: The Relation between the Friction and Viscoelastic Properties of Rubber. Proc. R. Soc. London, Ser. A 274(21), 21–39 (1963)

[2] Persson, B.N.J.: Theory of Rubber Friction and Contact Mechanics. Journal of Chemical Physics 115(8), 3840–3861 (2001)

[3] Klüppel, M., Heinrich, G.: Rubber Friction on Self-Affine Road Tracks. Rubber Chemistry and Technology 73, 578–606 (2000)

[4] Felhős, D., Xu, D., Schlarb, A.K., Váradi, K., Goda, T.: Viscoelastic Characterization of an EPDM Rubber and Finite Element Simulation of Its Dry Rolling Friction. Express Polymer Letters 2(3), 157–164 (2008)

[5] Pálfi, L., Goda, T., Váradi, K., Garbayo, E., Bielsa, J.M.: FE Prediction of Hysteretic Component of Rubber Friction (manuscript, 2009)

[6] Soós, E., Goda, T.: Numerical Analysis of Sliding Friction Behavior of Rubber. Materials Science Forum, 537–538, 615–621 (2007)

[7] Salant, R.F., Maser, N., Yang, B.: Numerical Model of a Reciprocating Hydraulic Rod Seal. Journal of Tribology 129, 91–97 (2007)

[8] Patir, N., Cheng, H.S.: An Average Flow Model for Determining Effects of Three-Dimensional Roughness on Partial Hydrodynamic Lubrication. Journal of Lubrication Technology 100, 12–17 (1978)

[9] Goda, T.: Numerical Modeling of Lubrication in Reciprocating Hydraulic Rod Seals. In: Proceedings of 6th Conference on Mechanical Engineering, on CD (2008) ISBN 978-963-420-947-8

Design of a Linear Scale Calibration Machine

Gyula Hermann

Budapest Tech, Bécsi út 96/B, H-1034 Budapest, Hungary
hermann.gyula@nik.bmf.hu

Abstract. This paper describes the modification of a length measuring machine for the semiautomatic calibration of linear scales with a pitch distance of a few micrometers. It consists of the following modules: a high resolution optics including a CCD camera, appropriate illumination, a motion system consisting two stacked linear stages with submicron resolution, responsible for moving the scale in front of the optics with a well defined speed. The paper presents the various pitch distance definitions. From these definitions simple algorithms are derived. The main point is to minimize the effect of optical distortion, diffraction and nonlinearity by using properly designed optics and illumination. The stage system carrying the scale is driven by piezomotor providing unlimited travel with nanometer resolution. The displacement of the scale is measured by an HP laser interferometer.

1   Introduction

The increasing application of micro- and nanotechnology emphasizes the importance of high resolution measurement, which in turn, increases the demand for calibration standards with submicron structures. Optically performed measurements thanks to their increased resolution and accuracy have gained importance. Various types of scales with periodic structures in one or two directions are relatively cheap and easy to use in the calibration of microscopes and other measuring instruments. For this purpose the pitch, the mean distance between the lines of such scales has to be known precisely.

Many kind of interferometric calibration systems for line scales have been built at many laboratories. Some laboratories modified commercial machines, usually length measuring machines. These calibration systems are designed for calibrating either one or two dimensional artifacts. Currently static line detection systems, mostly used in calibration laboratories, are to be replaced by systems using dynamic calibration technique.

One of the earliest paper [2] covering the subject describes the NIST length scale interferometer for measuring graduated length scales. It discusses in detail not only the machine and its operation, but elaborates also on the uncertainty, its

I.J. Rudas et al. (Eds.): Towards Intelligent Engineering & Information Tech, SCI 243, pp. 531–542.
springerlink.com   © Springer-Verlag Berlin Heidelberg 2009

sources, the required environmental conditions and how they are kept under control or compensated. Barakauskas and his co-authors investigate in their paper the effect of systematic and random component errors on the resulting accuracy of the calibration system. Design methods are introduced to increase motion accuracy. Also computational and active methods of geometric error compensation are described [1] [6]. Kaušinis paper addresses errors specific to dynamic line scale calibration caused by geometric and thermal deviations of the system components [5]. A 3D finite element model was used to investigate the thermal influence both on the scale and the machine structure. Meli's paper [7] presents another approach based laser diffractrometry, according to Littrow principle, resulting in picometer measuring uncertainty. The method is suitable for 2D gratings as well.

In recent years scanning probe microscopy emerged as a new technique to capture the distance between sub-sequent scale lines [8]. Atomic force microscope is the appropriate equipment for this purpose. However, if the lines forming the scale are covered, for example by a layer of glass, than this method is not applicable. Peng Xi etc. [9] solve this problem by applying non-contact near field detection of the measuring point. For the illumination of the scales monochromic light of a He-Ne laser was used. They claim to reach a resolution of ~50 nm.

A new linear measuring machine designed for the calibration of different standards, like step gauges, line scales and gauge blocks was described in the paper by Tae Bong Eom and Jin Wan Han [10]. The system consists of precision linear stage, an alignment system for the object to be calibrated, a module for detecting the measuring point and a laser interferometer. The measuring uncertainty for a 1000 mm target was 0.35 μm.

A good overview is given in [12] of the various calibration methods and equipments used by the major national metrology laboratories.

The paper by Druzuvec and his co-authors [3] discusses the effect of contamination in the calibration uncertainty model. Significant influence in calibration uncertainty budget is represented by the uncertainty of the line centre detection. The paper discusses different types of line scale contamination like dirt spot, scratches, line edge incorrectness and line intensity variations were simulated in order to test the error correction capability of the line centre detection algorithm.

To meet the requirements given in the subsequent paragraph, we have designed an additional carriage for the Zeiss measuring machine hereby extending it's capabilities for semi-automatic calibration of linear scales.

2   The Requirement

Graduated scales are made of various materials including steel, Invar, glass, glass-ceramics, silicon and fused silica. The cross sectional shape can be rectangular, H or U-form, or modified X called Tresca.

The machine should be able to accommodate and calibrate all these linear scales types with the following basic specifications:

- Measuring span: 400 mm;
- Resolution: approximately: 5 nm
- Measuring uncertainty: 50 nm
- Maximum measuring points: 0,5 mega points (400 mm/0,8 μm pitch)

The image of the lines should be captured optically. The structure has to be equipped with the appropriate number of high precision temperature sensors to provide sufficient data for determining the thermal distribution and compensating the effect of the resulting deformation.

The calibration machine is designed with the following concepts in mind:

- The Abbe's principle is satisfied,
- Moving table construction is employed on the existing length measuring machine,
- Laser interferometer with stabilized wavelength is used in a stabilized environment to minimize the effect of disturbances,
- Pitch detection both on the conventional way and on the base on Gabor filtering.

3 Overall Construction

The construction of the linear scale calibration system consists of three physically separated parts: the solid steel bed of the Zeiss length measuring machine and the carriage module, the optical microscope and image capturing module and the measuring system. The motion and the image evaluation is executed and coordinated by a personal computer. The carriage's position is determined by an HP heterodyne laser interferometer.

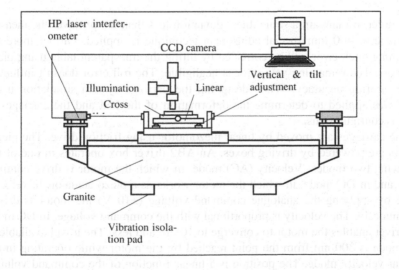

Fig. 1. The overall construction of the calibration machine

It is of vital importance that the distance between the microscope and the laser interferometers remain fixed during the scale measurement, because any relative motion between them will be seen as part of measured scale length, thus resulting in additional uncertainties.

4   The Motion System

The motion system consists of two linear stages - made of aluminum - mounted on top of each other. The straightness deviation of the guide rails causes substantial measurement error. While the carriage is in motion, angular errors can appear. Therefore, handlapping is performed to keep both the pitch and the yaw errors less than 0.5 arcsec.

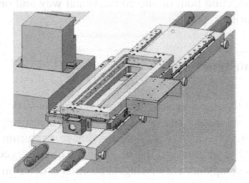

Fig. 2. The construction of the motion system

In order to maintain a straightness deviation less than 0.5 µm over the measuring range of 400 mm special adjustment technique is applied. Further more the remaining pitch error will be corrected by tilting the transparent table using piezo actuators. The remaining yaw error is negligible. The roll error doesn't influence the measuring accuracy. When designing the carriage system simulation using FEM was applied to determine the deformation of the bed and the carriages in various configurations.

The carriages are moved by linear piezomotor using friction drive. The piezomotors are powered by driving boxes. An AB2 driver box operates in one of the following two modes: Velocity (AC) mode, in which the motor is drive continuously and in DC mode, in which the motor works as a piezo-actuator. In velocity mode by applying the analogue command voltage (±10 V) the motor is driven continuously. The velocity is proportional with the command voltage. In DC mode the driver enables the motor to converge to 10 nm and less. The travel available in this mode is 300 nm from the point reached by the motor while operating in the regular velocity mode. The position is a linear function of the command voltage, with certain hysteresis and some asymmetry.

Fig. 3. Deformation of the motion system due to gravity

A table with flat surface is mounted on the top stage. They are linked together by flexures providing small frictionless linear motion. The distance between the two, are regulated by piezo actuators, resulting in a highly dynamic system. The aim of this additional construction is to enable the compensation for the tilt error. It also makes the correction of Abbe errors possible, to a certain extend. The angular errors are measured optically using techniques known from CD/DVD readers.

From the point of measuring uncertainty it is vital that the distance between the microscope and the interferometer remains fixed during scale measurement, because any relative motion between them will be seen as part of the scale length. The entire equipment rests on a vibration isolated concrete block and is housed in an environment where the temperature is kept 20°C±0.1°C.

The motion controller, implemented in software is working dual mode. The driver box is switched into velocity mode and a point 200 nm from the target position is approached following a trapezoidal velocity pattern. The stationary velocity, the acceleration and the deceleration distances can be introduced as control system parameters. Than the last 200 nm is covered in DC mode.

The parameters of the controller used in velocity mode were determined by mean of simulator taking into account the mechanical properties of the system.

5 Various Definitions of Pitch Distance

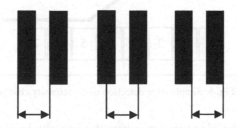

Fig. 4. Pitch distance definitions

Whereas the first two versions given in the picture can be measured directly, the third one requires more processing. The difficulty is generated by the line thickness variation and the straightness error of the edges. Contaminations of the scale also disturb pitch determination.

6   The Pitch Determination System

Linear scales are calibrated by moving the carriage and measuring it's displacement by the interferometers. The result of the measurement corresponds exactly to the scale pitch. The determining factor to this correspondence is the line centering process. The scale structures are captured optically. The optical system is assembled from various commercially available modules. The microscope consists of an objective, a manual control element for focusing, a tube lens, and a camera connection module.

The microscope with the CMOS camera is mounted on a beam fixed to the steel base. Digital measuring microscope enables precise estimation of the line edge quality and precise location of lines. To process the captured image data different approaches, described in the subsequent paragraphs are followed.

Optical measurement technology is confronted by the problem of having to achieve high accuracy with relatively wide surface sensors, which fall below the sensor dimensions by several multiples. As a rule, this problem is solved by blurring the edges of the test specimen across several pixels with the help of a fuzzy image. This technique thus makes possible the so-called sub-pixel edge detection. On this way precision measurements of up to a tenth of the pixel size (in some cases, even less) can be achieved. In Fig. 5, the basic principle of edge detection is given. Here the edge of an idealized test specimen is represented as defocused. It extends as a rising linear progression across several pixels (in this example, five).

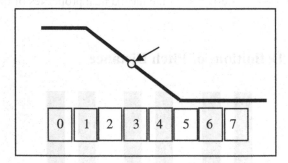

Fig. 5. Simple edge detection using intensity change

At the center of the edge function, where the intensity level is at 50%, lays the edge position to be detected. The pixels numbered 1, 2, ..., n, lie within the edge image, and receive intensities I_1, ...I_n. Let the pixels numbered 0 and n+1 be placed above and below the edge. Let the energy received be I_0 or I_{n+1}. The position of the edge p_{edge}, measured as the distance from the beginning of the pixel having the number 1, has the value

$$p_{edge} = p_{size}\left\{\frac{n}{2} + \frac{1}{n-1}\sum_{k=1}^{n}\frac{I_k - e_m}{I_{k-1} - I_k}\right\}$$
(1)

whereas p_{size} represents the pixel size and e_m is the intensity level for 50%. e_m is calculated from:

$$e_m = \frac{I_0 + I_{n+1}}{2}$$
(2)

This formula (1) is only one possibility for the detection of edges. There are several other formulas of equal value. Here, (1) stands as a representative for all edge detection formulas. Since the mathematical background is always similar, the statements made here about (1) apply to other edge detection algorithms as well.

With the help of edge detection, the position of the edge can be determined exactly, not only when the edge function is linear, but also that of all symmetrical edge progressions. The algorithm operates inaccurately, however, when the edge function is asymmetrical. In particular, the shift of the sensor relative to the edge in this case effects a displacement of the edge position.

In the first approach each horizontal line profile within the region of interest in the image is analyzed. The centre of the left and the right edge is used and the edge locations are determined with a moment based edge operator. A line is fitted through all these centers using only points within 2σ. The intersection of this fitted line with the reference line is used as the scale line position.

The second approach is based on the use of complex Gabor filter to obtain sub-pixel resolution:

$$h(x) = g(x)\left[\cos(\varpi_0 x) + j\sin\right]$$
(3)

where $g(x)$ is a Gaussian function.

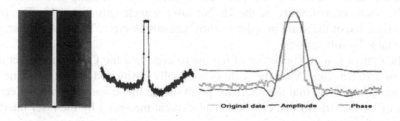

Fig. 6. Edge detection using Gabor filter

From the above figure it can be seen that the zero crossing of phase signal is the appropriate feature to find the line centre. Based on the detected lines centers, it is necessary to fit a straight line. In order to eliminate the effect of the outlier data points an M-estimator is used.

7 Illumination

As it was already mentioned from the optical point of view linear scales can be transparent (made of glass, silica or Zerodur) and opaque (made of various metals). Therefore the machine has to be equipped with both and backlight illumination.

For opaque (metal) scales diffuse axial illumination is envisaged. The lens looks through a beamsplitter that is reflecting light on the scale. Illumination is coaxial to the imaging system.

For transparent scales backlight illumination is applied. Backlighting is an excellent way to improve applications that need to measure edges, because silhouetting the object provides high contrast for improved edge detection while reducing unneeded surface detail. Normal backlights, however, can reduce the sharpness of the edge, because diffuse reflections occur from the broad area output and thus reduce the accuracy of gauging and inspection systems. Telecentric backlight illumination, is based on the same principles as telecentric measuring lenses, uses parallel rays, it avoids these problems. It also increases illumination compared to a standard backlight, allowing faster measurements.

8 The Measuring System

The pitch distance is the sum of the carriage displacement measured by the laser interferometer and the difference between the centre line of the camera and the detected line centre position in the camera coordinate system.

The HP interferometer operates in a $\frac{1}{4}\lambda$ counting mode with interpolation between the counts. Environmental parameter values of air temperature, barometric pressure, relative humidity and CO_2 content are measured and used to compute the refractive index of the air in the optical path, hereby compensating their effect.

The short term stability of the He-Ne laser wavelength is 1×10^{-8}. The laser beam is in line of the scale in order to eliminate cosine error. The scale is clamped to the table by soft springs.

The optical system consisting of the microscope and the CCD camera act as a kind of zero detector. After the scale is nominally positioned on the next line the carriage is moved so that the reference point on the line coincides with the center point of the camera. In this case the displacement measured by the laser interferometer is the pitch distance.

9 Measurement Uncertainty

Sources of measurement uncertainty:
- Uncertainty of the interferometer including environmental effects
- Misalignment
- Thermal expansion of the stage
- Thermal expansion of the scale

Some sources of uncertainty of the laser interferometer are intrinsic to this equipment and affect measurement accuracy. They are discussed in the subsequent paragraphs.

An interferometer generates fringes, equivalent to a fraction of the lasers wavelength, when displacement occurs between the measurement optics. When the wavelength changes fringes are generated giving an apparent displacement without actual displacement. The measurement error caused by wavelength deviation is proportional to the measured displacement, as it is specified in part-per-million of the nominal laser frequency. Assuming a wavelength accuracy of ±0.02 ppm, ensured by regular calibration to the national frequency standard, and a rectangular probability distribution for the laser uncertainty, its contribution to the combined standard uncertainty is:

$$u_w = \frac{\pm 0,02\, L}{\sqrt{3}} = \frac{\pm 0,008}{\sqrt{3}} \quad \mu m \qquad (4)$$

Optics non-linearity and electronics errors can be neglected as they are only a fraction of the wavelength error.

Uncertainties due to environmental effects are the influence of the atmospheric conditions on the interferometer accuracy and repeatability. As the wavelength of the laser is specified as the vacuum wavelength λ_v it is necessary to correct for the difference with the wavelength in the air λ_a. The index of refraction n of the air, whose value is dependent on the air temperature, pressure, relative humidity and CO_2 content, is given by the following equation:

$$n = \frac{\lambda_v}{\lambda_a} \qquad (5)$$

Using Edlen [4] equation and assuming standard and homogeneous air composition, 1 ppm error results from one the following conditions:
- 1°C air temperature change
- 2,5 mmHg change of the air pressure
- 80% change of the relative humidity

The thermometer and the pressure sensor built in the system have an accuracy of ±0.05°C and ±0.25 mmHg. The relative is measured with an accuracy of ±4%. It means that temperature; air pressure and humidity uncertainties are ±0.05 ppm,

±0.1 ppm and ±0.05 ppm. Assuming again rectangular probability distribution the standard uncertainties are 0.05 ppm/$\sqrt{3}$, 0,1 ppm/$\sqrt{3}$ and 0,05 ppm/$\sqrt{3}$. From this the standard uncertainty for the wavelength compensation is:

$$u_{comp} = L \frac{\sqrt{\left(\frac{0,05}{\sqrt{3}}\right)^2 + \left(\frac{0,1}{\sqrt{3}}\right)^2 + \left(\frac{0,05}{\sqrt{3}}\right)^2}}{\sqrt{3}} = L \frac{0,0707}{\sqrt{3}} \quad \mu m \quad (6)$$

Deathpath error results from the uncompensated length of the beam between the interferometer and the retroreflector in the zero position of the stage carrying the scale. The deadpath error d_e is the product of the deadpath distance d and the change of the laser wavelength during measurement:

$$d_e = d \frac{\Delta \lambda_a}{\lambda_w} \quad (7)$$

Cosine error is the sum of the misalignment of the scale relative to the carriage and the misalignment of the laser beam to the mechanical axis the motion. This results in a difference between the measured and the actual distances. The error is proportional to one minus the cosine of the misalignment angle. The cosine is approximately equal to $S^2/8L^2_{mm}$, where L_{mm} is the measured displacement in mm and S is the lateral offset of the returning beam in μm. For our case S, as a result of careful alignment, can be estimated to be about 100 μm over a measuring range of 400 mm. This results in a

$$\cos_{error}(ppm) = \frac{S^2}{8L_{mm}} = \frac{100^2}{8(400)^2} = \pm 0,00781 \quad \mu m \quad (8)$$

As the cosine error has a rectangular probability distribution its uncertainty is:

$$u_{cos} = \frac{0,00781}{\sqrt{3}} \quad \mu m \quad (9)$$

Correction for the thermal expansion of the stage is necessary in order to maintain the required measuring accuracy. The correction relates the displacement measurement results back to the temperature of 20°C. The uncertainty of the correction, consists of two components: uncertainty due to the error of the temperature measurement and uncertainty due to the deviation of the thermal expansion coefficient from the nominal value

The estimated uncertainty components are summarized in the following table.

Table 1. Estimated best measuring capability

Source of uncertainty	Probability distribution	Divisor	Standard uncertainty (μm)
Laser wavelength inaccuracy	Rectangular	$\sqrt{3}$	U_w=0,0046
Wavelength compensation	Rectangular	$\sqrt{3}$	U_{comp}=0,0163
Deadpath correction	Rectangular	$\sqrt{3}$	U_d=0,0006
Cosine error	Rectangular	$\sqrt{3}$	U_{cos}=0,0045
Carriage thermal compensation	Rectangular	$\sqrt{3}$	U_{tc}=0,0054
Combined standard uncertainty	Normal	---	U_r=0,0183
Expanded uncertainty	Normal	---	U_e=0,0367

Conclusions

The paper presents the development and design of a high precision carriage system, having nanometer resolution and a high resolution optical microscope responsible for capturing line data. The line data is processed by advanced filtering technique in order to determine the line centers. These two components enable dynamic calibration of scales, hereby reducing calibration time and costs.

Acknowledgments. The author gratefully acknowledges the support provided to this project by the National Science Foundation (OTKA) under the contracts no. T048850. The author also thanks Dr. Gy. Ákos, Professor Dr. Z. Füzessy, Dr. Cs. Sántha and Mr. K. Tomanyiczka for their valuable contribution.

References

[1] Barakauskas, A., Kasparaitis, A., Šukys, A., Kojelavičius, P.: Analysis of Geometrical Accuracy of Long Grating Scale Calibrator

[2] Bers, J.S., Penzes, W.B.: The NIST Length Scale Calibration Interferometer. Journal of Research of the National Standards and Technology 104(3), 225–252 (1999)

[3] Druzuvec, M., Acko, B., Godina, A., Weller, T.: Simulation of Line Scale Contamination in Calibration Uncertainty Model. Int. Journ. Simul. Model 7(3), 113–123 (2008)

[4] Edlen, B.: The Refractive Index of Air. Metrologia 2(2), 71–80 (1966)

[5] Jakštas, A., Kaušiniš, S., Flügge, J.: Investigation of Calibration Facilities of Precision Line Scales. Mechanika 3(53), 62–67 (2005)

[6] Kaušinis, S., Jakštas, A., Barakauskas, A., Kasparaitis, A.: Investigating of Dynamic Properties of Line Scale Calibration System. In: XVIII IMEKO World Congress Metrology for Sustainable Development, Rio de Janeiro, Brazil, September 17-22 (2006)

[7] Meli, F., Thalman, R., Blattner, P.: High Precision Pitch Calibration of Gratings Using Diffraction Interferometry. In: Proc. of the 1st Conf. on Precision Engineering and Nanotechnology, Bremen, May 1999, vol. 2, pp. 252–255 (1999)

[8] Misumi, I., Gonda, S., Kurosowa, T., Tanimura, Y., Ochiai, N., Kitta, J., Kubota, F.,
     Yamada, M., Fujiwara, Y., Nakayama, Y., Takamasu, K.: Comparing Measurements
     of 1D-Grating Samples Using Optical Diffraction Technique. In: CD-SEM and
     Nanometrological AFM, Proc. of the 3rd Euspen International Conference, Eindhoven,
     The Netherlands, May 26-30, 2002, pp. 517–520 (2002)
[9] Xi, P., Zhou, C., Luo, H., Dai, E., Liu, L.: Near Field Detection of the Quality of
     High-Density Gratings with Nanotechnology. In: Proceedings of SPIE, Nano and Mi-
     cro-Optics for Information Systems, vol. 5225, pp. 140–144
[10] Eom, T.B., Han, J.W.: A Multi-Purpose Linear Measuring System with Laser Inter-
     ferometer
[11] Theska, R., Frank, T., Hackel, T., Höhne, G., Lotz, M.: Design Principles for Highest
     Precision Applications
[12] WGDM-7 Preliminary Comparison on Nano-Metrology According to the Rules of
     CCL Key Comparisons. Nano 3 Line scale Standards. Final report. PTB Braun-
     schweig, August 29 (2003)

Engineering Objective-Driven Product Lifecycle Management with Enhanced Human Control

László Horváth and Imre J. Rudas

John von Neumann Faculty of Informatics, Budapest Tech
Bécsi út 96/B, H-1034 Budapest, Hungary
horvath.laszlo@nik.bmf.hu, rudas@bmf.hu

Abstract. Great advances in information systems and computer hardware, software, and communication technologies in their background motivated the authors in a work for human intent based and human acceptable automation of product modeling. In an advanced virtual space called as model space, increasingly complex products are described by information for engineering objects. Current product modeling in comprehensive product lifecycle management (PLM) systems suffers from quick development generated problems. Huge amount of relationships makes current product models less transparent. Decisions are not explained by their background in product models so that they can not be understood at applications. In this paper, the authors propose a contribution to solve the above problems by establishment of an extension to current product modeling that can be accessed through standard API. They outlined the problem and background in management of product information and conceptualized human thinking driven definition of product entities. Following this an extension to classical product modeling is proposed towards intelligent content, for information content driven control of conventional model entity generation, and for implementation in industrial PLM systems.

Keywords: Product lifecycle management, intelligent product modeling, information content in product model, control of product definition processes, tracking of changes.

1 Introduction

Engineering activities have been moving into computer systems for two decades. In the meantime, these activities have been integrated in product lifecycle management systems where efficient tools are available for any engineering activity, starting from the first sketch in a model space to finished recycling. Engineers define product in a model space where high number of parts, assemblies, analysis results, manufacturing plans, and other engineering objects are defined and related.

I.J. Rudas et al. (Eds.): Towards Intelligent Engineering & Information Tech., SCI 243, pp. 543–556.
springerlink.com   © Springer-Verlag Berlin Heidelberg 2009

Groups of engineers work on groups of engineering objects. Sometimes, several engineers work on the same group of engineering objects, concurrently. Concurrent engineering is organized to minimum innovation cycle so that results are applied at subsequent activity immediately. Engineers are in graphical dialogue with model creation processes and give and relate engineering object parameters. These are results of thinking process in course of which engineer consider intent of other engineers, the already decided engineering objects, and accepted and verified knowledge. Engineers can not communicate background of their decisions with modeling processes and thus background is not represented in product model. The only capability is definition of information, i.e. data and relationships.

The above characterized engineering realizes a high level of object modeling and is a result of development of product modeling, product information management, interoperability of systems, and human-modeling process communication. However, extensive attempts for the application of knowledge based methods in order to assistance of decision and improving of human-computer interaction have not resulted substantial advances in organized modeling of background of decisions. Consequently, while knowledge based methods are applied in relating of the object parameters successfully in the form of rule and check sets, neural networks, optimizing algorithms, etc., the answer is still for the question *what* and not for *how* and *why*. Human intent, motivation, and arguments are not recorded in the model and are not available for subsequent decision making. Work of engineer who revises decisions or continues work along a chain of decisions is not supported by representation of background of decisions.

The authors recognized the above problem as a definite need for new model entities that represent background of decisions during engineering activities. In order to achieve this, thinking process of engineers during engineering object decisions were analyzed. A new modeling was defined between the current information based modeling and dialogues on human communication surfaces.

Information-based product model in current engineering systems has reached a high level of its development. Practically any engineering object can be described and related from parts and their relationships to manufacturing, production and marketing of a product. In order to differentiate from its proposed extension and to consider it as a milestone, information-based product model is called as classical product model.

In this paper, the authors propose a contribution to solve the above problems by establishment of an extension to current product modeling that can be accessed through standard API. They outlined the problem and background in management of product information and conceptualized human thinking driven definition of product entities. Following this an extension to classical product modeling is proposed towards intelligent content, for information content driven control of conventional model entity generation, and for implementation in industrial PLM systems.

2 Problem and Background in Management of Product Information

In current industrial product lifecycle management (PLM) systems, product model includes a multilevel structure where engineering objects (EOs) are related and lower level objects are listed below a higher level object (Fig. 1). For example, a surface is followed by curves that was applies for its creation, constraints placed on the curves and on the surface itself, etc. This model is hard to survey because of the high number of object representations and relationships. Model entities and knowledge are represented as parameters and their relationships. Knowledge is also stored as information. In connection with human intent, the way to a decision is not included in the product model.

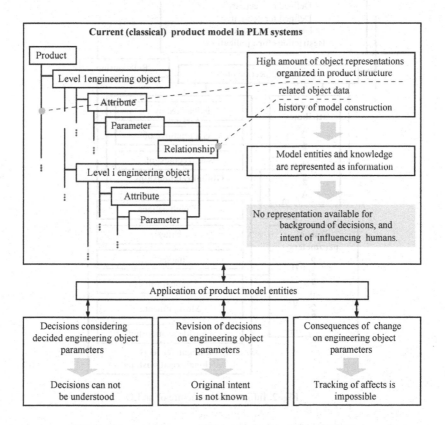

Fig. 1. Current product modeling: structure, entities and problem

The above model is representative for recent industrial product modeling. It is not suitable for efficient support of human decisions at application of product models in these systems. In order to support this statement, three applications are considered (Fig. 1). It was recognized, that EOs are placed in contextual chains in product structures. They can not be defined individually to match some specification.

Instead, they are defined in context of other engineering objects. When EOs in which context they were defined change, they change accordingly. Consequently, definition of an EO depends on other EOs and possible effects of decisions on those EOs also must be considered. The problem is that an engineer who considers early decided EOs in context of which a new EO must be defined, can not access to background of original decisions. Earlier decisions cannot be understood. The problem is the same at revision of parameters of EOs. Tracking of affects in contextual chains is practically impossible. Although contextual chains can be revealed in product models, representations are not available for original intent and for the way to decision in product model. Definition of a new EO can not be connected to earlier decisions organically in current product modeling.

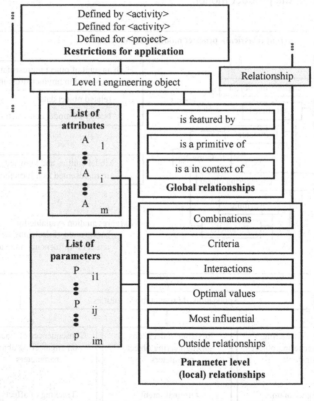

Fig. 2. Information structure of an EO

Even if background information would be recorded in product model, high number of EOs in product structure graph and increasingly huge number of unstructured EO relationships would make the related processing impossible. To solve this problem, a second recognition by the authors is existence of primary EOs in the contextual EO system. Decision on primary EOs including only primary parameters defines other parameters of primary EOs and a group of EOs those are associated to the primary EOs. This frame structure of EOs would make product model transparent if background representations would be available for human decisions. Before development of solution for this, it must be emphasized that automated decisions most represent humans with full responsibility for the product. Consequently, background representation must be relied on intent of humans who affect EOs.

On the way to a possible solution for the above problem, first step was analysis of information represented in an EO. On the basis of this analysis, an extended EO definition is developed (Fig. 2). EO is placed in one of the levels of product structure. It is characterized by its attributes. An attribute is described by one or more parameters. EO is connected to other EOs by relationship definitions. In other to make a first step towards better organized relationship definitions, the authors proposed to divide EO related relationships into local and global groups. Local relationships are defined on the level of EO parameters, while global relationships place EO in the structure of EOs. Local parameters record results of advanced parameter level analyses such as combinations and interactions of parameters, most influential parameters, etc. Global relationships record feature, primitive and contextual positions of an EO in a product structure. Finally, restrictions can be defined for the application of an EO in relation to activities and projects.

During the recent years, in efforts for better understanding of product model and improving decision assistance, high number of research results helped development of product modeling. Some of relevant works are cited below representing information modeling, change management, preservation of design intent during product data management, knowledge based product development, images as data sources of model representations, configuration and adaptation of products, and collaborative design process amongst different firms.

Authors in [1] describe a product information modeling framework in order to support the full range of PLM information needs. The framework is based on the NIST Core Product Model (CPM) and its extensions. It is intended to capture product, design rationale, assembly, and tolerance information to the full lifecycle.

In [2], it is emphasized that in redesign and design for customization of products are changed. During product development process, a change to one part of the product often results changes to other parts. An analysis of change behavior based on a case study in rotorcraft design is introduced. Mathematical models were developed to predict the risk of change propagation in terms of likelihood and impact of change. Likely change propagation paths and their impact on the delivery of the product were analyzed.

Authors of [3] emphasize importance of construction history, parameters, constraints, features, and other elements of design intent and suggest implementation of product model data exchange with the preservation of design intent, based on the use of newly published parts of the International Standard ISO 10303 (STEP).

Product data management (PDM) integrates and manages all the product objects. In [4], web-based PDM systems are reviewed. The PDM methodology is integrated with web architecture. Currently available PDM systems those have been integrated with web-technologies are reviewed.

In [5], a knowledge-driven collaborative product development (CPD) system architecture is proposed which will facilitate the provision of knowledge in product development. This work is motivated by a definite need by distance product development for information and knowledge in the place, time and format required.

Authors of [6] apply graph-theoretical formulation in order to bridge the gap between image and model representations in the area of automatic acquiring of generic 2D view-based model from a set of images.

Product line engineering is developing to create lines efficiently [7]. In order to achieve efficient reconfiguration, adaptation must be done at a generic level. In [7], an approach is described for analysis and specification of features that vary as a part of reconfigurations at runtime in case of software products.

The topic of [8] is creating a coherent set of ontologies to support a collaborative design process amongst different firms which develop mechatronic products is a challenge due to the semantic heterogeneity of the underlying domain models and the amount of domain knowledge that needs to be covered. In order to manage the complexity of the modeling task they separate the models into the foundational layer, the mechatronic layer consisting of three domain ontologies, one process model and one cross-domain model, and the collaborative application layer. For the development process, they employ a methodology for dynamic ontology creation, which moves from taxonomical structures to formal models.

3 Human Thinking-Driven Definition of Product Entities

Modeling of background of decisions on engineering objects makes record on concepts, methods, knowledge those are defined by humans who have an influence on EO definitions. Source of decision background is human thinking process while human uses experience and expertise. Starting from this approach, the authors analyzed and characterized products object modeling in recent PLM systems in [9]. They recognized primary role of human intent model and analyzed possibilities for relevant descriptions [10] and developed methods for organized description and behavior based definition of EOs [11].

The authors conceptualized modeling of background of decisions on EOs as information content (Fig. 3). Human is identified then gets access to modeling on the base of role. Following this, human defines intent for the subsequent activity in the process of product development. The base of human intent representation is characteristics of human intent and elements from human thinking process and partial decision points. Information content is defined in accordance with human intent and prepared for assistance of decisions on engineering objects. Decision making function generates adaptive activities to control processes those generate information for the description of EOs in classical product model.

Typical examples for elements of thinking process and interim decision points are shown in Fig. 4. In stage A of a thinking process human observes, perceives, and retrieves things about an EO or a group a EOs. During the stage *B*, think about EO, understands, considers, and inferences. Following this, human conducts experiments, verifies results, and then repeats stage *B*. At the end of thinking process interprets results. In the meantime, results are recorded at the end of each partial decision point. Result for any partial decision point must include content that can be processed into adaptive action that carries decision information for the control of definition of EO information. Fig. 4 shows a chain of interim decisions starting from a goal definition, placing the defined object in taxonomy, including earlier experience in product model and applying a procedure according to a selected algorithm and a method that defines the application of the selected procedure. A rule to help at the definition of a critical parameter, a dependence definition, and finally a range of the critical parameter within which the intent holder engineer accepts the computed result are additional partial decisions.

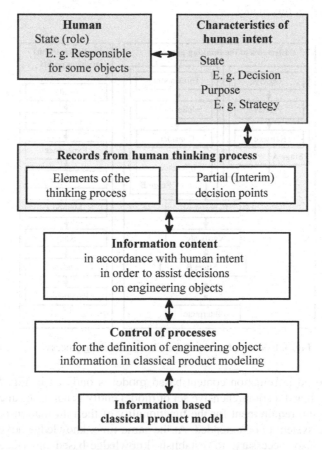

Fig. 3. Modeling of information content

4   Extension to Information-Based Product Modeling towards Intelligent Content

Captured and automatically operated intelligence in engineering systems is an old wish in product related engineering. The aim is to establish modeling functions in a modeling system in order to partly or fully substitute an engineer or a group of engineers at some decisions on EOs. In order to realize this aim, model must represent intent of all influencing humans at any conditions. Knowledge-based advising functions in current modeling systems, as it will be discussed later in chapter 5, use some simple knowledge definitions at the definition of EO parameters. However, these tools can not substitute but only assist engineer at decisions. Aim of the research that is reported in this paper is powerful assistance of engineers who apply model defined by other engineers. In the proposed modeling, engineer places information content in product model that is enough to make decisions at the application of this model.

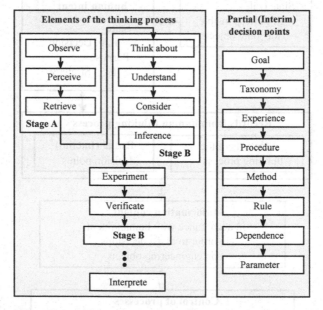

Fig. 4. Examples for record about human thinking process

The proposed information content-based model is outlined in Fig. 5. Product structure graph and a comprehensive set of model entity generation functions constitute minimum requirement for the functionality of the information-based product modeling system to be extended. At the same time, knowledge advisory functionality is also necessary to establish knowledge-based interface between information content and information-based sectors of a product model.

In order to allow a better surveying of object descriptions and their relationships than in the current information-based product models, a new multilevel structure of product information is created and applied. The purpose of this structure is to separate identification, application, associative connection, object description, and representation data. Adaptive actions communicate change decisions with this structure, coordinated in this structure, and executed from this structure. Advantage of this structure is that it organizes connections of EO descriptions (Fig. 2) with its application, relationships and representation. Besides a contribution to solving of the problem of unstructured relationships, this structure helps at the creation of instances from generic model entities.

A decision on a parameter or a parameter group of an EO or a group of EOs must be relied upon human intent. The question is that what representations are

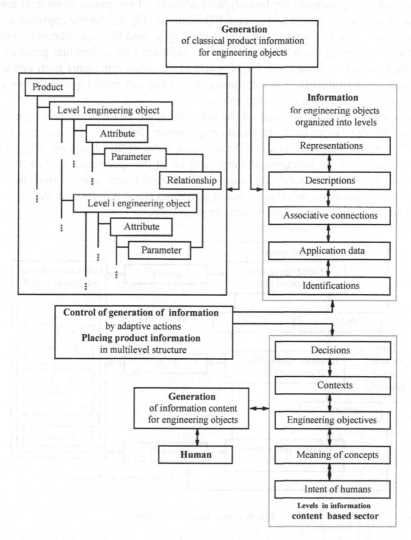

Fig. 5. Information-based product model extended with information content

necessary to process human intent representation into decision. The authors proposed application of new representations for meaning of concepts, engineering objectives, and contexts. Human defined intent and concept are applied at the definition of engineering objectives. An engineering objective is represented as a behavior and controls decision making on EOs taking contexts into consideration. Intuitive contributions to product development and model construction must be allowed integrating them into the above structure or placed in the model with a notice for responsibility of authorized person.

The next question in information content representations is that how contextual connections can be represented and handled. The selected solution is definition of contextual connections on the basis type of relations. Two groups of context entities are for content-content and content EO relations (Fig. 6). By the application of this method, pairs of content types and content types and EOs are related in order to achieve a correct model that can provide engineer with contextual generation and background of context. In Fig. 6, product structure represents both product structure graph in the classical information-based product model and the proposed multilevel representation of product information.

Few more words are devoted for the behavior space below. It includes specifications for EOs in order to represent engineering objectives. This approach is based on the recognition that any EO must serve some of the objectives. Objective definitions start from functions and represent all of the specifications. Functional structure description capability in recent information-based product modeling is necessary. It contains functions connecting engineering objects and thus it can serve as connection surface to engineering objectives.

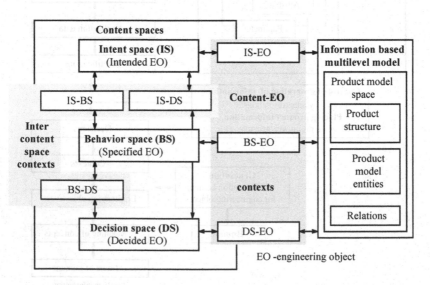

Fig. 6. Contextual connections

5 Information Content-Driven Control of Conventional Model Entity Generation

By the application of information content and multilevel structure of product information, surface for human dialogues is repositioned from the direct definition of EOs in classical product model to definition of human intent as initiation of information content processing chain. The original position of human as source of information for the control of EOs is taken by adaptive actions. The first question in this area is that where information content-based model can be interfaced with classical product model. Fortunately, recent developments in relating, grouping, and parameter definition capabilities of expert advising functions of classical information based product modeling offer a suitable surface for this purpose (Fig.7).

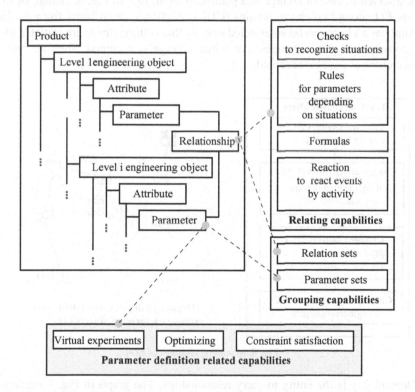

Fig. 7. Expert advising functions in classical product modeling

Main capabilities for relating EOs are rules, checks, reactions, and formulas. A rule is included in a product model for the generation of engineering object parameters depending on a situation. A check is defined to recognize a situation and advises modification of model information. A reaction is defined to react events by specified activity. Formulas establish relations amongst engineering object parameters.

Capabilities are available for the definition of parameters of engineering objects. Virtual experiments are planned and executed by performing computations of output parameters in case of different values of input parameters in defined ranges of analyzed parameters. Optimizing of parameters is performed by processes considering built-in or user specified algorithms. Grouping capabilities are for the definition of arbitrary sets of parameters and relations help to organize these entities according to engineering objective definitions.

When engineering objectives are available together with contextual connections for a set of EOs, decision on a parameter of an engineering object must be done considering all the affected parameters of all the affected EOs. In order to establish a suitable and feasible tracking of affects by a change at the definition of an engineering object, the authors introduced the concept of change affect zone (CAZ). CAZ is the subset of EOs in a product model EOs in which have a chace to be affected in case of change at a parameter of an EO. In Fig. 8, change of EO_x affects EO_s along two change chains (CHC). A change chain starts from the last EO that has a chance to be an affected one. In this context, there are initiated (IC) and consequence (CC) changes. An initial change is accepted when all consequence changes are also accepted.

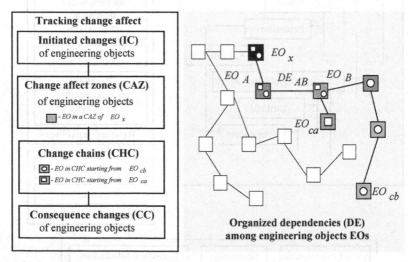

Fig. 8. Managing changes of EOs

Dependency is the entity to carry relationships. The graph in Fig. 8 organizes dependencies (DE) between pairs of engineering objects. For example, dependency DE_{AB} is defined between EO_A and EO_B. Two change chains are crossing at EO_B. Crossing connections are entities describing the actual adaptive actions and the crossing change chains together with the stratus of changes. Status of a change is changing during coordinating more or less interrelated decisions along change chains. It must be emphasized that this is not an automatic modeling process. Engineer is assisted by transparent structure and information content. On the other side, an advanced knowledge background record may allow for automatic decision making when responsible human can verify it. The feasibility of automation is not only scientific and technical problem. Secure and reliable product based on human intent is essential.

Status identifies the level of processing for a change of a parameter of an EO. There are two groups of change states. They are *under processing* and *concluded* ones. States are user defined according to local measurements. Under processing states may be, for example, *under discussion*, *under revision*, or *argued*. Concluded states are *accepted*, *decided*, or *executed*.

6 Implementation in Industrial PLM Systems

As it was stated above in this paper, the proposed modeling was defined as an extension to classical product modeling in professional PLM systems (Fig. 5). Implementation is outlined in Fig. 9. Extension to information-based product model represented in classical model space includes multilevel organization of information and three contextual information content spaces. The rest of extension is constituted by modeling functionality for information content. The programs covering this functionality can access modeling procedures, model data, and human interface through standard API. In other words, new elements of product modeling are built around the modeling in current PLM systems by using of open system functions in PLM systems.

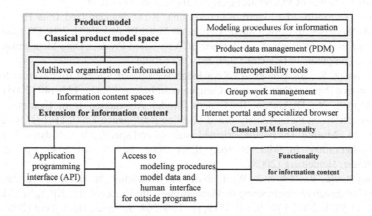

Fig. 9. Implementation of information content-based product model

Conclusions

In this paper, development of products together with other engineering activities during their lifecycle is discussed considering recent virtual environments. It was recognized, that recent PLM systems include product modeling that records and retrieves data for related engineering objects (EOs). This state-of-the-art modeling is defined as classical product modeling. The authors recognized that the main drawbacks in current product modeling are lack of background information on decisions on EOs at their application and high number of unstructured relationships among EOs. A development of current product modeling is proposed as a contribution to efforts to give new capabilities for the modeling systems in order to solve the above problems. Contributions by the proposed modeling are a multilevel organization of EO information, a new sector of product model for information content, and a new method for organizing change attempts from initiated changes to decided ones. In the meantime, effects of initiated changes are tracked in a change affect zone (CAZ), along change chains (CHC). One of the main objectives the reported work was to establish a human acceptable intelligent automation of product development that is suitable for engineers who are fully responsible for their work.

Acknowledgments. The authors gratefully acknowledge the grant provided by the OTKA Fund for Research of the Hungarian Government. Project number is NKTH-OTKA K 68029.

References

[1] Sudarsan, R., Fenves, S.J., Sriram, R.D., Wang, F.: A Product Information Modeling Framework for Product Lifecycle Management. Computer-Aided Design 37(13), 1399–1411 (2005)

[2] Clarkson, P.J., Simons, C., Eckert, C.: Predicting Change Propagation in Complex Design. Journal of Mechanical Design 126(5), 788–797 (2004)

[3] Kima, J., Prattb, M.J., Iyerc, R.G., Srirama, R.D.: Standardized Data Exchange of CAD Models with Design Intent. Computer-Aided Design 40(7), 760–777 (2008)

[4] Tony Liu, D., William, X.: A Review of Web-based Product Data Management Systems. Computers in Industry 44(3), 251–262 (2001)

[5] Rodriguez, K., Al-Ashaab, A.: Knowledge Web-based System Architecture for Collaborative Product Development. Computers in Industry 56(1), 125–140 (2005)

[6] Keselman, Y., Dickinson, S.: Generic Model Abstraction from Examples. IEEE Transactions on Pattern Analysis and Machine Intelligence 27(7), 1141–1156 (2005)

[7] Lee, J.J., Muthig, D.: Feature-oriented analysis and specification of dynamic product reconfiguration. In: Mei, H. (ed.) ICSR 2008. LNCS, vol. 5030, pp. 154–165. Springer, Heidelberg (2008)

[8] Damjanović, V., Behrendt, W., Plößnig, M., Holzapfel, M.: Developing Ontologies for Collaborative Engineering in Mechatronics. In: Franconi, E., Kifer, M., May, W. (eds.) ESWC 2007. LNCS, vol. 4519, pp. 190–204. Springer, Heidelberg (2007)

[9] Horváth, L., Rudas, I.J.: Modeling and Problem Solving Methods for Engineers, p. 330. Elsevier, Academic Press (2004)

[10] Horváth, L., Rudas, I.J.: Bringing up Product Model to Thinking of Engineer. In: Proc. of the, IEEE International Conference on Systems, Man, and Cybernetics, Singapore, pp. 1355–1360 (2008) IEEE Catalog Number: CFP08SMC-USB, ISBN: 978-1-4244-2384-2, Library of Congress: 2008903109

[11] Horváth, L.: Supporting Lifecycle Management of Product Data by Organized Descriptions and Behavior Definitions of Engineering Objects. Journal of Advanced Computational Intelligence and Intelligent Informatics, Tokyo 11(9), 1107–1113 (2007)

In-Situ Investigation of the Growth of Low-Dimensional Structures

Ákos Nemcsics

Institute for Microelectronic and Technology, Budapest Tech
Tavaszmező u. 17, H-1084 Budapest, Hungary
nemcsics.akos@kvk.bmf.hu

Abstract. In this work, MBE grown GaAs-based low dimensional structures are investigated. We are dealing with the realtime tracking of their growth with the help of RHEED. Particular behaviours of RHEED phenomenon are explained with the help of computing model. Some aspects to behaviour of RHEED and to the growth of superlattices, quantum dots and quantum rings are detailed here.

1 Introduction

A nanostructure is an object which has at least one dimension of intermediate size between molecular and microscopic structures. So, it is necessary to differentiate between the number of dimensions on the nanoscale. Nanostructured layers have one dimension on the nanoscale. Only the thickness of the layer is ranged beteween $0.1 - 100$ nm. They are two dimensional (2D) structures e.g. quantum wells (QWs), Super lattices (SLs). The objects which have two dimension on the nanoscale, they are one dimensional (1D) structures e.g. nano wires (NWs). The structures which have all three dimension on nano-scale, they are zero dimensional (0D) structures e.g. quantum dots (QDs). These nano-scale objects are called jointing as low-dimensional structures (LDS). The electronic structure of the LDS differs from the electronic structure in three dimensions (3D).

These properties service possibilities for new applications e.g. quantum computer. In the last time the fild of quantum information processing has experienced extremly rapid progress. The quantum computer is a device for computation that makes direct of use of quantum mechanical phenomena, such as superposition and entlanglement, to perform operations on data. The semiconductor technology provides the richest set of tools for construsting implementation of quantum-bit (qubit). E.g., two coupled QDs can serve as qubits.

LDS have been widely studied for applications also in optoelectronics. Until recently they had not considered as the absorbers in solar cells. However, it has recently been suggested that higher energy conversion efficiencies are possible if such LDS are incorporated into the active region of solar cells. The efficiency of the solar cell can enhance drastically with the using of different LDS (QW, NW, QD).

I.J. Rudas et al. (Eds.): Towards Intelligent Engineering & Information Tech., SCI 243, pp. 557–572.
springerlink.com   © Springer-Verlag Berlin Heidelberg 2009

Scientific and technological development has made it possible to grow materials with different properties onto each other, and this way we can build QWs, quantum islands and QDs, etc. – which leads to the possibility of creating novel devices and applications. The MBE (molecular-beam-epitaxy) is the nearly exclusive technique of growth the above mentioned LDS. The technology of LDS needs in-situ observation of the growth process, which is widely realized by reflection high-energy electron diffraction (RHEED) technique.

The growth of perfect crystal layers and LDS needs the control of epitaxy. For this control we need the knowledge and understanding of growth mechanism. The RHEED pattern and its intensity oscillation carry in-situ information for us about the crystal growth. The getting of growth information from the reflected – diffracted pattern, we need the understanding of the interaction between the grasing incident electron beam and the growing crystal surface.

We will briefly deal with the basic pieces of information carried by RHEED. After that we will investigate some special aspects of the RHEED behaviour e.g. damping and phase change of oscillation. Finally we will discuss the relation between the formation of different quantum objects and RHEED.

2   Growth of Low-Dimensional Structures by MBE

The evaporation under ultra-high vacuum (UHV) is a classical method for preparing thin film. Depending on conditions, the deposited films can also be crystalline. In this case the preparation method is called MBE. When the crystalline film grows on a substrate different from that of the deposited material, so the process called heteroepitaxy. MBE technology is based on the controlled interaction of beams of atoms and molecules on a heated crystalline surface under UHV conditions. MBE is the most versatile method for preparing well defined surfaces, interfaces, layer structures and different nanostructures of elemental- (Si, Ge) and compound semiconductors (GaAs, InAs, etc.). MBE method allows also growth of film structures with sharp doping profiles and different chemical composition (e.g. AlGaAs) different lattice constant (e.g. GaAs/InGaAs junction). Multilayer structures with alternating doping or alternating band gap can be growth. A new field of band-gap-engineering was made possible by the development of MBE. Forthermore, QW, QD and other quantum structure can be grown with the help of MBE.

Different MBE variants have been developed. They are solid-source MBE (SS-MBE), gas-source MBE (GS-MBE), metalorganic MBE (MOMBE) etc. The MBE chamber is equipped with pump system that gives base vacuum pressure in the range of 10^{-11} Torr. The UHV conditon is required to reduce the back ground contaminant. It can be hinder the contamination of the deposited structure. The UHV enviroment quarantee the collosion-free path of the atoms and molecules from the source to the target. In the UHV chamber, the growth of the LDSs can be in-situ monitored.

3   RHEED as a Tool for In-Situ Growth Characterization

3.1   The Static and Dynamic RHEED

RHEED is a widely used monitoring technique during MBE growth. The orientation, quality and reconstruction of the grown surface can be determined by the RHEED pattern. Compared to other in-situ investigations, the glancing-incidence-angle geometry of RHEED is ideally for the in-situ observation of growth process and furthermore has very high surface sensitivity. The penetration depth of electron beam into the surface, can be changed by the variation of the incidence angle. The intensity of the RHEED pattern oscillates under appropriate conditions during the growth process [1]. One period of these oscillations corresponds exactly to the growth of one complete monolayer (ML) in the case of the layer-by-layer growth mode. The growth rate, and the composition in the case of alloy materials can be determined with the help of RHEED oscillations [Nemcsics-96].

RHEED and its oscillations of intensity are very complex phenomena. This technique is a versatile tool for in-situ monitoring, in spite of the fact that we do not know many details of its nature. This oscillation is characterized by the period, amplitude, phase and damping of the oscillations, the behaviour at the initiation of growth, the recovery after growth and the frequency distribution in the Fourier spectrum of the oscillations. The origin and describing of oscillation were investigated by several authors [2-5]. Many properties and behaviours of the oscillations are not yet understood. For example, some of these problems are the different phases of the specular and nonspecular RHEED beams [6], and the varied behaviour of the oscillations in the case of III-V and II-VI materials [7]. There are still more interesting and open problems in the topic of the decay of intensity oscillations and of the inicial phase change, etc. Several authors have tried to describe these phenomena [8]. Several effects can be interpreted by the above mentioned geometrical desription. The oscillations in the case of two-dimensional growth mode, the disappearence of oscillations by step propagation, and the exponential decay of the oscillations can be explained by this geometrical description, too [8,9].

The RHEED is a suitable technique to investigate the binding properties on the surface. Furthermore, it is appropriate to distinguish special crystallographical directions, for which X-ray diffraction (XRD) is unable. This crystalographical direction are e.g. on the (001) surface, where the [110] and [1$\bar{1}$0] directions have different properties in crystalographical and in technological view, too. However the dynamic RHEED is approritate to determine and calibrate the growth parameters, such as e.g. growth rate and molecule flux [9, 10]. One of the interesting problems is the initial phase change of the oscillation, which is depend on the direction of the incident electron beam, and the technological parameter, too.

The amplitude and period of the initial swing of RHEED oscillations are different from what follows. Except for the first period, the measured decay of the oscillations fits well to an exponential function [9]. The incident electron beam impinges on the surface with low angle. If we change the incident or the azimuthal angle, the initial phase of the oscillations changes, too. For the phase change investigations, GaAs (001) serves as good model material [6, 8]. There are several models, which describe this phenomeon in broad range of incidence angle [11, 12]. These generally scattering based models take into account more MLs. These approaches describe qualitatively well the phase switch in the range of incident angle to about 4° but the fitting e.g. at low angle under 2° has some deviation from the measured curve.

3.2 Particular Behaviour of RHEED

The phase of the oscillation depends on the incident angle of the electron beam. However the very useful kinematic theory does not predict the phase shift of the oscillations, which depends on the condition of the electron beam. The contribution of inelastic processes such as Kikuchi scattering to the phase shift penomenon is not completely taken into account [6]. The RHEED phenomenon is partly reflection-like and partly diffraction-like. The effect of phase shift is described by the position of the minima of the oscillations. The behaviour of the minima and maxima of the oscillations can also be explained with a geometrical picture, which will be employed in this case. Because the specular spot is not a reflected beam, the interaction of the electron beam and the target surface must be described quantum mechanically. The glancing-incidence-angle electron beam touches the surface over a large area. The reflected-diffracted information obtained does not come from the whole area. The interaction between the surface and the electron beam exists only under special conditions, therefore we need to consider the surface coherence length (SCL) (w) [13]. The wave function after the interaction $|X_a\rangle$ is written as follows:

$$|X_a\rangle = W_B(r_1)W_T(r_2)|\chi_{\bar{a}}\rangle$$

where $W_B(r)$ and $W_T(r)$ are the wave packet of the electron beam and the interacting surface of the target, respectively. With the solution of this equation it can be shown that the SCL depends on the interaction potential between the incident electron and the target and depends less on the wave packet [13].

We can suppose that the SCL (w) is of the same order as the coherence length (Λ) of the beam. The energy of the electron beam is on the order of $E = 10$ keV, with a de Broglie wavelength $\lambda = 12.2 \cdot 10^{-12}$ m. The coherence length of the electron beam Λ is between 12.2 nm and 3.7 nm [14]. The spot size of the illuminating electron beam on the surface in the incident direction depends strongly on the incident angle. We can suppose that the SCL depends on the incident angle, too. The relation between the size of characteristic growth terrace (s) and the SCL (w) in the case of a polycrystalline surface was investigated [13]. This concept can be applied in this case. An estimate of the characteristic dimension of a growth terrace can be given from experiments. The terrace average width (s) and the migration length of Ga (l) depend on the substrate temperature. The RHEED oscillations are present if $l \leq s$ and absent if $l \geq s$. In our case the migration length is 7 nm because the substrate temperature is 580°C [8].

The binding energy on the (001) surface in the directions [110] and [1$\overline{1}$0] is not the same, which explains the dangling bonds. This anisotropy is manifested in the different growth rates. The growth rate in the [110] direction is larger than that in the perpendicular direction [15]. This anisotropy is apparent not only in the growth of the crystal (in other words composition of the crystal) but also in the etching (that is decomposition) of the crystal. The growth rates in the [110] and [1$\overline{1}$0] directions are different by more than a factor of two. This factor can be estimated with the help of the ratio of the etching rates of both directions.

We can suppose that the SCL and the average terrace width have commensurate dimensions at glancing-incidence-angles ($w \approx s$). This supposition seems right, because the touching length of the electron beam (also the SCL after our supposition) changes very abruptly at angles less than 1° and in this region the function $t_{3/2}/T$ is constant accordingly as $w > s$. The relation of SCL and average terrace width is changed with changes of the incident angle. If the incident angle increases, the SCL becomes smaller than the average terrace width ($w < s$), so and thus reflected-diffracted information comes from only a part of the average terrace.

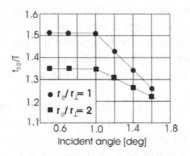

Fig. 1. The computed t3/2/T ratio vs. Incident angle in different crystallographical directions

For the calculation we used the polynuclear growth model in the two-dimensional case [5]. The simplified picture takes into consideration diffraction contributions only from the top most layer and the RHEED intensity is taken as proportional to the smooth part of the surface top layer [9]. The computational model is based on lattice node arrangement [16]. The surface area supplying the information decreases with the increasing incident angle of the beam. The different crystallographic directions mean different growth rates. Here the ratio $r_{[110]}/r_{[1\overline{1}0]}$ is estimated to be 2.4 at growth temperature (600 °C). The oscillations were computed for two different ratios of $r_{//}/r_{\perp}$, where $r_{//}$ and $r_{\perp}$ are the components of the growth rate in the observation direction (parallel with the electron beam) and the perpendicular direction, respectively. The calculated function of $t_{3/2}/T$ versus azimuthal angle in the two different directions is shown in Fig. 1.

The growth time for one complete monolayer in the two different directions is the same (T), but the phase is different ($t_{3/2}$) because of the anisotropic growth rate. These curves correspond with the measured data. If the SCL is larger than the average terrace width then the $t_{3/2}/T$ ratio remains constant (which constant value is determined by the $r_{//}/r_{\perp}$ ratio). If the SCL is smaller than the average terrace width, then the $t_{3/2}/T$ ratio decreases, too.

The behaviour of $t_{3/2}/T$ vs. incident angle was investigated under 1.8° glancing-incidence-angles. In real situations, the diffracted-reflected electron beam gets information not only from the topmost monolayer. A larger incident angle causes a larger penetration depth. The description of this phenomenon probably can be improved in either range by considerating more monolayers below the surface during the growth process. The calculated curves for the decay of the oscillations correspond very well with the measured data in the investigated range.

3.3   Some Aspects to the Deading of the RHEED Intensity Oscillations

Here, we investigate the decay of oscillation in the case of InGaAs growth on GaAs (001) surface. With the increase of In-content the decay of oscillation became faster. The least-squares-method was used to fit an exponential function to the decay of intensity. Except for the first period the measured decays fitted well to an exponential function. The first period is characterized not only by a period length different from that of the following oscillations, but also by an unusual intensity.

The exponential decay can be roughly modeled within a simplified picture, which only takes into consideration diffraction contributions from the topmost layer. We can suppose that at the start of the growth the surface of area A is perfectly smooth, i. e. $\Theta_0 = 1$, where Θ_0 is the coverage of the uppermost layer. In order to describe the increase of surface roughness during growth, we introduce a coverage reduction factor b, which gives the reduction in coverage from layer to layer. This factor may be regarded as a measure for the smoothness, therefore b can also be considered as epilayer quality ideality factor [9]. The first layer grown on the perfect surface is not complete, but has a coverage $\Theta_1 = b$ with $b < 1$, but close to unit. The next layer has $\Theta_2 = b^2$, or generally the nth layer:

$$\Theta_n = b^n$$

If in a first approximation the RHEED intensity I is taken as proportional to the smooth part of the surface top layer, the intensity of the perfectly smooth surface is given by $I_0 = c \cdot A$, where c is a constant which characterizes the diffraction power. After the growth of the first layer the intensity is $I_1 = b \cdot c \cdot A$, and generally after the nth layer:

$$I_n = b^n cA$$

A continuous description results from the replacing of n by $r \cdot t$ yields $I(t) = b^{rt} c \cdot A$, where r is the growth rate given by $r = 1/\tau_{InGaAs}(x)$ and t is the growth time. This can be written in the following form:
$$I = cA \exp(-t / \tau_d)$$
where τ_d is the decay time:

$$\tau_d = \frac{-1}{r \ln b}$$

with $\tau_d > 0$, since $b < 1$. In the fit of the experimental intensity decay the fit parameter was the coverage reduction factor, the growth rate being calculated from the period of the RHEED-oscillations. The growth rate increases with increasing composition x and therefore the decay time τ_d should decrease with x, if b is constant. The decay time constant is a function of composition, in agreement with the experimental results [9]. Calculating the coverage reduction factor from the experimentally observed decay time, one can see that b is not a constant with respect to composition x but decreases with increasing x. For this behaviour a qualitative explanation can be given: If the composition increases, the growth rate increases due to the preparation conditions. Therefore, the Ga and In atoms at the surface have less time for diffusion along the surface resulting in less perfect layers, i. e. in a smaller coverage reduction factor b. If a linear relation between b and $1/r$ is assumed, this results in:

$$\frac{b(x_i)}{\tau_{InGaAs}(x_i)} = \frac{b(x_j)}{\tau_{InGaAs}(x_j)}$$

where $b(x_i)$, $b(x_j)$ are the coverage reduction factors and $\tau_{InGaAs}(x_i)$, $\tau_{InGaAs}(x_j)$ the period lengths of RHEED-oscillations for composition x_i and x_j, resp. The decay time as a function of composition, i. e. $\tau_d(x) = -1/r(x)\ln(b(x))$, from the experimental results one can calculate $b(x)$. With $\tau_{InGaAs}(x) = 1/r(x)$ known from the RHEED-oscillations we obtain $b(x)/\tau_{InGaAs} = 0.30 s^{-1}$ within an accuracy of 10%. From this we may conclude that the model used here in spite of its simplicity gives a satisfying first order explanation of the RHEED-intensities observed in the experiment.

4   Some Aspects to the Production of the Heterojunctions and Nanostructures

4.1 Growth of Strained Heterostructures

The critical layer thickness (CLT) in the heteroepitaxy was investigated experimentally as well as theoretically [17]. Under real growth circumstances the measured CLT depends not only on misfit but on the growth parameters, too [18]. RHEED intensity oscillations are used for accurate determination of the threshold layer thickness in 2D growth mode which is technology dependent, too. The lattice mismatch also affects the behaviour of RHEED oscillations. The oscillation amplitude decreases rapidly during the InGaAs growth due to its strong dependence on the InAs mole fraction. This decrease also indicates the formation of the 3D growth mode.

The magnitude of intensity oscillations continously decreases during the growth process. With the increase of In composition not only the growth rate becomes faster but the decay of oscillations, too. With increasing composition x the decay time τ_d decreases. The measured value of the decay time constant is a function of composition $\tau_d(x)$, where the exponential function was fitted by the least-squares method. The fitted function is $\tau_d = a \cdot \exp(-x/b) + c$, where the parameters in our case are $a = 24.85$, $b = 0.103$ and $c = 4.1$ [9].

The measured values of the decay time constants depend on the composition. Most probably there is a relationship between the behaviour of the RHEED oscillation decay and CLT. The oscillation amplitude decreases rapidly during InGaAs growth due to the 3D growth which has a strong dependence on the InAs mole fraction. Not only the mismatch is responsible for this decay but the other technological conditions, too. The decay of oscillation amplitude exists even without mismatch e.g. at pure GaAs layer growth. The changes of the In content modify not only the mismatch but the conditions of growth (e.g. growth rate), as well. Therefore, the misfit and the technological parameters have a joint influence on the behaviour of amplitude decay.

The exponential function is a good approximation of the $I(t)$ function at every x values. Provided that both processes are independent from each other the phenomenon of decay at arbitrary x values can be described by two time constants as follows:

$$I(t) = A_0 \cdot \exp\left(-\frac{t}{\tau_d}\right) = A_0 \cdot \exp\left(\frac{-t}{\tau_M} + \frac{-t}{\tau_G}\right)$$

where τ_M and τ_G are the assumed time constants of the separated part effects which are responsible for the influence of misfit and the growth respectively. According to this model, the x dependent decay from misfit can be written as follows:

$$\frac{1}{\tau_M} = \frac{1}{\tau_d} - \ln\left(\frac{A_0}{A}\right) \cdot \frac{1}{t}$$

where the $\tau_d(x)$ and $\tau_M(x)$ time constants are functions of the composition x, and the $e(x)$ term contains the dependence from decay constants. In the case of pure GaAs growth ($x = 0$) there is no decay caused by misfit. It means that at $x = 0$ the reciprocal value of the decay time constant originates entirely from crystal growth phenomenon. Function $\tau_d(x)$ is known. The value of $1/\tau_M$ at $x = 0$ is zero. As a first approximation, the $1/\tau_M$ function describing the mismatch can be obtained by moving the $1/\tau_d$ function in negative ordinate direction. We can easily determine the $\tau_M(x)$ function from its reciprocal form. In order to be able to compare the calculated decay with the theoretical CLT we should change to layer thickness instead of period time because now we only have function of time vs. x. The grown thickness under one period of oscillation is one monolayer which corresponds to half of the lattice parameter. After evaluating the function of thickness vs. x instead of the function of time we can compare the calculated threshold thickness with the theoretical CLT. There is a good agreement between our calculation and the isotropic and anisotropic model [19].

As a first approximation it was supposed that the dependence of $1/\tau_G$ on the composition (x) is not significant. Although we have received similar course between the theoretical CLT and our calculated threshold layer thickness one can find difference between curves as well. The reason of the deviance can be in the fact that the τ_G also depends on the composition.

We can investigate the technological influence in the case of pure GaAs growth. We can increase the growth rate by increasing of the Ga flux. Let us see an example! At the Ga source temperature of 935 °C the growth rate is $r = 3.85$ 1/s ($\tau_{GaAs} = 2.61$ s) [3]. In this case the decay of oscillation can be described with decay constant $\tau_d = 15.3$ s, where -because of the homoepitaxy - τ_M is missing therefore τ_d and τ_G are equal. The above mentioned growth rate corresponds to the case of InGaAs growth ($\tau_{InGaAs} = 2.61$ s) $x = 0.13$ where ($T_{Ga} = 925$ °C and $T_{In} = 540$ °C). With the increase of the growth rate the decay constant decreases. In the case of InGaAs growth the increase of growth rate means simultaneous increase of InAs mole fraction, too. The more pronounced decay of oscillation can be traced back not only to the change of growth rate but to change in the In content as well. The course of RHEED intensity oscillations strongly depends on the observed material. The change of τ_G can be caused by the change of growth rate and/or change of In/Ga ratio. The supposed tendency of τ_d is supported by both above mentioned phenomena. The threshold thickness is defined as the thickness where the oscillation becomes weak enough. The supposed dependence of τ_G on the growth rate / composition seems valid because the deviance between CLT and the calculated data becomes less. The growth rate dependence was proved in the case of InGaAs growth by in-situ XRD measurement where the deposition temperature was 490 °C [20].

This description shows only the right tendency of the τ_G dependence. Reality is more complicated. We have not taken into account e.g. the material contrast and the different sticking coefficient of Ga and In. The CLT derived from this assumption shows relatively good agreement with CLT data from the literature. We can improve this agreement by supposing the composition / growth rate dependence of τ_G.

4.2 The Strain Induced Quantum Dot Growth

RHEED is also appropriate to observe the QD formation and to determine some parameters of QD. Nowadays, the production of self-assembled quantum structures has been intensively investigated for basic physics and device applications, too. It is very important to understand their growth process and the knowledge about their shape is particular significant. The archetypal system of these nano structures is the lattice-mismatched system such as InAs on GaAs, where the strain-induced process leads to the formation of QDs [21, 22]. Electronic structure of QD, which governs electronic and optical properties, depends on the shape [23]. In-situ information about the dot shape has been provided by RHEED. The chevron-shape spot appear in RHEED patterns from self-assembled InAs QDs grown GaAs (001) surface at $[0\bar{1}1]$ azimuth direction [24]. It is generally agreed that the essential driving force for coherent lattice mismatched QD formation is strain relaxation [25].

Besides the above treated 2D-3D transition [26], we can investigate the wetting layer (WL), characteristic times of the formation process [27], critical coverage of the layer, segregation of the components during QD formation [28], shape of the dot, change of the lattice mismatch and surface lattice parameter [29]. For example, the characteristic times of the QD formation can be determined with the help of pattern change and the intensity of RHEED picture. The shape of the formating dot can be investigated also by the pattern and by the so called chevron-shape spot of RHEED [29]. The RHEED gives more possibilities to investigate the parameters. For example, the 2D-3D transition can be investigated not only with the help of oscillation decay [27] but also with the help of change of diffraction pattern [29, 30] and chevron-shape spot [30].

The InAs/GaAs heteroepitaxial system has a lattice mismatch of ca 7% and under very specific conditions the growth mode follows a version of the Stranski-Krastanov (SK) mechanism, which results in the formation of coherent (dislocation-free) 3D islands, after the growth of between one and two MLs in a 2D layer-by-layer pseudomorphic mode. The 3D islands are usually referred to as self-assembled QDs. During the 2D-to-3D transition the island rapidly reach a saturation condition, with a rather narrow size (volume) distribution, where each growth conditions, principally deposition rate and substrate temperature. The essential driving force for coherent QD formation, after a WL has formed, is strain relaxation, whereby the energy gain from the increse in surface area via dot formation more than compensates the increase in surface area via dot formation more than compensates the increase in interfacial free energy.

In the case of InAs/GaAs QDs, the shevron-tails were attributed to perpendicular to the facets (reciprocal rod nornal to (113) and ($\bar{1}\bar{1}3$) facets because the angle between the two chevron-tails is about 55° [31]. For [1$\bar{1}$0] incident azimuth, an electron beam which enters via ($\bar{1}$13) facet and exits from the (1$\bar{1}$3) is refracted into the [00$\bar{1}$] direction. This gives rise to streaks perpendicular to the (001) surface. Alternative approach is based on refraction effect, which can explain the origin of RHEED pattern [29]. It was observed that refraction from inclined facets on small crystalites on a specimen surface should result in discrete displaced spots, not continuous streaks [31].

4.3 Development of Droplet-Epitaxial Quantum Dot

In the field of nanostructures, the self-assembly lattice matched quantum structures employing droplet epitaxy is an interesting and novel alternative to the establish technology of strain-driven QD formation [32, 33]. In the case of droplet epitaxial QD, two growth regimes depending on temperature were found and explained by the help of extended rate equation model [34]. It was the first to indicate the chevrons and their relation to the faceting in the case of droplet epitaxy [35]. The in-situ investigation of the evaluation of the quantum structures is very important to understand their growth kinetics.

The growth experiments were performed in a SS-MBE system on AlGaAs (001) surface. The sample temperature was 200 °C. The $\theta = 3.75$ ML Ga was deposited with the flux of 0.75 ML/s without As flux. The annealing was carried out at 350 °C temperature under the 5×10^{-5} Torr As pressure. The production of the QDs were tracked continuously in the direction of [1$\bar{1}$0] with the help of RHEED. In the case of initial stage of the surface, the RHEED pattern showed sharp streaks. After the Ga deposition, the pattern became diffuse on the RHEED screen. In the case of QDs, almost at the same time with the offering of As the RHEED pattern changed from diffuse to spotty. During the annealing phase, the pattern changed from spotty to spots with tails.

The sharp RHEED streaks disappear and the diffraction picture becomes diffuse during the Ga deposition. The deposited Ga is in liquid phase. The disappearance of the RHEED pattern originates from the appearance of the liquid phase (Ga droplet) on the surface. In the annealing phase of QD production, the RHEED pattern becomes from diffuse to spotty nearly simultaneously with the opening of As source. The sufficient As quantity (5×10^{-5} Torr) and the low temperature (200 °C) make the building-in (infiltration) of As in Ga phase that is the crystallization possible [36]. This process of infiltration takes about two-three minutes to the sharp chevron image. So, a crystallized shell comes into being on the Ga droplets. The appeared spotty RHEED pattern originates from the transmission electron diffraction.

The electron beam goes through the crystalline GaAs shell layers. It is observed, if there are crystallite formations or droplets on the surface, bulk scattering of the grazing beam can occur and the RHEED pattern may become dominated by spots rather than streaks due to transmission electron diffraction. The scattering from several planes modulate strongly the intensity along the reciprocal lattice rod. So, the streaks observed from two-dimensional surface are not observed when transmission dominates. For the transmission case, the reciprocal lattice is an array of points each broadened owing to the finite size of the scattering region. During the annealing, the As diffuses inside of droplets, while excess As builds in (infiltrates) in the shell [36, 37]. So, the droplet crystallizes trough slowly. At the same time, chevron-shape spot develops on the RHEED screen [35].

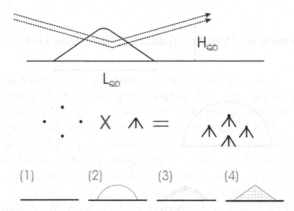

Fig. 2. Electron scattering on QD; Geomertical arrangement (upper part); Origin of the RHEED pattern (middle part); Temporal evaluation of the QD (lower part)

The side-facet angle of dropletepitaxial GaAs QD corresponds with the half angle between the two chevron-tails [35]. The angle between two RHEED streaks starting from same reciprocal lattice point is about 55°.

The observed RHEED pattern can be recognized by the product of square modulus of the separate Fourier-transform of the periodic constituents. In this case the intensity pattern is proportional with the product of intensity originated from transmission and from the pyramidal hut. A lucid explanation of the observed RHEED pattern and the temporal evaluation of QDs can be shown in Fig. 2.

4.4 Development of Droplet-Epitaxial Quantum Ring

Very recent experimental studies applying this droplet epitaxial technique demonstrate not only growth of QDs, but quantum rings (QRs) [38, 39] and even double QRs were grown [40]. Today is not yet available a theoretical description of the underlying growth mechanism. The growth experiments were performed also on AlGaAs (001) surface. The sample temperature was 300 °C. On the surface Ga was deposited as described before. During the annealing, the temperature remained the same (300 °C), but the As pressure was 4×10^{-6} Torr. The process were tracked also continuously in the direction of $[1\bar{1}0]$ with RHEED.

In the case of QR production, the course is the same to stage of Ga deposition. But after the Ga deposition, the change of the observed RHEED pattern is quite different. After the Ga deposition, the RHEED pattern becames diffuse. After the offering of As background of 4×10^{-6} Torr, the RHEED pattern develops slowly. The developed pattern contains in the middle a streak with a small spot and around elongated larger spots. The density of QDs and QRs are different, so the influence of the open (001) surface on the RHEED pattern is also different. The effect can be explained, similar like as in former case, with the appearence of the liquid Ga droplets on the surface. The annealing phase begins after the offering of As component with the pressure of 4×10^{-6} Torr, while the substrate temperature is 300 °C. This process takes about five minutes. During the annealing, the RHEED pattern consists of a streak with spots. The crystall structure and the form of the nano specimen influence together the RHEED pattern.

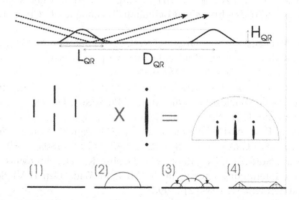

Fig. 3. Electron scattering on QR; Geometrical arrangement (upper part); Origin of the RHEED pattern (middle part); Temporal evaluation of the QR (lower part)

The volume of the QDs is large enough to receive transmission pattern during the electron scattering. The main lateral expansion L_{QD} and height H_{QD} of QD – according to the measurement – are 50 and 5 nm, respectively (see Fig. 2). The mean free paths of the electron Λ in GaAs between the crystal planes without collosion at the typical incidence angle of RHEED (about 2°) are less than 20 nm. So, in our case, there are several (ca. 9) lattice planes to receive transmission character. The situation in the QR is different. (The mean diameter of the principal ring D_{QR} is 60 nm.) The thickness of the ring L_{QR} - according to the measurement - is 20 nm and the height H_{QR} is 2 nm (see Fig. 3). In this case the measure of L_{QR} is comparable with Λ and the height H_{QR} consists only few planes. This is a marginal case where the dominate character of the diffraction pattern can be either transmission-like or reflection-like. The QR production takes long time during the annealing phase. The As background pressure is low here. So, the process of crystalization is slow. In this case, the liquid condition remains longer time, so there is enough time for the material transport, which leads to the formation of QR [41, 42]. Because, the crystalization occurs mainly after the transport process, only that can be demonstrated in the RHEED picture. Intermediate stages – because these happen mostly in amorpous state – provide the same diffuse RHEED pictures. The resulted RHEED picture is originated from the product of the reflection/transmission image and the diffraction image of the rotational-shaped object.

References

[1] Joyce, B.A.: In: Bauer, G., Kucher, F., Heinrich, H. (eds.) Two-Dimensional Systems: Physics and New Devices. Springer Series Solid State Sciences, vol. 67. Springer, Heidelberg (1986)

[2] Lent, C.S., Cohen, P.I.: Diffraction from Stepped Surfaces. Surf. Sci. 139, 121–154 (1984)

[3] Dobson, P.J., Joyce, B.A., Neave, J.H., Zhang, J.: Current Understanding and Application of the RHEED Intensity Oscillation Technoque. J. Cryst. Growth 81, 1–8 (1987)

[4] Clarke, S., Vvedensky, D.D.: Origin of Reflection High-Energy Electron –Diffraction Intensity Oscillations during Molecular-Beam Epitaxy: A Computational Modelling Approach. Phys. Rev. Lett. 58, 2235 (1987)

[5] Ishibashi, Y.: Intensity Oscillation of Reflection High-Energy Electron Diffraction (RHEED) by a Polynuclear Growth Model. J. Phys. Soc. Jap. 60, 3215–3217 (1991)

[6] Resh, J., Jamison, K.D., Strozier, J., Bensaoula, A., Ignatiev, A.: Phase of Reflection High-Energy Electron-Diffraction Intensity Oscillation during Molecular-Beam-Epitaxy Growth of GaAs (100). Phys. Rev. B 40, 11799–11803 (1989)

[7] Griesche, J., Hoffmann, N., Rabe, M., Jacobs, K.: Developments in the Use of RHEED for Interpreting Growth in the MBE of Wide Gap II-VI Semiconductors. Prog. Cryst. Growth 37, 103–121 (1998)

[8] Joyce, B.A., Neave, J.H., Zhang, J., Dobson, P.J.: Dynamic RHEED Techniques and Interface Quality. In: Larsen, P.K., Dobson, P.J. (eds.) Reflection High-Energy Diffraction and Reflection Imaging of Surfaces. NATO ASI Series B, Physics, vol. 188, pp. 163–183. Plenum Press, New York (1987)

[9] Nemcsics, Á., Olde, J., Geyer, M., Schnurpfeil, R., Manzke, R., Skibowski, M.: MBE Growth of Strained InGaAs on GaAs (001). Phys. Stat Sol (a) 155, 427–437 (1996)

[10] Heyn, C., Harsdorff, M.: Flux Control and Calibration of an As Effusion Cell in a Molecular Beam Epitaxy System for GaAs and AlGaAs with a Quadrupole Mass Spectrometer. J. Cryst. Growth 133, 241–245 (1993)

[11] Mitura, Z., Dudarev, S.L., Whelan, M.J.: Phase of RHEED Oscillation. Phys. Rev. B 57, 6309–6312 (1998)

[12] McCoy, J.M., Korte, U., Maksym, P.A., Meyer-Ehmsen, G.: Multiple-Scattering Evaluation of RHEED Intensities from the GaAs (001) 2x4 Surface: Evidence for Subsurface Relaxation. Phys. Rev B 48, 4721–4728 (1993)

[13] Beeby, J.L.: Theory of RHEED for General Surfaces. Surf. Sci. 80, 56–61 (1979)

[14] Nemcsics, Á.: The Initial Phase Shift Phenomenon of RHEED Oscillations. J. Cryst. Growth 217, 223–227 (2000)

[15] Horikoshi, Y., Yamaguchi, H., Briones, F., Kawashima, M.: Growth Process of III-V Compound Semiconductors by Migration-Enhanced Epitaxy. J. Cryst. Growth 105, 326–338 (1990)

[16] Nemcsics, Á.: Valuing of the Critical Layer Thickness from the Deading Time Constant of RHEED Oscillation in the Case of InAs/GaAs Heterojunction. Appl. Surf. Sci. 190, 294–297 (2002)

[17] Fitzgerald, E.A.: Dislocations in Strained-Layer Epitaxy: Theory, Experiment, and Applications. Mater. Sci. Rep. 7, 87–140 (1991)

[18] Grandjean, N., Massies, J.: Epitaxial Growth of Highly Strained InGaAs on GaAs(001): the Role of Surface Diffusion Length. J. Cryst. Growth 134, 51–62 (1993)

[19] Matthews, J.W.: Coherent Interfaces and Misfit Dislocations. In: Epitaxial Growth, Part B, vol. 559, pp. 273–280. Acad. Press, New York (1975)

[20] Gonzalez, M.U., Gonzalez, Y., Gonzalez, L.: In Situ Detection of an Initial Elastic Relaxation Stage during Growth of InGaAs on GaAs (001). Appl. Surf. Sci. 188, 128–133 (2002)

[21] Leonard, D., Krisnamurthy, M., Reaves, C.M., Denbaas, S.P., Petroff, P.M.: Direct Formation of Quantum-sized Dots from Uniform Coherent Islands of InGaAs on GaAs Surface. Appl. Phys. Lett. 63, 3203–3204 (1993)

[22] Bressler-Hill, V., Varma, S., Lorke, A., Nosho, B.Z., Petroff, P.M., Weinbergi, W.H.: Island Scaling in Strained Heteroepitaxy: InAs/GaAs(001). Phys. Rev. Lett. 74, 3209–3212 (1995)

[23] Stier, O., Grundmann, M., Bimberg, D.: Electronic and Optical Properties of Strained Quantum Dots Modelled by 8-Band kp Theory. Phys. Rev. B 59, 5688–5701 (1999)

[24] Zhang, K., Heyn, C., Hansen, W.: Ordering and Shape of Self-assembled InAs Quantum Dots on GaAs(001). Appl. Phys. Lett. 76, 2229–2234 (2000)

[25] Tersoff, J., LeGones, F.K.: Competing Relaxation Mechanisms in Strained Layers. Phys. Rev. Lett. 72, 3570–3573 (1994)

[26] Heyn, C., Hansen, W.: Desorption of InAs Quantum Dots. J. Cryst. Growth 251, 218–222 (2003)

[27] Krzyzewski, T.J., Joyce, P.B., Bell, G.R., Jones, T.S.: Wetting Layer Evolution in InAs/GaAs(001) Heteroepitaxy: Effects of Surface Reconstruction and Strain. Surf. Sci. 517, 8–16 (2002)

[28] Heyn, C., Hansen, W.: Ga/In- Intermixing and Segregation during InAs Quantum Dot Formation. J. Cryst. Growth 251, 140–144 (2003)

[29] Hanada, T., Koo, B.H., Totsuka, H., Yao, T.: Anisotropic Shape of Self-assembled InAs Quantum Dots: Refraction Effect Onspot Shape of Reflection High-Energy Electron Diffraction. Phys. Rev. B 64, 165307–165316 (2001)

[30] Belk, J.G., McConville, C.F., Sudijono, J.L., Jones, T.S., Joyce, B.A.: Surface Alloying at InAs/GaAs Interfaces Grown on (001) Surfaces by Molecular Beam Epitaxy. Surf. Sci. 387, 213–226 (1997)

[31] Pashley, D.W., Neave, J.H., Joyce, B.A.: A Model for the Appearence of Shevrons on RHEED Patterns from InAs Quantum Dots. Surf. Sci. 476, 35–42 (2001)

[32] Mano, T., Watanabe, K., Tsukamoto, S., Fujikoa, H., Ohshima, M., Koguchi, N.: New Self-Organized Growth Method for InGaAs Quantum Dots on GaAs(001) using Droplet Epitaxy. Jpn. J. Appl. Phys. 38, L1009-L1011 (1999)

[33] Lee, J.M., Kim, D.H., Hong, H., Woo, J.C., Park, S.J.: Growth of InAs Nanocrystals on GaAs(100) by Droplet Epitaxy. J. Cryst. Growth 212, 67–73 (2000)

[34] Heyn, C., Stemmann, A., Schramm, A., Welsch, H., Hansen, W., Nemcsics, Á.: Regimes of GaAs Quantum Dot Self-Assembly by Droplet Epitaxy. Phys. Rev. B 76, 75317–75324 (2007)

[35] Heyn, C., Stemmann, A., Schramm, A., Welsch, H., Hansen, W., Nemcsics, Á.: Facetting during GaAs Quantum Dot Self-Assembly by Droplet Epitaxy. Appl. Phys. Lett. 90, 203105–203114 (2007)

[36] Nemcsics, Á.: Behaviour of RHEED Oscillation during LT-GaAs Growth. Acta Polytechn. Hung. 4, 117–123 (2007)

[37] Mojzes, I., Sebestyén, T., Barna, P.B., Gergely, G., Szigethy, D.: Gallium Plus Metal Contacts to Gallium Arsenide Alloyed in an Arsenic Molecular Beam. Thin Solid Films 126, 27–32 (1979)

[38] Mano, T., Tsukamoto, S., Koguchi, N., Fujioka, H., Oshima, M.: Indium Segregation in the Fabrication of InGaAs Concave Disks by Heterogeneous Droplet Epitaxy. J. Cryst. Growth 227-228, 1069–1072 (2001)

[39] Gong, Z., Nin, Z.C., Huang, S.S., Fang, Z.D., Sun, B.Q., Xia, J.B.: Formation of GaAs/AlGaAs and InGaAs/GaAs Nanorings by Droplet Molecular Beam Epitaxy. Appl. Phys. Lett. 87, 93116–93124 (2005)

[40] Hwang, S., Nin, Z., Fang, Z., Ni, H., Gong, Z., Xia, J.: Complex Quantum Ring Structures Formed by Droplet Epitaxy. Appl. Phys. Lett. 89, 031921–031924 (2006)

[41] Mano, T., Watanabe, K., Tsukamoto, S., Fujioka, H., Oshima, M., Koguchi, N.: Fabrication of InGaAs Quantum Dots on GaAs(001) by Droplet Epitaxy. J. Cryst. Growth 209, 504–508 (2000)

[42] Liang, B.L., Wang, Z.M., Lee, J.H., Sablon, K., Mazur, Y.I., Salamo, G.J.: Low Density InAs Quantum Dots Grown on GaAs Nanoholes. Appl. Phys. Lett. 89, 43113–43114 (2006)

Monitoring of Heat Pumps

M. József Nyers and J. Lehel Nyers

Polytechnic Engineering College of Subotica, TERA TERM doo.
24000 Subotica, Vojvodina, Yugoslavia
jnyers@vts.su.ac.yu

Abstract. The aim of this paper is to stress the importance and necessity of monitoring heat pumps. The description and the analysis of individual activities in all the components of a heat pump are given from the point of view of monitoring problems. The components of the monitoring system are also analyzed, with respect to their operation and installation.

The schemes for installation and connecting the components of the monitoring system, depending on the strategy of the control system are given as well.

1 Introduction

A heat pump is a cooling device, operating on the principle of left-side cycle. It is used for transporting heat energy from lower to higher temperature levels by means of mechanic work or heat energy. In engineering practice, heat pumps with a compressor, that is those which operate by consuming mechanic work, have primarily been used. Devices with a compressor have a higher efficiency factor than absorption systems. With implemented systems that have a compressor the total efficiency factor reaches the value of 1: 3, 5.

2 Technical Description

Heating or cooling systems based on heat pump exchange heat energy with renewable sources, which are usually the outdoor air and groundwater.

The consumer of thermal energy is either a building or a technological process.

The system is comprised of three subsystems: the source, the transporting unit and the consumer of heat energy. A continual supply and removal of heat energy must be ensured for the continuous operation of the system. A flow of working mediums must exist in all three subsystems.

The analyzed water – water systems are made up of two water and one cooling cycle. The flow of working mediums is enabled by the pump in water circular flow and by the compressor in the cooling cycle. Working mediums are water and refrigerant, Freon.

I.J. Rudas et al. (Eds.): Towards Intelligent Engineering & Information Tech., SCI 243, pp. 573–581.
springerlink.com   © Springer-Verlag Berlin Heidelberg 2009

3   Monitoring Water – Water Heat Pumps

3.1   Introduction

The monitoring system is used for inspection of heat pump components by observing the relevant state parameters. Parameters show the state of the observed system components.

The monitoring system is at the highest hierarchical level, above the control system. In case the operation of the system is disturbed, the monitoring system through the control unit switches off the supply of energy into the system, thus preventing damage.

The lack of proper monitoring and intervention might lead to serious damage, and even explosion of the condenser, fracture of the compressor and freezing of the evaporator.

3.2   Monitoring the Condenser

Within the heat pump, the condenser is the most sensitive component in the system. Condensation is a self-regulating process, therefore without heat removal from the condenser the internal energy of the refrigerant and thus the pressure constantly increases. After a certain time, the pressure reaches a critical value and the weakest element of the transmission chain or the condenser fractures. A transmission chain is made up of a driving electric motor with a crankshaft, a piston rod with a pin, and the piston itself.

A flexible or copper tube which connects the compressor with the condenser is exposed to mechanical load resulting from the pressure in the refrigerant. It is desirable this tube to be the weakest component in the chain. If the pressure increases excessively and the tube ruptures, this leads to a damped expansion of the refrigerant from the system into the atmosphere.

With respect to explosion, shell-tube condensers are the most dangerous. A large amount of energy is stored in shell-tube condensers, expressed by p*V. The condenser shell, shield might drastically rupture so as to result in a sudden expansion of the refrigerant into the atmosphere, and the impact wave might cause great movement, collapse of the surrounding objects.

To eliminate this danger, it is necessary to connect the pressostat for monitoring pressure to the compressive branch of the compressor. The pressostat is set to the maximally allowed pressure in the refrigerant and if the working, operation pressure in the system reaches this value, it switches the system off.

Depending on the monitoring system, there are several ways of intervention.

3.3 Monitoring the Evaporator

In water-water systems, the evaporator is another component which should be monitored while in operation. The evaporator is filled with water, and if water does not flow it can freeze. Frozen water has a significantly greater volume, therefore the components of the evaporator are exposed to considerable mechanical load. If the resulting load is greater than the strength of the materials of the parts, it may result in fracture.

The above described situation occurs if the water in the evaporator does not flow. Water transports heat energy into the evaporator. Disturbances occur most frequently when the pump does not deliver water into the evaporator. Hardly ever does it happen that the flow of water is impeded due to clogging of the pipeline or the evaporator.

A thermostat with a sensor is most frequently used for monitoring evaporators. The thermostat directly measures water temperature in the evaporator and if the temperature reaches the minimally allowed value it switches the system off.

3.4 Monitoring the Compressor

Two factors should be taken into consideration when compressors are in question.

First of all, inside the compressor, the appropriate lubrication has to be ensured in all its assemblies.

With smaller size compressors, due to smaller dimensions, the satisfactory lubrication can be provided by immersing the crankshaft into oil. The crankshaft disperses oil in the casing of the compressor. This type of lubrication is fairly reliable. If oil freely returns to the compressor, it will certainly lead to lubrication. Only through the sight glass may the level of oil be visually monitored.

With larger compressors, transmission of oil is made possible by means of an oil pump. It is necessary to monitor the operation of an oil pump. The presence of oil in the appropriate assemblies is monitored indirectly through pressure in oil leaving the oil pump. A differential pressostat has been designed to monitor the pressure difference between oil leaving the oil pump and oil in the storage sump. If the difference in oil pressure is lower than is expected the pressostat switches off the entire system.

Secondly, each fracture in the compressor leads to a great decrease in the system capacity. The suction and compression valve, as well as the piston rod, fracture most frequently. It is possible to determine these damages only indirectly through suction pressure. If any kind of damage occurs in the compressor, the capacity decreases and the amount of transported refrigerant is reduced. The pressure in the evaporator increases due to the reduced amount of the refrigerant. It is possible to monitor damages by placing a pressostat into the suction branch of the compressor.

3.5 Monitoring the Electric Motors

Coils of driving electric motors should be protected from overload, dropout, and phase misbalanced. Without this kind of monitoring, wire insulation in the coil might burn through in case of excessive current. In order to prevent damage, the intensity of the current in the coil should be monitored continually. Electric motors should be monitored regardless of their type, whether they are single-phase or three-phase. If the intensity of the current exceeds the prescribed value, the monitoring device should switch the whole system off.

It is necessary to monitor the level and the difference in the level of voltage with three-phase electric motors. In case of insufficient voltage in the phases, the coils get damaged, they burn through. Devices with different executions have been designed for monitoring.

Their main purpose is to switch off the system, through the control phase, in case of insufficient current performances.

3.6 Monitoring the Amount of the Refrigerant

The amount of the refrigerant in the system is monitored visually through the sight glass, and possibly through sensors. Sensor monitoring is rarely applied since the executions of heat pumps are improving in quality, and it rarely happens that the refrigerant leaks. If it does happen the consequences are really unpleasant.

The heating power drastically decreases, therefore the device is constantly switched on. As the amount of the refrigerant decreases, the cooling intensity is gradually reduced as well as the lubrication of the cylinder inside the compressor. If the compressor operates under these conditions for a long time, the cylinder and the compressor head heats up and may even seize.

A sensor for monitoring the amount of the refrigerant in the system is called a pressostat. The pressostat is connected to the suction branch of the compressor. As the amount of the refrigerant decreases, the pressure in the suction branch drops as well. The pressostat records the level of the pressure, therefore if the level is lower than is prescribed the pressostat switches the system off.

3.7 Monitoring the Condition of Filters

The condition of filter is usually monitored indirectly, that is visually through the sight glass or through sensors by means of a pressostat. The symptoms indicating that the filter is clogged are completely the same as those indicating that the amount of the refrigerant is reduced. In the suction branch, the pressure drops in proportion with the clogging of the filter. A characteristic symptom that the filter is clogged is the temperature drop along the shield of the filter. The resulting difference in temperature can be determined by using a differential thermostat. One sensor is placed in the inlet, and the other in the outlet filter tube.

In practice, usually a single pressostat and sight glass are placed. The same instruments monitor the amount of the refrigerant and the clogging of the filter.

4 The Structure of the Heat Pump Monitoring System

The type of the monitoring system being used and the way it functions depend both on the strategy and the structure of the control system.

The complete monitoring of the heat pump is achieved if the structure of the monitoring system contains all the above mentioned components of monitoring.

The minimally required structure of the monitoring system must include monitoring the condenser, the evaporator and driving electric motors. If these components are not monitored, the heat pump must not be switched on.

It is also desirable to monitor all the other components, but it is not obligatory, since those kinds of disturbances do not cause drastic damages, failures in the system.

5 Types of Monitoring System Effects

The effect of the monitoring system depends on the control strategy.

In accordance with the control system, the monitoring system can be discrete and continual as well.

The mechanism of monitoring system effects is always the same, if disturbances occur, it switches off the whole heat pump system.
How this switching off is carried out depends on the type of the control system.

Discrete control can be strategically carried out directly and indirectly.

With **discrete indirect control system,** monitoring operates indirectly as well. In case of disturbances, the control system switches off the electromagnetic valve, the compressor continues to operate, therefore the pressure drops and the pressostat turns off the control phase. By switching off the control phase, the system stops.

With **discrete direct control system**, monitoring operates directly as well. In case of disturbances, monitoring system directly switches off the control phase.

Continual monitoring is based on the principle of semiconductors. The controller is a microprocessor with enormous possibilities. Control system sensors are adapted to the microprocessor and are used to replace discrete components, such as various pressostats and thermostats.

6   Discrete Components of the Monitoring System

Monitoring system for the discrete control system
If the controller is discrete, the components of the control system are discrete as well.

Discrete control system components are: a pressostat for low and high pressure, differential pressostat, thermostat, differential thermostat, sight glass, phase observation device, excessive current relay or thermistor.

The **pressostat for high pressure** should be connected to the discharge of the compressor. It detects the pressure in the compressive tube of the compressor and in the condenser. One and the same sensor monitors and protects the compressor and the condenser from mechanical overload. If the pressure exceeds the prescribed limit, it switches off the contactors through the control system and thus stops the supply of electric energy.

The pressostat for low pressure is connected to the suction of the compressor. It detects the pressure in the suction tube of the compressor.

If working pressure drops below the prescribed value, the pressostat switches off the contactors and thus stops the supply of electric energy. Working pressure decreases if: the amount of refrigerant is reduced due to lack of sealing, the filter is clogged, and the flow of water in the evaporator stops.

If working pressure increases above the prescribed value, another optional low pressure pressostat switches off the contactors. The rise in working pressure is caused by damages in the compressor, such as fracture of the valve, piston rod and the piston.

Differential pressostat is connected to the compressor. It detects the difference in oil pressure between the suction and compressive branch of the oil pump. It monitors the lubrication of the compressor. If the difference in pressure drops below the prescribed value, the pressostat switches the system off. Compressors without an oil pump do not require monitoring of this kind.

Thermostat. The sensor of the thermostat is placed on the shield of the evaporator. It detects the temperature of the evaporator. If the temperature drops below the prescribed value, the thermostat switches the system off. The fall in temperature is caused by either insufficient circulation of water in the evaporator or the complete lack of it. Insufficient flow is the result of damage in the pump or clogging of the evaporator, possibly of the tubing.

Differential thermostat. One sensor of the thermostat is placed on the inlet and the other on the outlet filter tube. The thermostat detects the difference in temperature between the inlet and outlet coolant. If the measured difference in temperature is greater than is prescribed, the differential thermostat switches the system off. The fall in temperature is caused by clogging of the filter.

Sight glass is built into the tube in front of the thermo expansion valve. It is used to monitor the condition of the refrigerant visually. If bubbles appear in the refrigerant, it is a sign that the e pressure has dropped in the suction branch. The causes are: lack of the refrigerant or clogging of the filter.

Intervention is manual. In case the monitoring is automatized, the low pressure pressostat replaces the sight glass.

Phase observation device is successively built into the control phase. It detects the level of voltage in the feed network. If the measured difference between the levels of voltage in the phases is greater than is prescribed, the phase observation device stops the control phase, thus stopping the whole system.

The excessive current relay is built into the power line for feeding the electric motor. It detects the intensity of the current which flows in the coils of an electric motor. If one of the currents has the intensity greater than is prescribed the relay stops the control phase and thus the whole system.

Thermistor is the sensor of the semiconductor and it is built into the coil of an electric motor. It replaces the excessive power relay. It detects the temperature in the coil. If the temperature exceeds the prescribed value, thermistor logic stops the control phase and thus the system.

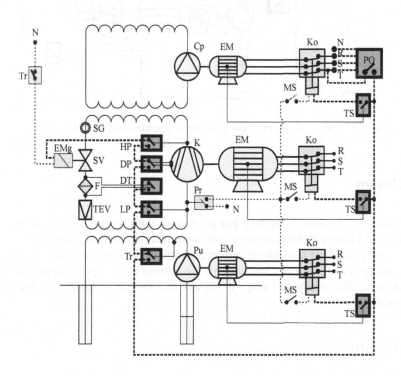

Fig. 1. Scheme of monitoring system in case of discrete indirect control strategy

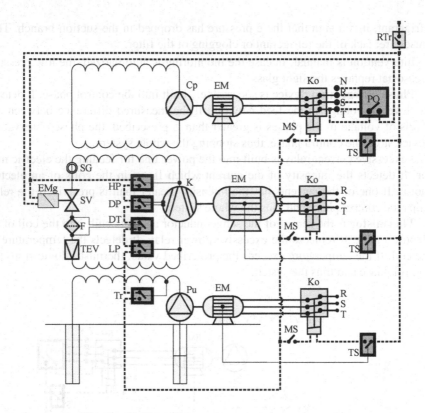

Fig. 2. Scheme of monitoring system in case of discrete direct control strategy

Conclusion

It is necessary to monitor the heat pump, at least with the minimum number of components. Lack of monitoring may lead to serious damage and fractures.

The principle of the monitoring system operation is always the same, in case of failure it directly or indirectly switches off the control phase. By switching off the control phase the supply of energy is cut off and the whole system of the heat pump stops.

Nomenclature

R,S,T – three phase
N - Null
Po – phase observer
RTr-inside air thermostat
Tr – water thermostat

EV – electric valve
TEV – thermo expansion valve
Cp – circulate pump
K – compressor
Pu – centrifugal pump

HP – high pressure pressostat
DP – different pressure pressostat
LP – low pressure pressostat
TS – thermistor switcher
DT – different thermostat
EM – electric motor

EMg – electromagnet
Ko – contactor
MS – manual switcher
SG – sight glass
F – Filter

References

[1] Nyers, J., Stoyan, G.: The Discrete Evaporator Model's Solution for Heat Pump by Means of Gauss-Newton Method. Bull. Appl. Maths, Balatonfüred, vol. 799/92 (1992)
[2] Nyers, J., Stoyan, G.: A Dynamical Model Adequate for Controlling the Evaporator of a Heat Pump. International Journal of Refrigeration 17(2), 101–108 (1994)
[3] Nyers, J.: Heat Energy Transfer between Heat Pump and Floor Heating in Steady Operation Mode. In: Hőerőgépek és környezetvédelem kongresszus, Proceedings, Tata, pp. 274–278 (1997)
[4] Njers, J.: Grejanje i hladjenje poslovne zgrade primenom toplotne pumpe sa podnim cevnim paketima, SMEITS Beograd (1998)
[5] Nyers, J.: Proceedings of International Conference on Comparison of a Traditional and a Microprocessor-controlled Heat Pump Control Strategy, Sweinfurt, pp. 141–146 (May 2001)
[6] Nyers, J.: International Conference on Evaporation Regulation a Heat Pump, Practical Aspects, Maribor, Slovenia (April 2000)
[7] Nyers, J.: The Diagnostics of Heat-Pumps with the Aplication of a Microprocessor. In: 20th International Conference on Science in Practice, Osijek, May 5-6 (2003)

The Nozzle's Impact on the Quality of Fabric on the Pneumatic Weaving Machine

István Patkó

Faculty of Light Industry and Environmental Protection, Budapest Tech
Doberdó út 6, H-1034 Budapest, Hungary
patko@bmf.hu

Abstract. The pneumatic weaving technology is a well known technology in the world. The quality of the fabric is sensitive to the flow condition of weaving machine. In this paper, I investigate the impact of the shape of nozzle on the fault of weaving technology. I did laboratory and industrial measurements. According to the results of them, I planed a new nozzle. In the industrial situation, I proved, the quality of fabric – which was produced by the new nozzle – had smaller faults.

1 Brief Description of Air Jet Looms

In air-jet looms, the weft is introduced into the shed opening by air flow.

The energy resulting from air pressure is converted into kinetic energy in the nozzle. The air leaving from the nozzle transfers its pulse to stationary air and slows down.

To this end, in order to achieve a larger rib width, V. Svaty developed in 1947 a confuser, which maintains air velocity in the shooting line.

The confuser drop wires are profiles narrowing in the direction of shoot, and they are of nearly circular cross section open at the top. These drop wires are fitted one behind the other as densely as possible. Therefore, they prevent in the shooting line the dispersion of air jet generated by the nozzle.

Fig. 1.1 shows the arrangement and design of the confuser drop wires applied in machines of the P type, as well as the arrangement schematic of weft intake. The nozzle (1) is secured to the machine frame, and the confuser drop wires (2) and the suction pipe (3) are fixed to the loose reed.

The confuser drop wires are profiles narrowing in the flow direction, and they have a conicity of 6°. These profiles (section A) may be made of metal (Fig. 1.1.a) or plastic (Fig. 1.1.b).

To be considered almost as a closed ring from the aspect of flow, a baffle plate of nearly circular cross section is placed on top of the latter.

I.J. Rudas et al. (Eds.): Towards Intelligent Engineering & Information Tech., SCI 243, pp. 583–592.
springerlink.com
© Springer-Verlag Berlin Heidelberg 2009

In the top part – in comparison with the metal confuser – they substantially reduce the air outflow, and therefore the reduction of air jet velocity will be smaller in the direction of shoot in the confuser drop wires.

The slay (1) is oscillated by a specific drive mechanism, to make sure that during the shoot, the swinging motion of confuser drop wires does not possibly influence the conditions of flow. This is because in case the displacement of the confuser drop wires is large during the shoot, the air flow conditions are unfavourable from the aspect of introducing the weft into the shed, and hence the warps may reach into the inner space of the confuser.

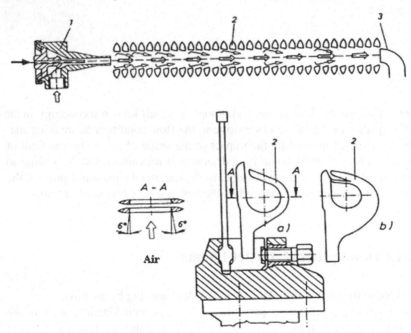

Fig. 1.1. Arrangement and design of the confuser drop wires applied in a type P machine

a/   metal confuser and its fixing,   b/ plastic confuser

In a type P machine, the nozzle is secured to the machine frame, while the confuser drop wires swing with the rib (Fig. 1.2).

During the shoot, the nearly stationary position of the slay is ensured by an eccentric articulated movement. In machine P 165 mentioned as an example, the introduction of the weft is carried out in 0.08 sec, four times in a second.

By the application of the confuser drop wires, a rib width of b=165 cm was achieved. This is where the name of loom type P 165 comes from. The confuser drop wires cover 75 to 85% of the rib width.

The design of the loom nozzle is shown in Fig. 1.3, indicating the velocity patterns of the air leaving the nozzle, in addition to the weft.

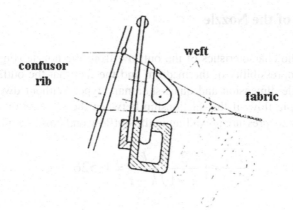

Fig. 1.2. Movement of the confuser drop wires

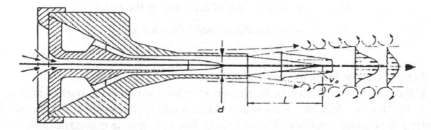

Fig. 1.3. Velocity distributions evolving behind the OEM nozzle

In front of the nozzle there is a yarn box, the function of which is to store one shoot of yarn. The box is designed in a way that the yarn can be removed from it almost without resistance.

One type of these boxes is the pneumatic yarn box (Fig. 1.4).

The pneumatic yarn box has a simple design, and its stores the yarn of specified length in a tube with a slow airflow, in the form of a loop. For introducing the weft, depending on the structure of the yarn, compressed air of 1.5 to 3.0 bar pressure is required.

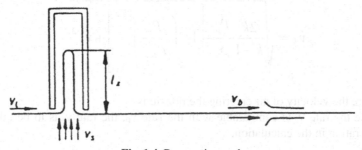

Fig. 1.4. Pneumatic yarn box

2   Testing of the Nozzle

To determine the characteristics of the outflow from the nozzle (Fig. 1.3), let us assume the compressibility of the medium and the fact that the outflow is of the isentropic, single-dimension and quasi-stationary type. Without dwelling on the laws of isentropic flow, if the relationship between the (P_0 environmental) pressure and the nozzle pressure prevailing in the outflow end cross section is

$$\frac{P_0}{P_{fuv}} < \left(\frac{2}{k-1}\right)\frac{k}{k-1} = 0.526 \tag{2.1}$$

then

P_0 is the pressure in the outflow (narrowest) cross section

P_{fuv} is the pressure of air entering the nozzle

k is the adiabatic coefficient for air

$k = 1.4$

In this case, the Mach number characterising the flow is: $M_a > 1$

If the outlet cross section is the narrowest cross section of the nozzle, then the air leaves the nozzle with a local sound velocity and this velocity is not exceeded under any pressure condition. The air exiting from the nozzle expands explosively at the moment of leaving the nozzle, and a complicated uncontrollable flow results. This flow will tear off the yarn end protruding from the nozzle or forces the yarn end on an undesired trajectory.

If the intention is to make sure that the velocity of outflow air is increased, then a widening extension is to be provided after the narrowest cross section. The outlet having an extension is called a Laval-nozzle. The role of the widening extension after the outflow aperture is to make sure that the air is able to expand – before escaping – to the pressure prevailing in the outflow cross section. Therefore, a well arranged flow pattern is generated at the outflow.

In this case the outflow velocity is:

$$v_0 = \sqrt{\frac{2k}{k-1}\frac{P_{fuv}}{\rho_{lev}}\left[1 - \left(\frac{P_0}{P_{fuv}}\right)^{\frac{k-1}{k}}\right] + v_f^{\,2}} \tag{2.2}$$

where the velocity of air entering the nozzle is v_f.

Since the rate v_f is comparable with the rate v_0, the rate v_f is to be taken into consideration in the calculation.

When examining the cross sections of the nozzle shown in Figure 1.3, the conclusion is that the cross section of the nozzle slightly narrows in the direction of the outflow. Therefore, the narrowest cross section exists in the plane of the outlet, where the pressure is equal to the ambient pressure. Furthermore, we shall indicate with an asterisk the characteristics prevailing in the narrowest cross section.

$P^*=P_0=1$ bar

$$P^*=P_0=1 \text{ bar}$$

P_0 is the ambient pressure

$$\frac{P_0}{P_{fuv}} < 0.526$$

from which

$$P_{fuv} \rangle \frac{P_0}{0.526} = \frac{1bar}{0.526} = 1.9bar$$

Consequently, in each nozzle pressure where

$$P_{fuv} \rangle 1.9bar$$

the air leaves the nozzle with a local sound velocity and expands explosively. In view of the operating conditions it can be determined that the condition $P_{fuv} > 1.9$ bar is met in 70 to 80% of the cases.

For these cases, we have designed such a nozzle, where the expansion of air takes place in the nozzle and then in the outflow cross section a larger outflow velocity v_0 – to be determined by 2.1 – can be achieved.

Examining the characteristic supply air pressures, it seems to be preferable to design two nozzle shapes for creating a new nozzle shape:

Shape I: the so-called parallel nozzle

                 for an operating pressure of $P_{fuv} = 1.5$ to 2 bar

Shape II: the so-called widening nozzle

                 for an operating pressure of $P_{fuv} > 2$ bar.

In the case of the parallel nozzle, we have replaced the existing narrowing cross section outflow stub by a parallel cross section stub.

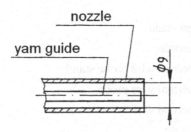

Fig. 2.1. Parallel nozzle

In the case of the <u>widening nozzle</u>, keeping the inner cross section (Ø D_b) of the OEM nozzle intact, we have fitted an extension widening like a Laval tube to replace the outlet stub.

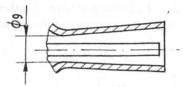

Fig. 2.2. Widening nozzle

In dimensioning the widening extension, the laws of single dimension isentropic flow have been applied.

Since the outflow cross section does not depend linearly on pressure P_{fuv}, we have sought such a dimensioning pressure, which is able to withstand with a relatively smooth outflow the nozzle pressure variations ranging between 2 and 3 bar.

On the basis of empiric experiments, we have determined from the dimensioning nozzle pressure the outlet cross section of the widening nozzle.

By cutting off the outlet stub of the OEM nozzle, both the widening stub and the parallel stub can be adjoined to the original nozzle body by a threaded joint.

3   Measuring Equipment

For measuring the velocities generated in the axis of the confuser array, we have assembled a test machine, which is also suitable for examining the factors that influence the velocity distribution in the axis of the confuser array. Therefore, we have assembled the test machine in a way that it simulates real life conditions as much as possible.

We have assembled the test machine from the spare parts of functioning machines, on the basis of the dimensions (positions) taken from these machines.

When designing the test machine, we have made sure that changing the positions of and replacing the components are possible.

The schematic diagram of test machine arrangement is shown in Figure 3.1, where:

1	Test stand
2	Recorder
3	Thermal wire probe stand
4	Thermal wire probe
5	Thermal wire velocity meter
6	Nozzle

7   Nozzle holder

8   Motion droppers

9   Movable fasteners

10   Air supply line ($\varnothing$ 8)

The air supply of the test machine has been controlled from a loom. In this way, it has been possible to model the movement of the weft under laboratory conditions.

The schematic arrangement of the test machine is shown in Fig. 3.1.

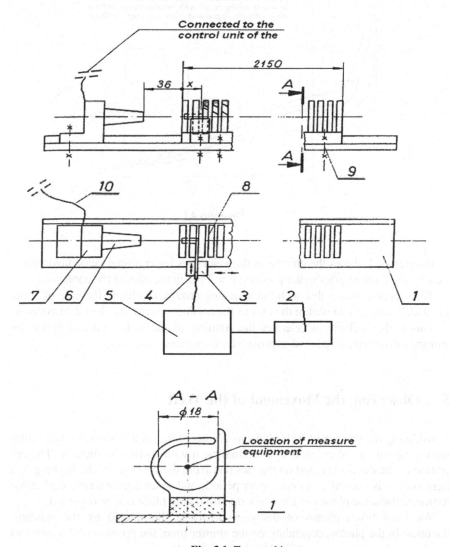

Fig. 3.1. Test machine

4 Laboratory Measurements

We have measured the flow velocities prevailing in the axis of the confuser array by a thermal wire velocity metering process on the test machine shown in Fig. 3.1, by simulating the operating conditions. In the course of measurement we have kept altering the nozzle (OEM nozzle and the widening/parallel nozzles developed by us)

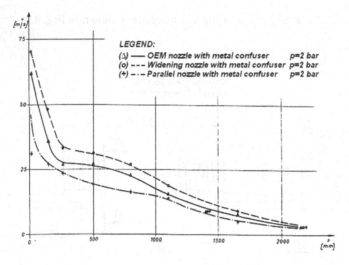

Diagram 4.1

Diagram 4.1 shows the change in the velocities (operational velocities) generated by the three nozzles under pressure P = 2 bar in the axis of the confuser.

The diagram proves that with the widening nozzle, a velocity increase of about 15 to 20% can be achieved in the axis of the confuser as against the OEM nozzles.

This higher velocity accelerates the running of the weft yarn and hence the looping of the weft is reduced as proven by factory experiments.

5 Observing the Movement of the Weft

Considering the process of weft intake as a quasi-stationary process, we have studied the movement of the weft by illuminating it with a synchro-nuctator. The operation of the device is based on the fact that from the main shaft, the flashing of a light source is controlled in each weft period, and in an appropriately dark environment, the same phases of a process rapidly taking place can be observed.

We have taken photos of the yarn positions illuminated by the synchronuctator. In the photos, depending on the shutter time, the position of the weft can be observed in one or more (5 or 6) subsequent weft periods.

On the receiving side, the behaviour of the weft can be well-observed in Pictures 1 and 2. It can be seen that when an OEM nozzle is applied, and the shooting is completed, the weft is looped, the shape of the loop varies and it frequently misses the suction aperture.

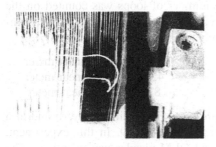

Picture 1

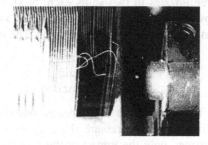

Picture 2

It is also a finding that – with a good approximation – the weft arrives at the receiving side always in the same loom position, which indicates that the weft velocity is the same.

When applying the widening nozzle, in association with the decreasing of yarn looping, the movement of the yarn is much more organised in the course of weft intake. The weft arrives at the receiving side without a loop and always enters the suction aperture (Pictures 3 and 4).

Picture 3

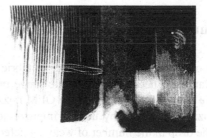

Picture 4

6   Weaving Experiments

To guarantee that the experiments are objective, the factory experiments have been carried out by the engineers of the weaving plant and they counted the number of weaving faults in one linear meter of the woven fabric.

Factory experiments were carried out in the weaving mill of Sopronkőhida Weaving Plant – in order to observe the looping of the weft yarn.

In the experiments – among others – the impact of the nozzle shape on weft looping was examined.

In examining the nozzle shape, the impact of the OEM nozzle and that of the so-called parallel and widening nozzles made by us was tested. The experiments were carried out on machine No. 429 of the weaving mill. The machine No. 429 was run with the same weft yarn and at the same reducer pressure with all the three nozzles in two shifts each. Next, the number of loops was counted on the woven fabric – in the right hand side 25 cm area of the fabric – and this was converted to one linear meter of fabric. The number of loops made by each nozzle:

OEM nozzle:	954 loops per linear meter
Parallel nozzle:	454 loops per linear meter
Widening nozzle:	82 loops per linear meter

This experiment was repeated one month later with a different yarn and at a different nozzle pressure (P_{fuv}) again on machine No. 429. In this experiment, however, only the impact of the widening and OEM nozzles was examined. The number of loops made by each nozzle:

OEM nozzle:	19 loops per linear meter
Widening nozzle:	6 loops per linear meter

The weaving experiments and pictures 3 and 4 prove confidently that by means of applying the widening nozzle developed by us, the number of weaving faults resulting from looping can be reduced by as much as 70 to 80%.

Summary

In order to improve the quality of fabric produced on a pneumatic loom, we have examined the impact of nozzle shape under laboratory and industrial conditions. We have found that when the OEM nozzle of the loom was replaced by the Laval nozzle developed by us, the improvement of fabric quality became remarkable. The drop in the number of weaving defects is 70 to 80%.

References

[1] Patkó, I.: Lamellák közötti áramlás tulajdonságainak meghatározása, Kandidátusi disszertáció, Budapest (1994)
[2] KMF Gépészeti Tanszék A P165 tipusú szövőgép vetülékbeviteli folyamatának fejlesztése. Kutatási jelentés (1982)
[3] Szabó, R.: Szövőgépek, Műszaki Könyvkiadó (1976)
[4] Alther, R.: Automatische Optimierung des Schusseintrages beim Luftdüsenweben. Kandidátusi disszertáció. ETH Zürich (1993)
[5] Lünenschloss, J., Wahhoud, F.J.: Das Eintragsverhalten verschidener Filamentgarme beim Idustriellen Luftweben. Textil Praxis Int. (1985)
[6] Patkó, I.: Weaving with Air Jet. In: Proc. of 6th International Symposium of Hungarian Researchers, Budapest, Hungary, November 18-19, 2005, pp. 401–412 (2005)

Part VI
Intelligent Systems and Complex Processes

A Comprehensive Evaluation of the Efficiency of an Integrated Biogas, Trigen, PV and Greenhouse Plant, Using Digraph Theory

Nicola Pio Belfiore[1], Umberto Berti[2], Aldo Mondino[2], and Matteo Verotti[1]

[1] Dept. Mechanics and Aeronautics, Sapienza Universita di Roma
via Eudossiana 18, 00184 Rome, Italy
belfiore@dma.ing.uniroma1.it

[2] GRUPPOFOR, S.r.l., Via di Trigoria, 191, 00128 Rome, Italy
info@gruppofor.eu

Abstract. This paper describes how directed graphs (*digraphs*) have been used for evaluating the total efficiency of a complex Plant that has been designed for the combined production of food, heat, cooling, and electrical energy. This task required three basic activities. Firstly, the overall plant scheme was studied in detail in order to identify the *elementary blocks*, each one having its own efficiency value. Each block corresponded to a single node of the *digraph*. Secondly, the system had to be considered as a whole. The development of the global scheme required the identification of all the *directed arcs* that were corresponding to the energy flows. Finally, a new algorithm for the evaluation of the overall efficiency was applied to the resultant *digraph*. This method allowed us to obtain the solution in algebraic symbolic form, automatically, provided that the digraph arc list was supplied to the developed PC code, together with the flow repartitions. The algorithm is quite robust and could also be used in systems with flow recirculation.

1 Introduction

During recent decades Society has been dealing with several important problems, the resolutions of which are essential for the survival of the human race. Among these are the quest for sustainable agricultural techniques and green energy. Related to energy production, the problem of carbon dioxide emission has come to global attention and the Kyoto Protocol has been signed by the representatives of most Countries. Modern Science and Technology have the answers to these requirements and some new concept Plants have been established all over the World to produce energy from renewable sources. The next step consists of integrating the production of both energy and food, synergically. Some endeavors are being developed in Central and Southern Italy, where new projects have been completed for the production of electrical energy, heat, cooling, and high quality vegetables from biological horticulture. Since the Plants are very complex it is not easy for investors to

I.J. Rudas et al. (Eds.): Towards Intelligent Engineering & Information Tech., SCI 243, pp. 595–605.
springerlink.com   © Springer-Verlag Berlin Heidelberg 2009

develop reliable business plans. Furthermore, realistic predictions on the economic balance can only be made after a detailed study of all the single parts of the complex mosaic and of the interactions between them. These drawbacks could be avoided if a preliminary study of the system efficiency has been performed. The basic idea that has been developed in this paper consists of approximating the physical plant as a set of arcs and nodes, the latter having characteristic values of the efficiency.

The efficiency of a simple system is quite easy to evaluate. For instance, the efficiency of a system composed of a series of single stations is the product of the efficiency values of the single stations. It is also well known that in systems that are composed of a number of machines which work in a parallel configuration the overall efficiency is the weighted average of the single efficiencies, taking as weights the corresponding input works. For more complex systems the evaluation of the efficiency becomes more complicated. In some cases, specific algorithms have to be conceived, especially where the energy circulates between some blocks. For systems that consist of a large number of elementary machines (blocks), the efficiency is not conceptually difficult to calculate, but it may become rather time consuming and complicated, with a significant increase in error occurrence in the calculations.

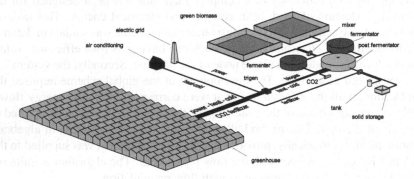

Fig. 1. An overall view of the Plant

In this paper a new algorithm, based on digraph theory, is applied to a complex Plant designed for the combined production of electricity, heat, cooling and fresh vegetables.

Elements of originality are believed to be both in the design of the overall Plant and in the algorithm that evaluates its efficiency.

2   A Survey on the Plant under Development

A project for a polygeneration Plant has been developed, after more than one year's research, and some areas of Central and Southern Italy have been identified as possible candidates for its actual establishment. The polygen system is composed of some basic stations that have been used *alone* for decades all over the world *as single production processes*:

- biogas production station;
- trigeneration station;
- greenhouse;
- photovoltaic field (integrated with the greenhouse);
- multipurpose buildings;
- flow and process control station.

2.1 Biogas Production

The system collects green biomass from the local agriculture and elaborates it in an anaerobic digester, where a series of processes takes place in order to allow microorganisms to break down biodegradable material in the absence of oxygen. The product of this process is biogas and fertilizer (see also [3], [5], and [6]). *If the biomass supply and the process are well controlled*, the digester neither emits pollution, nor radiations, nor any other dangerous matter in significant quantity, especially when compared with the other systems of energy production. Typically, biogas is composed of methane (around 50% or more) and carbon dioxide (from 35% to 50%), plus nitrogen (from 0% to 10%) and small quantities of hydrogen, hydrogen sulfide, and oxygen. *In fact, the only significant process waste is a high quality fertilizer.*

2.2 Trigeneration

After filtering and conditioning, the biogas is ready to be used to supply an internal combustion engine (see also [1]) that is specifically designed and optimized to run under a biogas supply regime. Its output shaft is mechanically connected to a generator. Most of the electrical energy that is being produced is conveyed to the grid with no particular problems. Other amounts are used within the Plant.

Since the internal combustion engine must be cooled, some heat can be regained from this operation. The amount of heat is quite large, that is to say it is about as great as the electrical energy produced. Hence, it is so important to retrieve energy that would be wasted otherwise. Therefore, as usual in the cogeneration Plants, the heat is extracted and used as much as possible. In our system, some of the recaptured heat is used as a source for processing and air conditioning. However, there is also a significant amount that can be used as a source of cooling. This is possible thanks to the application of *absorption cycle* refrigeration. The basic thermodynamic process is not a conventional thermodynamic cooling process which works with compressors. It is based on evaporation, where faster-moving hotter molecules carry heat from one material to another one that preferentially absorbs hot molecules (see [4] for theoretical aspects). Usually ammonia or lithium bromide salts are used.

2.3 Greenhouse

The trigeneration stage releases a great amount of heat that can be converted, totally or partially to cold, because the temperature of the cooling water is high enough for the absorption process to work. Either heat or cold can be used to condition the environment inside a greenhouse that is built within the Plant area. The greenhouse consists of a series of steel and glass constructions that are dedicated to vegetable production. Integrating the Plant with a greenhouse is very advantageous because this action generates a strong connection with the agricultural operators that provide the green biomass.

2.4 Photovoltaic Field (Integrated with the Greenhouse)

The above mentioned greenhouse can be regarded as a natural machine that is able to keep a certain quantity of irradiated solar energy. This characteristic is useful during wintertime. On the other hand, during the summer, the irradiated energy exceeds the needs for the natural growth of the vegetables. This occurrence is dramatically true in hot Countries, where farmers are used to adopting shading apparatuses. Therefore, the idea of using PV panels for ray interception arises as the natural way both to avoid excessive radiation and to get an increase of production of electrical power. The practical arrangement of this constructive solution is not easy and can be developed only with the assistance of an agro-technical project team. Finally, it is worth noting that the carbon dioxide conveyed within the biogas flow can be filtered and used as a fertilizer, as for the hydroponic crops (see also [2]).

2.5 Multipurpose Buildings

The availability of heat, cold, and electricity makes it possible to promote many kinds of commercial and social enterprises. Therefore, the same kind of constructions used for the greenhouses can be adopted as an area where several kinds of project may occur.

2.6 Flow and Process Control Station

The management of the overall system is a rather difficult task, because there are so many variables that need to be monitored and controlled. Therefore it is convenient to set up a central control room where a management team could observe the whole process and take the best decisions in order to improve security, avoid failures, and increase profit.

3 Description of the Algorithm

The method used in this paper was originally conceived for the evaluation of the efficiency of machines. However, the approach is more general and can be easily applied to any complex system that can be represented by means of flows that transmit energy amongst non ideal blocks.

Step 1. Definition of the directed graph G corresponding to the system
The initial stage consists in the definition of the elementary units that correspond to the single stations. Each station is modeled as a block which receives and transmits input and output energies. The balance of the single unit is not null because the output work is always less than the input. The ratio of the output work divided by the input work is referred to as the efficiency of the single block. One single block may receive energy from several blocks and may transmit energy to several other blocks.

Step 2. Addition of the virtual external world node
Apart from the n nodes that represent the blocks, another block is added to the set of vertices. This block, labeled as 1, represents the external world from which the power is originated to enter the system. Furthermore, it represents also the recipient of the power that exits from the system.

Step 3. Definition of the efficiency matrix E
Once the external block 1 has been defined, an efficiency matrix E can be introduced as a diagonal matrix whose generic $i-th$ element of the principal diagonal is equal to the efficiency of the $i-th$ block. Since the system regards the world as an external block, the element $E(1,1)$ has to be set as equal to the inverse of the overall system efficiency $\frac{1}{\eta_T}$.

Step 4. Evaluation of the Boolean adjacency matrix A of the directed graph
Matrix A is obtained by simply setting as *true* any element $i - j$ that corresponds to an arc whose tail and head vertex are, respectively, i and j.

Step 5. Definition of the Colored adjancency matrix M
The colored adjacency matrix M can be obtained by substituting the true value of the generic element $i - j$ with the label that corresponds to the same arc.

Step 6. Selection of the arc weights in M and definition of the set of n unknown
The works coming in from and going out to the external world are labeled with the symbols L_{mk} and L_{uz}, respectively, where k and z are the blocks involved in the corresponding energy transmission. The total energy coming out of the single j block is named e_j. If the block i transmits energy to two or more blocks, say, a, b, and c, the partition coefficients $p_{i,a}$, $p_{i,b}$, and $p_{i,c}$, are introduced, with

$$\sum_{j=a}^{c} p_{i,j} = 1, \tag{1}$$

in order to distinguish how much energy goes to the receiving blocks. For example, $e_i \cdot p_{i,a}$ represents the energy that block j has transmitted to node a.

Partition factors are known values, whereas energy e_i is unknown.

Step 7. Introduction of the balancing virtual edges in M
For each flow that transfers power from the external world block, one arc has to be added. Analogously, one other arc ought to be added for each flow that transfers power from any internal block to the external world.

Step 8. Determination of the output vector V
Introducing the N-dimensional vector Y, whose elements are all unit, the output vector is given by

$$V = M \cdot Y \tag{2}$$

Step 9. Determination of the reduced works matrix L
Matrix L is simply given by the equation

$$L = M \cdot E \tag{3}$$

Step 10. Evaluation of the exit reduced works vector X
The whole set of exit reduced works is then given by vector

$$X = L^T \cdot Y \tag{4}$$

Step 11. Definition of the set of n equations
Except for the 1-st rows, n equations are obtained by taking the rows from 2 to N of the set of equations

$$M \cdot Y - E^T \cdot M^T \cdot Y = \begin{bmatrix} 0 \end{bmatrix} \tag{5}$$

Step 12. Solution of the system of n equations
This procedure is performed in a closed form by using an algebraic manipulation programming language.

Step 13. Computation of the overall efficiency
The overall efficiency is given by solving the first equation of the system (eq:siste), where η_T is the only unknown.

3.1 Example of Application

The system represented in Figure 2 has been analyzed in order to better explain how the algorithm works. Following the above mentioned steps, matrix

$$
M = \begin{bmatrix}
0 & L_m & 0 & 0 & 0 & 0 & 0 \\
0 & 0 & e_2 p_{2,3} & 0 & 0 & 0 & e_2 p_{2,7} \\
0 & 0 & 0 & e_3 p_{3,4} & e_3 p_{3,5} & 0 & 0 \\
0 & 0 & 0 & 0 & 0 & e_4 & 0 \\
0 & 0 & 0 & 0 & 0 & e_5 & 0 \\
0 & 0 & 0 & 0 & 0 & 0 & e_6 \\
L_u & 0 & 0 & 0 & 0 & 0 & 0
\end{bmatrix}
\tag{6}
$$

is obtained at step 5. Then, at step 9 the reduced work matrix

$$
L = \begin{bmatrix}
0 & L_m \eta_2 & 0 & 0 & 0 & 0 & 0 \\
0 & 0 & e_2 p_{2,3}\eta_3 & 0 & 0 & 0 & e_2 p_{2,7}\eta_7 \\
0 & 0 & 0 & e_3 p_{3,4}\eta_4 & e_3 p_{3,5}\eta_5 & 0 & 0 \\
0 & 0 & 0 & 0 & 0 & e_4 \eta_6 & 0 \\
0 & 0 & 0 & 0 & 0 & e_5 \eta_6 & 0 \\
0 & 0 & 0 & 0 & 0 & 0 & e_6 \eta_7 \\
0 & 0 & 0 & 0 & 0 & 0 & 0
\end{bmatrix}
\tag{7}
$$

can be calculated.

Finally, the total efficiency is given at the last step 13, in symbolic form

$$
\eta_T = \eta_2 \eta_7 \left[p_{2,7} + p_{2,3}\eta_3\eta_6 \left(\eta_4 p_{3,4} + \eta_5 p_{3,5} \right) \right].
\tag{8}
$$

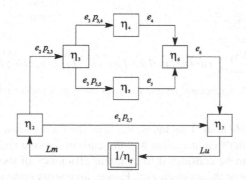

Fig. 2. A schematic representation of a system with $n = 6$ blocks (labels from 2 to 7) and $N = n+1$ total blocks (including the external world block 1)

4   Application of the Algorithm to the Actual Plant

Figure 3 shows a schematic representation of the whole system. Each block is represented by the value of its efficiency. The node numbers correspond to the blocks briefly described in Table 1. The numerical values of the efficiencies have been estimated by considering each single block, which has been regarded as a station that receives and transmits energy, not necessarily of a same kind (e.g. chemical energy may enters, while mechanical work exits). Therefore, each single value has been evaluated as the ratio of the output energy divided by the input energy. It is worth noting that when there is a demand for cooling energy, the heat flows have to be reversed because the valuable quantity is cold.

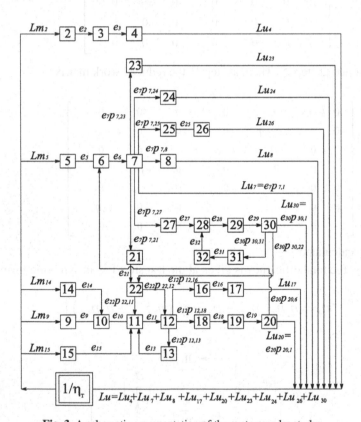

Fig. 3. A schematic representation of the system under study

The *external world node* 1 includes: the Sun, the Greenhouse interior, the electric grid, the customer's environment and the carbon dioxide tank. Finally, energy equivalences have to be considered to evaluate efficiency of those stations which transform the nature of the input energy. Hence, an energy equivalent content must be estimated for those input or output flows representing energy although not in conventional form (e.g. biomass, methane, rays of sunshine).

4.1 The System of Node Equations

Following the steps, the system of node equations can be found in a symbolic form completely automatically, as herein reported. The flow that exits from the block i is denoted by e_i, while the partition factor among nodes i and j is referred to as $p_{i.j}$:

$$Lm2 + Lm5 + Lm9 + Lm14 + Lm15 +$$

$$-\frac{e_7 p_{7.1}}{\eta_T} - \frac{e_{30} p_{30.1}}{\eta_T} - \frac{e_{20} p_{20.1}}{\eta_T} - \frac{Lu_4}{\eta_T} - \frac{Lu_8}{\eta_T} - \frac{Lu_{17}}{\eta_T} - \frac{Lu_{23}}{\eta_T} - \frac{Lu_{26}}{\eta_T} - \frac{Lu_{24}}{\eta_T} = 0$$

$$e_2 - Lm2\,\eta_2 = 0$$

$$e_3 - e_2\eta_3 = 0$$

$$Lu_4 - e_3\eta_4 = 0$$

$$e_5 - Lm5\,\eta_5 = 0$$

$$e_6 - e_{20}p_{20.6}\eta_6 - e_5\eta_6 = 0$$

$$e_7 p_{7.8} + e_7 p_{7.25} + e_7 p_{7.24} + e_7 p_{7.23} + e_7 p_{7.27} + e_7 p_{7.21} + e_7 p_{7.1} - e_6\eta_7 = 0$$

$$Lu_8 - e_7 p_{7.8}\eta_8 = 0$$

$$e_9 - Lm9\,\eta_9 = 0$$

$$e_{10} - e_9\eta_{10} - e_{14}\eta_{10} = 0$$

$$e_{11} - e_{22}p_{22.11}\eta_{11} - e_{10}\eta_{11} - e_{13}\eta_{11} - e_{15}\eta_{11} = 0$$

$$e_{12}p_{12.13} + e_{12}p_{12.18} + e_{12}p_{12.16} - e_{22}p_{22.12}\eta_{12} - e_{11}\eta_{12} = 0$$

$$e_{13} - e_{12}p_{12.13}\eta_{13} = 0$$

$$e_{14} - Lm14\,\eta_{14} = 0$$

$$e_{15} - Lm15\,\eta_{15} = 0$$

$$e_{16} - e_{12}p_{12.16}\eta_{16} = 0$$

$$Lu_{17} - e_{16}\eta_{17} = 0$$

$$e_{18} - e_{12}p_{12.18}\eta_{18} = 0$$

$$e_{19} - e_{18}\eta_{19} = 0$$

$$e_{20}p_{20.6} + e_{20}p_{20.1} - e_{19}\eta_{20} = 0$$

$$e_{21} - e_7 p_{7.21}\eta_{21} = 0$$

$$e_{22}p_{22.11} + e_{22}p_{22.12} - e_{30}p_{30.22}\eta_{22} - e_{21}\eta_{22} = 0$$

$$Lu_{23} - e_7 p_{7.23}\eta_{23} = 0$$

$$Lu_{24} - e_7 p_{7.24}\eta_{24} = 0$$

$$e_{25} - e_7 p_{7.25}\eta_{25} = 0$$

$$Lu_{26} - e_{25}\eta_{26} = 0$$

$$e_{27} - e_7 p_{7.27}\eta_{27} = 0$$

$$e_{28} - e_{32}\eta_{28} - e_{27}\eta_{28} = 0$$

$$e_{29} - e_{28}\eta_{29} = 0$$

$$e_{30}p_{30.31} + e_{30}p_{30.22} + e_{30}p_{30.1} - e_{29}\eta_{30} = 0$$

4.2 Solution of the System of Equations

Using a manipulating programming language the closed form solution can be found. The expressions of all the unknown flows can be found by solving the system of equations listed in the previous paragraph from the second to the last one. The value of the total efficiency can be found by solving the first equation with respect to η_T. The symbolic expressions are too long to be included in the paper.

However once the values of the single efficiencies and the partition factors are assumed, the expression of the total efficiency becomes similar to that of a system in a parallel configuration, for example,

$$\eta_T = \frac{0.11 L_{m2} + 0.31 L_{m5} + 0.27 L_{m9} + 0.27 L_{m14} + 0.25 L_{m15}}{L_{m2} + L_{m5} + L_{m9} + L_{m14} + L_{m15}}, \tag{9}$$

which has been obtained by assigning reasonable values to the single block efficiencies and by assuming an equal distribution of the output flows that exit form each block with multiple arc tails.

Table 1. List of correspondence between the blocks and the operating stations

Node	Block	Node	Block	Node	Block
1	World	12	Anaerobic Digester	23	Heat exchanger
2	Thin film PV	13	Slurry	24	Heat exchanger
3	BOS	14	Green biomass	25	Filtering
4	Inverter	15	Cattle dung	26	Carbon dioxide
5	Methane	16	Desiccator	27	Heat exchanger
6	Mixer	17	Fertilizer	28	Generator
7	Int. comb. engine	18	Filtering	29	Condenser
8	Alternator	19	Conditioning	30	Evaporator
9	Harvest	20	Distributor	31	Absorber
10	Biomass processor	21	Heat exchanger	32	Heat exchanger
11	Fermenter	22	Control		

Conclusions

In this paper an Integrated Biogas, Trigeneration, Photovoltaic and Greenhouse Plant has been presented. This complex system has been developed in order to deal with today's global economic crisis. In fact, many valuable products can be obtained from this Plant, such as electrical energy, heat, cold, liquid and gas fertilizer, and organic vegetables. Since this system is rather complex, a sophisticated algorithm has been conceived to evaluate its overall efficiency. The application of this algorithm shows that the typical expression of the efficiency consists of a weighted average of the energy amounts that are provided by the different sources (Sun and biomass). *Natural gas* can also be used when necessary. The numerical results show that the total *physical* efficiency ranges from about 11% to 31%. High losses are due to the Photovoltaic component. However, when economic implications are considered, it is clear that the PV gets more attention from the investor given that the daily fuel is cost free.

References

[1] Bari, S.: Effect of Carbon Dioxide on the Performance of Biogas-Diesel Duel-Fuel Engine. In: IV World Renewable Energy Congress, Denver, USA, June 1996, pp. 1007–1010 (1996)
[2] Jaffrin, A., Bentounes, N., Joan, A.M., Makhlouf, S.: Landfill Biogas for heating Greenhouses and providing Carbon Dioxide Supplement for Plant Growth. Biosystems Engineering 86(1), 113–123 (2003)
[3] Swaroopa, R.D., Krishna, N.: Ensilage of Pineapple Processing Waste for Methane Generation. Waste Management 24(5), 523–528 (2004)
[4] Tozer, R.M., James, R.W.: Fundamental Thermodinamics of Ideal Absorption Cycles. Int. J. Refrig. 20(2), 120–135 (1997)
[5] Yadvika, S., Sreekrishnan, T.R., Kohli, S., Rana, V.: Enhancement of Biogas Production from Solid Substrates Using Different Techniques—a review. Bioresource Technology 95(1), 1–10 (2004)
[6] Walla, C., Schneeberger, W.: The Optimal Size for Biogas Plants. Biomass and Bio Energy 32, 551–557 (2008)

Approximation of a Modified Traveling Salesman Problem Using Bacterial Memetic Algorithms

Márk Farkas[1], Péter Földesi[2], János Botzheim[3], and László T. Kóczy[4]

[1] Department of Telecommunications and Media Informatics, Budapest University of
Technology and Economics
Műegyetem rkp. 3-9, H-1111 Budapest, Hungary
farkasmark@gmail.com
[2] Department of Logistics and Forwarding, Széchenyi István University
Egyetem tér 1, H-9026 Győr, Hungary
foldesi@sze.hu
[3] Department of Automation, Széchenyi István University
Egyetem tér 1, H-9026 Győr, Hungary
botzheim@sze.hu
[4] Department of Telecommunications, Széchenyi István University
Egyetem tér 1, H-9026 Győr, Hungary
koczy@sze.hu

Abstract. The goal of this paper is to develop an algorithm that is capable to handle a slightly modified version of the minimal Traveling Salesman Problem in an efficient and robust way and produces high-quality solutions within a reasonable amount of time. The requirements of practical logistical applications, such as road transportation and supply chains, are also taken into consideration in this novel approach of the TSP. This well-known combinatorial optimization task is solved by a bacterial memetic algorithm, which is an evolutionary algorithm inspired by bacterial transduction. A new method is also proposed to deal with the time dependency in the cost matrix. The efficiency of the implementation, including time and space constraints, is investigated on a real life problem.

Keywords: Traveling Salesman Problem, time dependent fuzzy costs, eugenic bacterial memetic algorithm.

1 Introduction

In every aspect of life, productivity and extensive cost-reduction are two vital factors that have to be taken into consideration.

Logistics can be regarded as an art of delivering every goods to their intended destinations by using up as narrow resources (transfer capacity, cost and time) as possible. The primary objective is to satisfy the needs of every customer without sacrificing the quality of the service or wasting limited resources (time and money respectively).

I.J. Rudas et al. (Eds.): Towards Intelligent Engineering & Information Tech., SCI 243, pp. 607–625.
springerlink.com   © Springer-Verlag Berlin Heidelberg 2009

In the field of logistics, as in every optimization task, the size of the search space is often so enormous that widely available or practical methods could not cope with the task of providing quality solutions efficiently any longer. Either the performance of the algorithm is not adequate, or the quality of the solutions is not satisfactory. One such hard problem, which is also derived from logistics and serves as a foundation of many other applications, is the famous Traveling Salesman Problem.

1.1 The Traveling Salesman Problem

The Traveling Salesman Problem (TSP) is a nice example of combinatorial optimization problems in the field of logistics. The problem is loosely modeled after an average traveling salesman whose job is to visit each city at least (and preferably at most) once while the traveling cost or time must be kept as low as possible.

The salesman starts his/her journey from the company headquarters and after the journey he/she must return to this place. Given a list of cities and the distances between them the goal is to find the shortest possible tour that contains each city exactly once and returns to the first city. [2]

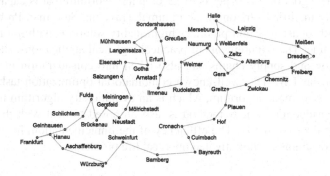

Fig. 1.1. An example of a Traveling Salesman Problem on German cities

Although one might think this problem is solely theoretical, even in its purest formulation it has several applications, such as planning, logistics, and the manufacture of microchips. The slightly modified traveling salesman may also appear as a part of larger problems, for example genome sequencing.

In the former example the concept of city represents DNA fragments and the distance is defined as a similarity measure between DNA fragments. Additional constraints in these applications may render the problem considerably harder. [1]

1.2 Combinatorial Optimization Problems

In order to formalize the Traveling Salesman Problem an introduction to the combinatorial optimization problems is necessary. An optimization problem is a computational problem in which the task is to find a solution in the feasible region which has the minimum (or maximum) value of the objective or measure function. In this context feasible solution is the set of all possible solutions of an optimization problem. The measure or objective function is a function associated with an optimization problem which determines how good a solution is.

Definition 1.1. *Optimization problem:* An A optimization problem can be formally defined as a quadruple: [3]

$$A = (I, f, m, g) \tag{1}$$

- I is a set of instances of A
- Given an instance x, $f(x)$ denotes the set of feasible solutions where S is the set of solutions.

$$f : I \to \mathrm{P}(S), \mathrm{P}(S) := \{s | s \subseteq S\} \tag{2}$$

- The measure or objective function is an m function which given an instance f assigns a positive real number to a feasible solution y according to its goodness.

$$m : I \times S \to \mathbb{R}^+ \tag{3}$$

- g is the goal function which is either *min* or *max*.

Definition 1.2. *Optimum solution:* y is the optimum solution of optimization problem A regarding instance x if y is the best solution amongst the set of feasible solutions. Here the term best is defined against the measure m and the goal function g. This function maps an instance x to an optimum solution. [3]

$$opt_A : I \to S, \forall x \in I, y \in f(x):$$
$$y = opt_A(x) \leftrightarrow m(x, y) = g\left\{ m\left(x, y'\right) \middle| y' \in f(x)\right\} \tag{4}$$

Combinatorial optimizations are optimization problems where the set of feasible solutions is discrete.

1.3 Traveling Salesman Problem as an NP-Hard Task

The minimal Traveling Salesman Problem can be regarded as a graph problem
with edge weights [13]

$$G_{TSP} = \left(V_{cities}, E_{conn} \right)$$
$$V_{cities} = \{v_1, v_2, \ldots, v_n\}, E_{conn} \subseteq \{(v_i, v_j) \mid i \neq j\} \tag{5}$$
$$d : V_{cities} \times V_{cities} \to \mathbb{R}, C := \left[d(v_i, v_j) \right]$$

An instance in the TSP problem is a set of interconnected cities (G_{TSP}) and a
distance (d) for each pair of cities. The set of feasible solutions to an instance contains
all possible permutations of the vertices in the graph, i.e. all possible directed
Hamiltonian cycles in G_{TSP}. The distance measure is defined as a sum of distances
between adjacent cities in the tour. The goal is to select a directed Hamiltonian
cycle with minimal total length amongst all possible directed Hamiltonian cycles
in the instance graph:

$$I_{TSP} = (G_{TSP}, C), \forall x \in I_{TSP} :$$
$$f_{TSP}(x) = \left(v_{\pi(1)}, v_{\pi(2)}, \ldots, v_{\pi(v(G_{TSP}))} \middle| \pi : [1 \ldots v(G_{TSP})] \to [1 \ldots v(G_{TSP})] \right) \tag{6}$$
$$\forall x \in I_{TSP}, y \in f(x) : m_{TSP}(x,y) = C_{\pi(v(G_{TSP})),\pi(1)} + \sum_{i=1}^{v(G_{TSP})} C_{\pi(i),\pi(i+1)}$$

The corresponding decision problem (the decidable question whether a directed
Hamiltonian cycle exists in a G graph whose total length is smaller than a given
m_0) is known to be an NP-complete problem as a polynomial time many-to-one
reduction (Karp reduction) exists, here the TSP will mark the corresponding decision
problem of the minimal TSP optimization problem: [21]

Theorem 1.1. The TSP is an NP-complete decision problem

Proof. It is obvious that the TSP is an NP problem, as a graph and a possible directed
Hamiltonian cycle with a total length smaller than m_0 certifies that the problem
lies in the NP complexity class according to the Witness theorem. Therefore it
is enough to find a Karp-reduction to a well-known NP-complete decision problem.
H denotes the decision problem whether a directed/undirected Hamiltonian
cycle exists in a G graph. H is known to be NP-complete. [14] H can be regarded
as a special subset of the TSP where every edge-weight is 1 and m_0 is equal to the
number of vertices in the G undirected graph. Therefore such a mapping is a polynomial
time many-to-one reduction, a Karp-reduction. The corresponding decision
problem is NPO-complete. [20]

1.4 The Modified Traveling Salesman Problem

In the field of logistics it is essential to solve the minimal Traveling Salesman Problem in a fast and efficient (close to optimal) manner. Some minor modifications will be introduced to the original TSP. These alterations make the problem considerably harder to solve than the traditional metric TSP but it suits the needs of logistical applications better.

Let us take a bustling metropolitan area anywhere around the World as an example. The city center which is also a place where people make ends meet and do business is encompassed by increasing layers of industrial and residential areas. In traditional city designs where residential and business/industrial areas are separated people have to travel great distances on a daily basis. This leads to rush hours and eventually inevitable traffic snarls and inequalities. Instead of the geological distance it is more convenient to apply time as a cost function. While traveling through a metropolitan area, the distance remains identical regardless the date of departure, the elapsed time greatly depends on it. That means two identical routes may have different cost depending on the date of departure.

This cost function can no longer be considered a metric as neither symmetry, nor triangle inequality hold true anymore. As the triangle inequality is not valid anymore, shortcuts may exist: Often it is more rewarding to travel on a longer but less time consuming route. Sometimes trailer trucks have to wait at one place during holidays due to government restrictions and regulations. These inevitable cost increases have to be taken into account as well. Even multiple visits in the same city might be reckoned if it decreases overall cost. Thus, a further parameter is introduced into our cost function: the time.

It is obvious at the first glance that this model suits the needs of logistics better. The proposed optimization problem is not harder than the original TSP, albeit a lot of approximation algorithms and heuristics are based on assumptions which are no longer valid in this context.

Without the loss of universality let us assume that time is always measured from the date of departure and time dependent cost functions are chosen according to this date. (Different cost functions belong to trucks departing in the morning and different ones to evening departures)

Based on the model defined in section 1.3 a new method is introduced to calculate time dependent continuous cost matrices: [8]

$$I_{TSP} = (G_{TSP}, C'), C' := \left[d\left(v_i, v_j, t \right) \right], \forall x \in I_{TSP} :$$

$$f_{TSP}(x) = \left(v_{idx(1)}, v_{idx(2)}, \ldots, v_{idx(n)} \middle| idx : [1 \ldots n] \rightarrow [1 \ldots v(G_{TSP})], n \geq v(G_{TSP}) \right)$$

$$\forall x \in I_{TSP}, y \in f(x) : m_{TSP}(x, y) = \sum_{i=1}^{n-1} C^{t_i}_{idx(i), idx(i+1)} + C^{t_n}_{idx(n), idx(1)} \quad (7)$$

$$t_1 := 0, t_{i+1} := t_i + \tau \left(C^{t_i}_{idx(i), idx(i+1)} \right)$$

The set of solutions are series of vertices (tours) where each vertex occurs at least once in the tour. The measure of a solution is defined as an aggregated cost function of the graph cycle.

The time is calculated using the recursive formula above, where the date of departure from the next vertex (city) in the tour is equal to the date of departure from the previous city plus the cost of travel between the previous and next vertices transformed to time. Tau is a transformation which transforms cost to time. This allows us to define cost as an abstract principle, which may not have physical meaning at all.

Function idx is a function which maps tour stages to cities. This function is surjective as its values span its whole co-domain.

1.5 Main Concepts behind Evolutionary Algorithms

- Gene: A functional entity that encodes a certain property
- Allele: The value of a gene
- Genotype: The combination of an individual's alleles
- Phenotype: All properties of an individual
- Individual: Chromosome, a candidate solution to the problem
- Population: Group of individuals
- Generation: Peer (contemporarily existing) individuals
- Fitness function: The measure of the individuals' goodness
- Evolution: Development and advance of the population
- Selection: The survival of certain individuals
- Crossover: Combination of two distinct individuals' chromosome
- Mutation: Random change in the chromosome
- Convergence: The approximation of phenotypes to the optimum

1.6 Bacterial Evolutionary Algorithms

Bacterial evolutionary algorithms [19] are based on bacteriologic adaptation and form a subset of evolutionary algorithms [11]. It is widely inspired by evolutionary ecology, which tries to observe the adaptation of living organisms in the context of their environment. The main idea behind evolutionary ecology is that in a heterogeneous case, there is no single individual that fits the whole environment. One needs to reason at the population level. [4]

In bacterial evolutionary algorithms the bacterial mutation, cloning, and gene transfer replace the selection, crossover, and mutation operators known from traditional genetic algorithms. The method was inspired by the gene transfer (transduction) of bacterial populations. Transduction is the process by which DNA is transferred from one bacterium to another by a virus. It also refers to the process whereby foreign DNA is introduced into another cell via a viral vector. [10]

Bacterial evolutionary algorithms may have better convergence properties when applied on certain problems than corresponding genetic algorithms. In many cases less generation is sufficient to reach a quasi-optimal solution and due to the presence of clones in bacterial mutation the population can be smaller. [6]

1.7 Memetic Algorithms

Evolution is not exclusive to biological systems as any complex system that fulfills the traits of an evolving system (shows the principles of inheritance, variation and selection) is subject to evolution. Therefore the term meme was introduced as the basic unit of cultural transmission, or imitation (unit of imitation in cultural transmission) by Richard Dawkins in his book the Selfish Gene. [7]

Meme in the context of global optimization algorithms can be regarded as an inheritable problem specific knowledge. Memetic algorithms are the synthesis of a local and a population based global search. [17]

There are two different approaches to memetic algorithms. According to the genetic algorithm community memetic algorithms are a special form of genetic algorithms that are extended with local hill-climb methods. In this interpretation memetic algorithms are hybrid genetic algorithms. [9, 17] Nevertheless this is a simplistic viewpoint since generally speaking, memetic algorithms are rather some form of population based global metaheuristics (not always GA) where individuals perform heuristic local search in the problem domain. This search is based on some particular problem specific information (meme). [15]

Algorithm 1.1. *Memetic algorithm:*

- *Population construction:* The population is formed from individuals that represent a feasible solution to the optimization problem. Individuals can be constructed randomly or using a given heuristics.

- *Local search:* Each individual performs local search on the problem domain (e.g. Monte Carlo Simulated Annealing)

- *Interaction:* Individuals interact with each other within the population. This can be either a competitive interaction, which is any competitive act or race between the members of the population or a cooperative one. Generally speaking, the aim of the cooperation is to share information amongst the members of the population. An example for cooperation in the case of bacterial memetic algorithms would be the gene transfer phase.

- Repetition of the above 2 steps until the exit conditions are fulfilled. This step is often amended with the measure of diversity and other population characteristics.

As one might observe the above loose definition of memetic algorithms adequately describes almost every evolving competitive and cooperative systems where members of a population search a common goal. Memetic algorithms are therefore not limited to the application of linear, genetic representations. As a consequence of this, the later conceptualization is more allowing than the traditional one.

Mainly due to the introduction of per individual local search and the concept of the memes, memetic algorithms tend to find high quality solutions faster than their traditional evolutionary counterparts.

2 Formulating a Solution for the Modified TSP

In the following a method will be proposed, which is based on bacterial memetic algorithms [5] and it is capable to solve the modified Traveling Salesman Problem described earlier. As mentioned before, metaheuristics only give us a problem independent framework but there are certain problem specific parts in every metaheuristics that need to be chosen in accordance with the problem. Characteristics of the heuristics greatly depend on these problem specific parts. For example in the case of bacterial memetic algorithms the implementation of a proper bacterial mutation and gene transfer operators, which suit the needs of the problem itself, is essential to obtain correct behavior and assist the interchange of genetic information within the population. The bacterial mutation provides the sufficient hill-climbing properties.

The first step in the development of evolutionary algorithms is often the encoding the problem into a genetic representation. Appropriate fitness or cost function also plays an important role.

Every metaheuristics has some ubiquitous parameters that are also somewhat problem dependent. E.g. cool down factor in the case of simulated annealing and gene transfer count in the case of bacterial memetic algorithms. Choosing these parameters wrongly often leads to the premature convergence or heavy disruption of the population in case of the generation based metaheuristics.

In memetic algorithms the other problem dependent part is the local search itself, which drives individuals in the population towards a local optimum using problem specific knowledge. The local search algorithm shall be adopted to work on linear genetic representations. Local search algorithms must also be deterministic as stochastic behavior is part of the evolutionary algorithm itself.

2.1 Encoding of the Modified Traveling Salesman Problem

An obvious linear genetic representation of our modified Traveling Salesman problem is the following:

A tour in an arbitrary graph can be expressed as a sequence of vertex indices. If every vertex has an associated index (possibly an integer value), that is a bijective mapping exists between the set of vertices in a graph and the set of integers between $[0, n)$ then every tour can be transformed into a sequence of integers and each sequence of integers has a corresponding tour in the graph.

Every tour in a graph starts from the same city (this is indexed with 0); therefore the start city is not present in the genetic representation of the tour. Thus with the exception of 0, every integer between $[0, v(G_{TSP}))$ appears at least once in the linear genetic representation. If segment multiplier is set to one then the length of each chromosome is equal to the number of vertices $- 1$ (as the initial city is not counted); thus every integer between $[0, v(G_{TSP}))$ appears at least and at most once, this solves the original TSP problem where each city is visited only once.

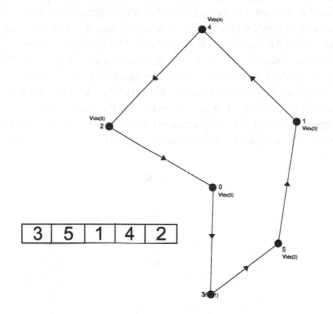

Fig. 2.1. The encoding of a tour

2.2 Creating the Initial Population

The creation of the initial population can be crucial in reaching high quality solutions quickly. In order to achieve this goal some easily implementable heuristics are brought in, which can find a better than average tour within a very short time. These heuristics guarantee that some individuals with above the average fitness are always present in the initial population. This is an eugenic element in the algorithm. [22]

Method 2.2.1. *Random creation:* During the initial construction all but a few individuals are created randomly. The length of the chromosome is also chosen randomly. If the segment multiplier is 1 then n equals to the number of vertices in the graph otherwise it is a random number between $(v(G_{TSP}), m_{seg}\, v(G_{TSP})]$ where m_{seg} stands for the segment multiplier. First of all the presence of all cities in the random tour must be ensured therefore a random permutation is created, which has a length of $v(G_{TSP})$ - 1. After that random integers between $[0, v(G_{TSP}))$ are inserted into random places in the chromosome $n - v(G_{TSP})$ times. This method guarantees that at least a major part of the population is distributed uniformly in the search space. As mentioned earlier this is essential in order to attain the apt genetic variance of the population.

Method 2.2.2. *Nearest neighbor (NN) heuristics:* The nearest neighbor heuristic is responsible for the creation of one individual in the population if it is enabled. It is a deterministic heuristics with a predictable result thus it is not necessary to create more than one individual using this method. The algorithm recursively takes the nearest neighbor from the set of remaining cities in the graph. Initially this set contains all cities except the start city (0). The term nearest is not necessary defined over a metric space but rather means the city that costs the least to travel to from the currently visited city. If a city is chosen as the next visited city then it is immediately removed from the set of remaining cities and it is appended to the sequence of already visited cities. The algorithm stops when the set of remaining cities is empty.

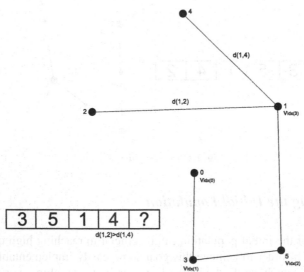

Fig. 2.2. The NN heuristics

$$\pi(0) := 0, \pi(i+1) := \min_{\forall k \neq \pi(j), 0 \leq j \leq i} \left\{ d\left(v_{\pi(i)}, v_k, t_i\right) \right\} \tag{8}$$

Method 2.2.3. *Secondary nearest neighbor (SNN) heuristics:* The secondary nearest neighbor heuristics, as its name might suggest always chose the second least costly city after each consecutive step in the recursion. The rest of the algorithm is the same as it was in the nearest neighbor algorithm.

Method 2.2.4. *Alternating nearest neighbor (ANN) heuristics:* The alternating nearest neighbor heuristic can be regarded as a combination of the above two methods as it takes the nearest and second nearest cities in alternating order after each step. This method produces an above average result with a certain amount of discrepancy, which is useful for genetic variety. All three methods can be used in our implementation to construct deterministic individuals.

2.3 Calculating the Aggregated Cost of a Tour

The distance matrix plays an important role in the algorithm. This matrix is an ordinary matrix where each item can be a scalar, a fuzzy value or a function of time.

When the distance matrix contains solely scalar values the corresponding optimization task is an asymmetric TSP problem (or symmetric TSP if the matrix is symmetric itself). Each matrix item stores the constant cost of travel between the adjacent cities.

Distances can be defined as a fuzzy value. Any matrix item can be a triangle membership function.

The distance between two arbitrary cities can be expressed as a function of time. In this case the cost of travel between these cities depends on the time of departure from the source city. The time elapsed between the traveling is calculated using a transformation function. This can be anything that can meaningfully convert cost into time. To simplify our model we chose metric distance as a universal cost of travel and travelers perform linear motion with a constant average velocity between each city.

$$C^t := \left[d_{i,j}(t) \right]$$

$$m_{TSP}(x,y) = \sum_{i=1}^{n-1} C^{t_i}_{idx(i),idx(i+1)} + C^{t_n}_{idx(n),idx(1)}$$

$$t_1 := 0, t_{i+1} := t_i + \frac{C^{t_i}_{idx(i),idx(i+1)}}{\left| \bar{v}_{idx(i),idx(i+1)} \right|}$$

(9)

Cost evaluation is a linear function with respect to the number of vertices in the graph. Despite that this evaluation can be rather time consuming but this greatly depends on the functions embedded inside the continuous distance matrix.

2.4 Bacterial Mutation

Bacterial evolutionary algorithms use bacterial mutations instead of the mutations implemented in standard genetic algorithms. [18] The bacterial mutation is an operator performed on every single individual in the generation.

First, the bacterium is replicated in N_{clone} instances. Chromosomes are divided into fixed length but not necessarily coherent segments. A randomly chosen i^{th} segment of the chromosome is altered in the clone bacteria (mutation) but the same segment in the original bacterium is left intact.

After that, each clone bacteria including the original one are evaluated according to the fitness function and the ith segment of the fittest bacterium is transferred into the other bacteria.

This process – mutation, evaluation, selection, and replacement – is consecutively applied on each segment of the chromosome. At the end of the bacterial mutation the most fit clone bacterium is selected and the others are destroyed. As a result of the mutation a more or equally fit bacterium is created. [6]

Initially N_{clone} clones are created from the original bacterium. At the beginning these clones share exactly the same chromosome. Chromosomes are divided into segments.

Method 2.4.1. *Loose segment mutation:* The key difference between our implementation and traditional bacterial evolutionary algorithms is that segments are not necessarily cohesive, alleles from one segment can come from anywhere within the chromosome and do not need to be adjacent in the chromosome. Therefore a segment is an ordered collection of indices that point to alleles in the chromosome. Permutation provides a very convenient way to achieve such a behavior: A permutation of the vertex indices can be split into coherent segments. These segments store indices to alleles in the original chromosome.

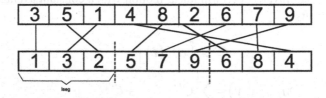

Fig. 2.3. Incoherent segments

Method 2.4.2. *Coherent segment mutation:* In this case, as opposed to the loose segment mutation, segments are cohesive and alleles from one segment need to be adjacent in the chromosome as well. The chromosomes are divided into coherent segments where each allele in one segment is adjacent in the chromosome. Segment creation is a very simple procedure: the chromosome of each bacterium is split into segments with a predefined length. The rest of the algorithm is the same as the loose segment mutation.

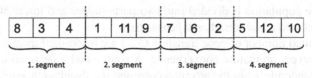

Fig. 2.4. Coherent segment

Algorithm 2.1. *Bacterial mutation on a segment:* Bacterial mutation on a segment is basically a random permutation in the case of each clone. N_{clone} number of $l_{segment}$ long permutations are made:

- In the original bacterium the segment is left intact.
- In one bacterium the segment is reversed.
- Other segments are permuted randomly.

After the bacterial mutation was performed on the segment the bacteria are evaluated according to the fitness function and the fittest bacterium is selected. In our implementation the rest of the clones are destroyed and new clones are derived from the fittest bacterium after each subsequent per segment mutation. When the segment length variance is allowed the length of each segment may vary with a given probability after every successive segment mutation. That means after reordering the alleles in a specific segment, random alleles (which appear at least one more time elsewhere in the chromosome) may disappear from the segment –thus shortening the chromosome- or random alleles (including the 0 allele) may be inserted somewhere into the segment.

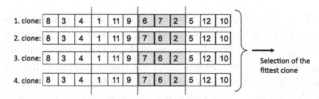

Fig. 2.5. Bacterial mutation

The time complexity of the mutation is $O(N_{clones}\ n^2)$ and space requirement is $O(N_{clones}\ n)$.

2.5 Gene Transfer

The gene transfer is a phenomenon when a part of the source bacterium is transferred into the chromosome of the destination bacterium. The gene transfer helps to spread advantageous genetic material within the population and it is based on the assumption that fitter individuals have more advantageous coherent chromosome segments than less fit ones have. [6] First and foremost members of the population are sorted in a descending order according to their fitness measure. After that the whole population is divided into two parts (upper and lower 50%) a good (advantageous) and a bad (disadvantageous) part.

Only a limited number of gene transfers (N_{inf}) are performed in each generation in order to prevent premature convergence. A source and a destination bacterium are selected randomly from the advantageous and the disadvantageous part of the population. Only the destination bacterium is altered and straight after the gene transfer it is reinserted into the ordered population.

As a consequence of the gene transfer a randomly selected segment with a pre-defined length (this is different to the segment length introduced in the mutation operator) is transferred into the target bacterium from the source bacterium. This pre-defined segment length will be referred as $l_{transfer}$.

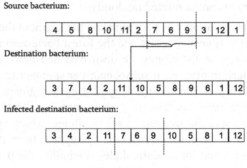

Fig. 2.6. Gene transfer

Algorithm 2.2. *Gene transfer:* A random source segment is selected from the source bacterium (see Fig. 2.6.) (7,6,9) and this segment is transferred to the target bacterium (between 11 and 10). In the target bacterium this segment is inserted into the chromosome at a given offset (after the fourth allele). If the offset generation is enabled then the destination offset can be different from the source offset and it is chosen randomly. Otherwise the destination offset is the same as the source offset. The alleles in the inserted chromosome must be deleted from everywhere besides the target chromosome. (E.g. the first allele in the transferred segment, which is found at the second place in the destination chromosome and therefore in the chromosome of the infected bacterium 7, is deleted from its original place. 4 immediately follows 3 in the infected chromosome.)

At the beginning of the population based gene transfer, the individuals are sorted according to their fitness measure. The fitness calculation requires $O(N_{idv} v(G_{TSP}))$ consecutive distance calculation and the sorting requires on average $O(N_{idv} \log N_{idv})$ comparison. After each gene transfer the infected bacteria is reinserted into the sorted population. It is also required to recalculate the fitness measure of the target bacteria. (*n* distance calculation and N_{idv} comparison) This has the time complexity of $O(N_{inf}(n + N_{idv}))$. The aggregated time complexity of the gene transfer is:

$$O\left(N_{idv} \cdot \left(v\left(G_{TSP}\right) + \log\left(N_{idv}\right)\right) + N_{inf} \cdot \left(n + N_{idv}\right)\right) \tag{10}$$

2.6 Local Search

The local search techniques are crucial parts of every memetic algorithms and they can be viewed as methods that rely on vital problem specific knowledge. In the case of memetic algorithms local search algorithms are responsible for the improvement of the candidate solutions in the population. Usually these algorithms are not sophisticated enough to find the optimum solution alone so they are combined with black box global optimization methods (metaheuristics) such as the bacterial evolutionary algorithms. A local search algorithm starts from a candidate solution and then iteratively moves to a neighbor solution. This is only possible if a neighborhood relation is defined on the search space. [12]

In the case of the TSP there are multiple different neighborhoods defined on a candidate solution and generally a local optimization with neighborhoods that involve changing up to k components of the solution is often referred to as k-opt. [12, 16]

Method 2.6.1: *2-opt local search:* The 2-opt local search is a very simple algorithm that can improve candidate solutions by exchanging edge pairs in the original graph. The algorithm works on the metric TSP but with certain constraints it can be adapted to a non-metric TSP as well. Edge pairs *(AB, CD)* are iteratively taken from the graph and the following inequality is examined: $|AB| + |CD| > |AC| + |BD|$. If the inequality above is true then edge pairs are exchanged; *AB* and *CD* edges are deleted from the graph and *AC* and *BD* edges are inserted instead. One of the sub-tours between the original edges is reversed.

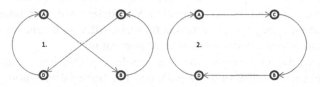

Fig. 2.7. A graph before and after a 2-opt step

As mentioned before non-metric TSPs are treated otherwise. The inequality above does not guarantee that the modified tour will have lower aggregated cost because cost function is not necessarily symmetric. After the edge exchange both the original and the modified tours are evaluated according to their fitness measure and only the less costly tour is kept.

The 2-opt local search, which deterministically improves a tour in a metric TSP problem, is only a random algorithm in the case of non-metric TSPs. The 2-opt local search has a worst-case time complexity of $O(2! \, n^2)$ in naive array-based implementations.

Method 2.6.2: *3-opt local search:* As one may have observed in the 2-opt local search, which comprised an edge pair exchange, the 3-opt local search algorithm works on edge triples. The selected edge triples are removed from the tour, which thus falls into 3 distinct sub-tours. Besides the original edge order there are two possible ways to reconnect these sub-tours (the other possibilities can be obtained as a result of consecutive 2-opt steps). The output of a 3-opt step is always the less costly tour.

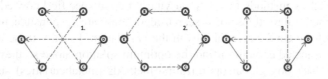

Fig. 2.8. A possible 3-opt move

The 3-opt local search has a worst case time complexity of $O(3! \; n^3)$ in naive array-based implementations.

2.7 *Putting the Parts Together*

First and foremost the tour construction heuristics construct N_{idv} random and maximum 3 deterministic tour by using one of the deterministic tour construction method described in 2.1.2.

The population is driven by a bacterial evolutionary algorithm. A bacterial mutation is performed on every member of the population and the intermediate population is ranked according to the fitness value of the individuals. The sorted population is divided into two parts which serve as the source and the destination of the gene transfers. After the bacterial mutation N_{inf} consecutive gene transfer is performed on random source and destination bacteria. After each gene transfer the altered bacterium is reinserted into the ordered population.

The memetic part of the algorithm is the problem specific local search, which is performed on each member of the population. The bacterial mutation, gene transfer and the local search together result in the next generation of the population. The bacterial evolutionary algorithm drives the entire population towards the global optimum while problem specific local optimization algorithms, such as the 2-opt and the 3-opt local search methods, improve candidate solutions locally.

3   Possible Applications

3.1 *PCB Drilling*

An important real-life application of the minimal TSP optimization is the PCB (Printed Circuit Board) drilling. In every aspect of production activities, productivity and cost reduction are two crucial factors. During PCB manufacturing multiple holes are drilled into the PCB material, these holes can serve as soldering points or interconnects between different layers of the circuit board. In a complex PCB there can be thousands of soldering points or interconnects. In order to reduce processing time and related costs the total length of the tool path must be kept as short as possible.

3.2 VLSI Signal Distribution

Another similar application of the minimal TSP is also from the field of electronics: The signal distribution inside a VLSI (Very Large Scale Integrated) circuit between several terminals. A proper signal distribution network greatly contributes to the lower power consumption of the VLSI circuits; therefore it is essential to find an optimum solution. The description of this problem is beyond the scope of this paper but our example case comes from the famous VLSI related TSPLIB graphs.

The example network was taken from a "classical" reference instance TSP (www.tsp.gatetech.edu, xqf131). The total length of the optimal tour is 564. The quality of the solutions produced by the algorithm greatly depends on the parameters of the bacterial memetic algorithm. The more generations and individuals we have the more precise solutions are produced however computation time increases as well. As already mentioned earlier, evolutionary algorithms transform a certain optimization problem into another one. In this case the optimization task is to find the optimal parameters that result the best possible solutions.

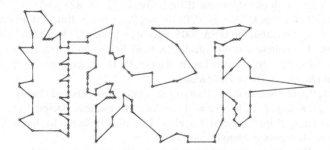

Fig. 3.1. A possible solution to the xqf131 VLSI network

$$N_{gen}=100, N_{idv}=40, l_{mutation}=4, l_{transfer}=50, N_{clones}=12, N_{inf}=12$$

Using the parameters above a shortest tour with a length of 566.42 was obtained (15 test runs were conducted with an average of 567.59 and standard deviation of 1.27) which is only 0.4% worse than the optimal tour of the corresponding TSP problem.

Standard deviation is adequately low, which means that these results are easily reproducible, and the average is also very close to the global optimum. These factors make the algorithm applicable and practical in real life scenarios.

The calculation of the optimal tour in the xqf131 network took approximately 4-5 minutes on an Intel Core2 Duo T7400, 4GB RAM workstation with a peak memory consumption of 20280 kB.

The scope of future research activity is to adapt the algorithm to the Vehicle Routing Problem (VRP) and to set general rules that can help to achieve the most efficient parameters of the bacterial memetic algorithm according to the size and other features of the given TSP graph.

Acknowledgments. This paper was supported by a Széchenyi University Main Research Direction Grant 2009, the National Scientific Research Fund Grant OTKA T048832 and K75711.

References

[1] Applegate, D.L., Bixby, R.E., Chvátal, V., Cook, W.J.: The Traveling Salesman Problem, A Computational Study. Princeton University Press, Princeton (2006)

[2] Applegate, D.L., Bixby, R.E., Chvátal, V., Cook, W.J.: Solving Traveling Salesman Problems. Retrieved from: The Traveling Salesman Problem (2008), http://www.tsp.gatech.edu/

[3] Ausiello, G., Crescenzi, P., Gambosi, G., Kann, V., Marchetti-Spaccamella, A., Protasi, M.: Complexity and Approximation: Combinatorial Optimization Problems and their Approximability Properties. Springer, Heidelberg (1999)

[4] Baudry, B., Fleurey, F., Jézéquel, J.-M., Le Traon, Y.: Automatic Test Case Optimization: A Bacteriologic Algorithm. IEEE Software 22, 76–82 (2005)

[5] Botzheim, J., Cabrita, C., Kóczy, L.T., Ruano, A.E.: Fuzzy Rule Extraction by Bacterial Memetic Algorithm. In: IFSA 2005, Beijing, China, pp. 1563–1568 (2005)

[6] Botzheim, J.: Intelligens számítástechnikai modellek identifikációja evolúciós és gradiens alapú tanuló algoritmusokkal. Budapest: Ph.D. dissertation, Technical University of Budapest (2007) (in Hungarian)

[7] Dawkins, R.: The Selfish Gene. Oxford University Press, Oxford (1989)

[8] Földesi, P., Kóczy, L.T., Botzheim, J., Farkas, M.: Eugenic Bacterial Memetic Algorithm for Fuzzy Road Transport Traveling Salesman Problem. In: International Symposium of Management Engineering (2009)

[9] Goldberg, D.: Genetic Algorithms in Search, Optimisation, and Machine Learning. Addison-Wesley, Reading (1989)

[10] Griffiths, A.J., Miller, J.H., Suzuki, D.T., Lewontin, C.R., Gelbart, M.W.: Gene Transfer in Bacteria and their Viruses. Retrieved from National Center for Biotechnology Information (2000), http://www.ncbi.nlm.nih.gov/books/ bv.fcgi?rid=iga.section.1363

[11] Holland, J.H.: Adaption in Natural and Artificial Systems. The MIT Press, Cambridge (1992)

[12] Hoos, H.H., Stutzle, T.: Stochastic Local Search: Foundations and Applications. Morgan Kaufmann, San Francisco (2005)

[13] Kann, V.: Minimum Traveling Salesperson. Retrieved from Network Design (2000), http://www.nada.kth.se/~viggo/wwwcompendium/node69.html
[14] Karp, R.: Reducibility Among Combinatorial Problems. In: Symposium on the Complexity of Computer Computations, p. 14. Plenum Press (1972)
[15] Kauffman, S., Levin, S.: Towards a General Theory of Adaptive Walks on Rugged Landscapes. Journal Theory of Biology 11, 128 (1987)
[16] Lin, S., Kernighan, B.W.: An Effective Heuristic Algorithm for the Traveling-Salesman Problem. Operations Research 21, 498–516 (1973)
[17] Moscato, P.: On Evolution, Search, Optimization, Genetic Algorithms and Martial Arts: Towards Memetic Algorithms. Caltech Concurrent Computation Program, Tech. Rep., California (1989)
[18] Nawa, N.E., Hashiyama, T., Furuhashi, T., Uchikawa, Y.: Fuzzy Logic Controllers Generated by Pseudo-Bacterial Genetic Algorithm. In: IEEE Int. Conf. Neural Networks (ICNN 1997), Houston, pp. 2408–2413 (1997)
[19] Nawa, N.E., Furuhashi, T.: Fuzzy System Parameters Discovery by Bacterial Evolutionary Algorithm. IEEE Tr. Fuzzy Systems 7, 608–616 (1999)
[20] Orponen, P., Mannila, H.: On Approximation Preserving Reductions: Complete Problems and Robust Measures. Helsinki: Technical Report C-1987-28, Department of Computer Science, University of Helsinki (1987)
[21] Rónyai, L., Ivanyos, G., Szabó, R.: Algoritmusok. Budapest: TypoTex (1998) (in Hungarian)
[22] Ye, J., Tanaka, M., Tanino, T.: Eugenics-based Genetic Algorithm. IEICE Transactions on Information and Systems E79-D(5), 600–607 (1996)

Towards Intelligent Systems with Incremental Learning Ability

Tomas Reiff and Peter Sinčák

Center for Intelligent Technology (CIT)
Dept. of Cybernetics and Artificial Intelligence, FEI TU of Košice
Letná 9, 042 00 Košice, Slovak Republic
peter.sincak@tuke.sk, www.ai-cit.sk

Abstract. The paper deals with the great challenge in building intelligent systems and it is a creation of human made systems with ability to learn incrementally. The approach is agent based and speed of learning is very high. The paper presents the concept of the starting project which is now underway in CIT and have ambition to create a strong project proposal towards domain oriented intelligent system with strong ability of incremental learning, knowledge fusion and sharing for its agents. We are ready to launch a server client approach and any community can join and make its own contribution in this project.

Keywords: universal incremental learning system, Agents, knowledge, fusion of knowledge, pattern recognition, neural networks, fuzzy systems, evolutionary systems.

1 Introduction

The Human-made intelligent system is **a dream** of the mankind and still it is very big problem of its definition and basic concept of these systems. Basically we can state that the industrial revolution around us is underway and AI and related fields are trying to find its position and contribution with aim top bring a new better technology and provide investors higher profit and better conditions for human life and future of humanity.

Generally we can state that for any process made by human we need certain amount of Intelligence (even we do not know what it is) but generally we can state that

$$GI = HI + MI \tag{1}$$

I.J. Rudas et al. (Eds.): Towards Intelligent Engineering & Information Tech., SCI 243, pp. 627–638.
springerlink.com   © Springer-Verlag Berlin Heidelberg 2009

Where **GI** is Global Intelligence necessary for accomplishing a process, **HI** is Human Intelligence necessary for accomplishing a process and **MI** is Machine Intelligence necessary for accomplishing a process. Generally proportion between **HI** and **MI** tells something about the **Autonomity** of the process. If we do assume that **GI** is always constant **1** then we do scale all the parameters in definition interval <0,1> that means that **HI** and **MI** are between <0,1> we can state that the Autonomity is high if **HI** is converging to **0** and **MI** is converging to **1**. Based on this assumtion we define Autonomity Measure of Intelligence **AMI** as follows:

$$AMI = HI / MI \tag{2}$$

that means that **AMI** has a definition interval of **<0 , LN)** (LN is large number), **HI** is between **<0,1>**, **MI** is between **<0,1>** . So based on that AMI describes a "fully" Autonomous system if **AMI** is **0**. The AMI should be linked to **MIQ** which is very difficult to describe and for sure it MUST be domain oriented. The problem is **"WHO"** will define **HI** and **MI** and I think it must be only made by human or community which define HI and MI by observation of the process. A good example of changing **AMI** is piloting a large plane 20 years back and now. **AMI** is completely **different** and is a nice example how things are changing towards **AMI equal to 0**.

These ideas and tools are important to implement. There is a strong believe that 4 main stream of technologies will affect this goal and these are: **nanotechnology, Information technology, Cognitive technology and Biolotechnology**. Convergence of these technologies should be the way how to create an Intelligent System and bring these systems into everyday use.

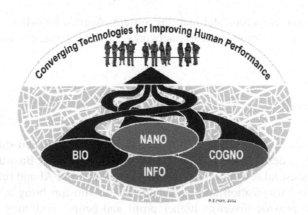

Changing the societal "fabric" towards a new structure
(upper figure by R.E. Horn)

Fig. 1. Importance of the technological convergence

2 Multiagent Sophisticated System

Our lab is setting up a new project for creating a Multiagent Sophisticated System (MASS) where we want to make an attempt for creating a prototype which will be able to fulfill basic requirements for conditions as follows:

- Agent-based concept with server-client approach where client is an Agent and server is a "Master of Agents". We do not like "full democracy in Agent Systems
- Each Agent will be a source for information which will be stored in "Masters – knowledge base" (MKB) with fast access technology
- Each Agent will be able to use a common MKB for its own "life" and activities
- Conflicts will be handled by very care and each conflict will be solved either by human or by preference rule base which is created by human
- Interactive technology is an important part of MASS
- We will start our project in domain – it will be much easier to handle a specific problem using domain oriented problems.
- The candidates for these domains are related to projects and funding for this MASS.

There are number of problems in this concept which can be summarized by the following questions:

1   What kind of Pattern recognition systems for "real" discrimination between classes we will use?
2   What kind of feature space we will use and will it be static or dynamical?
3   What kind of answers will the MASS provide to the user – crisp or fuzzy?
4   What if the user observes the error of MASS – how will the MASS create the feedback and "correction" in this situation?
5   What will be a form of knowledge database and what kind of technology will be used for information fusion
6   Can we use Grid related architecture in this research problem for server concept?

These questions are very difficult and are very domain related. The pilot project is made by MSc students and it is related to various playing card recognition using simple WEB camera. We will set up a server and put public a client downloadable from world/wide and we will ask people show to a client any playing card in front of the WEB camera.

The system will be a pilot system for creating a speed up learning using many agents worldwide. We do hope that we will be able to proof the basic concept of the system and proceed for more domains and more complicated and complex problems.

Currently we are working **on toy problem** regarding the playing cards and bank notes. **Our Dream** is to create a System which will be able recognize any playing card using WEB camera. We plan to set up the Web page of the system in December 2009 and offer to the World Wide community a client download and participate on World Wide learning of playing Cards. This toy problem can be a nice example of the system and also can reveal number of problems regarding to the global goal of building Intelligent Systems for Technological Applications.

3   MASS and SIFT for Pattern Recognition

Multi-Agent Sophisticated System (MASS) is considered as an approach or policy how to create families of Intelligent Systems. In this paper we describe MASS and its utilization in image pattern recognition methods.

The development of image matching by using a set of local interest points can be traced back to the work of Moravec ([1], 1981) on stereo matching using a corner detector. The Moravec detector was improved by Harris and Stephens ([2], 1988) to make it more repeatable under small image variations and near edges. Harris ([3], 1992) also showed its value for efficient motion tracking and 3D structure from motion recovery, and the Harris corner detector has since been widely used for many other image matching tasks. While these feature detectors are usually called corner detectors, they are not selecting just corners, but rather any image location that has large gradients in all directions at a predetermined scale.

The ground-breaking work of Schmid and Mohr ([4], 1997) showed that invariant local feature matching could be extended to general image recognition problems in which a feature was matched against a large database of images. They also used Harris corners to select interest points, but rather than matching with a correlation window, they used a rotationally invariant descriptor of the local image region. This allowed features to be matched under arbitrary orientation change between the two images. Furthermore, they demonstrated that multiple feature matches could accomplish general recognition under occlusion and clutter by identifying consistent clusters of matched features.

The Harris corner detector is very sensitive to changes in image scale, so it does not provide a good basis for matching images of different sizes. Earlier work of Lowe ([7], 1999) extended the local feature approach to achieve scale invariance. Work also described a new local descriptor that provided more distinctive features while being less sensitive to local image distortions such as 3D viewpoint change. This method was named Scale Invariant Feature Transform (SIFT). Later Lowe ([8], 2004) described how SIFT could be used in object recognition. The recognition part of this work is based on this but some methods were modified. Because SIFT was very successful, its approach was modified resulting to similar methods like PCA-SIFT ([9], 2004), FastAproximatedSIFT ([10], 2006) and SURF ([11], 2006).

4 MASS and SIFT for Pattern Recognition

As was mentioned before, this master thesis builds on [8] and it is using original SIFT method invented by Lowe. However some changes were made in the additional approach which is needed when using SIFT in object recognition (result of this approach is a transformed rectangle boundary around each recognized object according to its pose). These changes will be described in the following parts after which the system MASS will be described as well.

This work is using distance ratio between the first and second closest neighbor as well as Lowe's approach. The difference is that first and second closest neighbors could represent the same object. This provides better results especially with objects which contains common parts like playing cards. Also the distance ratio used in this work is 0.7 instead of Lowe's 0.8.

Because this part was not described clearly in [8], some implementation details are provided but the point of this part is the same. If $I(\theta_i, \sigma_i, x_i, y_i)$ are information from descriptor in test image and $M(\theta_m, \sigma_m, x_m, y_m)$ are information from matched database descriptor, then for rotation r, scale s, translation t_x in x axis and t_y in y axis could be written the following:

$$r = \theta_i - \theta_m,$$

$$s = \frac{\sigma_i}{\sigma_m}, \tag{3}$$

$$t_x = x_i + s(y_m \sin(r) - x_m \cos(r)),$$

$$t_y = y_i - s(x_m \sin(r) + y_m \cos(r)),$$

where r is from interval $<0; 2\pi)$ and σ_i, σ_m are absolute keypoint scales counted by formula for the absolute keypoint scale σ_a:

$$\sigma_a = \frac{\sigma}{s_o}, \tag{4}$$

where σ is refined keypoint scale and s_o is the relative size of the octave in which the keypoint was found (2, 1, 0.5, 0.25 ...).

Error e is used instead of half the parameters error range from Hough transform as was used by Lowe. The following formula is used to count error for all matched descriptors in every cluster:

$$e = MAX\left(\frac{|u_c - u|}{sw}; \frac{|v_c - v|}{sh}\right), \tag{5}$$

where $[u_c \; v_c]^T$ is counted location of the test image descriptor with use of train image descriptor location and affine parameters, $[u \; v]^T$ is the real location of test image descriptor, s is scale (counted from affine parameters), w is the width of trained object and h is the height. Only the matched descriptors with error less than 0.03 remain for re-solution of affine parameters.

After we have final affine parameters for remaining clusters, we will sort the clusters according to count of its members. The cluster with the biggest member count will be the best cluster in the first step and all other clusters which contain some matched descriptors from this best cluster or descriptor matches set, made from already removed clusters, are removed. In the next step, the best cluster will be the second best cluster and we will repeat this till the last remaining cluster (obviously it will be the worst cluster that remained). After this, multiple clusters of the same object should be removed.

Problem is that there could be a special case especially with objects like playing cards. Playing cards has marks which are the same in four or two corners. Therefore we can still have clusters which belong to the same object but consist of different descriptor matches (the difference is only that they are in other corner, but same). To avoid having doubled frame which differs only in opposite orientation, the intersection count of descriptor matches is used. If there are less than 3 descriptor matches outside of intersection, the worse cluster is removed.

Everything from the previous part was implemented as a plugin for MASS. This part will describe MASS and plugin types which are used in this system.

MASS could be described as multi-agent, incremental, plugin system with client-server architecture. With plugins for object recognition it is possible in parallel manner for many clients around the world to learn various objects or to recognize them. Gained knowledge will be stored on server which will host the object recognition setup in MASS.

Fig. 2 describes the plugin system of MASS. Red plugin types are required and together they represent the smallest possible system. Green plugin types are optional. As you can see there can be more filter and processor plugins in the system and processor plugins can by placed whether on client or server side. Flow plugin is special because the same plugin is placed on both sides and these sides are communicating together. Every plugin type will be now briefly described.

Flow is required part of the system. This plugin type is managing the client-server communication. If some user input is needed, form is included in the plugin.

Output is required part of the system. Its task is to provide and react on results provided by server. The reaction could be manipulation with some connected robot or device.

Input is optional part of the system. It provides input data from devices like webcams, microphones, sensors...

Filter is optional part of the system. It is changing its input, which can be from Input or other Filter plugin. The purpose is similar to Filters known from image or audio processing.

Processor is optional part of the system. It can be placed whether on client or server side of the system. Processor should change the input data to data which can be used in classification, clustering or other types of Handler plugin tasks.

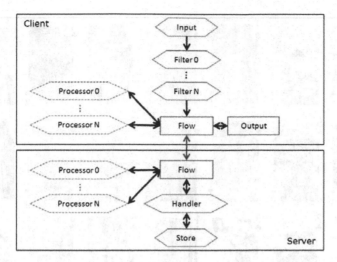

Fig. 2. MASS plugins system

Handler is optional part of the system. This plugin should have all the functionality required for manipulation with data in database. Classification, clustering and other similar operations should be implemented in this type of plugin.

Store is optional part of the system. Here should be implemented everything related to data storage. This could be implemented all by authors or they can implement link to SQL or similar database system. Moreover also internet can be considered as some sort of database.

Chosen domain for experiments was playing cards. Because the result of this object recognition is not only recognition of object but also its frame, correct recognition needs to be defined.

If recognized object class agrees with the real class and object surface which is not covered by frame of recognized object or part of this frame which surface is not covering the object do not exceed one fourth of the object surface, then given recognition is considered as correct.

Experiments are using 21 different playing cards (Fig. 3). These cards were trained with use of regular webcam under various illuminations. Because the count of descriptors for cards was various, two categories of cards were created to setup better test scenes. First category is strong cards with more than 200 descriptors and the second is weak cards with less than 201 descriptors.

To test the quality of SIFT, we will have 3 different scales and 3 different view angles for the test images (Fig. 4). Together it forms 3*3=9 categories and each will have 5 test images. First scale was chosen to be 1 and the next scales increase randomly. Cards count is scale dependent and for first scale there will be 2 cards (1 weak, 1 strong), for second scale 3 cards (1 weak, 2 strong) and last scale will have 6 cards (2 weak, 4 strong). First view angle is the same as training angle and other angles are random.

Fig. 3. Objects used in experiments

Fig. 4. Examples of test images

All test images contain together 165 cards. Results were that 149 cards were correctly recognized, 13 cards were not found and 4 cards were recognized incorrectly. The reason why this is 166 is that from 4 incorrectly recognized cards has 3 cards only bad frame and 1 card was recognized in spite of it was not there. You can see result examples on Fig. 5 also with this extra card.

From the results were made two charts (Fig. 6). First is showing correct recognition, not found and incorrect recognition percentage by all test image categories. The second chart is only showing percentage in all categories.

It would be good to mention facts about speed of training and recognition. Average training time was little less than 3 seconds and after all cards were trained, database had 5599 descriptors. Recognition took about 6 seconds in average.

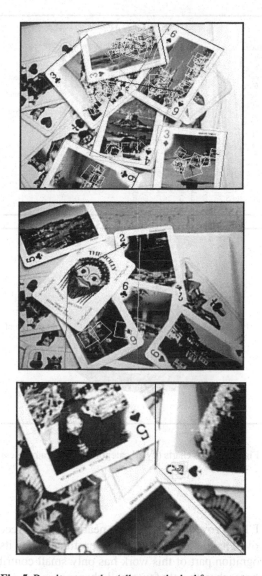

Fig. 5. Results examples (all correctly, bad frame, extra card)

Finally Fig. 7 shows examples of all correctly recognized cards. No doubt that object recognition done in this thesis could be implemented to run faster, but try to imagine that somebody will show you 21 cards each for 3 seconds and then you will have 6 seconds to recognize and bound them like computer did on these examples. It could be a problem.

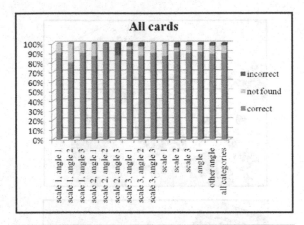

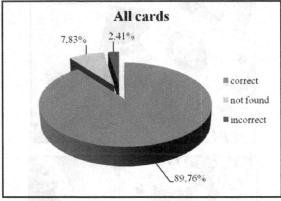

Fig. 6. Results charts (by all categories, in all categories)

Conclusion

Briefly said, SIFT has proved its quality by nearly 90% correct recognition rate and MASS has proved its legitimacy and future profitability by its essence.

The object recognition part of this work has only small contribution because it is using Lowe's approach with only few small changes.

The main contribution lies in the system MASS. With this system it is really easy to implement, test or develop various methods for various tasks. This makes the research easier and faster. In case of object recognition in MASS, it is possible to start building a huge object database. This database could be used by autonomous systems for better understanding of their surroundings. With this information they could be able to analyze more complex situations resulting in better behavior in their environment.

Future work could be improving object recognition approaches especially part 2.3 where instead of descriptor matches count, another criterion could be used to rate clusters. For example we can use surface of descriptor matches presence. Another possibility is to work on MASS to release final core which could be used as a framework for many artificial intelligence tasks.

Fig. 7. Examples of all correctly recognized cards

Acknowledgments. Research supported by the National Research and Development Project Grant 1/0885/08 "Learnable Systems based on Computational Intelligence" 2008-2010.

References

[1] Moravec, H.: Rover Visual Obstacle Avoidance. In: International Joint Conference on Artificial Intelligence, Vancouver, Canada, pp. 785–790 (1981)

[2] Harris, C., Stephens, M.: A Combined Corner and Edge Detector. In: Fourth Alvey Vision Conference, Manchester, UK, pp. 147–151 (1988)

[3] Harris, C.: Geometry from Visual Motion. In: Blake, A., Uille, A. (eds.) Active Vision, pp. 263–284. MIT Press, Cambridge (1992)

[4] Schmid, C., Mohr, R.: Local Gray Value Invariants for Image Retrieval. In: IEEE Trans. on Pattern Analysis and Machine Intelligence, pp. 530–534 (1997)

[5] Lindeberg, T.: Detecting Salient Blob-like Image Structures and their Scales with a Scale-Space Primal Sketch: a Method for Focus-of-Attention. International Journal of Computer Vision, 283–318 (1993)

[6] Lindeberg, T.: Scale-Space Theory: A Basic Tool for Analyzing Structures at Different Scales. Journal of Applied Statistics, 224–270 (1994)

[7] Lowe, D.G.: Object Recognition from Local Scale-Invariant Features. In: International Conference on Computer Vision, Corfu, Greece, pp. 1150–1157 (1999)

[8] Lowe, D.G.: Distinctive Image Features from Scale-Invariant Keypoints. International Journal of Computer Vision (2004),
http://www.cs.ubc.ca/~lowe/papers/ijcv04.pdf

[9] Ke, Y., Sukthankar, R.: Pca-Sift: A More Distinctive Representation for Local Image Descriptors (2004),
http://www.cs.cmu.edu/~yke/pcasift/pca-sift-cvpr-04.pdf

[10] Grabner, M., Grabner, H., Bischof, H.: Fast Approximated SIFT (2006),
http://www.icg.tu-graz.ac.at/pub/pubobjects/pdf/
fastapproximatedsift.pdf

[11] Bay, H., Tuytelaars, T., Van Gool, L.: SURF: Speeded Up Robust Features (2006),
http://www.vision.ee.ethz.ch/~surf/eccv06.pdf

[12] Brown, M., Lowe, D.G.: Recognizing Panoramas (2003),
http://www.cs.ubc.ca/~mbrown/papers/iccv2003.pdf

[13] Se, S., Lowe, D.G., Little, J.: Global Localization Using Distinctive Visual Features. In: International Conference on Intelligent Robots and Systems, Lausanne, Switzerland (2002)

[14] Kuniyoshi, Y.: University of Tokio, Japan (2008),
http://www.liveleak.com/view?i=00b_1207503927

[15] Nikolakis, G., Tzovaras, D., Strintzis, M.G.: Object Recognition for the Blind (2004),
http://www.ee.bilkent.edu.tr/~signal/defevent/papers/
cr2056.pdf

[16] Witkin, A.P.: Scale-Space Filtering. In: International Joint Conference on Artificial Intelligence, Karlsruhe, Germany, pp. 1019–1022 (1983)

[17] Koenderink, J.J.: The Structure of Images. Biological Cybernetics, 363–396 (1984)

[18] Mikolajczyk, K.: Detection of Local Features Invariant to Affine Transformations, Ph.D. thesis, Institut National Polytechnique de Grenoble, France (2002)

[19] Brown, M., Lowe, D.G.: Invariant Features from Interest Point Groups. In: British Machine Vision Conference, Cardiff, Wales, pp. 656–665 (2002)

Systems Engineering Approach to Sustainable Energy Supply

István Krómer

Institute for Electric Power Research, Budapest, Hungary
i.kromer@veiki.hu

Abstract. The world faces no greater challenge in the $2f^t$ Century than arresting the rapidly increasing accumulation of greenhouse gases in the atmosphere that cause climate change risks.

Despite an enormous amount of efforts done to reduce the uncertainties surrounding the danger of climate change, there is a little progress toward a global understanding about how to tackle the global warning disaster. It appears that none of the number of options considered to date to meet carbon targets is ample to address this extremely difficult problem.

This report is devoted to the analysis of interactions between the major uncertainties in forecasting the development of world energy needs, the impacts of the changing supply paradigms and the enabling technologies' R&D needs.

Obviously, the energy sector is looked at from a holistic perspective in view of the fact that a coordinated and comprehensive modernization framework is needed to achieve the main goals. The articulate vision of the major future energy supply issues outlined in the report helps to create a common understanding and a shared purpose of the numerous stakeholders regarding the most important and urgent development needs.

1   Introduction

Climate change is a global problem facing everyone in the 21st Century due to greenhouse gas emissions from human activity, the most significant of which, in quantitative forms, is carbon dioxide produced by burning fuels. CO_2 emissions have increased by more than 20% over the last decade. If the future is in line with the present trends, CO_2 emissions, oil and natural gas demand will continue to grow over the next decades. Stabilizing emissions at today's level would cause the concentration of CO_2 to continue to rise. Stabilizing the concentration of CO_2 implies that the global net CO_2 emissions to the atmosphere must peak and than decline year after year.

The scientific community state that limiting climate change risk will require movement to a very different energy system. The focus of the near term should be on preparation for dramatic transformations in the energy systems through

I.J. Rudas et al. (Eds.): Towards Intelligent Engineering & Information Tech., SCI 243, pp. 639–651.
springerlink.com   © Springer-Verlag Berlin Heidelberg 2009

technology experimentation, exploration and development. The key reason to develop and deploy advanced energy technologies is to control the cost of stabilizing greenhouse gas concentration.

While the current global economic crisis could make long term actions more difficult, it could also provide an unexpected impetus. If wisely allocated, funds invested in economic recovery can help address climate change which also advancing "green technologies" and industries what can lead to a new wave of economic growth.

The actual energy strategy making seems to be a veritable Tower of Babel. There are different groups of interest whose efforts collide in developing a new strategy. It is the task of systems engineers to provide the necessary linkages that enable to function these desperate actors as a strategy making community. In general, a systems approach involves defining objectives for the system and then looking at the ways they might be achieved by exploiting the potential synergies between elements of the system or finding new ways to overcome existing barriers.

A major issue in climate-change abetment decisions is the considerable uncertainty surrounding the problem including not only the scientific understanding of the processes driving climate change and the risk of possible catastrophic effects but also the economic values to be linked to these impacts.

This report is devoted to the analysis of interactions between the major uncertainties in forecasting the development of energy use, the change of supply paradigm and the trends of socially optional R&D investments.

2 Energy Supply Constraints

Accurate long-range forecasting of even the most basic energy data is difficult. Over the past several decades, a large portion of energy forecasts and relevant strategic recommendations turned out to be inaccurate and mistaken [1]. The gap between actual and forecasted energy consumption generally increase over time. However it is worth noting, a number of elements of past forecasts were correct and may well persist into the future, they include the following:

- The gradual electrification of the economy will continue and the economy will become more electricity intensive.
- Natural gas will continue to increase its share in primary fuel-mix, although it is uncertain how long this trend will last.
- Environmental protection will remain a high priority and global warming will remain a major concern.
- World population will continue to increase primarily in the developing countries. They will account for an increasing share of world energy consumption. Therefore future energy and environmental technology development will be increasingly dictated by developing nations' needs.

- Renewables will continue to increase in importance and account for a larger share of energy mix, however, their overall contribution will remain modest.

Ascertaining the likely energy trends and parameters for the world over the next decades remains a crucial exercise with major economic, environmental and political consequences.

Even the most sophisticated energy forecasts are strongly influenced by events and trends of the time of the forecasts.

The total energy demand growth is influenced by three major factors: the increase of world population, the economical growth and the decrease of energy intensity.

Fig. 1 shows the predicted impact of these factors on the energy consumption by 2030. China and India account for over the half of incremental energy demand to 2030 while the Middle East emerges as a major new demand center.

According to population projections, world population will continue to grow until around 2050. The fastest rates of world population growth (above 1,8%) occurred briefly during the 1950s then for a longer period during the 1960s and 1970s.The 2008 rate of growth has almost halved since its peak of 2,2% per year, which was reached in 1963.

As to the influence of the growth of population, it is interesting to note, that one toe/capita/year seems the minimum energy needed to guarantee an acceptable level of living despite many variations of consumption patterns and lifestyles across countries. The influence of per capita energy consumption on the human well-being begins to decline somewhere between 1 and 3 toe per inhabitant.

Advanced studies stated that with 1 toe per capita energy developing countries can provide any standard of life ranging from the present low level to a level as high as the industrialized regions in the mid and late 1970s for the majority of the population [3]. Large improvements in living standards can be achieved by shifting from traditional, inefficiently used, non-commercial fuels to modern energy carriers. As an example, the efficiency of commercial energy use is about 25% or more whilst the efficiency of non-commercial energy use for cooking purposes range around 10%.

The key to improving well-being without a major increase in primary energy consumption is the modernization and increased end-use efficiency in the use of fuels and end-use appliances.

Economic growth is closely related to growth in energy consumption because higher the economic growth more energy is used. Statistical studies found that the relationship between growth in energy use and gross domestic product growth became weaker after the first energy crisis, but later it seems to have become stronger again in the most developed countries. The rate of growth in world GDP to 2030 reflects the rising weight in the world economy of the non-OECD countries, where growth will remain the fastest. An IEA published world average is about D Energy = 0.5 D GDP. This value was taken in constructing Fig. 1.

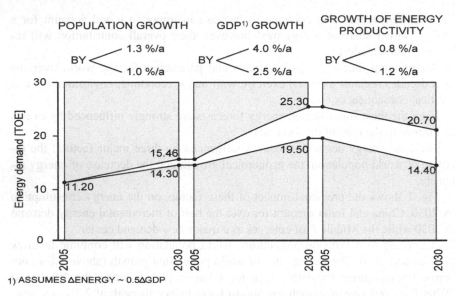

Fig. 1. Three factors determine growth in energy demand [2]

It is possible to decouple energy consumption and economic growth to some extent. More efficient use of energy may compensate the influence of economic growth and reduce energy consumption. Energy intensity is the ratio of total primary energy supply to GDP. There is an inverse relationship between energy productivity and energy intensity. Energy intensity improvements have provided enormous reductions in energy needs over the last fifty years. On a world level, energy intensity dropped at a rate of 1.8% per annum from 1990 to 2000 and 1.4% per annum from 2000 to 2006. The decrease can be explained mostly by the economic growth and growing energy needs of China. The reductions of energy intensity were the result of a combination of improvements in the efficiency of energy consuming devices, changes in the energy intensity of industrial processes and shifts in consumer demand away from energy intensive products toward lower energy intensity products and services. A clear decline in the energy intensity of developing countries can be observed as foreign investment increases. The reason for that is the use of modern technologies that came with foreign technologies, a leapfrogging over the old fashioned traditional technologies in use in these countries.

While there are sufficient conventional (hydro-carbon based) sources of energy to meet even these projections of future demand, the environmental consequences would be unacceptable. To accommodate the large increase in demand for energy without major environmental consequences, it is necessary to expand the choice of energy sources and at the same time increase the efficiency of energy use.

The world's future energy system, satisfying the growing energy needs will evolve from today's energy system. During its more than 150 years history, the global energy system continues to diversify. Even fuels whose market shares decline show trends of expanded deployment as the energy use grows overall.

Enhancing the energy supply security will require diversifying the fuel mix as well. There is no single answer as to the optimum fuel mix option, that will be determined by national circumstances. Interesting to note that almost all of the increase in fossil energy production will occur in non OECD countries.

It is clear that the low carbon energy must expand in the future energy mix. The key factor in expanding the choices of affordable and acceptable energy sources as well reducing energy needs is the development of new technologies. Recent research shows that aggressive development of a full portfolio of advanced technologies could reduce CO_2 emission levels by about 45% in 2030. The results reveal that a combination of new and existing technologies will be required.

A major topic in some recent forecasts concerns the future market potential for distributed generation and decentralized small energy systems.

3   New Supply Paradigm

Providing for energy consumption by means of local energy sources is becoming increasingly competitive with centralized supply. Over decades, centralized systems have provided the potential for efficient resource allocation and generated substantial economics of scale in the process of building and operating reliable energy conversion plants and transportation systems. Electricity was originally generated by burning coal in the city centres, delivering electricity to nearby buildings and recycling the waste heat to make steam to heat the same buildings. Over time, coal plants grew in size, facing the pressure to locate them far from the population because of their pollution. Later hydro, oil, natural gas and nuclear energy have subsequently been introduced as primary energy sources for large scale power plants. Eventually, a huge grid was developed and the power industry built the generation facilities far from users. These conditions prevailed from 1910 through 1970 and it was widely recognized that central generation was optimal because it delivered power at the lowest cost versus other alternatives.

During the last decades, a number of events have highlighted the vulnerability of the current centralized energy supply infrastructure. Unilateral decisions by primary energy exporting or transit countries can have dramatic consequences. The disputes on natural gas prices during the winter of 2005/2006 and 2008/2009 led Russia to cut its gas supply to Ukraine and consequently to Europe.

A great part of the centralized electricity supply system is approaching the end of its lifetime. The recent major blackouts highlighted how the complexity and vulnerability of the electricity grid can cause widespread social and economic troubles.

Large central power plants using fossil fuels produce large amount of greenhouse gases. It is evident that shifting from the centralized energy supply system to small-scale decentralized systems, where energy production and consumption are usually tightly coupled, might improve reliability and security of supply through using more distributed energy sources. New problems emerging by the mass introduction of dispersed generation can be solved by developing a new intelligent energy system.

Although most distributed energy systems are not new in a technological respect, they are receiving increased attention today because of their ability to provide a wide range of services: heat and power, peak power, demand reduction, backup power, improved power quality and ancillary services to the power grid.

The development on harvesting distributed renewable energy resources has to focus on the improvement of energy capture and conversion efficiencies, and reducing investment cost.

A system approach to the local renewable energy resources is a promising concept addressing distributed generation as an alternative solution for technical, economical and environmental constraints of conventional power system.

Radical proposals to replace existing grid infrastructure, given its wide deployment, high capital costs and well-understood technologies, are unlikely to succeed. A model of infrastructural coexistence and service displacement would be applied. There are radical changes that can be done near-term to exploit information technology to improve the reliability, visibility and controllability of existing grid and to support demand side participation in load reduction. Longer term, the Internet suggests alternative organizing principals for a 21st Century grid offering a model of infrastructural co-existence and service displacement. The same approach can yield a new architecture for local energy generation and distribution that leverages the existing energy grid, achieves new levels of efficiency and robustness. The power delivery systems have the potential to make the kinds of leap in capabilities and cost efficiencies happened in telecommunications and most other industries during the past decades.

The future grid will seamlessly integrate many types of electrical generation and storage systems with a simplified interconnections process analogous to "plug-and-play". Interconnections standards will enable a wide variety of generation and storage options.

The residential sector is an important model area for efficiency improvements, and large scale deployment of dispersed energy resources as it accounts for 22 percent of global energy consumption.

Consumers in residential sector need as cheap energy as possible but like their traditional "supplied" status too. Because of lack of skills this group needs plug-and-play solutions. A major part of the energy use is for space and water heating. There have been continued technical improvements in the efficiency of the large household appliances. However, these improvements have been offset by increases in the use, numbers and size of large appliances and the growing number of smaller, mostly electronic devices. In Hungary, only 13% of the total final household energy consumption is provided by electricity. The major part (87%) of the residential energy consumption produced by burning gas or oil directly and this burden would not be influenced at all by an increased efficiency of electricity use. But, advances in building design could make huge savings in energy use for space heating.

A more efficient way of using energy of local fuels is to burn them in combined heat and power generating (CHP) units to provide both heat and electricity. Using CHP concept to supply the needs of households entails either installing a separate unit in each develling or constructing a local heat distribution network [4].

The cogeneration concept leads to improved energy utilization by burning less fuel and reducing harmful emissions. Considering the limited transportability of heat decentralised, dispersed supply systems based on locally available alternative resources open a new field for innovative applications of cogeneration options.

The most extensively available local alternative sources for CHP applications are carbon-based: energy crops, agricultural and forestry wastes and municipal wastes. Although, their use produces carbon dioxide emissions, this is largely compensated by the carbon dioxide removed from the atmosphere by growing vegetation in a planned rotation or in the case of municipal wastes by preventing more damaging emissions of methane. Recently questions have been raised by a variety of authors about the indirect effect of bio energy production on deforestation rates, crop prices and non-CO_2 greenhouse gas emissions.

Biomass for energy use can be produced and converted efficiently into more convenient forms such as pellets, gases, liquids and electricity over a range of application forms. Direct combustion remains the most common technique for deriving energy from biomass for both heat and electricity. Biomass CHP systems provide efficiencies as high as 80%. Beyond biomass, ground source heat solar heat and photovoltaic have the potential to be widely implemented for household purposes.

Ground source heat pumps extract the energy from the sun. A few meters below the ground, the earth keeps a constant temperature of around 11-12 C° through the year. Ground source heat pumps can transfer the heat from the ground into a building to provide space heating and in some cases preheating domestic heat water. The heat pumps use electricity to pump the heat and they generate considerably more energy (300-400%) in heat than they consume in electricity.

Heat pumps, which are already used in parts of Continental Europe, are favoured over biomass for delivering low carbon heat as biomass supply is limited and the transport of large volume of biomass into urban environments is problematic.

Solar thermal panels use the energy from the sun for heating too. They have been found to be most cost-effective when they produce a larger part of the household's average hot water needs. Recently, the emergence of combined systems can be observed. The combined solar systems integrate the hot water supply and the space heating in one system.

Solar photovoltaic has the most potential for cost reduction, performance improvements and material breakthrough, which could lead to a dramatic change in the use of solar rooftops.

The best mix of renewable energy technologies at a site depends on renewable energy resources, technology characterization, various incentives and financial parameters.

Before analyzing the best technology mix options, a short review of a few non-renewable based dispersed generation options is necessary. It must be mentioned that the most popular DG technologies are based on natural gas.

Small gas-turbines: a few decades ago they worked mainly as emergency sources. Nowadays having higher efficiency – first of all with CHP option – they are used for continuous production too.

Reciprocating engines: the traditional diesel emergency units are able for continuous operation, and their reliability is high, but because of the high price of diesel oil this DG solution loses its grip. At the same time the natural gas fuelled reciprocating engines with CHP option are very popular. The high combined efficiency, relative low emission rate and the reliable operation bolster up the popularity.

Micro-turbines: they are featured with simple but relative reliable mechanical design – single shaft, air bearing, and permanent magnet generator – and high rotating speed. Solid state inverter is used to produce high quality power. CHP option is available. They are usually fuelled by natural gas. Hydrogen and cleaned bio-gas are also applicable.

Stirling engines: a rediscovered technology. The working material is heated up externally. The environmental impact is determined by the external fuel and heating units. In the cleanest version the sun shine is used for heating. Their efficiency is very high, it is near to the theoretical value of Carnot cycle.

Fuel cells: emerging technology. The units may be fuelled by hydrogen or natural gas. Practical applications are mostly based on natural gas. Wide power range, high efficiency, CHP possibility are the most important features. Main disadvantage is the short lifetime.

A secure, clean and economically sound future smart electricity system will be flexible enough to integrate a spectrum of local resources thus combining the benefits of centralized and distributed systems. A growing strategic portion of the supply demand gap will be fulfilled by low emissions, efficient generation located close to loads. A smart grid would use digital technology to collect, communicate and react to data making the system more efficient and reliable. The intelligent grid needs a flexible power grid communication architecture that provides a common service platform for disseminating power grid status information to the power producers, end-users and other participants.

Just like the internet, the electricity grid will be interactive for both power generation and power consumption actors. Enabled by smart metering, electronic control technologies, modern communication means and the increased awareness of costumers, local energy supply management will play a key part in meeting emission reduction and efficiency improvement targets while facilitate the fulfilling of the growing demands.

4 Technology Development Risks

To reduce the potential threat of climate change, many scientists and policy makers envision long scale reductions in CO_2 emissions. The development of cheaper and more efficient technologies will be critical for reducing costs and increasing the social and political visibility of substantial greenhouse gas emissions reductions over time. Although, efforts that employ currently available technologies can slow the growth of emissions, considerable technology advancement is needed to reduce the cost of near-term emissions reduction options and much larger reductions required to stabilize concentrations (Fig 2). Development of a portfolio of advanced technologies is the most promising approach to reduce CO_2 emissions [10]. The goal of reserving the global concentration of CO_2 while satisfying the continued growth of energy consumption present a huge but vital challenge for the research, development and demonstration efforts.

USD / tC (2000 USD)

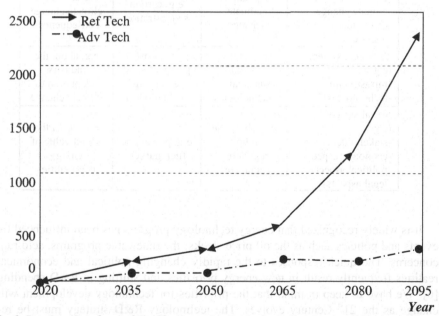

Fig. 2. Development of a portfolio of advanced technologies is the most promising approach to reduce CO_2 emissions [10]

Achieving the necessary technological advances requires investments in a number of technologies at different stages of development. An investment portfolio to develop technologies needs to include a broad range of options to be able to deal effectively with future uncertainties. An uncertainty matrix is shown in Table 1 in order to allow for capturing uncertainties in energy forecasting in a broader and systematic way [5].

The matrix provides a categorization of uncertainties by the type of knowledge relationship and the objective of uncertainty. It distinguishes between three types of knowledge relationships that are assumed to be established among an actor and an object. Those are unpredictability, incomplete knowledge and multiple knowledge frames. Each of the three knowledge relationships can refer to different objects of uncertainty within the natural, technical and social systems. The use of uncertainty matrix offers a new and more structured way of approaching the issue of uncertainty.

Table 1. Uncertainty matrix

Object \ Source of uncertainty	Incomplete knowledge	Unpredictable behaviour	Conflicting knowledge frame
Natural system (e.g. -climate change -ecosystem -resources)	e.g. about climate change	e.g. carbon sequestration	e.g. about the climate change consequences
Technical system (e.g. -infrastructure -technologies)	e.g. about emerging sustainable technologies	e.g. cost perspective of emerging technologies	e.g. about the industry adaptation to climate change
Social system (e.g. -stakeholders -economic aspects -political aspects -legal aspects)	e.g. about the future regulatory framework	e.g. power and fuel markets	e.g. about the social value of energy externalities

It is widely recognized that energy technology progress has been influenced by events and policies such as the oil price sleeks, the renewable programs, acid rain concerns. The quick responses to the rapidly changing political and economical realities frequently result in new energy priorities and shifting in R&D spending [6]. We have to keep in mind that the priorities for technology development will change as the 21st Century evolves. The technology R&D strategy must be reviewed on a regular basis to incorporate new information. A part of the information, such as fuel costs and availability and climate change motivated policies are uncertainties that will guide the priorities but are not directly controlled by the R&D efforts. These uncertainties will shape the critical R&D portfolios in the future (Fig. 3) It be can seen that increasing risk of climate damages will have a large impact on investments in risky R&D programs. Rising natural gas and other primary fuel prices push the economy to shift toward a high-technology base with steady increases in adoption of energy-efficient technologies.

Technology forecasts often err by confusing what is technically feasible in an engineering sense with what is likely to be deployed in the future [5]. We have long decades of optimistic forecasts for various technologies and concepts those have not achieved significant market penetration. For example, the future contribution of solar energy and renewable has been consistently misforcast for the last decades. Although many advocates have great expectations for renewable energy, all of the major studies indicate that the contribution of certain renewable will remain limited for at least the next 10 to 20 years. Photovoltaic, fuel cells, and other technologies have been predicted to become economically viable within short time for a long time. These technologies appear to hold great promise for the future and they may someday become viable.

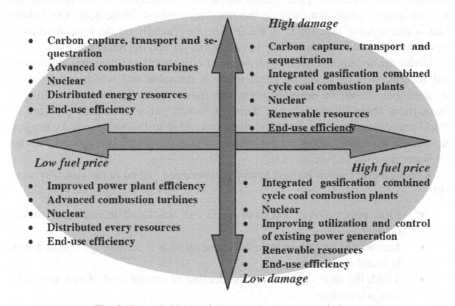

High damage

- Carbon capture, transport and sequestration
- Advanced combustion turbines
- Nuclear
- Distributed energy resources
- End-use efficiency

- Carbon capture, transport and sequestration
- Integrated gasification combined cycle coal combustion plants
- Nuclear
- Renewable resources
- End-use efficiency

Low fuel price

High fuel price

- Improved power plant efficiency
- Advanced combustion turbines
- Nuclear
- Distributed every resources
- End-use efficiency

- Integrated gasification combined cycle coal combustion plants
- Nuclear
- Improving utilization and control of existing power generation
- Renewable resources
- End-use efficiency

Low damage

Fig. 3. Future R&D portfolios meeting key uncertainties

However, the experience of the past decades suggests that it is not without considerable risk to suggest that these technologies may be 5 years away from being commercially viable. The current methods for assessing technology readiness or maturity does not capture the risk involved in adopting a technology. A holistic approach to technology risk should include not only the readiness level of a given technology but also dampening factors that could lead to reconsideration of adopting a given technology even one that is of highest maturity [7]. Examples of dampening factors may be obsoletion and leapfrogging. The reasons for obsolation are many and this retiring factor has to be addressed for evaluating a given technology.

The risk of obsoletion occurs when a better technology is on the horizon in the near term or a technology is outdated. Recently the concept of leapfrogging is being used in the context of sustainable development as a theory of development which may accelerate development by skipping less efficient, more expensive or more polluting technologies and move directly to more advanced ones. The adoption of solar energy technologies in developing countries could be an example, where the mistakes made by industrialized countries in creating a fossil fuel based energy infrastructure is not repeated, but they can jump directly into the solar energy era.

It is possible that there are additional dampening factors that could also introduce some level of risk in choosing a high maturity technology. There is no single, general effect of technological change on the costs of abatement. It is possible that technological change can increase the marginal costs of abatement. The bottom line is that optimal investment in R&D depends crucially on the type of technology considered and the uncertainly in both damages and the R&D process.

Under the uncertainly about both eventual climate-related damages and technical success, it is unclear how much R&D is desirable and which categories of technologies should be developed. On one hand, there is a value to waiting to learn more about climate change damages which put a downward pressure on R&D investment, on the other hand there is a value to investing in risky projects in order to learn more about the probability of success, which put upward pressure on financing risky technologies [8].

A few generic lessons learned from the experience gained in the past fifty years [9]:

- First, policy driven intensive R&D can accelerate technological development,
- Second, the growth of installed capacity of new technology results a drop in initial costs,
- Third, the time required for the decline of capital cost of new technologies is uncertain,
- Fourth, the role of private R&D can be very effective in reducing the costs of a new technology but its focus is weak on long-term objectives under market conditions.

The rate and magnitude of technological development is a major component in estimating the reduction potential of future anthropogenetic carbon emissions. There are two reasons why technology is important for climate change analysis. Firstly, it is the application of fossil fuel technologies that has caused the anthropogenic contribution to climate change. Secondly, a change to a low carbon society will require wide-spread development and mass deployment of new, low carbon technologies. Consequently, we must develop a foresight capability to scope out the complexity and uncertainties of the future environment in which technology will be developed.

Conclusions

The projected growth in the world in the next decades implies an important increase in energy needs. Unsustainable pressures on the environment and on natural resources are inevitable if energy demand remains closely coupled with economic growth and if fossil fuel use is not reduced.

Decentralized energy systems can significantly contribute to increase the overall efficiency of existing energy systems, decrease the dependency of fossil fuel resources and consequently to reduce CO_2 emissions.

Achieving the necessary technological advances to mitigate the climate change risks requires investments in a great number of technology development programs. Technology development depends not only on the technological opportunities that currently exist, but also on evolving opportunities created by the ever-changing economical, political and scientifique environment.

References

[1] Bezdek, R.H., Wendling, R.M.: A Half Century of Long-Range Energy Forecasts. Journal of Fusion Energy 21(3/4), 155–172 (2002)

[?] Drenckhahn, W., Pyo, I., Riedle, K.: Global Energy Demand and its Constraints. VGB Power Tech, 1-2, pp. 28–34 (2009)

[3] Goldenberg, J.: Energy and Human Well-Being. UNDP World Energy Assessment (2001)

[4] Chaudry, M., Ekanayake, J., Jenkins, N.: Optimum Control Strategy for a MicroCHP Unit. Int. Journal of Distributed Energy Resources, 265–280 (October-December 2008)

[5] Bruguach, M., et al.: Towards a Relational Concept of Uncertainty. In: MOPAN Conference, Leuven, Belgium (2007)

[6] Krómer, I.: Managing the Uncertainties in Power Engineering R&D Planning. In: 6th Biennial International Workshop Advances in Energy Studies, Graz University of Technology, pp. 345–350 (2008)

[7] Valerdi, R., Kohl, R.J.: An Approach to Technology Risk Management. In: MIT ESD Symposium, Cambridge, MA (2004)

[8] Baker, E., Adu-Bonnah, K.: Investment in Risky R&D Programs in the Face of Climate Uncertainty. Energy Economics (2006) doi: 10.1016/j.eneco.2006.10.003

[9] Capros, P., Mantzos, L., Voyonkas, E.L.: Technology Evolution and Energy Modelling: Overview of Research and Findings. Int. Journal of Global Energy Issues (1), 1–32 (2000)

[10] Clarke, L., Calvin, K., Edmonds, J.A., Kyle, P., Wisw, M.: Technology and International Climate Policy, The Harvard Project on International Climate Agreements (December 2008), http://www.belfercenter.org/climate

Fuzziness of Rule Outputs by the DB Operators-Based Control Problems

Márta Takács

Budapest Tech
Bécsi út 96/B, H-1034 Budapest, Hungary
takacs.marta@nik.bmf.hu

Abstract. After a short introduction about type-2 FS and basics of Mamdani type fuzzy approximate, a possible influence of the fuzziness of fuzzy sets involved in approximate reasoning model is given. In the reasoning process the observed rule in the rule base system is weighted by the measure of fuzziness at the highest point of the common area of the rule premise and system input. The basic operators in the fuzzy approximate reasoning system are the distance-based (DB) operators.

1 Introduction

In fuzzy control systems the typical fuzzy approach is as follows: for measured fuzzy input data, and given rule base system an output signal should be generated to reach the better state of the controlled system using fuzzy approximate reasoning method. However, applications of type-1 fuzzy systems (FS) are used to approximate random, imprecise or incomplete data or to model an environment that is changing in an unknown way with time [1], [2]. Based on the facts, that there is a paradox of type-1 fuzzy systems which can be formulated as the problem that the membership grades are themselves precise real numbers, L. A. Zadeh introduced type-2 and higher-types FS in 1975 [3]. Type 2 fuzzy sets give a model to handle the fuzziness of fuzzy values, which is very often left out by the control problems. For many applications the data generating system is known to be time-varying but the mathematical description of the time-variability is unknown, the measurement noise is non-stationary and the mathematical description of them unknown. In those cases the results of type-1 fuzziness give imprecise boundaries

I.J. Rudas et al. (Eds.): Towards Intelligent Engineering & Information Tech., SCI 243, pp. 653–663.
springerlink.com
© Springer-Verlag Berlin Heidelberg 2009

of FS-s [4]. For that reason many control mechanisms based on type-2 fuzzy sets or type-2 fuzzy logic and inference, called a type-2 (T2) fuzzy system, are investigated [5], [6], [7].

Since the introducing of the fuzziness of fuzzy values [4], there are periods of different intense activity in type-2 fuzzy set theory and applications investigation. In recent years we can find in [1], [2] a summary of basic terms of T2 fuzzy systems, and there are several published results related to the reasoning methods based on T2 fuzzy systems [8], [9].

In the same period there are practical motivations for the introduction of new operators on fuzzy sets, like uninorms. In many applications related to multicriteria decision making problems the aggregation of variables is one of the key issues. Such situations can be modelled by uninorms [10], [11], [12], and distance-based operators [13]. Expanding uninorm based reasoning methods with type-2 fuzzy set theory an effective method can be constructed to handle applications, where type-1 fuzziness of uncertain linguistic terms are used and have a non-measurable domain of FS-s, or results imprecise boundaries.

The question is raised: how should the fuzziness of fuzzy sets and fuzzy operators be effected on the fuzzy approximate reasoning process? Applying the uninorm operators and distance-based operators with the changeable parameter e in fuzzy approximate reasoning systems is done taking into consideration that the underlying notions of soft-computing systems are flexibility and the human mind. The choice of the fuzzy environment must support the efficiency of the system, and it must comply with the real world. This is more important than trying to fit the real world into the inflexible models.

Maximum distance-based operator as one from the family of distance based operators is an idempotent, commutative, associative, left continuous, increasing operator on each place of $[0,1] \times [0,1]$, and as it was proven, it is a uninorm with the unit element e [14]. As a uninorm it is suitable for refined respond of system behavior. Changing the operator parameter e in some applications it is possible to improve the effectiveness of fuzzy controller more than 30% [15]. Furthermore, the next dimension of fuzziness, the type-2 fuzzy set representation presents an opportunity to measure the reliability of fuzzy memberships.

For that purpose in the paper at first the basic type-2 fuzzy set terms, as well the 3D represented distance-based operator results are given. It offers an opportunity to introduce approximate reasoning model based on type-2 fuzzy sets and uninorms, especially on maximum distance based minimum operator. According to the Mamdani type fuzzy approximate reasoning basic idea, a possible influence of the fuzziness of fuzzy sets involved in approximate reasoning model is given. The basic idea is that in the reasoning process the observed rule in the rule base system is weighted by the measure of fuzziness at the highest point of the common area of the rule premise and system input.

2   Type 2 Fuzzy Sets

Type-2 and higher-types fuzzy sets (FS) eliminate the paradox of type-1 fuzzy systems which can be formulized as the problem that the membership grades are themselves precise real numbers. Type-1 fuzzy set (T1 FS) has grade of membership that is crisp, whereas a type-2 fuzzy set (T2 FS) has grades of membership that are fuzzy, so T2 FS are „fuzzy-fuzzy" sets.

The 2-D representation of fuzzy membership of fuzzy sets is the footprint representation of uncertainty (FOU), which is given with the uncertainty about left and right end point of the left side of the membership function, and with the uncertainty about left and right end point of the right side of the membership function.

Let be $x \in X$ from the universe of basic variable of interest. Let be $MF(x)$ the T1 fuzzy membership function of the fuzzy linquistic variable or other fuzzy proposition. Functions $UMF(x)$ and $LMF(x)$ are functions of the left-end and right-end point uncertainty. In a fix point x' of the universe X it is possible to define so called vertical slices of the uncertainty, describing it for different possibilities of the $MF_i(x)$ functions ($i=1,2,..N$), included in the shading of FOU.

In this case, for example if we have Gaussian primary membership function (MF), very often the uniform shading over the entire FOU means the uniform weighting, possibilities. T2 FS with FOU representation and uniform possibilities on FOU is called interval type-2 FS (IT2 FS).

The second way is to use 3D representation, where the F1 fuzzy set $A(x)$ is represented in the domain xOy, and in the third dimension for every crisp membership value $A(x)$ of the basic variable x a value of possibility (or uncertainty) is given as the function $MF(x,A(x)) = \mu(x, A(x))$. It is the embedded 3D T2 FS (Fig. 1). On Fig. 1 the value $\mu(x, A(x))$ has a Gaussian distribution value to represent the uncertainty of the type 1 fuzzy set (T1 FS). This uncertainty is the lowest at the kernel of the T1 FS.

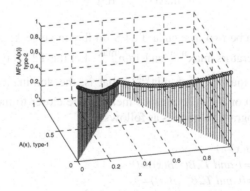

Fig. 1. 3D T2 FS

In Fig. 2 the value $\mu(x, A(x))$ is a random value from the interval [0,1].

3   Distance-Based Operators on Type 2 Fuzzy Sets

The maximum distance minimum operator with respect to $e \in [0,1]$ is defined as

$$\max_e^{min}(x,y) = \begin{cases} x, & if\ d(x,e) > d(y,e) \\ y, & if\ d(x,e) < d(y,e) \\ \min(x,y), & if\ d(x,e) = d(y,e) \end{cases} \tag{1}$$

The minimum distance maximum operator with respect to $e \in [0,1]$ is defined as

$$\min_e^{max}(x,y) = \begin{cases} x, & if\ d(x,e) < d(y,e) \\ y, & if\ d(x,e) > d(y,e) \\ \max(x,y), & if\ d(x,e) = d(y,e) \end{cases} \tag{2}$$

It can be proven by simple computation that if the distance of x and y is defined as $d(x,y) = |x - y|$ then the distance-based operators can be expressed by means of the min and max operators as follows [14]:

$$\max_e^{min} = \begin{cases} \max(x,y), & if\ y > 2e - x \\ \min(x,y), & if\ y < 2e - x \\ \min(x,y), & if\ y = 2e\text{-}x \end{cases} \tag{3}$$

$$\min_e^{max} = \begin{cases} \min(x,y), & if\ y > 2e - x \\ \max(x,y), & if\ y < 2e - x\ . \\ \max(x,y), & if\ y = 2e\text{-}x \end{cases} \tag{4}$$

Let $A(x)$ and $B(x)$ be two triangular T1 FS on the universe X, and let e be a fix parameter of the operators $\max_e^{min}$ and $\min_e^{max}$. Let the T2 FS type represent the result of operation $\max_e^{min}$ and $\min_e^{max}$, that is, the dominance of one of the fuzzy operands, $A(x)$ or $B(x)$. Dominance means to be closer to the given parameter (level) e. The program solution is as follows:

if abs(A(x)-e)<abs(B(x)-e)
    then T2A(x,A(x))=1 and T2B(x,B(x))=0;
    else T2A(x,A(x))=0 and T2B(x,B(x))=1;
end.

It is easy to observe, that at the third dimension we clearly see the result of the $\max_e^{\min}(A(x), B(x))$ operation, and at the basic plain we have the representation of the $\min_e^{\max}$, which is basically dual pair of the result $\max_e^{\min}(A(x), B(x))$.

It is shown that using this calculus a new, T2 representation of the operation result can be achieved. In this case a uniform uncertainty level exists (Fig. 2).

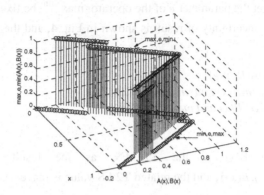

Fig. 2. T2 as result of the DBO

Let now the parameter e be changeable just as the change of the basic variable x. It can be a random value from the interval [0,1] or for example a measured parameter of the uncertainty at the current value of x.

The program solution is as follows:

```
if abs(A(x)-randome)<abs(B(x)-randome)
    then T2A(x)=1 and T2B(x)=0;
    else T2A(x)=0 and T2B(x)=1;
end.
```

With the changeable parameter e it is not anymore a uniform level of uncertainty, but it still shows the dominance of one of the operands $A(x)$ or $B(x)$. The T2A and T2B values are from the set of {0,1}. (Fig. 3)

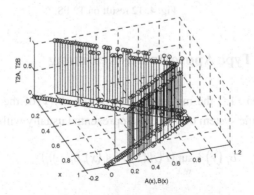

Fig. 3. T2A and T2B values with random e

3.1 T2 Result on T2 Operands

Let $T2A(x, A(x))$ and $T2B(x, B(x))$ be T2 fuzzy sets, with random fuzziness. Let us find $\max_e^{\min}(T2A, T2B)$, calculating it on the domain $\max_e^{\min}(A(x), B(x))$, taking into account the fuzzy values $T2A$ and $T2B$ of the fuzzy membership functions $A(x)$ and $B(x)$. Let the parameter e of the operator $\max_e^{\min}$ be fixed now.

The value of uncertainty of result is shown in Fig. 4, and the program solution is as follows:

if abs(A(x)-e)<abs(B(x)-e)
   then maxeAB(x)=A(x) and maxe(x)=T2A(x);
   else maxeAB(x)=B(x) and maxe(x)=T2B(x)
end.

where $\text{maxeAB}(x)$ is the T1 FS, as the result of the operation $\max_e^{\min}(A(x), B(x))$, and the related T2 FS value is $\text{maxe}(x)$.

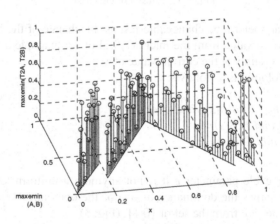

Fig. 4. T2 result on T2 FS

4   Mamdani Type Approximate Reasoning

In Mamadani-based FLC the model the rule output $B_i'(y)$ of the ith rule if x is A_i then y is B_i in the rule system of n rules is represented usually with the expression

$$B_i'(y) = \sup_{x \in X}(T(A'(x), T(A_i(x), B_i(y)))) \tag{5}$$

where $A'(x)$ is the system input, x is from the universe X of the inputs and of the rule premises, and y is from the universe of the output. For a continuous associative t-norm T, it is possible to represent the rule consequence model by

$$B_i'(y) = T\left(\sup_{x \in X} T(A_i(x), A'(x)), B_i(y) \right)$$ (6)

The consequence (rule output) is given with a fuzzy set $B'(y)$, which is derived from rule consequence $B(y)$, as an upper-bounded, cutting membership function derived from the of the $B(y)$. The cut,

$$DOF_i = \sup_{x \in X} T(A_i(x), A'(x))$$ (7)

is the generalized degree of firing level of the rule, considering actual rule base input $A'(x)$, and usually depends on the covering over $A(x)$ and $A'(x)$, i.e. on the *sup* of the membership function of $T(A'(x), A(x))$.

Rule base output, B'_{out} is an aggregation of all rule consequences $B_i'(y)$ from the rule base. As aggregation operator usually S conorm fuzzy operator is used.

$$B'_{out}(y) = S(B_n'(y), S(B_{n-1}'(y), S(...., S(B_2'(y), B_1'(y)))))).$$ (8)

The crisp FLC output y_{out} is constructed as a crisp value calculated with a defuzzification method.

4.1 Approximate Reasoning Based on Distance-Based Operators

Because of the properties of the distance-based operators [14], it was unreasonable to use the classical degree of firing, to give expression to coincidence of the rule premise (fuzzy set A_i), and system input (fuzzy set A'), therefore a degree of coincidence (*Doc*) for these fuzzy sets has been initiated. It is nothing else, but the proportion of area under membership function of the modified entropy-based intersection of these fuzzy sets, and the area under membership function of their union (using *max* as the fuzzy union). There are several types of reasoning methods based on those similarity measures of two fuzzy sets [14].

The rule output fuzzy set (B_i') should achieved as a cut of rule consequence (B_i) with *Doc*.

$$B_i'(y) = \max(B_i(y), Doc_i)$$ (9)

where Doc_i is the degree of coincidence, and gives expression to coincidence of the rule premise (fuzzy set A), and system input (fuzzy set A') in the ith rule of the rule system:

$$Doc_i = \frac{\int\limits_X T_e^{\min}\left(A_i(x), A'(x)\right)dx}{\int\limits_X \max\left(A_i(x), A'(x)\right)dx}.$$ (10)

The FLC rule base output is constructed as crisp value, calculated from associative using of the t-conorm on all rule outputs Bi'(y) [14].

5   Approximate Reasoning in T2 Environment

Let us assume, that we have T2 fuzzy sets as the rule premise of the rules and as the system input, described respectively by

$$MFA_i(x, A_i(x)) = \mu(x, A_i(x))$$ (11)

and

$$MFA'(x, A'(x)) = \mu(x, A'(x)),$$ (12)

where the fuzziness represent the uncertainty of T1 fuzzy membership values of these sets.

Let take into the consideration the fuzziness of these fuzzy sets. Calculating the rule output it is possible to introduce weightiness at the ith rule proportionally with the fuzziness of rule premise and rule input in the considered rule. The gain value G_i (weightiness of the observed i^{th} rule) can be calculated as the maximum fuzziness at $x_i^* \in X$, where

$$DOF_i = \sup_{x \in X} T\left(A_i(x), A'(x)\right) = T\left(A_i\left(x_i^*\right), A'\left(x_i^*\right)\right)$$ (13)

regarding $MFA_i(x, A_i(x))$ and $MFA'(x, A'(x))$, i.e.

$$G_i = max\left(MFA_i(x, A_i(x)), MFA'(x, A'(x))\right)$$ (14)

The weighted i^{th} rule output is calculated by the

$$B_i'(y) = G_i * T\left(DOF_i, B(y)\right).$$ (15)

If we apply the distance-based operators in the reasoning process, the gain G_i can be the maximum of the fuzziness at the level e, where e is the unit element of the operator.

5.1 Example

Applying Matlab Fuzzy toolbox, and using Gaussian distribution for the representation of the fuzziness of the rule premise and system input, based on the basic operators of min and max, the control surfaces are presented:

- In Fig. 6 without weighted rule outputs,
- In Fig. 7 with weighted rule outputs.

The investigated system is a MISO system, usually used in the case of control problems with two inputs on the universe $X_1 \times X_2$ (for example *error* and *error changes*), shown on Fig. 5.

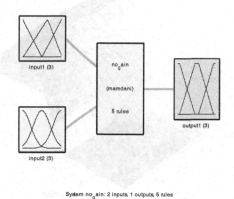

System no_gain: 2 inputs, 1 outputs, 5 rules

Fig. 5. The investigated system

Using gains G_i a significant result was achieved. There are areas of the universe of inputs $X_1 \times X_2$, where the control surface represents a stable, almost constant output, thanks to the rule outputs weighted by the level of fuzziness of the fuzzy outputs. It can have positive influence on the stability of the system.

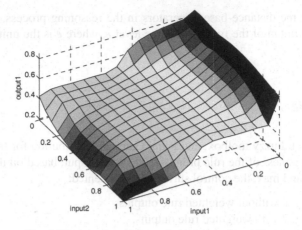

Fig. 6. The investigated system's control surface without gains

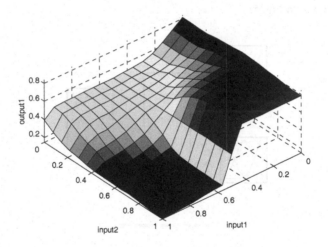

Fig. 7. The investigated system's control surface with gains

Conclusions

After a short introduction about type-2 FS, distance-based operators and Mamdani
type reasoning method, a possible calculation for T2 results of T2 operands was
given based on $\min_e^{\max}$ operator. It is of primary importance in all preliminary
works in order to later construct inference mechanism based on this family of
fuzzy operators and T2 fuzzy logic. Furthermore a possible influence of the fuzziness of fuzzy sets involved in approximate reasoning model was given, based on
Mamdani type fuzzy approximate reasoning. In the reasoning process the observed

rule in the rule base system is weighted by the measure of fuzziness at the highest point of the common area of the rule premise and system input. The control surfaces show that the gained reasoning process returns more powerful control than the classical type of reasoning process.

Acknowledgment. The research was partially supported by the Hungarian Scientific Research Project OTKA T048756 and by the project "Mathematical Models for Decision Making under Uncertain Conditions and Their Applications" financed by Vojvodina Provincial Secretariat for Science and Technological Development.

References

[1] Mendel, J.M.: Type-2 Fuzzy Sets: Some Questions and Answers. IEEE Neural Network Society (August 2003)

[2] Mendel, J.M., John, R.I.B.: Type-2 Fuzzy Sets Made Simple. IEEE Transactions on Fuzzy Systems 10(2), 117–127 (2002)

[3] Zadeh, L.A.: The Concept of a Linguistic Variable and its Application to Approximate Reasoning – 1. Information Sciences 8, 199–249 (1975)

[4] Ozen, T., Garibaldi, J.M.: Investigating Adaptation in Type-2 Fuzzy Logic Systems, Applied to Umbilical Acid-Base Assessment. In: Proc. the 2003 European Symposium on Intelligent Technologies (EUNITE 2003), Oulu, Finland, July 2003, pp. 289–294 (2003)

[5] Sepulveda, R., Castillo, O., Melin, P., Rodriguez-Diaz, A., Montiel, O.: Experimental Study of Intelligent Controllers Under Uncertainty Using Type-1 and Type-2 Fuzzy Logic. Information Sciences 177, 2023–2048 (2007)

[6] Wu, D., Tan, W.W.: Genetic Learning and Performance Evaluation of Type-2 Fuzzy Logic Controllers. Submitted to IEEE Trans. on Systems, Man, and Cybernetics—Part B: Cybernetics (2005)

[7] Castro, J., Castillo, O., Mellin, P.: An Interval Type-2 Fuzzy Logic Toolbox for Control Applications. In: Proc. IEEE FUZZ Conference, London, UK, July 2007, pp. 61–66 (2007)

[8] Thiele, H.: On Approximate Reasoning With Type-2 Fuzzy Sets. In: Proc. IPMU Conference, Annecy, France, July 1-5, 2002, pp. 355–362 (2002)

[9] Wang, T., Chen, Y., Tong, S.: Fuzzy Reasoning Models and Algorithms on Type-2 Fuzzy Sets. International Journal of Innovative Computing, Information and Control ICIC International 4(10), 2451–2460 (2008)

[10] Yager, R.R., Rybalov, A.: Uninorm Aggregation Operators. Fuzzy Sets and Systems 80, 111–120 (1996)

[11] Klement, E.P., Mesiar, R., Pap, E.: Triangular Norms. Kluwer Academic Publishers, Dordrecht (2000)

[12] Yager, R.R.: Uninorms in Fuzzy System Modeling. Fuzzy Sets and Systems 122, 167–175 (2001)

[13] Rudas, I.J.: Evolutionary Operators: New Parametric Type Operator Families. Fuzzy Sets and Systems 23, 149–166 (1999)

[14] Takacs, M.: Approximate Reasoning in Fuzzy Systems Based On Pseudo-Analysis, Phd Thesis, Univ. of Novi Sad (2004)

[15] Takacs, M., Pozna, C.: Quick Comparison of the Efficiency of Fuzzy Operatios Used in FLC. In: Proc. of SISY 2006, pp. 73–80 (2006)

Fine Structure Constant – A Possibilistic Approach

Péter Várlaki*, Robert Fullér**, and Imre J. Rudas***

  * Budapest University of Technology and Economics, Hungary
    varlaki@kme.bme.hu
 ** IAMSR, Åbo Akademi University, Joukahaisenkatu 3-5, FIN-20520 Åbo, Finland
    robert.fuller@abo.fi
*** Budapest Tech, Bécsi út 96/B, H-1034 Budapest, Hungary
    rudas@bmf.hu

Abstract. The paper discusses some unusual interpretations of the fine structure constant (FSC) taking into account of our earlier results and the concept of "undetached observer" introduced by Pauli. A joint possibilistic-probabilistic approach and a two-fold "controlling-observing equations" are proposed for a new reinterpretation of FSC using specific bivariate possibility distribution as a hypothetically proper concept of FSC.

1   Introduction

In our earlier papers we discussed some interesting potentially fruitful "results" of the Pauli-Jung correspondence concerning mainly the problem of the fine structure constant. [37-38]. Since Heisenberg's early work, [18] many authors have dealt in detail, with the different scientific, physical, psychological and historical aspects of the cooperation of these two Great Minds of the 20th Century (e.g. [12, 13, 15]).

Our early consideration mainly focuses on the germs of the background control and system theory connected strongly with the interpretation of fine structure constant and number archetype 137. [20, 21] On the basis of this approach we introduced a new, very simple mathematical formula, a so-called "controlling-observing equation" for computation and symbolical interpretation of the fine structure constant and Number-archetype 137. [8] Concerning the characteristic epistemological problem of FSC we may cite the next concise "evaluation".

"The mystery about (the FSC) it is actually a double mystery. The first mystery – the origin of its numerical value $\alpha \approx 1/137$ – has been recognized and discussed for decades. The second mystery – the range of its domain – is generally unrecognized." [11]

I.J. Rudas et al. (Eds.): Towards Intelligent Engineering & Information Tech., SCI 243, pp. 665–679.
springerlink.com   © Springer-Verlag Berlin Heidelberg 2009

Concerning the last "statement" our conjecture is that for the interpretation of the FSC a "joint" probabilistic-possibilistic approach can be realized, which is corresponding to the Pauli's interpretation on the role of probability in physics [17][1]

After the critical evaluation of the Bernoulli's theorem, from the point of view of the undetached observer and the anticipated "theory of possibility", it seems to be, as Pauli would propose a complementary treatment of the so called objective ("Kolmogorovian") and subjective (e.g. Keynesian, von Mises and de Finetti-like, etc.) probabilities. From the point of view of the question for "probability vs. possibility", in the measurement practice every probability distribution "automatically" generates a possibility variables on the limits of the distributions. Contrary, as we shell show, the possibility concepts under certain general conditions can be derived from the properties of the uniform probability distribution. This approach in a natural way rises up the problem of the undetached observer. [12, 15]

Reflecting to the opinion of Bohr on the "detached observer" (1955) Pauli underlined the significance of the concept of the undetached observer in the future of physics:

"Dear Bohr, ... under your great influence it was indeed getting more and more difficult for me to find something on which I have a different opinion then you. To a certain extent I am therefore glad, that eventually I found something: the definition and the use of the expression 'detached observer'... I consider the impredictable change of the state by a single observation in spite of the objective character of the results of every observation and notwithstanding the statistical laws for the frequencies of repeated observation under equal conditions to be an abandonment of the idea of the isolation (detachment) of the observer from the course of physical events outside himself." [12]

Accepting Pauli's opinion on the partially undetached observer we propose that the theoretical and experimental results of the FSC are naturally and properly different as marginal cases originated from an output of an unknown but hypothetically (in principle) identifiable bilinear system. Applying the possibilistic and probabilistic approach the joint (bivariate) possibility distribution of the FSC can be concerned as an output possibility distribution of a possibilistic (fuzzy) control system similarly to the random interpretation of the FSC. In this last case as we mentioned above the output would be treated by the bivariate probability distribution of the output of a bilinear stochastic system.

---

[1] "In this purely mathematical form, Bernouilli's theorem in thus not as yet susceptible to empirical test. For this purpose it is necessary somewhere or other to include a rule for the attitude in practice of the human observer, or in particular the scientist, which takes account or the subjective factor as well, namely that the realization, even on a single occasion, of a very unlikely event is regarded from a certain point on as impossible in practice. Theoretically it must be conceded that there is still a chance, different from zero, of error; but in practice actual decision are arrived at in this way, in particular also decisions about the empirical correctness of the statistical assertions of the theories of physics or natural science. At this point one finally reaches the limits which are set in principle to the possibility of carrying out the original programme of the rational objectivation of the unique subjective expectation." [17]

Therefore, it means that the searching an "average", as looking only one value of the FSC from our point of view hopeless and possible biased approach. Our endeavor is looking for a control and system-like approach (see still e.g. [14]) for the joint treatment of the probabilistic and possibilistic analysis taking into consideration the necessary role of the partially undetached observer as an essential part of the hypostatized unknown but principally identifiable system. This approach forms a quite different angle comparing with the usual characteristic interpretations of FSC. Contrary with these approximations we postulate a possibilistic, probabilistic relationship between the obtained results of the two kinds (concerning the experimental and theoretical computations) of the partially undetached observers.[2] To anticipate, we shall stress that our approach is basically empirical and heuristic and it concerns the questions of discovery rather than that of philosophical legitimacy.

2 A Simple Controlling-Observing Interpretation of the Fine Structure Constant

The *fine structure constant* is a ratio which characterizes the "amount" of the electromagnetic (mutual) effect (independent of the selection of the dimension) and can be found in the description of the fine structure of hydrogen spectrum:

$$\alpha = \frac{e^2}{4\pi\varepsilon_0 hc}$$

where e is the elementary charge of electron, c is the speed of light, h is the Planck-constant and ε_0 is the vacuum permittivity. On the other hand, the value of fine structure constant can be calculated from the direct measurements too However, the specific "value of the concept" obtained from the measurements depends upon the type (and accuracy) of the concrete quantum-electrodynamic model. [20]

Without knowing the "accepted", probably, most accurate values (considering just the 137.03… value of FSC), the following formula for the "general" (synchronistic) definition of the fine structure constant was proposed in 1983 by Stanbury [6]:

$$\alpha^{-1} = 4\pi^3 + \pi^2 + \pi = \pi\left(4\pi^2 + \pi + 1\right) = 137,0363037\ldots = \alpha^{-1}(\pi)$$

It can be seen that this formula is simple, general, self-expressive and aesthetically also neat. Furthermore, besides π, sufficiently – according to certain old "scientific" (alchemical) and traditionally hermeneutical rules – it consists only of the first four integer numbers. The first three numbers (as powers) have some "generative characteristics", but the fourth one (i.e. 4) with certain topological characteristics (as a multiple) also meets the usual "symbolic demands".

---

[2] The possibilistic approach is in complete harmony with Pauli's opinion because contrary to the fruitful possibility description of the relationships between very uncertain variables and events, the probabilistic approach (average, correlation etc.) can not properly characterize the „possible" connections among the improbable events (i.e. events with very low probabilities).

Therefore, it is able to symbolize the completeness or perfectness according to the mentioned hermeneutical principles. On the other hand, the first three integer numbers appear in generative way as powers of π, while the fourth one, the '4', hints at a topological structure (as a multiple) satisfying the usual Jungian interpretations, as well [20, 21].

The essence of the interpretations is that the tetragonal substitution of π can be 4 as outside the square-measure or 2 as the inside square-measure of the "generative circle" with unit radius.

Applying the previous quaternary substitutive "interpretations" the following natural (integer) structure numbers can be obtained

$$\alpha^{-1}(4) = 4.4^3 + 4^2 + 4^1 = 4.64 + 16 + 4 = 256 + 16 + 4$$

$$\alpha^{-1}(2) = 4.2^3 + 2^2 + 2^1 = 4.8 + 4 + 2 = 32 + 4 + 2$$

$$\frac{\alpha^{-1}(4)}{2} - 1 = 137 < \alpha^{-1}(\pi) < 138 = \frac{\alpha^{-1}(4)}{2}$$

It is also well-known, according to the latest research on the basis of the more precise astronomical measurements, that the empirical value of the fine structure constant probably depends on time too [22].[3]

It can be seen that this formula and value can probably be considered as a crucial orienting parameter which is of course, not the empirical fine structure constant in itself.[4]

On the basis of Stanbury's note Sherbon suggested the next interesting and important corrected formula [5]:

$$\alpha^{-1} \cong 4\pi^3 + \left(1 - \pi_r^{-2}\right)\pi^2 + \pi \cong 137.03599916$$

where π_r is the harmonic of π radians = 180 [40]. It can be seen that this formula and value can be considered as a further step to a hypothetical controlling expression. However on the basis of the revised interpretation of the famous "World Clock vision" of Pauli as the "spontaneous symbolic expression" of the fine structure constant „observed" partly unconsciously by Pauli – following Sherbon "corrective formula" – we may proposed a new mediator number archetype related to the fine structure constant. [8]

$$\alpha^{-1}(\pi) = 4\pi^3 + \pi^2 + \pi - \frac{\delta}{1-\delta}\frac{\delta}{1+\delta},$$

---

[3] Not concerning the details, hypothetically we can say that circa in the "middle of the age" of the Universe, the value of the fine structure constant was approximately 137.0368... although the actual accepted value is roughly 137.0360... [22] (The introduced formula of α^{-1} has the value of 137,0363037... that occurred in the given time interval of the Universe).

[4] It is also obvious, that from the point of view of the life, the value of the FSC can not change arbitrarily. Were its value is very different, carbon atoms would not be stable and organic life, as we know, would not be possible. This evidence increasing underlines the significance of 137 as an integer and, at the same time, as a mediator or controlling number. The latter seems to be a perfect manifestation for the mutual concept of the arithmetic and geometric number archetype and at the same time as a central number archetype of the Self (Selbst). [20]

$$\delta = \frac{2\pi}{360}, \quad \alpha^{-1}(\pi) = 4\pi^3 + \pi^2 + \pi - \sum_{n=1}^{\infty}\left(\frac{2\pi}{360}\right)^{2n}$$

The value $\alpha^{-1}(\pi) = 137.0359990656$ obtained from this formula is practically equivalent to the proposed result of the fine structure constant (137.035999070 and 137.035999084) obtained by measurements in 2007 and 2008, respectively. [10, 24] On the generalization of the above formula we may introduce the next "Controlling-Observing Equation" of the Fine Structure Constant:

$$\alpha^{-1}(\pi,x) = 4\pi^3 + \pi^2 + \pi - \frac{\delta}{1-\delta}\frac{\delta}{1+\delta}, \quad \delta = \frac{2\pi}{x}$$

$$\alpha^{-1}(\pi,x) = 4\pi^3 + \pi^2 + \pi - \sum_{n=1}^{\infty}\left(\frac{2\pi}{x}\right)^{2n}$$

This general formula expresses that there are different courses and rhythms for the circulation of the circle with unit radius. If the x is equal to 32 the value of the expression is 136.99..., it means that from practical point of view x = 32 (33) generalizes the number archetype 137. In the second parts of our earlier paper we gave manifold interpretations for the controlling and observing character of this equation. [8]

The selection of the 360 = 36 × 10 seems to be very natural for us. In the famous vision of Pauli the number 32 in the given context and the "spiritual environment" generate the number 360. The 36 is a pair of 32 in Pauli thinking (see his preoccupation of 32-36 and 137 [20]). The ten (10 = 3 + 7) is a permanent factor in the dreams and their interpretations of Pauli. The 36 is a really strong number archetype in a Jungian sense of the word. It has a mediator role because it has as objective importance as the first composed natural number ($6^2 = 36$, $3 \times 12 = 36$ = 4 x 9 =$2^2 \times 3^2$ = 36). From other side the 36 has some evidently „human aspects", as well. Symbolically the $1-\partial$ and $1+\partial$ can express an oppositional property for the two circulations using the step length of the rhythm of circulation. This oppositional character of delta can be interpreted in a symbolic mathematics according to Pauli, as an "orthogonal, perpendicular" relationship. In the last part of this paper we shell generalize the above controlling-observing equation of FSC for the interpretation of the complex plane according to Pauli's further "observations"[5]. [15, 17]

---

[5] Some physicists have the opinion that the FSC is only an esthetical category, which can be actually everywhere bypassed in the modern physics. According to this view (obviously rejecting the idea found in Pauli's famous study of Kepler) it can be considered like Kepler's real physics, i.e. his mathematical laws about the orbit of planets, relates to his idea about eternal harmony. Others are constructing unexplainable and uninterpretable complex equations of all kinds in order to approximate the most recent measurement results of FSC. Our endeavor falls between the "physical negation" and the playfulness and is related to a new such a formula partly derived from Pauli's imagination which beyond the physics also carries productive and well interpretable meaning, and perhaps may give a creative impulse for the future physics as well.

3 Some Basic Concepts of "Possibility Number Theory" (Joint Possibility Distributions and Copulas)

3.1 Concept of the Possibility Distributions

A fuzzy number A is a fuzzy set $\mathbb{R}$ with a normal, fuzzy convex and continuous membership function of bounded support. The family of fuzzy numbers is denoted by F Fuzzy numbers can be considered as possibility distributions [4]. A fuzzy set C in $\mathbb{R}^2$ is said to be a joint possibility distribution of fuzzy numbers $A, B \in F$, if it satisfies the relationships [1]

$$\max\{x \mid C(x, y)\} = B(y) \text{ and } \max\{y \mid C(x, y)\} = A(x)$$

for all $x, y \in \mathbb{R}$. Furthermore, A and B are called the marginal possibility distributions of C. Let $A \in F$ be fuzzy number with a γ-level set denoted by $[A]^\gamma = [a_1(\gamma), a_2(\gamma)]$, $\gamma \in [0,1]$.

A function $f : [0,1] \to \mathbb{R}$ is said to be a weighting function if f is non-negative, monotone increasing and satisfies the following normalization condition

$$\int_0^1 f(\gamma) d\gamma = 1.$$

If $A, B \in F$ are non-interactive then their joint membership function is defined by $C = A \times B$. It is clear that if $x \in [A]^\gamma$ and $y \in [B]^\gamma$ then $(x, y) \in [C]^\gamma$.

3.2 Copulas

Copulas characterize the relationship between a multidimensional probability function and its lower dimensional margins. A two-dimensional copula is a function $C : [0,1]^2 \to [0,1]$ which satisfies [1]

- $C(0, t) = C(t, 0) = 0$
- $C(1, t) = C(t, 1) = t$ for all $t \in [0,1]$

- $C(u_2, v_2) - C(u_1, v_2) - C(u_2, v_1) + C(u_1, v_1) \geq 0$ for all

  $u_1, u_2, v_1, v_2 \in [0,1]$ such that $u_1 \leq u_2$ and $v_1 \leq v_2$.

Equivalently, a copula is the restriction to $[0,1]^2$ of a bivariate distribution function whose margins are uniform on $[0,1]$. The importance of copulas in statistics is described in the following theorem.

Theorem (Sklar, [3]): Let X and Y be random variables with joint distribution function H and marginal distribution functions F and G, respectively. Then there exists a copula C such that

$$H(x, y) = C(F(x), G(y)),$$

moreover, if marginal distributions, say, $F(x)$ and $G(y)$, are continuous, the copula function C is unique. Conversely, for any univariate distribution functions F and G and any copula C, the function H is a two-dimensional distribution function with marginals F and G. Thus copulas link joint distribution functions to their one-dimensional margins and they provide a natural setting for the study of properties of distributions with fixed margins. For further details, see Nelsen [2]. For any copula C we have,

$$W(u, v) = \max\{0, u + v - 1\} \leq C(u, v) \leq M(u, v) = \min\{u, v\}.$$

In the statistical literature, the largest and smallest copulas, M and W are generally referred to as the Fréchet-Hoeffding bounds.

3.3 Joint Possibility Distributions

Let C be a joint possibility distribution of fuzzy numbers $A, B \in F$. The f-weighted measure of possibilistic covariance between $A, B \in F$, (with respect to their joint distribution C), defined by [1], can be written as [1]

$$Cov_f(A, B) = \int_0^1 Cov(X_\gamma, Y_\gamma) f(\gamma) d\gamma,$$

where X_γ and Y_γ are random variables whose joint distribution is uniform on $[C]^\gamma$ for all $\gamma \in [0,1]$.

For uniform distributions on $[a_1(\gamma), a_2(\gamma)]$ and $[b_1(\gamma), b_2(\gamma)]$ we look for a joint distribution defined on the Cartesian product of this level intervals (with uniform marginal distributions on $[a_1(\gamma), a_2(\gamma)]$ and $[b_1(\gamma), b_2(\gamma)]$): this joint distribution can always be characterized by a unique copula C, which may depend also on γ.

3.4 Non-interactive Fuzzy Numbers

The product copula $\Pi(x, y) = xy$ corresponds to non-interactive fuzzy numbers (with zero covariance). [1]

If A and B are non-interactive, i.e. $C = A \times B$. Then $[C]^{\gamma} = [A]^{\gamma} \times [B]^{\gamma}$, that is, $[C]^{\gamma}$ is rectangular subset of $\mathbf{R}^2$ for any $\gamma \in [0,1]$. Then X_{γ}, the first marginal probability distribution of a uniform distribution on $[C]^{\gamma} = [A]^{\gamma} \times [B]^{\gamma}$, is a uniform probability distribution on $[A]^{\gamma}$ (denoted by U_{γ}) and Y_{γ}, the second marginal probability distribution of a uniform distribution on $[C]^{\gamma} = [A]^{\gamma} \times [B]^{\gamma}$, is a uniform probability distribution on $[B]^{\gamma}$ (denoted by V_{γ}) that is X_{γ} and Y_{γ} are independent. So,

$$cov(X_{\gamma}, Y_{\gamma}) = cov(U_{\gamma}, V_{\gamma}) = 0,$$

for all $\gamma \in [0,1]$, and, therefore, we have

$$Cov_f(A, B) = \int_0^1 cov(X_{\gamma}, Y_{\gamma}) f(\gamma)\gamma = \int_0^1 0 \times f(\gamma)\gamma = 0.$$

If A and B are non-interactive then $Cov_f(A, B) = 0$ for any weighting function f. (see Fig. 1)

The upper Fréchet-Hoeffding bound of copulas, $M(x, y) = \min\{x, y\}$ corresponds to interactive fuzzy numbers with correlation 1 (see Fig. 2)

The lower Fréchet-Hoeffding bound of copulas $W(u, v) = \max\{0, u + v - 1\}$ corresponds to interactive fuzzy numbers with correlation -1 (see Fig. 3).

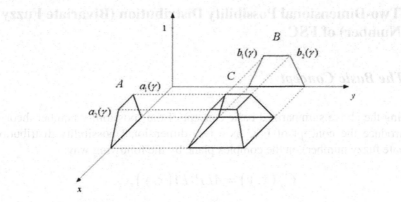

Fig. 1. The case of non-interactive fuzzy numbers A and B.

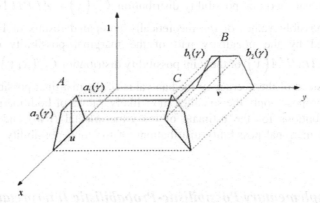

Fig. 2. The case of $\rho_f(A,B) = 1$.

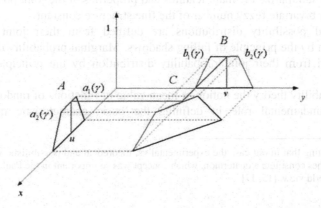

Fig. 3. The case of $\rho_f(A,B) = -1$.

4   Two-Dimensional Possibility Distribution (Bivariate Fuzzy Number) of FSC

4.1 The Basic Concept

Applying the above summarized basic concept of the "possibility number theory" we introduce the concept of FSC as a two dimensional possibility distribution (bivariate fuzzy number) on the complex plane by the following way:

$$C_\alpha(x, y) = ALPHA(x, y),$$

where, x is the real and y the imaginary part for the possible experimentally and theoretically computed values of the FSC.

Here the set of estimated values obtained from the measurements is characterized by the real marginal possibility distribution $C_\alpha(x) = ALPHA(x)$ meanwhile the possible values of the theoretically computed results of FSC can be characterized by the imaginary part of the marginal possibility distribution $C_\alpha(y) = ALPHA(y)$ of the joint possibility distribution $C_\alpha(x, y)$.

Now, instead of the average or median value of the marginal possibility distribution we propose apply the so called most likely values of both marginal possibility distributions for the estimate of the experimental (real) and theoretical (imaginary) marginal possibility distributions of the joint possibility distribution of FSC.

4.2 Complementary Possibilistic-Probabilistic Interpretation of the "Double-Faced" Concept of the FSC[6]

Now let us summarize the main features and properties of the joint possibility distribution or bivariate fuzzy number of the fine structure constant.

Marginal possibility distributions are defined from their joint possibility distribution by the principle of falling shadows. Marginal probability distributions are defined from their joint probability distribution by the principle of falling integrals.

In probability theory the notion of mean value of functions of random variables plays a fundamental role in defining the basic characteristic measures of

---

[6] It is interesting that in our case the experimental vs. theoretical and possibilistic vs. probabilistic approaches constitute a quaternion, which concept was so important in the Pauli scientific and spiritual world view. [15, 17]

probability distributions. For instance, the measure of covariance, variance and correlation of random variables can all be computed as probabilistic means of their appropriately chosen real-valued functions.

In our proposal we equip each level set of a possibility distribution (represented by a fuzzy number) with a uniform probability distribution, then apply their standard probabilistic calculation, and then define measures on possibility distributions by integrating these weighted probabilistic notions over the set of all membership grades. These weights (or importances) can be given by weighting functions. Different weighting functions can give different (case-dependent) importances to level-sets of possibility distributions.

The possibilistic mean value of a possibility distribution is nothing else but the weighted average of the probabilistic mean values of the respective uniform distributions on the level sets of that possibility distribution.

The measure of possibilistic variance is nothing else but the weighted average of of the probabilistic variances of the respective uniform distributions on the level sets of that possibility distribution.

The possibilistic covariance between marginal possibility distributions of a joint possibility distribution is defined as the weighted sum of probabilistic covariances between marginal probability distributions whose joint probability distribution is (supposed to be) uniform on each level-set of the joint possibility distribution.

The measure of possibilistic correlation between marginal possibility distributions of a joint possibility distribution is defined as their possibilistic covariance divided by the square root of the product of their possibilistic variances.

In other words, possibilistic correlation represents an average degree of interaction between marginal distributions of a joint possibility distribution as compared to their respective variances.

If the marginal possibility distributions of a joint possibility distribution are independent (or non-interactive) then their correlation coefficient is equal to zero, and their joint possibility distribution is nothing else but the Cartesian product of the marginal possibility distributions. In this case the level-sets of the joint possibility distribution are rectangulars. That is, any level set of the joint possibility distribution is the Cartesian product of the level sets of the marginal possibility distributions.

If marginal possibility distributions of a joint possibility distribution are completely positively correlated, that is their correlation coefficient is equal to one, then the level-sets of the joint possibility distribution are simple line segments; and any level set of the second marginal distribution is a positive affine transform of the corresponding level set of the first marginal distribution.

If marginal possibility distributions of a joint possibility distribution are completely negatively correlated, that is their correlation coefficient is equal to minus one, the level-sets of the joint possibility distribution are simple line segments; and any level set of the second marginal distribution is a negative affine transform of the corresponding level set of the first marginal distribution.

In probability theory copulas link joint distribution functions to their one-dimensional margins and they provide a natural setting for the study of properties of distributions with fixed margins.

The importance of copulas in statistics is described in the previous part of the paper. Note, in particular, that a copula is itself a continuous bivariate probability distribution with uniform margins. [1]

For any copula K as we have seen, we have,

$$W(u,v) = \max\{0, u+v-1\} \le K(u,v) \le M(u,v) = \min\{u,v\}.$$

In the statistical literature, the largest and smallest copulas, M and W are generally referred to as the Fréchet-Hoeffding bounds. It is clear that the lower Fréchet-Hoeffding bound of copulas is the Lukasiewitz triangular norm, and the upper Fréchet-Hoeffding bound of copulas is the minimum triangular norm.

The upper Fréchet-Hoeffding bound of copulas, $M(x,y) = \min x,y$ is linked to interactive marginal possibility distributions with correlation 1. The lower Fréchet-Hoeffding bound of copulas $W(u,v) = \max 0, u+v-1$ is linked to interactive marginal possibility distributions with correlation -1. The product copula $\Pi(x, y) = xy$ is linked to non-interactive marginal possibility distributions.

5   The Two-Dimensional Possibility Distribution for the Pauli's Equation of FSC[7]

Applying the results from the previous chapters we introduce the joint possibility distribution of FSC by the following way:

$$C_\alpha(x, yi) = ALPHA(x, yi) = ALPHA\left(\alpha_\pi^{-1}(x), \alpha_\pi^{-1}(yi)\right),$$

where $\alpha_\pi^{-1}(x)$ is the real version of the "Pauli's equation" for the experimental results meanwhile $\alpha_\pi^{-1}(yi)$ is the imaginary part for the theoretical result of FSC, respectively. Applying the "Pauli's equation" we obtain the following formulas

$$\alpha_\pi^{-1}(x) = 4\pi^3 + \pi^2 + \pi - \frac{\delta_x}{1-\delta_x}\frac{\delta_x}{1+\delta_x}, \quad \delta_x = \frac{2\pi}{x}$$

$$\alpha_\pi^{-1}(x) = 4\pi^3 + \pi^2 + \pi - \sum_{n=1}^{\infty}\left(\frac{2\pi}{x}\right)^{2n}$$

---

[7] The "concept" of „Pauli equation" on the basis of the Stanbury's note was initiated by Sherbon [5].

$$\alpha_{\pi}^{-1}(yi) = 4\pi^3 + \pi^2 + \pi + \frac{\delta_y}{1-\delta_y}\frac{\delta_y}{1+\delta_y}, \quad \delta_y = \frac{2\pi}{yi}$$

$$\alpha_{\pi}^{-1}(yi) = 4\pi^3 + \pi^2 + \pi + \sum_{n=1}^{\infty}\left(\frac{2\pi}{yi}\right)^{2n}$$

Here we rise up to hypothesis that the most likely value of the real marginal possibility distribution of FSC (for experimental results) $\alpha_{\pi}^{-1}(x)$ can be obtained when x=360. From other side according to our conjecture the most likely value of the imaginary marginal possibility distribution of FSC (for theoretical results) $\alpha_{\pi}^{-1}(yi)$ can be obtained when y=360, i.e.

$$\alpha_{\pi}^{-1}(x=360) = 4\pi^3 + \pi^2 + \pi - \frac{\delta_x}{1-\delta_x}\frac{\delta_x}{1+\delta_x}, \quad \delta_x = \frac{2\pi}{360}$$

$$\alpha_{\pi}^{-1}(x=360) = 4\pi^3 + \pi^2 + \pi - \sum_{n=1}^{\infty}\left(\frac{2\pi}{360}\right)^{2n}$$

$$\alpha_{\pi}^{-1}(y=360) = 4\pi^3 + \pi^2 + \pi + \frac{\delta_y}{1-\delta_y}\frac{\delta_y}{1+\delta_y}, \quad \delta_y = i\frac{2\pi}{360}$$

$$\alpha_{\pi}^{-1}(y=360) = 4\pi^3 + \pi^2 + \pi + \sum_{n=1}^{\infty}\left(i\frac{2\pi}{360}\right)^{2n}$$

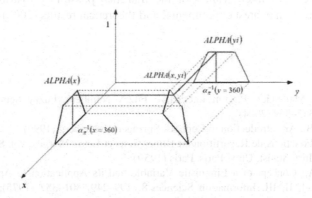

Fig. 4. The illustration of 'twin' interactive fuzzy numbers of FSC with possibilistic correlation 1

Consequently the FSC as number archetype is a two dimensional possibility distribution or in other words a bivariate fuzzy number, which is realized by its two marginal possibility distributions. As we have seen there is a complementary probabilistic interpretation of the two dimensional possibility distribution trough a suitable undistinguishable bivariate uniform probability distribution. In other words in the complementarity concept of the possibilistic and probabilistic approaches there is a final undistinguishability between the two dimensional possibility and bivariate uniform probability distributions.

More specifically we apply the most likely value of both marginal possibility distributions for the characterization of the experimental and theoretical results of FSC. One of these is related to the experimental and the other one is related to the theoretical result of FSC.

In our cases for the most likely values of the two marginal possibility distributions from the above formulas we obtain

$$\alpha_\pi^{-1}(x = 360) = 137,0359990656... \text{ and}$$

$$\alpha_\pi^{-1}(y = 360) = 137,035999251...$$

The latest experimental result (2008) and the latest theoretical result (2008) have the following values [9, 10]:

$$\alpha_e^{-1} = 137,035999084... \text{ and } \alpha_t^{-1} = 137,035999252...[8]$$

The above interpretations and results are illustrated in Fig. 4.

Conclusions

A new possibilistic-probabilistic approach was proposed for the interpretation and evaluation of the experimental and theoretical results of the FSC. On the basis of this approach a bivariate possibility distribution was applied on the complex plane for the possibilistic interpretation of the so called "Pauli's equation". We have shown that the most likely values of the marginal possibility distributions are almost the same as the latest experimental and theoretical results (2008) of FSC[9].

References

[1] Fullér, R., Majlender, P.: On Interactive Fuzzy Numbers. Fuzzy Sets and Systems 143, 355–369 (2004)
[2] Nelsen, R.B.: An Introduction to Copulas. Springer, Heidelberg (1999)
[3] Sklar, A.: Fonctions de Répartition àn Dimensions et Leurs Marges, vol. 8, pp. 229–231. Publ. Inst. Statist. Univ. Paris, Paris (1959)
[4] Zadeh, L.A.: Concept of a Linguistic Variable and its Application to Approximate Reasoning. I, II, III, Information Sciences 8, 199–249, 301–357 (1975); 9, 43–80 (1975)
[5] Sherbon, M.A.: Constants of Nature, Dynamics of Time, Publication ID: 103-446-592, p. 10 (2008)
[6] Stanbury, P.: The Aleged Ubiquity of Pi. Nature 304, 11 (1983)
[7] Várlaki, P., Nádai, L., Bokor, J.: Number Archetypes and "Background" Control Theory Concerning the Fine Structure Constant. Acta Polytechnica Hungarica 57, 71–104 (2008)

---

[8] As we mentioned earlier the experimental results in 2006 and 2007 are cca around 137.03599907. [24]

[9] The epistemological and hermeneutical interpretations of the problem can be found in [25].

[8] Várlaki, P., Nádai, L., Bokor, J., Rövid, A.: Controlling-Observing Interpretation of the Fine Structure Constant. In: Proc. of the IEEE 13th International Conference on Intelligent Engineering Systems, Barbados, pp. 61–71 (2009)

[9] Schönfeld, E., Wilde, P.: Electron and Fine Structure Constant II. Metrologia 45(3), 342 (2008)

[10] Hanneke, D., Fogwell, S., Gabrielse, G.: New Measurement of the Electron Magnetic Moment and the Fine Structure Constant. Physical Review Letters 100, 120801 (2008) arXiv:0801.1134v1

[11] Gregor, M.H.M.: The Power of Alpha, p. 69. World Scientific, Singapore (2007)

[12] Szabó, G.: Life of Pauli and his Role in the History of Science. In: Internation Symposium in Memoriam Wolfgang Ernst Pauli, Budapest, March 27 (2009)

[13] Atmanschpacher, H., et al. (eds.): Der Pauli–Jung-Dialog und seine Bedeutung für die moderne Wissenschaft. Springer, Berlin (1995)

[14] Bokor, J., Nádai, L., Rudas, I.: Controllability of Quantum Bits – from the von Neumann Architecture to Quantum Computing. In: 3rd Intl. Conference on Computational Intelligence and Intelligent Informatics, Agadir, Morocco (2007); On CD-ROM

[15] Lindorff, D.: Pauli and Jung: The Meeting of Two Great Minds. Quest Books (2004); Meier, C. (ed.): Atom and Archetype: The Pauli/Jung Letters, pp. 1932–1958. Routledge, London (2002)

[16] Michaletzky, G., Bokor, J., Várlaki, P.: Representability of Stochastic Systems. Akadémiai Kiadó, Budapest, Hungary (1998)

[17] Pauli, W., (Enz, C., Meyenn, K.V. (eds.)): Writings on Physics and Philosophy. Springer, Heidelberg (1994)

[18] Heisenberg, W.: Wolfgang Paulis Philosophische Auffassungen (in Ztschr). für Parapsychologie und Grenzgebiete der Psychologie III(2/3), 127 (1960)

[19] Várlaki, P., Bokor, J., Nádai, L.: Historical Background and Coincidences of Kalman System Realization Theory. In: 5th IEEE Int. Conference on Computational Cybernetics, Gammarth, Tunisia (2007); On CD-ROM

[20] Várlaki, P., Nádai, L., Bokor, J.: Number Archetypes and "Background" Control Theory Concerning the Fine Structure Constant. Acta Polytechnica Hungarica 5, 71–104 (2008)

[21] Várlaki, P., Nádai, L.: Background Control and Number Archetype in Perspective of the Pauli–Jung Correspondence. In: Proc. of Symp. on System and Control Theory, Budapest Univ. of Technology and Economics, pp. 195–229 (2009)

[22] Webb, J.K., Murphy, M.T., Flambaum, V.V., Dzuba, V.A., Barrow, J.D., Churchill, C.W., Prochaska, J.X., Wolfe, A.M.: Further Evidence for Cosmological Evolution of the Fine Structure Constant. Phys. Rev. Lett. 87(9), 091, 301 (2001)

[23] Várlaki, P., Nádai, L., Bokor, J.: Numbers and System Representations in Perspective of the Pauli-Jung Correspondence. In: CD-ROM of the 6th IEEE Symposium on Applied Machine Intelligence and Informatics, Herlany, Slovakia (2008)

[24] Gabrielse, G., Hanneke, D., Kinoshita, T., Nio, M., Odom, B.: New Determination of the Fine Structure Constant from the Electron g Value and QED. Phys. Rev. Lett. 97 (030802) (21.07.2006); 030802, doi:10.1103/PhysRevLett.97.030802

[25] Varlaki, P., Rudas, P.: Twin Concept of Fine Structure Constant as the "Number Archetype" in Perspective of the Pauli-Jung Correspondence, Part I: Observation, Identification and Interpretation. Acta Polytechnica Hungarica 6 (2009)

Observer-Based Iterative Fuzzy and Neural Network Model Inversion for Measurement and Control Applications

Annamária R. Várkonyi-Kóczy

Integrated Intelligent Systems Japanese-Hungarian Laboratory
Dept. of Measurement and Information Systems
Budapest University of Technology and Economics
Magyar tudósok krt. 2, H-1117 Budapest, Hungary
koczy@mit.bme.hu

Abstract. Nowadays model based techniques play very important role in solving measurement and control problems. Recently for representing nonlinear systems fuzzy and neural network (NN) models became very popular. For evaluating measurement data and for controller design also the inverse models are of considerable interest. In this paper, different observer based techniques to perform fuzzy and neural network model inversion are presented. The methods are based on solving a nonlinear equation derived from the multiple-input single-output (MISO) forward fuzzy model simple by interchanging the role of the output and one of the inputs. The utilization of the inverse model can be either a direct compensation of some measurement nonlinearities or a controller mechanism for nonlinear plants. For discrete-time inputs the technique provides good performance if the iterative inversion is fast enough compared to system variations, i.e., the iteration is convergent within the sampling period applied. The proposed method can be considered also as a simple nonlinear state observer which reconstructs the selected input of the forward (fuzzy or NN) model from its output using an appropriate strategy and a copy of the fuzzy or neural network model itself. Improved performance can be obtained by introducing genetic algorithms in the prediction-correction mechanism. Although, the overall performance of the suggested technique is highly influenced by the nature of the non-linearity and the actual prediction-correction mechanism applied, it can also be shown that using this observer concept completely inverted models can be derived. The inversion can be extended towards anytime modes of operation, as well, providing short response time and flexibility during temporal loss of computational power and/or time.

Keywords: inverse modeling, inverse control, iterative methods, fuzzy modeling, neural network models, anytime techniques, genetic algorithms, observers, singular value decomposition.

I.J. Rudas et al. (Eds.): Towards Intelligent Engineering & Information Tech., SCI 243, pp. 681–702.
springerlink.com   © Springer-Verlag Berlin Heidelberg 2009

1   Introduction

Model-based schemes play an important role among the measurement and control strategies applied to dynamic plants. The basically linear approaches to fault diagnosis [1], optimal state estimation [2] and controller design [3] are well understood and successfully combined with adaptive techniques (see. e.g. [4]) to provide optimum performance. Nonlinear techniques, however, are far from this maturity or still are not well understood. Although, there is a wide variety of possible models to be applied based on both classical methods [5] and recent advances in handling [6] information, up till recently practically no systematic method was available which could be offered to solve a larger family of nonlinear control problems. The efforts on the field of fuzzy and neural network (NN) based modeling and control however, seem to result in a real breakthrough also in this respect. With the advent of adaptive fuzzy controllers very many control problems could be efficiently solved and the model based approach to fuzzy controller design became a reality [7]. Using model based techniques in measurement and control also the inverse models play a definite role [4].

Recently different techniques have been published (see e.g. [7]-[11]) for inverting certain fuzzy models, however, exact inverse models can be derived only with direct limitations on the fuzzy models applied. In this paper, an alternative approach is investigated which is based on the quite general concept of state observation widely used in measurement and filtering applications. The key element of this concept is to force a model of a physical system to "copy" the behavior of the system to be observed (see Fig. 1). This scheme is the so called observer structure which is a common structural representation for the majority of iterative data and/or signal processing algorithms. From Fig. 1, it is obvious that here the inversion of dynamic system models is considered.

Traditionally, the observer is a device to measure the states of dynamic systems having state variable representation. According to our proposition, however, these states can be regarded as unknown inputs, and therefore their "copy" within the observer as the result of model inversion. In this scheme, the fuzzy models appear as static nonlinear Multiple Input Single Output (MISO) mappings from the state variables to the system output, i.e., represent the output equation. A copy of this output equation is also present within the observer, and in this case observer dynamics is simply due to the iterative nature of the algorithm.

At this point it is important to note that the inversion is not unique if more than one input is considered, i.e., from one observer input value more than one output is to be calculated. In this paper only unique inversions are investigated, therefore

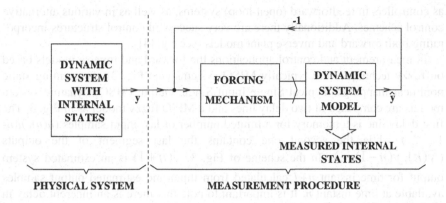

Fig. 1. The observer concept

the calculation of one controllable input is regarded based on the desired output and uncontrollable input values. Although, we also point out the generalization of the inversion for more then one input.

The paper is organized as follows. The possible role of inverse fuzzy models and the main features of the explicit inversion methods are described in Section 2. Section 3 presents the observer based iterative inversion technique. Section 4 is devoted to the genetic algorithm and its application within the inversion procedure. In Section 5 the inversion scheme is extended to be able to operate in anytime modes. Section 6 details the extension to neural network models and finally, in Section 7 illustrative examples are presented.

2 Inverse Fuzzy Models in Measurement and Control

Nowadays solving measurement and control problems involves model-integrated computing. This integration means that the available knowledge finds a proper form of representation and becomes an active component of the computer program to be executed during the operation of the measuring and control devices. Since fuzzy models represent a very challenging alternative to transform typically linguistic a priori knowledge into computing facilities therefore it worth reconsidering all the already available model based techniques whether a fuzzy model can contribute to better performance or not.

The role of the inverse models in measurements is obvious: observations are mappings from the measured quantity. This mapping is performed by a measuring channel the inverse model of which is inherent in the data/signal processing phase of the measurement. In control applications inverse plant models are to be applied as controllers in feedforward (open-loop) systems, as well as in various alternative control schemes. Additionally there are very successful control structures incorporating both forward and inverse plant models (see e.g. [4]).

In measurement and control applications the forward and inverse models based on fuzzy techniques are typically MISO systems (see Fig. 2) representing static nonlinear mappings. Typical Single Input Single Output (SISO) dynamic system models are composed of two delay lines and a MISO fuzzy model as in Fig. 3. The first delay line is a memory for a limited number of last input samples ($u(n)$, $u(n-1)$, …) while the second one contains the last segment of the outputs ($\hat{y}(n), \hat{y}(n-1),...$). In the scheme of Fig. 3, $\hat{y}(n+1)$ is an estimated system output for time instant $n+1$ calculated from input and estimated output samples available at time instant n. It is important to note that there is an inherent delay in such and similar systems since the model evaluations take time. The (partial) inverse of such a dynamic system is also a SISO scheme (see Fig. 4) which is in correspondence with Fig. 3 except the role of the input and output is transposed. The input $r(n)$ is a sample of the reference signal and a predicted value of the input will be the model output. It is obvious from these figures that if the forward fuzzy model is available only the inverse (nonlinear static) mapping must be derived.

There are different alternatives to perform such a derivation. One alternative is to invert the fuzzy model using the classical regression technique based on input-output data. To solve this regression problem, iterative algorithms can also be considered. The result of such a procedure is an approximation of the inverse, and the accuracy of this approximation depends on the efficiency of the model fitting applied. In this case, however, the inverse is not necessarily a fuzzy model in its original sense.

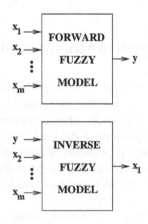

Fig. 2. Forward and inverse model

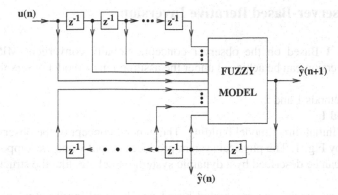

Fig. 3. Dynamic system model

Recently very interesting methods have been reported for exactly inverting certain type of fuzzy models [7]. For the case of the standard Mamdani fuzzy model [12] with singletons in the rule consequents exact inverse can be derived. Obviously the general conditions of invertibility must be met: the forward fuzzy model should be strictly monotone with respect to the input which is considered to be replaced by the output. If the forward model implements a noninvertible function it must be decomposed into invertible parts and should be inverted separately. There are, however, several other prerequisites of the exact inversion concerning both antecedent and consequent sets, rule base, implication and T-norm, as well as the defuzzification method. The construction of the inverse proved to be relatively simple at the price of strong limitations.

There are other promising approaches ([8], [10]) where the inversion is solved on linguistic level, i.e. rule base inversion is performed. These techniques can accommodate various fuzzy concepts but must be combined with fuzzy rule base reduction algorithms if the number of the input sets is relatively high.

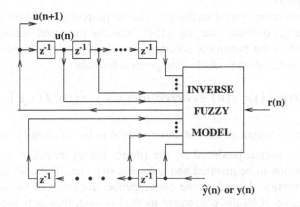

Fig. 4. Inverse model

3   Observer-Based Iterative Inversion

Theorem 1 Based on the observer concept, globally convergent MISO fuzzy model inversion can be achieved under the assumption of model observability.

Proof

See Methods 1 and 2.

Method 1

Step 1 Initial direct model building. The general concept of the observer is represented by Fig. 1. The physical system produces output y and we suppose that its behavior can be described by a dynamic system model e.g. like the structure given in Fig. 3.

The fuzzy model can be created based on observed input-output pairs of the system, probably with support of human expert knowledge. This system description becomes the inherent part of the measurement procedure and is forced to behave similarly to the physical system. A more detailed description of this idea for static nonlinear fuzzy model is given by Fig. 5. As an example, here a three-input one-output fuzzy model can be seen.

Step 2 Iteration. Repeat steps 3-5 till convergence or stopping criteria is reached.

Step 3 Estimation of the output-error based cost function. Input x_1 is considered unknown and therefore to be observed via comparing the output of the physical system and that of the model.

Step 4 Evaluation of the adaptation (forcing) mechanism.

Step 5 Correction of the unknown model input estimation with the help of step 3. If the correction (forcing) mechanism is appropriate the observer will converge to the required state and produce the estimate of the unknown input. The strength of this approach is that this iterative evaluation is easy to implement, e.g., using standard digital signal processors. The complete system of Fig. 5 can be embedded into a real-time environment, since the necessary number of iterations to get the inverse can be performed within one sampling time slot of the measurement or control application.

For the correction several techniques can be proposed based on the vast literature of numerical methods (see e.g. [13]) since the proposed iterative solution is nothing else than the numerical solution of a single variable nonlinear equation. The iteration is based on the following general formula

$$\hat{x}(n+1) = \hat{x}(n) + correction[y - \hat{y}, \hat{x}(n), f(), \mu]   \tag{1}$$

where y is the output of the unknown system to be inverted, $\hat{y}$ denotes the estimation of the output produced by the (direct fuzzy) model, $f(\ )$ stands for the nonlinear function to be inverted and μ for the step size. The convergence properties of the inversion depend on the convergence properties of the applied correction technique, i.e. if locally convergent method is used, then only local convergence of the iteration can be ensured. Although, after the convergence of a step, in most of the cases, the output remains within a predictable distance of the previous output if the input change is under a limit and vice versa.

One of the simplest solutions is if Newton iteration is applied. In this case, (1) has the form of

$$\hat{x}(n+1) = \hat{x}(n) + \mu \frac{(y - \hat{y})}{f'(\hat{x}(n))} \tag{2}$$

where $f'(\ )$ denotes the derivative of the $f(\ )$. This latter must be evaluated locally using simple numerical technique:

$$f'(\hat{x}(n)) \cong \frac{f(\hat{x}(n) + \Delta) - f(\hat{x}(n) - \Delta)}{2\Delta} \tag{3}$$

The computational complexity of this iterative procedure depends mainly on the complexity of the forward fuzzy model itself. It is anticipated, however, that after the first convergence, if the input of this observer changes relatively smoothly, then in the majority of the cases only a few iterations will be required to achieve an acceptable inverted value (see also [S49]).

In principle the proposed method can be generalized even for two or more (forward model) inputs (see Fig. 5). For this generalization, however, it must make clear that to calculate two or more inputs at least the same number of ("measured") outputs is required. With static nonlinear MISO models this is not possible. If the fuzzy model in Fig. 5 is replaced by a dynamic, state variable model, where the "inputs" will correspond to the state variables, following the state transitions new outputs can be "measured" and as many values can be collected as required to the iterative solution of the multi input problem. This idea obviously requires the "observability" [2] of the states and the application of state variable models instead of the schemes of Figs. 4 and 5 [14]. The complexity of the iterative technique to be applied in such cases is under investigation.

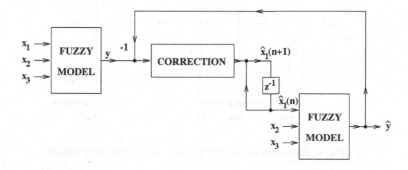

Fig. 5. Block diagram of the iterative inversion scheme for a three input one output case

4   Genetic Algorithms for Fuzzy Model Inversion

Unfortunately, the simple Newton iteration may fail if the nonlinear function represented by the fuzzy model has multiple minima. This is because gradient-based techniques work using only locally available information. If multiple minima may occur global search techniques are to be considered, however usually with a burden of high complexity.

4.1 The Multiple Root Problem

Another problem is the case of multiple roots. Fig. 6 illustrates these situations. Here a function $y=f(x_1,x_2,\ldots)$ is shown with fixed x_2, … input values. The calculation of the inverse means to find the corresponding roots x_{10i} ($i=1,\ldots$) for a given $y=y_0$. Unfortunately, in general it is a hard problem to decide which root equals (or approximates) the actual fuzzy model input. At our present knowledge the only thing we can state is that, having the correspondence, small variations of the input should result in small deviation of the estimate. Intuitive techniques based on Taylor series expansion may provide proper orientation, but the general solution is still an open issue.

Since the situation indicated in Fig. 6 is not necessarily typical, there are intervals with one-to-one correspondence between input and output. Based on these values the root decision problem can be solved.

It is important to note at this point that if multiple roots may occur in a region, and there is no proper hint available to find the appropriate one, then it is unavoidable to search for all the roots and make the decision afterwards.

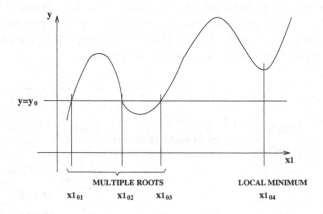

Fig. 6. Illustration for the multiple roots-local minimum problem

4.2 Genetic Algorithm-Based Inversion

Genetic algorithms (GAs) [15], [16] are optimization methods which search in the entire space to find the global optimum, and they are quite resistant against the problem of local minima. The classical versions of GAs are applied to solve discrete optimization problems, however, by applying new forms for representation and for genetic operators more and more continuous optimization problems are solved based on GAs.

In the continuous optimizations the chromosomes should store, instead of the traditional binary strings (as representation of numbers, e.g. [17]), one or more real numbers. Here the so-called direct real coding is used, (see e.g. [18]) which is a more efficient representation and it is immune against several unfortunate effects [16], [18].

The presented methods apply real-coded representation. The chromosome contains basically one real number x that is the candidate for x_1. The fitness-function $ff(x)$ is derived from the difference $d(x) = |f(x) - y|$ as $ff(x) = M - d(x)$ where, $M = $ max $d(x)$ + min $d(x)$, max and min are evaluated over the entire population.

The initialization of the population and the mutation are performed in a similar way. The population is derived from an interval I_n which is updated in every generation. The operators work on random values having uniform distribution over this interval.

If the tracing mode is applied, then the initial population comes from interval I_0. This interval is around the value x_0 with radius r which is derived from the previous x_1 and Δx_1 values: $x_0 = x_1 + \Delta x_1$, $r = \Delta x_1$. As the generations go on, I_n is grown as follows: $I_{n,\min} = I_{0,\min} - nv_1$, $I_{n,\max} = I_{0,\max} + nv_1$, where $v_1 = 0.5\Delta x_1$.

The crossover operator is the same in both the global search and the tracing mode methods. It uses linear interpolation of the function, deriving from parent values x_A, x_B and from $f(x_A)$, $f(x_B)$: $x_{C0} = \dfrac{x_A f(x_B) - x_B f(x_A)}{f(x_B) - f(x_A)}$. The final children are taken from an interval around x_{C0}, with radius $r = 0.1 \min(|x_A - x_{C0}|, |x_B - x_{C0}|)$.

The GAs use roulette wheel selection. The probability of the crossover is 0.8, the mutation works with 0.05. The population size is 10. The terminating condition depends on the error of the best value: $d(x_b) / |f'(x_b)| < E$ where $f'(x_b)$ is estimated by the nearest x to x_b, E is the estimation error limit.

In our experience the performance of the GAs for iterative fuzzy model inversion is quite acceptable, but its computational complexity is relatively high.

In the followings, as a low complexity global search technique, a new hybrid method is presented which is based on the combination of some low complexity local search technique (like Newton iteration) and globally convergent genetic algorithms.

4.3 The New Globally Convergent Hybrid Inversion Method

To decrease the complexity of the inversion, a combined technique has been elaborated which is basically a Newton method, but if it fails, it is switched to the GAs for one iteration.

Method 2

Our assumption concerning operation is that of associated with the observer mechanism, i.e., a tracking mode, where the iteration is based on a "good", previous estimate, since the inputs and the output do not change drastically. The introduction of genetic search can be justified (1) at the beginning, as we start the iterative procedure, and try to find the first estimate of the input, (2) during continuous operation, when gradient based search techniques fail due to local minima. In other cases the simple Newton iteration might be acceptable.

The major steps of the combined method are as follows:

Genetic algorithm:

    Evaluate genetic algorithm

    Switch to Newton iteration

Newton iteration:

    Set the input variables, evaluate the fuzzy model $y = f( )$, $y(n)$

    Evaluate equation (3)

    IF $|y - \hat{y}(n)| / |f'(\hat{x}(n))| \langle E$, where $E > 0$ is the estimation error limit,

    THEN stop

    ELSE IF $|f'(\hat{x}(n))| \langle \delta$, where $\delta > 0$ is a small number, OR IF the number of

        iterations is larger than a predefined value N, THEN switch to the genetic algorithm

    ELSE Evaluate equation (2)

    $n = n + 1$

    Switch to Newton iteration

The advantageous, low complexity figures of the proposed hybrid method has been proved by simulations (see also [19]).

5 Fuzzy Inversion in Anytime Systems

5.1 Anytime Systems

In computer-based monitoring, diagnostic, control, etc. systems the operations should be performed under prescribed response time conditions. It is an obvious requirement to provide enough computational power but the achievable processing speed is highly influenced by the precedence, timing and data access conditions of the processing itself. It seems to be unavoidable even in the case of extremely careful design to get into situations where the shortage of necessary data and/or processing time becomes serious. Such situations may result in a critical breakdown of the monitoring, diagnostic, and/or control systems. The concept of "anytime" processing tries to handle the case of too many abrupt changes and their consequences in larger scale embedded systems. The idea is that if there is a temporal shortage of computational power and/or there is a loss of some data the actual operation should be continued to maintain the overall performance "at lower price", i.e., information processing based on algorithms and/or models of simpler complexity should provide outputs of acceptable quality to continue the operation of the complete system. The accuracy of the processing will be temporarily lower but possibly still enough to produce data for qualitative evaluations and supporting decisions. Consequently "anytime" algorithms [20] provide short response time and are very flexible with respect to the available input information and computational power. The expectations against such algorithms are: Low complexity, changeable, guaranteed response time/computational need and accuracy, known error.

Iterative algorithms are popular tools in anytime systems, because their complexity and computational time-need can be flexibly changed according to the temporal conditions. Unfortunately, the usability of iterative algorithms is limited because adequate evaluation methods can not always be found.

Besides the iterative algorithms, a wide-range of other types of computing methods/algorithms can be used in anytime systems by using modular architectures (see Fig. 7). The applicability of this technique is more general, however, it needs some extra planning and considerations.

Using this latter solution, each module of the system offers several implementations (having different attribute-values) for a certain task. The units within a given module have uniform interface (same inputs, outputs and solve the same problem) but differ concerning their computational need and accuracy. In the knowledge of the current conditions (tasks to complete, achievable time/resources, needed accuracy, etc.), an expert system can choose the adequate configuration, i.e., the units which will be used. This means the optimization of the whole system, instead of individual modules, i.e., it can be more advantageous to reduce the computational complexity and accuracy of some parts of the systems and rearrange the resources to others offering at the moment more important services.

5.2 SVD-Based Complexity Reduction of Fuzzy Inference Systems

Although fuzzy tools are very popular in engineering systems, their usability is limited by their exponential increasing computational complexity. There is no

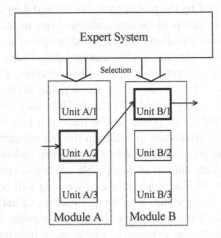

Fig. 7. Anytime system with a modular architecture

universal method to estimate the necessary number of antecedent fuzzy sets to achieve a desired accuracy. In practice, the number of antecedent fuzzy sets is usually overestimated resulting in large and redundant rule-bases which require a lot of unnecessary computing. A complexity-reduction method can help in the elimination of this redundancy, however, in a lot of cases the achieved complexity reduction is still not enough for the realization.

From the lot of known complexity reduction techniques the SVD-based reduction technique [21] seems very advantageous in anytime systems. It can be used in a wide range of fuzzy systems and it makes possible not only the automatic elimination of the redundancy of the rule-base, but also further, non-exact reductions with easily computable complexity and inaccuracy. It is worth mentioning, that the SVD-based reduction finds the optimum, i.e., minimum number of parameters which is needed to describe the system.

Definition 1 (SVDR): The SVD based complexity reduction algorithm is based on the decomposition of any real valued $\underline{\underline{F}}$ matrix:

$$\underline{\underline{F}}_{(n_1 \times n_2)} = \underline{\underline{A}}_{1,(n_1 \times n_1)} \underline{\underline{B}}_{(n_1 \times n_2)} \underline{\underline{A}}_{2,(n_2 \times n_2)}^T \qquad (4)$$

where $\underline{\underline{A}}_k$, $k=1,2$ are orthogonal matrices ($\underline{\underline{A}}_k \underline{\underline{A}}_k^T = \underline{\underline{E}}$), and $\underline{\underline{B}}$ is a diagonal matrix containing the λ_i singular values of $\underline{\underline{F}}$ in decreasing order. The maximum number of the nonzero singular values is $n_{SVD} = \min(n_1, n_2)$. The singular values indicate the significance of the corresponding columns of $\underline{\underline{A}}_k$. Let the matrices be partitioned in the following way:

$$\underline{\underline{A}}_k = \left| \underline{\underline{A}}^r_{k,(n_k \times n_r)} \quad \underline{\underline{A}}^d_{k,(n_k \times (n_k - n_r))} \right| \text{ and } \underline{\underline{B}} = \left| \begin{array}{cc} \underline{\underline{B}}^r_{(n_r \times n_r)} & 0 \\ 0 & \underline{\underline{B}}^d_{((n_1 - n_r) \times (n_2 - n_r))} \end{array} \right|,$$

where r denotes "reduced" and $n_r \le n_{SVD}$.

If $\underline{\underline{B}}^d$ contains only zero singular values then $\underline{\underline{B}}^d$ and $\underline{\underline{A}}^d_k$ can be dropped: $\underline{\underline{F}} = \underline{\underline{A}}^r_1 \underline{\underline{B}}^r \underline{\underline{A}}^{rT}_2$. If $\underline{\underline{B}}^d$ contains nonzero singular values, as well, then the $\underline{\underline{F}}' = \underline{\underline{A}}^r_1 \underline{\underline{B}}^r \underline{\underline{A}}^{rT}_2$ matrix is only an approximation of $\underline{\underline{F}}$ and the maximum difference between the values of $\underline{\underline{F}}$ and $\underline{\underline{F}}'$ equals

$$E_{RSVD} = \left| \underline{\underline{F}} - \underline{\underline{F}}' \right| \le \left(\sum_{i=n_r+1}^{n_{SVD}} \lambda_i \right) \underline{\underline{1}}_{(n_1 \times n_2)} \tag{5}$$

5.3 Anytime Fuzzy Models

With the help of the (HO)SVD-based complexity reduction fuzzy systems can be operated in an anytime mode [22]. This operational form can be advantageous in model inversion, as well. For anytime fuzzy model inversion the forward model within the inversion scheme has to be replaced by the appropriate truncated model corresponding to the actual circumstances, i.e., the temporarily available amount of computational resources and time. The computational need of the inversion based on the reduced forward model is directly proportional to the computational complexity of the used model, thus the necessary model reduction can be computed.

Since convergence can be ensured in the observer scheme, the accuracy of the inverted model will be determined by the accuracy and the transfer characteristics of the forward model. This latter can be approximated locally using e.g. the simple numerical technique in equation (3).

The steps of preparing a fuzzy model to be able to work in anytime inversion are, as follows:

Method 3

Step 1 First a practically "accurate" fuzzy system is to be constructed. For the determination of the rule-base, expert knowledge can be used. Further improvement can be obtained by utilizing training data and some learning algorithm. In this step there is no need to deal with the complexity of the obtained model.

Step 2 In the second step, by applying (HO)SVD-based complexity reduction algorithm, a reduced but "accurate" model can be generated. (In this step only the redundancy is removed.)

Step 3 The SVD-based model can be used in anytime systems either by applying the iterative transformation algorithm developed for PSGS fuzzy systems [23] or in the more general frame of the modular architecture.

In the first case, the transformation can be performed off-line, before the anytime operation starts (i.e. it does not cause any additional computational load on the system) and the model evaluation can be executed without knowing about the available amount of time. The newest output will always correspond to the in the given circumstances obtainable best results.

In the second case, based on the SVD transformed model of the system, further non-exact variations of the rule-bases of the model must be constructed. These models will differ in their accuracy and complexity. For anytime use, an alternative rule-base is characterized by its complexity and its error that can be estimated by the sum of the discarded singular values.

The different rule-bases form the different units realizing a given module (Fig. 7).

Step 4. During the operation, an intelligent expert system, monitoring the actual state of the supervised system, can adaptively determine and change for the units (rule base models) to be applied according to the available computing time and resources at the moment. These considerations need additional computational time/resources (further reducing the resources). On the other hand, because the inference algorithm within the models of a certain module is the same, only the rule-bases – a kind of parameter set – must be changed resulting in advantageous dynamic behavior. One can find more details about the intelligent anytime monitor and the algorithmic optimization of the evaluations of the model-chain in [24].

6   Inversion of Neural Network Models

The presented inversion technique can be generalized to invert neural network models, as well [25]. When we refer to "inversion" of a neural network for the acquisition of certain input parameters, we are actually referring to a constrained inversion. That is to say, given the functional relationship of the NN input to the output, we have the forward and want to have the inverse relationships (see Fig. 8), where $x_1, x_2, \ldots, x_n$ is the input set of the forward neural network model and $\underline{y}$ is the output vector of the model. Fig. 8b illustrates the inverse NN model, where output vector $\underline{x}^s$ is composed of a subset of the input variables of the forward model. Given the forward relationship described in a neural network form

$$\underline{y} = f(\underline{x}^s, \underline{u}) \tag{6}$$

where $\underline{x}^s$ stands for the unknown environmental parameters we wish to obtain, and $\underline{u}$ denotes a vector of the known environmental and system parameters ($\underline{x} = \underline{x}^s \cup \underline{u}$), we wish to find an inverse mapping

$$\underline{x} = g(\underline{y}, \underline{u}).\tag{7}$$

For the inversion, Method 2 can be used, however with taking into consideration that we need as many input as many output is to be inverted, i.e. e.g. in case of a multiple-input-single-output (MISO) system we have to use as many output values as input of the inversion scheme as many input parameters are to be inverted [22].

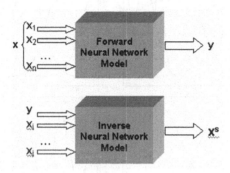

Fig. 8. Forward and inverse neural network models

$$\underline{x} = \{x_1,...,x_n\},\ \underline{x}^s \subseteq \underline{x} \equiv \{x_1,...,x_n\} \setminus \{x_i,...,x_j\}$$

7   Illustrative Examples

In this Section, to illustrate the proposed iterative inversion method, three simple examples are presented. The first one is a three input one output forward fuzzy model. The purpose of this model is to describe the frequency, sound intensity and age dependency of the human hearing system [26]. The output of the model is the sound intensity felt by the person, i.e., the subjective intensity. The inversion of this fuzzy model might be interesting if the input sound intensity is to be estimated from known subjective intensity, frequency, and age information.

The inputs are represented by Gaussian shape membership functions: five-five sets for the frequency and the input intensity, respectively, and another three to represent the age. The characterization of the output is solved by five triangular shape sets (see Fig. 9). The rule base of the model consists of 49 rules which is a linguistic equivalent of the widely known, measurement-based plots describing the human hearing system (see Table 1). Fig. 10 shows the obtained surface as a function

of frequency and sound intensity at age of 60 years. The inverse was calculated by two methods. The proposed iterative gradient method was operated with
$\mu=1$ and termination condition at $E = 10^{-4}$ or 50 iterations. The error surface
given in Fig. 11 is defined as the absolute difference of the original and the calcu-
lated values of the input variable objective intensity ($|x - \hat{x}(n)|$). The number of
iterations can be kept at a relatively low level as it is illustrated by Fig. 12. The inversion was carried out also by using GA based inverse search. The obtained results (an average of ten independent runs) for the error surface and for the number
of evaluations are shown in Figs. 13-14.

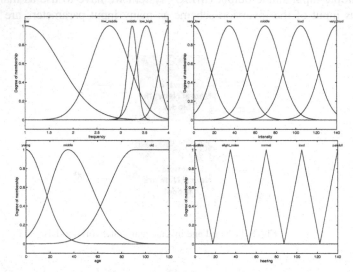

Fig. 9. Input and output fuzzy sets for the frequency, intensity, age, and hearing

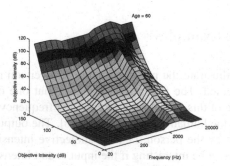

Fig. 10. The obtained surface as a function of frequency and sound intensity at age of 60 years

In the second example (Ex2) the inversion is carried out over a relatively different surface containing several local minima. The five-five Gaussian shape sets
for the input variables and the five triangular shape output sets of the two input
one output forward fuzzy model are shown on Fig. 15. The rule base of the model
consists of 25 rules as is given by Table 2.

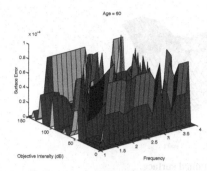

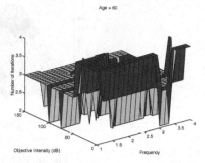

Fig. 11. The error surface (grad. method)

Fig. 12. The number of iterations (grad method)

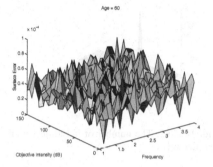

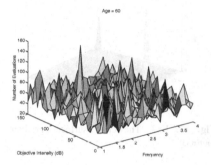

Fig. 13. The error surface (GA method)

Fig. 14. The number of evaluations (GA method)

The obtained surface can be followed on Fig. 16. As it is illustrated by Fig. 17 the gradient based iteration suffers from the local minima (see also Fig. 18 for the number of iterations). The GA based iteration avoids this problem (see Figs. 19-20).

To keep the computational complexity on a lower level the proposed combined method can be applied. Fig. 21 shows the obtained error surface, while Fig. 22 the number of evaluations.

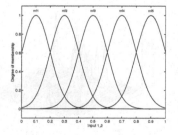

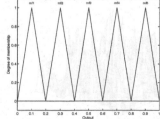

Fig. 15. Ex2: input fuzzy sets for x1 and x2(left); output fuzzy sets (right)

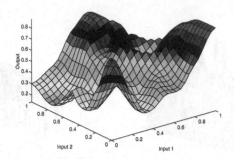

Fig. 16. Ex2: The obtained surface

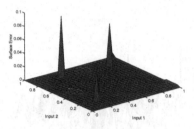

Fig. 17. Ex2: The error surface (gradient method)

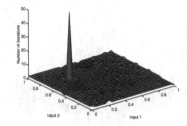

Fig. 18. Ex2: The number of iterations (gradient method)

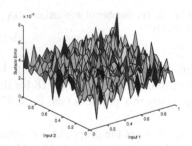

Fig. 19. Ex2: The error surface (GA method)

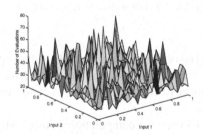

Fig. 20. Ex2: The number of evaluations (GA method)

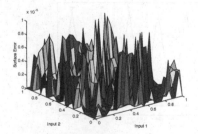

Fig. 21. Ex2: The error surface (combined method)

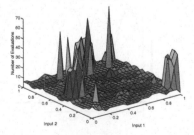

Fig. 22. Ex2: The number of evaluations (combined method)

Table 1. Rules of hearing fuzzy-model

Hearing (Age = Young)		Objective intensity				
		Very low	Low	Middle	Loud	Very loud
Frequency	Low	H1	H1	H2	H3	H5
	Low middle	H1	H2	H3	H4	H5
	Middle	H2	H3	H4	H5	H5
	Low high	H1	H2	H3	H4	H5
	High	H1	H1	H2	H3	H5

Hearing (Age = Middle)		Objective intensity				
		Very low	Low	Middle	Loud	Very loud
Frequency	Low	H1	H1	H1	H2	H5
	Low middle	H1	H1	H2	H3	H5
	Middle	H1	H2	H3	H4	H5
	Low high	H1	H1	H2	H3	H5
	High	H1	H1	H1	H2	H5

Hearing (Age = Old)		Objective intensity				
		Very low	Low	Middle	Loud	Very loud
Frequency	Low	H1	H1	H1	H2	H5
	Low middle	H1	H1	H2	H3	H5
	Middle	H1	H1	H3	H4	H5
	Low high	H1	H1	H2	H3	H5
	High	H1	H1	H1	H2	H5

H1 = non-audible, H2 = slight noise, H3 = normal, H4 = loud, H5 = painful

Table 2. Rules of example 2

Output		Input 2				
		mf1	mf2	mf3	mf4	mf5
Input 1	mf1	mf5	mf1	mf3	mf2	mf5
	mf2	mf1	mf3	mf2	mf4	mf5
	mf3	mf3	mf1	mf5	mf2	mf4
	mf4	mf1	mf3	mf4	mf5	mf2
	mf5	mf2	mf3	mf5	mf4	mf1

Example 3 shows the inversion of the dynamic neural network model of the Anti-lock Brake System (ABS). The block scheme of the ABS system is illustrated in Fig. 23. The input are the slip (the ratio of the velocities of the vehicle and the wheel, i.e. (vehicle speed - wheel speed) / vehicle speed): the desired slip (0.2), and the actual slip and the output are the speed of the vehicle and the wheel or the actual slip (see Fig. 23). The system works in the following way: The actual slip is compared to the desired slip and based on the result the brake pressure is controlled according to the estimated maximum of the allowable brake pressure. The speed of the vehicle and the speed of the wheel together with the adhesion (μ) can also be determined from the slip. (From the actual brake pressure we can obtain the actual brake torque and from this latter the wheel speed.)

The aim of the inverse system is to reconstruct the operation of the ABS in time, i.e. having the values of the speed of the car and that of the wheel we are able to determine the time diagram of the braking pressure of the ABS.

In Fig. 24 the speed of the vehicle can be followed, while Fig. 25 shows the time dependence of the wheel speed. In Fig. 26 the time dependence of the slip (effected by the braking process) is illustrated. The accuracy of the inversion is quite acceptable: The relative mean square error of the estimated braking pressure or torque (evaluated by the inverse system) compared to the actual value remains always less then 1%.

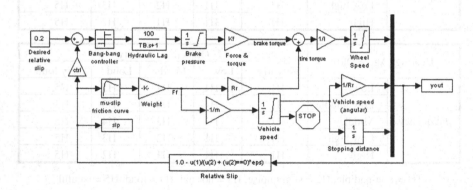

Fig. 23. Block diagram of the Anti-lock Brake System

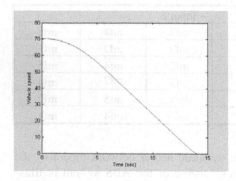

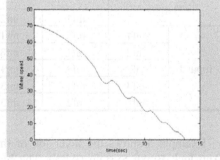

Fig. 24. Time dependence of the speed of the vehicle

Fig. 25. Time dependence of the speed of the wheel

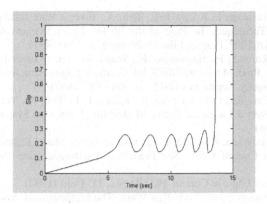

Fig. 26. Time dependence of the slip

Conclusions

In this paper an "on-line" iterative technique has been proposed to solve the inversion of fuzzy and neural network models for measurement and control applications. The derivation of this iterative technique is related to the state observer concept which proved to be very successful in the interpretation of the different techniques applied on this field. Additionally a step toward completely inverted models can also initiated by introducing state variable dynamic models combined with fuzzy logic based components. The proposed inversion scheme can be operated in anytime systems, as well, if as forward model SVD-based fuzzy or NN model is used.

Acknowledgments. This work was sponsored by the Hungarian Fund for Scientific Research (OTKA K 78576).

References

[1] Patton, R., Frank, P., Clark, R.: Fault Diagnosis in Dynamic Systems. Prentice Hall International (UK) Ltd., Englewood Cliffs (1989)
[2] Anderson, B.D.O., Moore, J.B.: Optimal Filtering. Prentice-Hall, Inc., Englewood Cliffs (1979)
[3] Boyd, S.P., Barratt, C.H.: Linear Controller Design, Limits of Performance. Prentice Hall, Inc., Englewood Cliffs (1991)
[4] Widrow, B., Walach, E.: Adaptive Inverse Control. Prentice-Hall, Englewood Cliffs (1996)
[5] Billings, S.A.: Identification of Nonlinear Systems – A Survey. IEE Proc. 127(6), Pt. D, 272–284 (1980)
[6] Klir, G.J., Folger, T.A.: Fuzzy Sets, Uncertainty, and Information. Prentice-Hall International, Inc., Englewood Cliffs (1988)
[7] Babuska, R., Sousa, J., Verbruggen, H.B.: Model-Based Design of Fuzzy Control Systems. In: Proc. of the Third European Congress on Intelligent Techniques and Soft Computing EUFIT 1995, Aachen, Germany, pp. 837–841 (1995)

[8] Baranyi, P., Bavelaar, I., Kóczy, L.T., Titli, A.: Inverse Rule Base of Various Fuzzy Interpolation Techniques. In: Proc. of the 7th Int. Fuzzy Systems Association World Congress, IFSA 1997, Prague, June 25-29, pp. 121–126 (1997)

[9] Baranyi, P., Korondi, P., Hashimoto, H., Wada, M.: Fuzzy Inversion and Rule Base Reduction. In: Proc. of the 1997 IEEE Int. Conf. on Engineering Systems, INES 1997, Budapest, Hungary, September 12-15, pp. 301–306 (1997)

[10] Baranyi, P., Bavelaar, I.M., Babuska, R., Kóczy, L.T., Titli, A., Verbruggen, H.B.: A Method to Invert a Linguistic Fuzzy Model. Int. Journal of Systems Science 29(7), 711–721 (1998)

[11] Batur, C., Shrinivasan, A., Chan, C.-C.: Inverse Fuzzy Model Controllers. In: Proc. of the American Control Conf., San Francisco, California, June 1993, pp. 772–774 (1993)

[12] Jager, R.: Fuzzy Logic in Control, PhD Thesis, TU Delft (1995)

[13] Acton, F.S.: Numerical Methods That Work, The Mathematical Association of America, Washington D.C (1990)

[14] Várkonyi-Kóczy, A.R., Péceli, G., Dobrowiecki, T.P., Kovácsházy, T.: Iterative Fuzzy Model Inversion. In: Proc. of the IEEE Int. Conference on Fuzzy Systems, FUZZ-IEEE 1998, Anchorage, Alaska, USA, May 5-9, vol. 1, pp. 561–566 (1998)

[15] Whitley, D.: A Genetic Algorithm tutorial, Technical report, Colorado State University (1993)

[16] Goldberg, D.E.: Genetic Algorithms in Search, Optimization and Machine Learning. Addison-Wesley, Reading (1989)

[17] Ko, M.-S., Kang, T.-W., Hwang, C.-S.: Function Optimization Using an Adaptive Crossover Operator Based on Locality. Engineering Applications of Artificial Intelligence 10(6), 519–524 (1997)

[18] Djurisic, A.B., Elite: Genetic Algorithms with Adaptive Mutations for Solving Continuous Optimization Problems – Application to Modeling of the Optical Constants of Solids. Optics Communications 151, 147–159 (1998)

[19] Várkonyi-Kóczy, A.R., Álmos, A., Kovácsházy, T.: Genetic Algorithms in Fuzzy Model Inversion. In: Proc. of the 8th IEEE Int. Conference on Fuzzy Systems, FUZZ-IEEE 1999, Seoul, Korea,, August 22-25, vol. 3, pp. 1421–1426 (1999)

[20] Zilberstein, S.: Using Anytime Algorithms in Intelligent Systems. AI Magazine 17(3), 73–83 (1996)

[21] Yam, Y.: Fuzzy Approximation via Grid Point Sampling and Singular Value Decomposition. IEEE Trans. on System, Man, Cybernetics 27, 933–951 (1997)

[22] Várkonyi-Kóczy, A.R., Takács, O.: Anytime Extension of the Iterative Fuzzy Model Inversion. In: Proc. of the 10th IEEE Int. Conference on Fuzzy Systems, FUZZ-IEEE 2001, Melbourne, Australia, December 2-5, pp. 976–979 (2001)

[23] Takács, O., Várkonyi-Kóczy, A.R.: Iterative-Type Evaluation of PSGS Fuzzy Systems for Anytime Use. In: Proc. of the 2002 IEEE Instrumentation and Measurement Technology Conference, IMTC/2002, Anchorage, USA, May 21-23, pp. 233–238 (2002)

[24] Várkonyi-Kóczy, A.R., Samu, G.: Anytime System Scheduler for Insufficient Resource Availability. International Journal of Advanced Computational Intelligence and Intelligent Informatics (JACIII) 8(5), 488–494 (2004)

[25] Várkonyi-Kóczy, A.R., Rövid, A.: Observer-based Iterative Neural Network Model Inversion. In: Proc. of the 14th IEEE Int. Conference on Fuzzy Systems, FUZZ-IEEE 2005, Reno, USA, May 22-25, pp. 402–407 (2005)

[26] Moore, B.C.J.: An Introduction to the Psychology of Hearing, 3rd edn. Academic Press, London (1995)

Role of the Cyclodextins in Analytical Chemistry

Zoltán Juvancz[1], Judit Némethné-Katona[1], and Róbert Iványi[2]

[1] Institute of Environmental Engineering, Budapest Tech
Doberdó út 6, H-1034 Budapest, Hungary
juvancz.zoltan@rkk.bmf.hu, katona.judit@rkk.bmf.hu
[2] Cyclolab Ltd., H-1525, P.O.B. 435, Budapest, Hungary
ivanyi@cyclolab.hu

Abstract. This paper is dealing with the analytical usage of cyclodextrins and their other applications (e.g. pharmaceutical, food and cosmetics) are out of scope. Their fluorescent, sensory and chromatographic (solid phase extraction, gas chromatography, high performance liquid chromatography, capillary electrophoresis) applications are detailed. Special emphasizes is taken for the chiral recognition feature of cyclodextrins.

1 Introduction

This article reviews the present status of the applications of cyclodextrins (CDs) in chemical analyses. CDs are frequently applied as sensors, fluorescence enhancement agents, adsorbers in sample collections processes, purification agents in sample pretreatment and as separation agents in instrumental analyses [1]. CDs are also applied frequently as control release agents in pharmaceutical administrations, solubilizers and stereo specific catalysts in synthetic chemistry and remediation improving materials in environmental protection [1], however theses applications are out of scope of this paper.

I.J. Rudas et al. (Eds.): Towards Intelligent Engineering & Information Tech., SCI 243, pp. 703–718.
springerlink.com   © Springer-Verlag Berlin Heidelberg 2009

2 Characterization of Cyclodextrins

2.1 Physical and Chemical Characterization of Cyclodextrins

CDs are cyclic oligosaccharide molecules (Figure 1), which comprise six, seven or eight D(+)-glucopyranose units, assigned the Greek letters, α, β and γ, respectively [1]. The glucose units join together with α-1,4-glycosidic linkages.

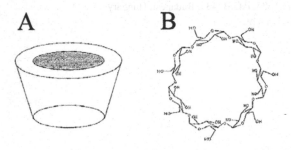

Fig. 1. Shape and chemical structure of β-cyclodextrin. A, shape of β-cyclodextrin; B, chemical structure of β-cyclodextrin. The primary hydroxyl groups are on lower rim and the secondary hydroxyl groups on the upper rim.

The CDs have truncated cone shapes (Figure 1a) having axial cavities with primary hydroxyl groups (in the C(6) positions) around its narrower rim and, secondary hydroxyl groups (in the C(2) and C(3) positions) on the opposite, wider rim (Figure 1b). Continuous H-bond interactions exist among the secondary hydroxyl groups along the wider rim, which give a rigid structure. The main physical parameters of CDs are summarized in Table 1. The CDs are solid molecules. CDs decompose higher temperature (caramelization) without melting.

Table 1 Some characteristic data of cyclodextrin molecules [1]

Cyclodextrin type	Number of glucose units	Cavity (nm)		Molecular weight	Solubility in water (g/L, 25° C)
		Diameter (upper)	Depth		
α	6	0.57	0.78	972	145
β	7	0.78	0.78	1135	18.5
γ	8	0.95	0.78	1297	232

The CDs can include molecules in their cavities creating host-guest complexes. A tight fitting between of host molecule and cavity of the CD results in a stabile of inclusion complex. A deeply immersed host molecule creates more stable complex, than other guest molecule, which is immersed in less extent in the cavity of CD. Increasing the cavity size of CDs, they can include bigger and bigger molecules. The α-CD produces stabile complexes with benzene and monosubstituted benzene derivatives. The β-CD is ideal complexing agent for *meta* disubstituted benzene derivatives. Finally the γ-CDs are ideal partner for polyaromatic-hydrocarbons (e.g. pyrene, chrysene).

The CD molecules offer hydrophobic environment in their cavities. The rims and outer surfaces of CDs show hydrophilic characters, offering H-bond and electron donor-acceptor interactions.

The hydroxyl groups of CDs can substitute various other chemical groups (e.g. alkyl, hydroxypropyl, acyl, carboxymethyl, phosphate, sulfate, amine hydroxyl-amine, etc.). The substitution of CDs results in changes in sizes and shapes of their cavities resulting in the shifting their complexation properties comparing with their mother (native) compounds. Moreover the substituents offer new interaction possibilities too.

Generally, substitutions of functional group of CDs produce big number of molecules differing from each other in the number of substituents, and the positions of substituents. The methylation of β-CD can produce several thousands compounds differing from each other in their substitution rates and patterns. Members of such CD derivatives mixtures show also deviations in their physical - including complexation- and chemical characters. The variety of molecules further increases, if more than one type of derivatizations are managed with CDs.

An alternative synthesis way is the single isomeric derivatives, where the products have uniform, only one structure [2]. These single isomeric CD derivatives show well-defined characteristics, with excellent batch to batch reproducibility.

2.2 Isomer Selectivity of Cyclodextrins

The most useful property of CDs is their ability to recognize differences between isomeric molecules [1].

The CDs can include the isomers in different extent, producing different stability complexes with the isomers. CDs can create strong complexes only those molecules, which shapes and size fit well to the cavity of CDs. The appropriate steric arrangement of functional groups of the guest is also necessary for the strong interaction of functional group of CDs. These requirements offer ability to use CDs for making differences between isomers. In several occasions, only one isomer can create tight fitting complex with CDs. The isomers have different shape therefore the isomers can interact differently with CDs. For example, only the *para* isomer of disubstituted benzene isomers can fit well to the cavity of α-CD. The *meta* and *orto* isomers too broad to immerse deeply into the cavity of α-CD (Figure 2). It means the α-CD creates strongest complex with *para* isomers and weaker ones with *orto* and *meta* isomers.

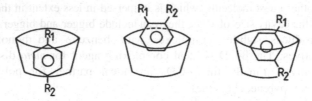

Fig. 2. Different inclusion of disubstituted benzene isomers into the cavity of α-cyclodextrin. Only the *para* isomer can deeply immerse into the cavity of α-cyclodextrin. The *meta* isomers immerse less extent than *para*, and the *orto* isomers least. The inclusion energies are follow: *para > meta > orto*.

The CDs are a very effective agent for differentiation of constitutional and stereoisomers. Constitutional isomers have same molecular formula, but the atoms have different connectivity (e.g. branched chain, anellation, *orto-meta-para*). Stereoisomers are compounds, which made up of the same atoms connected by the same sequence of bonds, but having different three dimensional structures [3, 4]. The stereoisomers can be *cis-trans* isomers, enantiomers, diastereomers. Enantiomers (enantiomer pairs, optical isomers) are mirror images of each others, but they cannot be superposed on their mirror image [3, 4]. Enantiomers are chiral, having asymmetry in their structure. Members of an enantiomer pair relate to its mirror image as left and right hands. An asymmetrically substituted carbon atom results in a asymmetric chiral molecule, creating enantiomer pair with its mirror image (Figure 3).

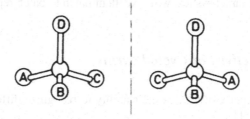

Fig. 3. Symbols of enantiomer pair molecules. The central atoms have the same substituents :a, b, c, and d. The two molecules are asymmetric (chiral) and they can not be superimposed, because every substituent is different of central atoms. Two molecules are mirror image (enetiomers) of each other.

Enantiomers are identical chemical and physical properties in symmetric environment. In the asymmetric environment, however, the enantiomers act differently. Since many molecules in the bodies of living beings are enantiomers themselves, there is often a marked difference in the effects of members of an enantiomer pair on living beings, including human beings. Thalidomide (Contergan) pharmaceutical, a mixture of enantiomer molecules, in which one enantiomer produces a desirable antiemetic effect, whereas the other enantiomer is toxic and produces a teratogenic side-effect [5]. From 1956 to 1962, approximately 10,000 children in Africa and Europe were born with severe malformities, including phocomelia, because their mothers had taken thalidomide during pregnancy. Enantiomeric mixture of Robitussin, an over-the-counter cough medicine caused lethal case. The useful enantiomer dextromethorphan worked well, but its enantiomer pair, levomethorphan caused lethal overdose effect, because it have deposited in patient [6]. These cases and several other disturbances of application enantiomeric mixture of pharmaceuticals forced the authority to introduce enantiomer pure medications [7, 8]. It means a charial drug can contain less than 0.1% of its optical isomer in pharmaceutical products. The requirement of enantiomer pure agrochemicals are also introduced some compounds, because those materials act differently in the environment [9]. For example, some chiral pollutants show enantioselective carcinogenesis.

The CDs are chiral molecules, therefore they act as chiral selective agents. One β-CD molecules has 35 asymmetric atoms, thus it has several possibility for chiral selective interactions. The broad spectra of chiral selectivity of a CD molecules make them the one of most popular chiral selectivity agents [1, 11-14].

Olefins and related structures can also be stereoisomers. If the two high priority ligands lie on the same side of the double bond, the system isomer is *cis*, but if they are on opposite sides, the isomer is *trans*.

Diastereomeric molecules have more than one asymmetric chiral center, they are stereoisomers, but they are not mirror images each other. The physical and chemical properties of diastereomer molecules show deviation from each other even in a symmetric environment [7, 8].

3   Selected Application Fields of Cyclodextrins in Analytical Chemistry

3.1   The Application of Cyclodextrins in Fluorescence Spectroscopy

Fluorescence spectroscopy is a type of electromagnetic spectroscopy which analyzes emitted light from a sample [15]. It involves using a beam of light (shorter wavelength), that excites the electrons in the analyte molecules and causes them to emit light (longer wavelength) of a lower energy. CDs have been found to remarkably enhance fluorescent and phosphorescent emissions of several included molecules [1, 16]. Generally, fluorescence phenomenon is stronger in apolar environment than in water media, because the oxygen of water acts as quenching agents. The CDs shield the included fluorescence molecules from the quenching effect of water in their apolar cavity. Moreover the guest molecules are prevented from collisional deactivation in the cavity of CD [1]. More than 500 times enhancement was reported in the fluorescent emission using CD. The emission maximums of analytes are also influenced by inclusion, with a shift towards shorter wavelengths.

It is not necessary to include the whole guest molecules for fluorescent enhancement. Only those parts have to include, which are responsible for fluorescence emission.

The sensitivity of detection of mycotoxins can improve with their CD complexes [18]. Several members of mycotocins have native fluorescence emission, which enhance and shift the complexation with CDs. In this way, the detection of aflatoxins was possible under nanomol concentration range. Carbamate pesticides have weak fluorescence emissions in water, but their CD complexes show strong emission, achieving μg/ml detection limit from water sample [19].

Phosphorence analysis have been done using CD complexes of polyaromatic hydrocarbons (PAH) [17]. The PAHs do not show phosphorence in water. They produce, however, intensive phosphorence emission, if the PAHs cocomplexed with a heavy atom containing molecule (e.g. CH_2Br_2) in the cavity of CDs. Such method is appropriate to determine concentration of PAHs even in 10^{-13} Mol concentration range.

3.2 Cyclodextrin-Based Sensors

A sensor is a device that measures a physical quantity and converts it into a signal which can be read by an observer or by an instrument [20].

Several publications deal with the application of CDs as substance specific highly sensitive sensors [1, 21]. The inclusion complex formation result in changing physical properties (e.g. electrode potential, spectral band, oscillation frequency) of fixed CD or tested media. A compound of interest, having strong complexation energy can insert the originally included compound from the cavity of CD.

Environmental early warning sensors have been constructed based on CDs [21]. Polyvinylchloride polymer containing fixed CD is used as reversible sensor for detection of bisphenol A in 7x10-8 mol/L range [22].

Phenols and nitro-phenols were selectively sensed with CDs containing electrode using cyclic voltametric method [23]. CD-coated membrane is appropriate for determination of organic pollutants by piezoelectric effect [21]. The insertion of a fluorescent compound from cavity of CD, causes changing of microenvironment of fluorescent marker molecules resulting in results in significant shift in its fluorescent spectra. In many cases, the displaced marker chemically linked to the CD achieving a reversible sensor. Such a device has been constructed to attach the fluorophor inidolizine to CD for the determination of benzene, toluene, phenol, p-cresol in 0.00025 0.01 millimol range [24].

Several publications are dealing with also CD-based chiral selective sensors (e.g. amino acids, enflurane, carvone, methotrexate) [25].

3.3 The Chromatographic Applications of CDs

3.3.1 Basics of Chromatography

Chromatography is a separation method in which the components are distributed between two phases, one of which is stationary (stationary phase) while the other (the mobile phase) moves in a definite direction [26].

The outstanding role of CDs is generally acknowledged in those methods which are using capillary columns (gas chromatography, supercritical fluid chromatography and capillary electrophoresis), but their role also significance in other modes of chromatography (liquid chromatography, solid phase extraction and thin layer chromatography) [13].

The advantages of chromatographic methods are followings in analytical chemistry:

- The chromatography is one of the most frequently used analysis methods. The well chosen chromatographic parameters creates a disturbance free retention window, where the compounds of interest do not co elute with matrix compounds.
- Chromatography offers determination of trace impurities without disturbance of other major constituents.
- The chromatographic methods have a broad range of linearity (10^3- 10^7).
- Chromatography makes possible the exact determination of trace components even in very complex matrices.
- Chromatography is a fast method. The literature has referred analysis under one sec, but the longest runs are completed within one hour.
- A chromatography analysis requires small amount of sample (fg – mg).
- Determination of several compounds during one analysis is well accustomed in chromatography.
- Chromatographic instruments can couple to structure identification instruments (e.g. MS, IR, NMR, ORD).
- Chromatographic instruments offer well-automated analyses even 24 hours in a day.

The advantages of chromatography are most obvious in chiral selective chromatography, where the CDs are applied frequently [13]. Namely, the well separated members of isomer pair are determined independently from each others. In this way even 0.01% minor isomer can be determined exactly without disturbance of 99.91% major isomer.

To manage a chiral chromatographic separation, either the mobile phase or the stationary phase must contain chiral selective agents. In chromatography CDs and their derivatives are one of the most frequently chiral agents [10-13, 28-30]. They are used as part of chiral stationary phases or as chiral mobile-phase additives in solution form. The compound which forms the more stable diastereoisomeric association will be more retained, whereas the opposite enantiomer will form less stable diastereoisomeric association i.e. will elute first.

The capillary column using chromatographic methods apply mostly CDs. These techniques are very efficient to compensate the moderate selectivity of separation agents. The high efficient separations makes possible the separation of compounds having only 0.1 kJ/M interaction energy difference with the separation agents. [27].

3.3.2   Application of Cyclodextrins in Solid Phase Extraction

Solid phase extraction (SPE) can be used to isolate and concentrate analytes of interest from a wide variety of matrices, including urine, blood, water samples, beverages, soil, animal tissue, and consumer products [31]. In the first step of separation, the solutions of compound of interests are poured to adsorber, and the adsorber retains selectively the compounds from dilute solution. Several disturbing components can wash out during this process. The following step is the retained molecules can elute with a small volume strong solvent from adsorber. CDs, bonded solid surface, enable to extract certain components from solutes [32, 33].

β-CD molecules have been bonded to acryl amide polymer matrix, achieving better than 95% recovery for ursolic acid. It seems the best field for CD-based solid phase extraction is the steroid analysis [32]. A β-CD containing absorbent could retain 77 steroid from urine with 82-112% recovery.

3.3.3   Application of Cyclodextrins in Gas Chromatography

The CDs containing stationary phases can use for separation constitutional and stereoisomers [1, 11, 137, 138, 140] in gas chromatographic practice. Recently, overwhelming part of GC separations are done on long (5-100 m) capillary columns. The selectivity values of CD containing stationary phases are moderate, therefore high separation power is required for the separations. The shape selective interactions represent only a few per cent of the working chromatographic interactions in gas chromatography. Generally, gas chromatography uses CD derivatives, because a stationary phase must be in liquid state. The solid native CDs decompose instead of melting in higher temperature. On the other hand, several CD derivatives have liquid state, few of them in room temperature [11]. More than 50 CD derivatives have been used for stationary phase in the following forms: molten state, dissolved in an achiral matrix or bonded chemically to silicone polymer. The up to date CD containing stationary phases consist of a selective CD and a silicone polymer matrix giving high efficiency [13].

The disubstituted benzene isomers (*orto, meta, para*) have different shapes, therefore their inclusion ability can differ significantly from each others (Figure 2)

[1]. The retention orders of isomers depends on the used CD derivatives and types of substituent groups.

The determination of carcinogenic benzene toluene, xylene isomers ethybenzene (BTEX) aromatic compounds is important task of environmental analysis. CD containing stationary phases are excellent for this purpose (Figure 4) [34, 35].

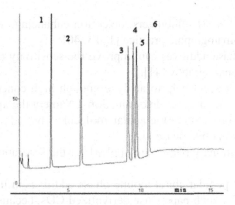

Fig. 4. Gas chromatographic analysis of BTEX compounds on permethy-β-CD containing stationary phase. Abbreviations: 1, benzene; 2, toluene; 3, *para*–xylene; 4; *meta*-xylene; 5, ethylbenzene; 6, *orto*-xylene. Conditions: GC/MS instrument, Shimadzu QP5000; column, 25 m x 0.25 mm FSOT; stationary phase, Cydex-B (0.25 μm); carrier, He (30 cm/sec); split ratio, 200; temperature program 55°C (2min)→2°C/min→100°C.

CD-containing stationary is highly recommended as first choice for gas chromatographic chiral separation [11, 13]. The literature survey show that more than 90% of the chiral GC separations have been done on a CD-based stationary phase.

The CDs can separate every class of enantiomers, even the functionless hydrocarbons [11]. Broad spectra of enantiomers having pharmaceutical interests have been separated with CD containing stationary phases in GC [11, 35]. Only CD-based stationary phases are appropriate for separation of chlorinated hydrocarbon pesticides (e.g. DDT, lindane, heptachlor, toxaphene) [36]. Enantiomer selective degradation of environmental pollutants, which refer the time and condition of pollutions, can also observed with CD-based selector [37]. The flavor and flagrancy industry is frequently use CD-based gas chromatographic separation to determine the ratio of enantiomers [38]. Such analysis can recognize the adulteration of perfumes. Using two dimensional gas chromatographic system, several hundred component can be separate in one analyses [39]. Such systems use frequently GC-based chiral column as second dimension.

3.3.4 Application of Cyclodextrins in High Performance Liquid Chromatography

The high performance liquid chromatography (HPLC) is the most frequently used chromatographic method in analytical chemistry. This techniques use short (10-25 cm) packed columns with moderate efficiency, requiring high selectivity value for the total resolutions of peaks.

The CDs are used in this method many times as selectors as stationary phases as well mobile phase additives [1, 13]. The CDs are applied as shapes selective and solubilizer agents. Their popularity partly comes from their low background signal in UV range, because the UV is the most frequently used detector in liquid chromatographic.

The CDs are applied as solubilizers, detection enhancing, and shape selective agents in liquid chromatographic practice [1, 13, 40].

CDs as mobile phase additives can improve the sensitivity of fluorescence detection in liquid chromatography [41].

CDs as solubilizer agent help to analyze enough high concentration for compounds of interest for their exact determinations. Namely the applied water based mobile phases can poorly solve the apolar molecules, but adding CDs their concentration increase in mobile phase [42].

The CD-based stationary phases are applied for the separation of Polyaromatic Hydrocarbons [43].

CDs are the most popular chiral mobile phase additives in liquid chromatography [44, 45]. Majority such papers use derivatized CDs, because they have better solubility than native ones in water based media. Last but not least the CD-based mobile phase additives are cheaper and having lower UV absorbance values than several other type chiral selectors.

The liquid chromatographic practice uses the amylose and cellulose based chiral stationary phases in the majority of enantiomer separations, because these phases have very high chiral selectivity compensating the moderate efficiency of liquid chromatographic columns [13]. In spite of the minor importance of CD containing stationary phase in chiral liquid chromatography, more than 5600 enantiomer pairs have been separated in CD-based stationary phases [13]. Not only the native CDs, but several derivatized have been fixed to silica surface and applied [40]. Some CD derivatives (e.g. carbamoylated, naphthylethylcarbamoilated) show unique chiral separation feature. They have three different types chiral selectivity according to the mobile phase: normal, reverse and polar organic types [40].

3.3.5   Application of Cyclodextrins in Capillary Electrophoresis

Capillary electrophoresis (CE) is the most dynamically developing branch of analytical chemistry. Advantages of capillary electrophoresis include very efficient, fast analyses, fast method developments, low material consumption and simplified sample preparation from biological matrices [46, 47].

The CDs have versatile functions in capillary electrophoresis: detection enhancers, solubilizers, electro dispersion reducers and shape selective (e.g. *cis-trans*, annellation, chiral) selective separation agents [14, 30]. Generally, the CDs are dissolved in the buffer and they act as pseudo-stationary phases.

The CDs have been applied as good solubilizers for hydrophobic compounds in capillary electrophoresis practice too [48]. The concentration apolar materials can increase and unwanted adsorption decrease, because high percentages of apolar molecules exist as CD complex. In this way an apolar materials reach enough high concentration to establish its trace impurities in water based capillary electrophoresis system.

The CDs enhance the fluorescence detectability as was mentioned earlier. This property of CD is also applied in capillary electrophoresis practice [49, 50]. Mycotoxin zarelonon has been determined from maze with 5 ng/g quantization limit using methylated-β-CD as mobile phase additive.

The small inorganic ions show very different mobility feature. Analyzing inorganic ions with various charge and electro negativity result in bad efficiency, because the broad spectra of ions do not match the conductivity of background electrolyte. The addition of CD can, however, reduce the differences of migrations of ions [51, 52]. Namely the mobility of CD complexes much less differ each other than the mobility of ions themselves. as was demonstrated in the case of Cl^-, Br^-, SO_4^{2-}, NO_3^-, NO^{2-}, J^-, F^-, HPO_3^{2-}.

The CDs improve the mobility differences of various isomers, because their inclusion depend on not only their polarity but their shape too. Namely the mobility of analytes change according they are in free or included states. In this way the CDs can separate *cis-trans* isomers [48], anellation isomers of polyaromatic hydrocarbons [53]. Capillary electrophoresis is appropriate for separation of environmentally important 25 phenolic compounds using 1 millimol sulfobutylether-β-cyclodextrin [54]. A sensitive method, achieving 20 µg/L detection limit, has been developed for the determination of explosives and their metabolites with capillary electrophoresis mass spectrometry on line coupling applying sulfobutylether substituted β-CD [55].

The most important application of CD is chiral separations in capillary electrophoresis [12-14, 30 56].

According to the literature, more than two thirds of chiral separations have been made with CD chiral selectors in capillary electrophoresis [13].

The low analysis temperature and broad variability of conditions (pH, ionic strength of buffer, types and concentration of selectors and additives) offer good, finely tunable selectivity. The low material consumption allows the application of expensive chiral selectors (such as sulfate or amine substituted CDs). Capillary electrophoresis E offers much more advantages than other chromatographic modes in chiral analysis. For example the counter current, or partial filling, method applies selectors with migration direction towards the injection point, so the selectors do not disturb UV detection or MS measurements [30, 57].

The capillary electrophoresis uses native CDs as well derivatives CDs.

Recently the charged CDs become more and more popular. Using oppositely charged CDs and analytes, less than 1 millimol selector can result in a total separation of enantiomers, because the analytes show opposite migration directions in complexed and free states. This opposite increments of migration directions result in "increased column length". In a certain concentration range the multicharged selector can produce infinite resolution [58].

Moreover, the charged CD can separate neutral and charged enantiomers, but neutral CDs can separate only charged ones. Overwhelming part of charged CD derivatives are statistically substituted. For the shake of good reproducibility, however, single isomer derivatives CDs have been also introduced [2]. Several pyrethroid acid enantiomers have been separated efficiently with such a basic CD derivatives (Figure 5). The selectivity of charged Cds is welldemonstrated that 1.75 millimol acid derivatized sulfobutylether-β-CD well separates the basic baclofen enantiomers with 0.1 µg/L detection limit from formulated drug [57].

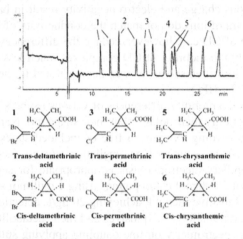

Fig. 5. Simultaneous chiral separation of six pyrethroid acid using basic single isomer cyclodextrin derivate. Conditions: column, 50 cm x 0,05 mm id uncoated fused silica tubing; chiral selector, 15 millimol monoamino-permethylated-β-cyclodextrin; background buffer, 40 millimol Britton-Robinson buffer; pH, 6.5; potential +30 kV; detection, 200 nm UV; temperature 15°C.

Applying dual separation system – buffer containing two CD selectors – becomes more and more popular [59]. Some combinations show synergetic selectivity improvements. Mixture of sulfobuthylether-β- CD and hydroxypropyl-β-CD is appropriate for validated chiral separation of hydroxychloroquine and its metabolite from urine in 10-100 ng/mL range [2].

CDs play role in capillary electrochromatography (CEC) too [61].

In electrochromatographic technique, the mobile phase is forced through the packed columns by the electroosmotic flow in spite of the pressure-driven flow of regular liquid chromatography [62].

The microfluidic chip technology is also used in capillary electrophoresis. To explore bioorganic signatures on Mars, a chip form capillary electrophoresis instrument has been developed [63]. A small channel has been cut into the chip as channel. CD based chiral selector will analyze the enantiomer ratios of amino acid analyses. The microchip allows very fast analyses. For example, the naproxen enantiomers have been separated within 1 min on such a device [64].

4 Mechanistic Considerations of the Broad Chiral Selectivity of Cyclodextrins

The overwhelming part of chiral separations use a CD selector, therefore a deeper insight is taken for reasons of their protruding chiral recognition characters. CDs can separate enantiomers with any functional groups. Even saturated branched aliphatic hydrocarbons have been separated on CD-based stationary phase in GC [11]. There are no strict requirements for the structure of analytes for a successful chiral separation with CD-based chiral selectors [1, 13]. The broad chiral recognition feature of CDs is based on following phenomena:

- They have numerous chiral centers – five in every glucose units. In a glucose unit, every chiral center has different orientations and they show different distances from their neighboring atoms. Moreover, the shapes of the glucose units do not repeat themselves from units to unit, so β-CD has 35 different chiral recognition sites.
- Some substitutions (such as hydroxypropyl and naphthyl ethylcarbamoyl) add more chiral centers to CDs, broadening further their recognition spectra.
- Most CD derivatives (randomly substituted) are not homogeneous products, but consist of a large number of isomers, which differ from each other in the numbers and the positions of the substitutions, and almost every isomer has different chiral recognition abilities. Even migration reversals of enantiomers have been reported as depending on the degree of sulfate substitution of β-CD [65].
- CDs can change their shape to interact intimately with analytes, the so-called "induced fit mechanism [1, 13]. The "induced fit" interactions produce a further increase in the chiral selectivity spectra of CDs.
- The ionizable CDs can change their selectivity spectra according to their ionization states, as was observed in the case of phosphated CD [66].
- Some CD derivative change chiral recognition feature, according to the mobile phase [40].

The great variety of chiral centers and induced fit produce the multimodal characteristics of CDs, which can yield more than one mode of interaction with an enantiomer pair [27]. In several occasions, the inclusion is one of the key interactions of chiral recognition, but in other occasion the inclusion does not play role in chiral recognition [13].

Conclusion

CDs are versatile sensors, fluorescence enhancement materials, adsorbers and separation agents.

Their main advantage is their broad chiral recognition characters. CDs are frequently applied in the analyses of pharmaceutical, flavor, food industry and environmental protections.

Acknowledgments. The NKTH-OTKA No. NI-68863, N KFP0 0947-00949/2005 and NKA 1-00150/2007, NaturSep grants are gratefully acknowledged.

References

[1] Szejtli, J., Osa, T. (eds.): Comprehensive Supermolecular Chemistry. Cyclodextrins, vol. 3. Pergamon, Oxford (1997)

[2] Ivanyi, R., Jicsinszky, L., Juvancz, Z.: Chiral Separation of Pyrethroic Acids with Single Isomer Permethyl Monoamino β-Cyclodextrin Selector. Electrophoresis 22, 3232–3236 (2001)

[3] Mislow, K.: Molecular Chirality. Top Stereochem 22, 1–82 (1999)

[4] Welch, C., Szczerba, T., Perrin, S.: Review of Stereochemistry. Regis Technologies Inc. (2000)

[5] Teo, S.K., Colburn, W.A., Tracewell, W.G., et al.: Clinical Pharmacokinetics of Thalidomide. Clin Pharmacokinet 43, 311–327 (2004)

[6] Rouhi, M.A.: Chirality at Work. Chemical and Eng. News 81, 56–61 (2003)

[7] Policy Statement for the Development of New Stereoisomeric Drugs (1992) Food and Drug Administration, Department of Health and Human Services, USA, Fed. Regist. 57 (May 2, 1992)

[8] Ali, I., Aboul-Enein, H.Y., Ghanem, A.: Enantioselective Toxicity and Carcinogenesis. Current Pharmaceutical Analysis 1, 109–125 (2005)

[9] Liu, W., Gan, J., Schlenk, D., Jury, W.A.: PNAS 102, pp. 701–706 (2005)

[10] Ali, I., Aboul-Enein, Y.: Chiral Pollutants: Distribution, Toxicity and Analyis by Chromatography and Capillary Electrophoresis. Wiley, Chicester (2003)

[11] König, W.A.: Gas Chromatographic Enantiomer Separation with Modified Cyclodextrins. Hüthig, Heidelberg (1992)

[12] Fanali, S.: Enantioselective Determination by Capillary Electrophoresis with Cyclodextrins as Chiral Selectors. J. Chromatogr. A 875, 89–122 (2000)

[13] Juvancz, Z., Szejtli, J.: Role of Cyclodextrins in Chiral Selective Chromatography. Trac. 21, 379–387 (2002)

[14] Juvancz, Z., Kendrovics, B., Ivanyi, R., Szente, L.: The Role of Cyclodextrins in Chiral Capillary Electrophoresis. Electrophoresis 29, 1701–1712 (2007)

[15] Lakowicz, J.R.: Principles of Fluorescence Spectroscopy. Springer, Wiesbaden (2006)

[16] Szente, L., Szejtli, J.: Non-Chromatographic Analytical Use of Cyclodextrins Analyst 123, 735–741 (1998)

[17] Scypinski, S., Love, C.L.J.: Room-Temperature Phosphorescence of Polynuclear Aromatic Hydrocarbons in Cyclodextrins. Anal. Chem. 56, 322–327 (1984)

[18] Maragos, C.M., Appell, M., Lippolis, V., et al.: Use of Cyclodextrins as Modifiers of Fluorescence in the Detection of Mycotoxins. Food Additives and Contaminants: Part A 25, 164–171 (2008)

[19] Pacioni, N.A., Veglia, A.V.: Determination of Poorly Fluorescent Carbamate Pesticides in Water, Bendiocarb and Promecarb, Using Cyclodextrin Nanocavities and Related Media. Anal. Chim. Acta. 583, 63–71 (2007)

[20] Fraden, J.: Handbook of Modern Sensors. Springer, Wiesbaden (2003)

[21] Fenyvesi, E., Jicsinszky, L., Kolbe, et al.: Ciklodextrin-tartalmú szenzorok, mint korai figyelmeztető rendszerek. NKFP-3/020/2005 project (2007)

[22] Wang, X., Zeng, H., Wei, Y., Lin, J.-M.: A Reversible Fluorescence Sensor Based on Insoluble Beta-Cyclodextrin Polymer for Direct Determination of Bisphenol A (BPA). Sens. and Actuators, B: Chem. B 114, 565–572 (2006)

[23] Chen, E.T.: U.S. Patent h26, 582, 583 (2003), http://patft.uspto.gov/netacgi/

[24] Becuwe, M., Landy, D., Delattre, F., et al.: Fluorescent Indolizine-β-Cyclodextrin Derivatives for the Detection of Volatile Organic Compounds. Sensors 8, 3689–3705 (2008)

[25] Szeman, J.: Application of Cyclodextrins in Enantioselective sensors. Cyclodextrin News 21(4), 1–4 (2007)

[26] Cazes, J., Scott, R.P.W.: Chromatography Theory. CRC, Boca Raton (2002)

[27] Juvancz, Z., Petersson, P.: Enantioselective Gas Chromatography (Review) J. Mirocol. Sep. 8, 99–114 (1996)

[28] Juvancz, Z., Markides, K.E.: Enantiomer Separation with SFC, a Promising Possibility of Chromatography. LC-GC International, 44 (1992)

[29] Schneiderman, E., Stulcap, A.M.: Cyclodextrins: a Versatile Tool in Separation Science (review). J. Chromatogr. B 745, 83–102 (2000)

[30] Chankvetadze, B.: Capillary Electrophoresis in Chiral Analysis. Wiley, Chichester (1997)

[31] Thurman, E.M., Mills, M.S.: Solid Phase Extraction, Principle and Practice. Wiley, Chicester (1998)

[32] Liu, H., Liu, C., Yang, X., et al.: Uniformly Sized Cyclodextrin Molecularly Imprinted Microspheres Prepared by a Novel Durface Imprinting Technique for Ursolicacid. Anal. Chim. Acta. 628, 87–94 (2008)

[33] Moon, J.-Y., Jung, H.-J., Moon, M.H., et al.: Inclusion Complex-based Solid-Phase Extraction of Steroidal Compounds with Entrapped Cyclodextrin Polymer. Steroids 73, 1090–1097 (2008)

[34] Campos-Candel, A., Llobat-Estellés, M., Mauri-Aucejo, A.: Comparative Evaluation of Liquid Chromatography Versus Gas Chromatography Using a β-Cyclodextrin Stationary Phase for the Determination of BTEX in Occupational Environments. Talanta 78, 1286–1292 (2009)

[35] Juvancz, Z., Grolimund, K., Schurig, V.: Pharmaceuticasl Applications of Bonded Cyclodextrin Stationary Phase. J. Mirocol. 5, 459–468 (1993)

[36] Vetter, W., Schurig, V.: Enantioselective Determination of Chiral Organochlorine Compounds in Biota on Modified Cyclodextrins by Gas Chromatography (Review). J. Chromatogr. A 774, 143–175 (1997)

[37] Buerge, I.J., Poiger, T., Müller, M.D., Buser, H.-R.: Enantioselective Degradation of Metalaxyl in Soils: Chiral Preference Changes with Soil pH. Environ. Sci. Technol. 37, 2668–2674 (2003)

[38] Bicchi, C., D'Amato, A., Rubiolo, P.: Cyclodextrin Derivatives as Chiral Selectors for Direct Gas Chromatographic Separation of Enantiomers in the Essential Oil, Aroma and Flavour Fields. J. Chromatogr. A 843, 99–121 (1999)

[39] Liu, B., Chen, C., Wu, D., Su, Q.: Enantiomeric Analysis of Aanatabine, Nornicotine and Anabasine in Commercial Tobacco by Multi-Dimensional Gas Chromatography and Mass Spectrometry. J. Chromatogr. B 865, 13–17 (2008)

[40] Cyclobond Handbook, Advanced Separation Technology. Whippany (1997)

[41] Shimada, K., Komine, Y., Miamura, K.: High-Performance Liquid Chromatographic Separation of Bile Acid Pyrenacyl Esters with Cyclodextrin-Containing Mobile Phase. J. Chromatogr. 565, 111–118 (1991)

[42] Shimada, K., Hirakata, K.: Retention Behavior of Derivatized Amino Acids and Dipeptides in High-Performance Liquid Chromatography Using Cyclodextrin as a Mobile Phase Additive. J. Liq. Chromatogr. 15, 1763–1771 (1992)

[43] Panda, S.K., Schrader, W., Andersson, J.T.: β-Cyclodextrin as a Stationary Phase for the Group Separation of Polycyclic Aromatic Compounds in Normal-Phase Liquid Chromatography. J. Chromatogr. A 1122, 88–96 (2006)

[44] Bielejewska, A., Lukasik, B., Duszczyk, K., Sybilska, D.: Cyclodextrins as Chiral Additives for Chromatographic Separation of Some Mandelic Acid Esters. Chem. Anal. 47, 419–427 (2003)

[45] Szeman, J., GanzkerK: Use of Ccyclodextrins and Cyclodextrin Derivatives in High-Performance Liquid Chromatography and Capillary Electrophoresis. J. Chromatogr. 668, 509–517 (1994)

[46] Khaledi, M.G.: High Performance Capillary Electrophoresis. Wiley, Chicester (1998)

[47] Prost, F., Thorman, W.: Enantiomeric Analysis of the Five Major Monohydroxylated Metabolites of Methaqualone in Human Urine by Chiral Capillary Electrophoresis. Electrophoresis 22, 3270–3280 (2001)

[48] Juvancz, Z., Urmos, I., Klebovich: Capillary Electrophoretic Separation of Clomiphene Isomers Using Various Cyclodextrins. J. Cap Electrophoresis 3, 181–190 (1996)

[49] Maragos, C.M., Appell, M.: Capillary Electrophoresis of the Mycotoxin Zearalenone Using Cyclodextrin-Enhanced Fluorescence. J. Chromatogr. A 1143, 252–257 (2007)

[50] Tanaka, Y., Naruishi, N., Nakayama, Y., et al.: Development of an Analytical Method Using Microchip Capillary Electrophoresis for the Measurement of Fluorescein-Labeled Salivary Components in Response to Exercise Stress. J. Chromatogr. A 1109, 132–137 (2006)

[51] Stathakis, K., Cassidy, R.M.: Control of Relative Migration of Small Iinorganic and Organic Anions with Cyclodextrins in Capillary Electrophoresis (CE). Can J. Chem. 76, 194–198 (1998)

[52] Masar, M., Bodor, R., Kaniansky, D.: Separations of Inorganic Anions Based on their Complexations with α-Cyclodextrin by Capillary Zone Electrophoresis with Contactless Conductivity Detection (CE) J. Chromatogr. 834, 179–188 (1999)

[53] Szolar, O.H.J., Brown, R.S., Luong, J.H.T.: Separation of PAHs by Capillary Electrophoresis with Laser-Induced Fluorescence Detection Using Mixtures of Neutral and Anionic β-Cyclodextrins. Anal. Chem. 67, 3004–3010 (1996)

[54] Groom, C.A., Luong, J.H.T.: Sulfobutylether-β-Cyclodextrin-Mediated Capillary Electrophoresis for Separation of Chlorinated and Substituted Phenols. Electophoresis 18, 1166–1172 (1997)

[55] Groom, C.A., Halasz, A., Paquet, L., et al.: Detection of Nitroaromatic and Cyclic nitramine Compounds by Cyclodextrin Assisted Capillary Electrophoresis Quadrupole Ion Trap Mass Spectrometry. J. Chromatogr. A 1072, 73–82 (2005)

[56] Chankvetadze, B.: Enantioseparations by Using Capillary Electrophoretic Techniques, The Story of 20 and a Few More Years. J. Chromatogr. A 1168, 45–70 (2007)

[57] Desiderio, C., Rossetti, D.V., Perri, F., et al.: Enantiomeric Separation of Baclofen by Capillary Electrophoresis Tandem Mass Spectrometry with Sulfobutylether-β-Cyclodextrin as Chiral Selector in Partial Filling Mode. J. Chromatogr. B 875, 280–287 (2008)

[58] Rudaz, S., Le Saux, l., Prat, J., Gareil, P., et al.: Ultrashort Partial-Filling Technique in Capillary Electrophoresis for Infinite Resolution of Tramadol Enantiomers and its Metabolites with Highly Sulfated Cyclodextrins. Electrophoresis 25, 2761–2771 (2004)

[59] Lorin, M., Defepe, R., Morin, P., Ribet, J.P.: Quantification of Very Low Enantiomeric Impurity of Efaroxan Using a Dual cyclodextrin System by Capillary Electrophoresis. Anal. Chim. Acta 592, 139–145 (2007)

[60] Oliveira, M.A.R., Cardoso, C.D., Bonato, P.S.: Electrophoresis 28, 1081–1091 (2007)

[61] Wistuba, D., Schurig, V.: J. Sep. Sci. 29 1344–1355 (2006)

[62] Bartle, K.D., Myers, P.: Capillary Electrochromatography. RSC, London (2001)

[63] Skelley, A.M., Mathies, R.A.: Chiral Separation of Fluorescamine-labeled Amino Acids Using Microfabricated Capillary Electrophoresis Devices for Extraterrestrial Exploration. J. Chromatogr. A 1021, 191–199 (2003)

[64] Guhen, E., Hogan, A.-M., Glennon, J.D.: High-Speed Microchip Electophoresis Method for Separation of (R,S)-Naproxen. Chirality 21, 292–298 (2009)

[65] Lurie, I.S., Odenal, N.G., McKibben, T.G., Casale, J.F.: Effects of Various Anionic Chiral Selectors on the Capillary Electrophoresis Separation of Chiral Phenethylamines and Achiral Neutral Impurities Present in Illicit Methamphetamine. Electrophoresis 19, 2918 (1998)

[66] Juvancz, Z., Markides, K.E., Jicsinszky, L.: Enantiomer Separation of Disopyramide with Capillary Electrophoresis Using Various Cyclodextrins. Electrophoresis 18, 1002–1007 (1997)

An Intelligent GIS-Based Route/Site Selection Plan of a Metro-Rail Network

András Farkas

Faculty of Economics, Budapest Tech
Tavaszmező út 17, H-1084 Budapest, Hungary
farkas.andras@kgk.bmf.hu

Abstract. The goal in any route/site selection problem is to find the best location with desired conditions that meet predetermined selection criteria. In such a process, manipulation of spatial data of multiple criteria is essential to the success of the decision making. Conventional geographic information systems (GIS), while recognized as useful decision support technologies, do not provide the means to handle multiple decision factors. A GIS-based system, however, appropriately combined with spatial multi-criteria decision analysis (SMCA) techniques may become to be capable of transforming the geographical data into a decision. A hierarchical decision tree model is developed to join and prioritize the diverse engineering, economical, institutional, social and environmental objectives. An illustrative study of a metro-rail route/site selection process of an urban transportation project is also presented.

Keywords: Route/site selection, metro-rail network, geometric information system.

1 Introduction

Route/site selection of different transportation facilities is usually considered one of the most challenging parts of urban planning projects. This process requires the consideration of a comprehensive set of factors that meet desired goals and balancing of multiple objectives in determining the suitability of a particular area. The selection of a rail-route, or a site for a station, involves the complex treatment of critical, sometime contradicting factors drawing from engineering, economic, social, and environmental disciplines. Respect for legislation and public awareness of environmental issues make the final choice of suitable locations for facilities increasingly complicated, particularly when the facilities may have an adverse impact on protected areas and residents of a metropolis. Conventional decision support techniques lack the ability to simultaneously take into account a variety of such aspects. More importantly, the traditional multi-criteria decision making (MCDM) techniques have been non-spatial. However, in a real life situation it can

I.J. Rudas et al. (Eds.): Towards Intelligent Engineering & Information Tech., SCI 243, pp. 719–734.
springerlink.com
© Springer-Verlag Berlin Heidelberg 2009

hardly be assumed that the entire study area is spatially homogenous, because the evaluation criteria used to vary across space. Also, some of the important factors that add to the difficulty of the proper selection include the existence of a great number of feasible options within a sought territory, multiple objectives, intangible attributes, diversity of interest groups, lack of quantitative measures of the factors' impact, uncertainties regarding government/municipality influence on the selection process through legislations and different interests regarding permissions and construction [10].

Over the recent decades, geographic information systems (GIS) have emerged as useful computer-based tools for spatial description and manipulation. Although, they are often described as decision support systems (DSS), there have appeared some disputes regarding whether the GIS decision support capabilities of such systems are sufficient [7]. From an architectural aspect, an intelligent GIS must be able: (i) to integrate the expert's knowledge and the GIS to facilitate the site suitability analysis in determining various suitability constraints for generating and displaying composite suitability maps in the GIS; (ii) to employ a multi-criteria decision tool to accommodate trade-offs among the multiple and conflicting decision criteria in determining a preferred location; (iii) to design flexible feedback loops throughout the decision-making process to allow the user to revise the intermediate decisions by examining the consequences and to gradually narrow the solution space to reach a desired solution; and (iv) to develop an efficient graphic user interface by which the modules in the system are seamlessly linked [8]. If these requirements are incorporated in the system and user-specified routines through generic programming languages are adding also into its existing set of commands, then it is classified as a fully integrated GIS-DSS [3, 14]. In such a computerized system, a dynamic iteration of the data flow between the decision tool and the GIS according to the user's need can be performed by the user interface to interact with either system through graphical menu-based tools [1]. This paper will discuss and describe the use of such an intelligent GIS-DSS to support the transportation planning and modeling activities of a metro-rail network.

2   The Use of GIS for Urban Transportation Planning

Transportation and land-use planning are key elements of urban developments. Planning of any public transit project should begin with the recognition of an existing or projected need to meet the growing demand in the future. In trying to implement such a long-term investment governments/municipality councils are faced with several problems:

-    How best to serve the city's growing population, based on the forecasted increase of the population in a long run;
-    Compliance with regulations, which require metropolitan areas to reduce reliance on the car and reduce vehicle kilometer of travel per capita during the next 15 to 25 years;

   – Improving accessibility to employment, education and non-work activities
     in a situation where traffic congestion is expected to get worse during the
     plan period;
   – Analysis of all feasible transportation alternatives (possible transportation
     modes) simultaneously, by the help of computer simulation.

To address these and other issues, it is apparent that traditional transportation
modeling approaches (e.g., when transportation model data are aggregated to traffic analysis zones that represent average values for the factors inside the zones)
are not sufficient to meet today's needs to plan and build metro-rail networks.
Thus, urban travel system improvements should extend modeling capabilities.
Also, in order to collate and manage the data inputs to the model in the level of detail specified, more accurate information and data locating are required. These
may include [19]:

- Disaggregate data collection using household as unit of analysis.
- Activity sites, e.g. employment places or institutions analyzed as single units.
- Travel patterns modeled involving one or more trips as part of "activities
  chain".
- Development of indicators to measure and evaluate pedestrian access to transit.
- Mixed-use measure of jobs-housing balance.

On a metropolitan or a regional basis, professional departments should be responsible to gathering base year data and producing forecasts to demographic,
employment and land-use data for travel modeling and managing the data with an
integrated database and GIS. Such a system is operated through region-wide base
maps and associated databases maintained using, e.g. Arc/Info GIS programming
and spatial analysis, mapping, spatial database design and database management
since geography is a major factor in urban growth management. Spatial and attribute data are stored and managed on a central server by using a database and SDE
(Spatial Data Engine). These organizations maintain an average of 75-100 data
layers of demographic, employment, environmental and transportation data for the
entire region or city and continually improve the base map and attribute data and
produce a CD-ROM of the latest datasets usually every quarter [19]. The data on
the CD-ROM are formatted as shape files and can therefore be read directly by
Arc/View GIS as a data viewer of the Arc/Info coverage. Categories of such an
urban GIS data bank are: boundary, census, environment, land-use, places, streets,
tax lots, transit, water, etc. GIS is then used to collate and manage a variety of data
inputs and outputs.

Travel behavior is a complex phenomenon. Nevertheless, statistical modeling
techniques have been developed that are able to establish the significant factors
that effect travel demand. These variables include income, household size, age of
household members, access to autos, employment and non-employment activities
and accessibility to activity centers. Household survey data captures the range of
data that is required to perform these types of analyses. Household surveys are the
foundation from which most travel demand models are constructed. GIS can be

utilized to geocode the location of households and the activity centers. This provides a precise geographic dataset of the distribution of trip origins and destinations together with the location specific characteristics of household or employment site. With these data it is possible to perform accurate spatial analyses of trip generation and trip distribution factors [19]. For example, distance or travel time is known to be an important factor in impeding travel: the nearer or shorter the destination the more likely one is to travel to it compared to a more distant location. GIS is able to measure very accurately these types of factors, using actual network distance from each site, to produce an average value for a group of points. Thus, using geocoded locations improves the accuracy of the model data and the modeling process in digital map forms.

One of the key objectives is to accomplish mixed land-use developments that encourage closer proximity between households and employment sites. Experience has shown that more even jobs-housing balance reduces dependence on the automobile (which is more attractive for longer commute journeys). Thus jobs-housing balance promotes travel choice for transit, bicycling and walking journeys. Complimentary to the mixed land-use policy is to encourage transit use rather than auto modes. Mode choice models have shown that if transit is to compete with the auto it must offer at least comparable travel times and accessibility to destination activities [19]. The bus services are planned to feed into the metro-rail stations and the light rail stops, thereby improving suburban access to transit along the metro-rail and the light rail corridors. This in turn will help attract new commercial development to these corridors.

An important component of transit trip making is pedestrian accessibility to the bus stops, metro-rail or light rail stations. Pedestrian accessibility is often absent from transit modeling activities, in part because zone based measures are not able to determine this factor. In addition, transportation models do not have the capability of analyzing local pedestrian networks in the way that GIS can, like the measurement of pedestrian accessibility to transit services. This consists of factors to examine such as the location of street crossings, quality of sidewalks and ease of walking (slope of the streets). Most recently, a more sophisticated index has been developed using an interval scale of measurement [19]. The higher the score the more pedestrian accessible or pedestrian friendly is the neighborhood toward transit.

3   Integrated GIS and Spatial Multi-criteria Evaluation

Spatial multiple criteria evaluation (SMCE) is based on multiple attribute decision analysis techniques and combines multi-criteria evaluation methods and spatio-temporal analysis performed in a GIS environment [13, 17]. A world widely known MCDM procedure is usually employed for the evaluations, called the analytic hierarchy process (AHP), to identify the most suitable site in the decision phase [15]. The performance assessment of an option in one or more criteria at a point in time can be described by a defined set of maps. Therefore, the spatial decision problem can be visualized as a two or three dimensional table of maps, or map of tables, which has to be transformed into one final ranking of the alternatives [16]. Such a GIS-based system implements two-directional integration and starts from the GIS module generating the appropriate data for the counter-part decision support module. Latter module performs the required decision making operations, and ultimately uses resultant data to add to the existing attribute data set in GIS and creates maps based on this newly created information.

SMCE partially utilizes Herwijnen's model of spatial multi-criteria analysis [2]. In a SMCE, the decision alternatives are the series of maps, and the criteria are the pixels (basic units for which information is explicitly recorded) or polygons in the maps. The spatial aggregation is first applied to attribute maps, after which the aggregate effects are evaluated and ranked [17]. Thus, spatial multi-criteria decision analysis (SMCA) is a process that combines and transforms geographical data (the input) into a decision (the output). This process consists of procedures that involve the utilization of geographical data, the decision maker's preferences and the manipulation of data and preferences according to specified decision rules. For ranking of the alternatives, the evaluation table of maps has to be transformed into one final ranking of alternatives. The ranking of the alternatives could be different, since the decision makers, i.e. the groups of stakeholders, may have conflicting interests as they represent dissimilar perspectives. Most recent real world applications of the integrated GIS/SMCA approach to different route/site selection problems can be found in the excellent works of Eldrandaly et al [2], Keshkamat [11], Keshkamat et al. [12] and Sharifi et al. [16, 17, 18].

Keeney [9] distinguished two major approaches in MCDM: (i) the alternative-focused and (ii) the value-focused approach. The alternative-focused approach starts with development of alternative options, specification of values and criteria, then, the evaluation of each option is done. The value-focused approach considers the values as the fundamental component in decision analysis. Therefore, first, it concentrates on the specification of values (value structure), then it develops the values feasible options and evaluates them with respect to the predefined value and criteria structure. This implies that the decision alternatives should be generated in a way that values specified for a decision situation are best met. Hence, the order of thinking is focused on what is desired, rather than the evaluation.

In the context of route/site selection of urban transportation facilities the value-focused approach has many advantages over the other [18]. To implement this, for an urban transportation project like a metro-rail system is, a top-down decision analysis process is proposed to define the goal, the objectives and their related indicators for the facilities. This hierarchical decision tree model is presented in Figure 1. In the decision making phase, a consulting team, technical committee members, designers, investors, local authority officials and public representatives are involved as the basis for development and evaluation of the project. The various elements of this criteria structure are briefly described as follows:

Goal and Objectives: The goal of this framework is to identify an effective public mass transportation system for a metropolitan area integrated with an efficient land use so that it meets the present and long-term socio-economic and environmental requirements of the residents of the marked territory. This goal can be achieved if the following objectives are met:

Economic objective: Economic objective seeks to maximize feasible economic return on investment from the system. A number of criterion is used to measure how well an option performs on each indicator, e.g., benefit/cost ratio, first year return, internal rate of return, net present value, construction cost and operation cost, as well as minimizing land/real estate acquisition (expropriation of property), intensification of existing land use and maximizing the potential of the location.

Engineering Objective: This objective looks at three main concerns that are the efficiency of the system, the construction issues and the effective use of the network for work and non-work travels. The criteria used to measure the extent to which such achievements are met by the transit route or facility options are the following:

• Efficiency is measured by examining the minimum number of transfer, (whereby an alternative with excessive transfer will score low for this criteria) A transit option which contributes to a reduction in travel time compared to time spent on the roads and provides a close-to-optimal convenience for pedestrian access and links to other local and commuter transportation modes, and, also an effective connection of housing jobs, retail centers, recreation areas is beneficial and will score high.

• From the construction perspective, alternatives that have rail routes passing through high demand areas like high-density built-up areas, commercial, industrial and institutional areas, will score high for this criterion. This aspect, however, particularly if it is accompanied by poor geological conditions at a route/site option, conflicts with a low construction cost requirement. To build metro-line stations, the commonly used construction modes are: open-cast construction (just below grade, building pit is beveled or secured by walls, requires large construction areas, more flexibility in design); bored-piled and cover-slab construction with or without inner shell (bored-piled wall, generates column free space, reduces surface interruption); diaphragm wall and cover-slab construction (excavation after diaphragm and cover-slab are constructed, multi-story basement structure, structure growths from top to downwards); mine tunneling construction (extremely deep situation, use shot crete, but cracks and leakages are not avoidable).

• Engineering characteristics and alignment are evaluated with respect to the measures/attributes constituting the geological environment (including soil mechanics, intrusive rock structure, stratification, etc.); hydro-geological conditions (including underground water-level, chances of inrush, perviousness, locations of permeable or impermeable layers, chemical and physical characteristics of underground water and their effects on the built-in architectural structures) and geotechnics (rock boundaries, response surfaces, geographic configuration). A special focus should be given to safety, therefore, the recognition and control of risk factors are of utmost importance (water intrusion, gas explosion, earth quake).

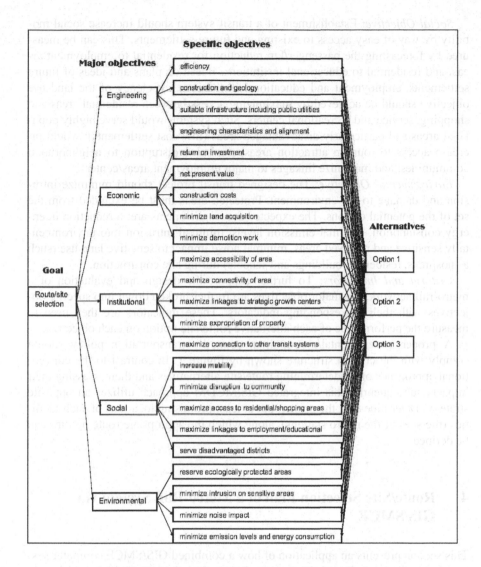

Fig. 1. Hierarchy of goals, objectives criteria and indicators

• Infrastructure involves the careful examination/analysis of overground building up, the suitability of the existing public utility network and the required overground organization to be made before the construction works are started.

Institutional Objective: This objective measures the match between the transit system and spatial policies of the government/urban municipality, e.g. to maximize interconnectivity to existing public transport systems; to maximize linkages to strategic growth centers (as designated/proposed in local plans), to provide good linkages among urban centers and suburban railway networks, airports, long-distance bus stations, park and ride lots as well as to minimize land acquisition.

Social Objective: Establishment of a transit system should increase social mobility by way of easy access to existing and future settlements. This can be measured by forecasting the passenger/km reduction for residential to employment areas, and residential to educational institutions. Based on plans and ideas of future settlements, employment and educational institutions, efficiency of the land use objective should be achieved by maximizing access between residential areas and shopping, service and recreational centers. Such systems would serve highly populated areas and particularly disadvantaged areas (low cost settlements); would increase access to tourism attraction areas; minimize disruption to neighborhood communities; and maximize linkages to major employment areas/centers.

Environmental Objective: The designed transit project should minimize intrusion and damage to the environment. Protected areas must be excluded from the set of the potential options. The expected accomplishments are: a reduction in energy consumption, minimal emission levels, minimal intrusion into environmentally sensitive and reserved areas, minimal noise impact to sensitive land use (such as hospitals, residential buildings and schools) during site construction.

Criteria and Indicators: To further support the design and evaluation of a metro-rail network, the major objectives are further broken down into specific objectives with their corresponding indicators. These indicators are then used to measure the performance of each alternative route/site option on each objective.

A proper governmental/metropolitan council's transportation policy should comply with the criteria structure shown by Figure 1. In contrast to the conventional approaches of predetermining route/site alternatives and then assessing their impacts subsequently, this integrated GIS/MCDM approach utilizes an opposite strategy. Determine first the proper, but at least promising locations of such facilities (the sites of the metro stations), along which the appropriate route options can be defined.

4   Route/Site Selection Plan of a Metro-Rail System via GIS/SMCE

This section presents an application of how a combined GIS/SMCE computer system can assist the design of alternative solutions for urban transit zone locations in a given metropolitan area. As is usual in many countries, spatially referenced data (with geometric positions and attribute data) are rarely available in a direct way. Therefore, the author has chosen a built-in database from the ILWIS (Integrated Land and Water Information System) library [6], which has been developed by the International Institute for Aerospace Survey and Earth Sciences (ITC), Enschede, The Netherlands. ILWIS is a Windows-based remote sensing and GIS software which integrates image, vector and thematic data in one powerful package on the desktop. In this study, Release 3.4 is applied (as an open source software as of July 1, 2007) which contains a strong SMCA module [5].

4.1 Study Area

The study area is Cochabamba city, a fast growing center located in the Andean region of Bolivia with a fast growing population of approximately 550000. The city is located at an elevation of about 2600 meters above sea level in a large valley on the alluvial fans at the foot of steep mountains. The city's northeastern side area is occasionally subjected to landslides, soil erosion and heavy flashfloods. Hence, from a perspective of urban development, the improvement of its transport infrastructure is of utmost importance, however, topographical and geological characteristics do form quite serious considerations to build a metro-rail system.

4.2 Geographic Data

Spatial data includes field collected data and GIS datasets (which consist of data derived by remote sensing from satellite imagery and/or field measurements) Attribute data are partly based on actual measurements, but, for the most part, are elicited from judgments, and, thus, they are fictive. To display geographic data (spatial and attribute data) on screen or in a printout, digitized vector maps (point, segment and polygon maps) and raster maps are used in a conveniently chosen visual representation form. Each map should contain the same coordinate system and georeference. In a raster map, spatial data are organized in pixels (grid cells). Pixels in a raster map all have the same dimensions. A particular pixel is uniquely determined by its geographic coordinates expressed in Latitudes (parallels) and Longitudes (meridians). With the help of a map projection, geographic coordinates are then converted into a metric coordinate system, measuring the X and Y directions in meters (UTM). This way a very high degree of accuracy is reached.

4.3 Description of Data Sets

The geographic area of the planned metro-rail project (network system) is given by the polygon map "Cityblock" and is shown in Figure 2. (The skewness of the chart is due to the north-pole orientation of the map.) This map has a total of 1408 blocks (polygons). To each of these polygons an identifier code is assigned. Block attributes are the geometric area in square meters; the prevailing land use type, i.e. residential (city blocks used primarily for housing), commercial (city blocks containing malls, supermarkets, shops, banks, hotels, etc.), institutional (such as schools, universities, hospitals, museums, governmental offices), industrial (buildings dedicated to industrial activities, storages), recreational (including protected areas, parks, sport fields), existing transport facilities (railway stations, bus stations, taxi services, public parking lots), airport, water (including lakes and rivers) and vacant (blocks that are not used for any urban activity); the codes of city districts and population (number of persons living or using a city block).

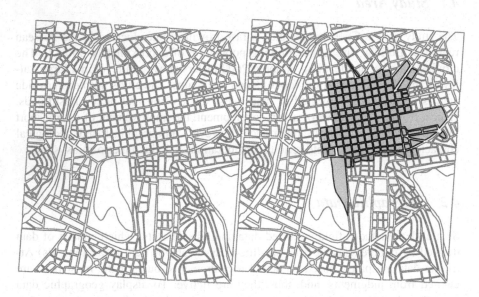

Fig. 2. Polygon map "Cityblock"   **Fig. 3.** The embedded polygon map "Center"

4.4 Identifying the Objectives/Criteria

As a simplified illustration of the site selection problem, consider the central part of the city only. This dependent polygon map "Center" has 137 blocks and its location is shown by the shaded area that is added to the layer "Cityblock" as it is depicted in Figure 3. Its block attributes include the following specific objectives (with their computed or estimated numerical data) for each polygon:

C1 = engineering characteristics and geological soil structure (rocks) [% scale],

C2 = ecological suitability [% scale],

C3 = connectivity index [m] (inversely converted into an interval scaled score),

C4 = population density [number of people/area-hectare], and

C5 = projected construction costs [mi$].

In the course of the aggregation to calculate the values of the composite attributes, of these criteria, C2 represents a spatial constraint that determines areas which are not at all suitable (these areas will get a value of 0 for that pixel in the final output); C1, C3 and C4 are criteria representing spatial benefits that contribute positively to the output (the higher the values are, the better they are with respect to those criteria); and C5 represents a spatial cost factor that contributes negatively to the output (the lower the value is, the better it is with respect to that criterion).

4.5 Processing of Raster Datasets

The raster layers were derived by applying an appropriate GIS raster processing method to the vector maps. These generated raster maps contain the data sets required for the SMCE. ILWIS requires all raster overlays to have the same pixel size. In this study, a pixel size of 20.00 meter was chosen to rasterize all vector layers.

4.6 Weighting of Criteria

Weights of the major objectives of the hierarchical decision model of Figure 1 were generated by an expert group formed of 5 transportation engineers, 3 mechanical engineers and 2 economists using the pairwise comparison matrix of the AHP. In real life problems, obviously, more groups of stakeholders must be requested. Our results, therefore, will not represent the positions involved organizations and civil members take and are only indicative. Still, we attempted to illustrate the deviations rising to the surface in the views of the different stakeholders' groups through evaluation. The inconsistency measures, μ_i, of the pairwise comparison matrices generated by the committee members varied between 0.023 and 0.042.

4.7 Spatial Multi-criteria Assessment

For the major objectives, their embedded factors and constraints together with their attached weights a criteria tree was built in ILWIS for three different project policies (equal vision, engineering vision, economic vision). In such an application of SMCA, each criterion is represented by a map. Due to the different units of measurement, standardization of all criteria should be carried out using an appropriate method ("Attribute", "Goal", or "Maximum") depending on the given factor and data characteristics. As a result, all the input maps are normalized to utility values between 0 (not suitable) and 1 (highly suitable). The completed criteria tree constructed in ILWIS is exhibited in Figure 4 for the engineering vision. In this study we selected only one specific objective from each set of the five sets of the major objectives as it can be seen from Figure 1.

  This process resulted in output maps for each of the policy visions, showing the suitable locations of metro-rail stations in the inner part of the city. As an example, the suitability maps of the single objectives (criteria) and the composite suitability map for this metro-rail station problem are shown in Figure 5 for the engineering vision. In these raster maps, areas of low suitability (valued 0 or close to 0) are symbolized by dark colors, while areas of highest suitability (valued 1 or close to 1) by hell colors. For true color interpretation, the reader is referred to the

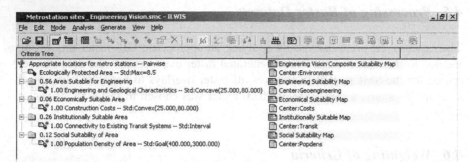

Fig. 4. The ILWIS screenshot of the criteria tree for identifying suitable locations

web version of this paper. The pixel information catalog contains the utility values in numerical terms for every pixel. We remark that the pixel information is invariant within a particular polygon (city block), since the functionality of these blocks can be regarded to be homogenous.

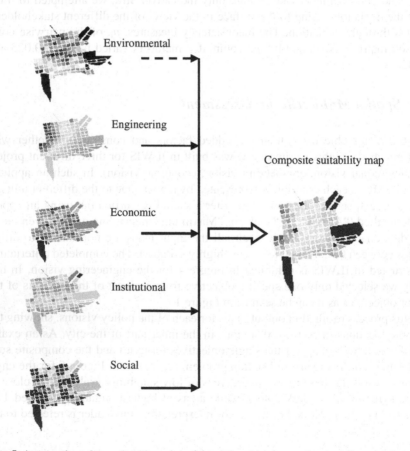

Fig. 5. Aggregation of the suitability maps of the major objectives to the overall composite map

4.8 Designing Alternative Metro-Rail Paths

In this step of the planning process the assessment of appropriate metro-rail routes is performed. First the processing of the raster datasets is extended to all other city blocks (beyond the blocks contained by the "Center" raster map), then, the output suitability maps for the polygon map "Cityblock" are generated. A careful analysis of the resulted maps for suitable locations of metro-rail stations enabled us to design proper pathways leading between the two major transit zones of the city from the origin node (South Railway Station) to the destination node (North Railway Station). These corridors, which span more than one block in the polygon map "Cityblock", are indicated by the shaded areas in Figure 6 for the engineering vision. It was required also to keep ourselves to the technical requirements, i.e. to the track building and vehicle engineering standards and specifications (e.g., feasible length and radius of transition curves, possible slope of the tracks, etc.), when these corridors were mapped out. As is displayed in Figure 7, three metro-rail routes for potential metro line alternatives have been identified (Blue Line, Red Line and Green Line). By further investigating the values of the multiple factors at different pixels within these three corridors the final locations for the metro-rail stations were fixed. Thus, a rough feasibility plan of this metro-rail network project was completed as is shown in Figure 7.

Fig. 6. Corridors of the metro-rail routes   **Fig. 7.** Feasibility plan of the metro network

4.9 Network Analysis and Evaluation of the Alternative Metro-Rail Routes

Effectiveness and efficiency of both construction and operation of a particular route are mostly determined by the embedded stations along that route. Therefore, it is reasonable to measure the extent to which an average suitability of the stations along a given route contributes to these characteristics. Introducing the mean spatial utility measure of a given metro-rail route as

$$\text{MSU}_i = \frac{\sum_{j=1}^{N} u_j}{N}, \qquad i = 1,2,...,M, \tag{1}$$

where u_j is the utility (suitability index) of the pixel (raster cell) underlying the jth site (metro station) along the ith route, N is the number of the selected sites along the ith route, M is the number of the alternative route options. To form the conventionally used measure in transportation problems called impedance, we should compute the complementary of each value of MSU_i and multiplying them by the total length of the routes [11, 12]. Hence the impedance of a route within the metro network system yields

$$\Omega_i = (1 - \text{MSU}_i) \cdot L_i, \qquad i = 1,2,...,M, \tag{2}$$

where L_i is the length of the ith route option (the length of the ith polyline). The higher the value of the impedance Ω_i is, the greater the costs associated with that route and/or the lower the benefits attained by it. Thus, the best route option is obtained by

$$\Omega^* = \min_i \{\Omega_i\}, \qquad i = 1,2,...,M. \tag{3}$$

The multiple criteria evaluation of the metro-rail network developed had been carried out based on the performance of each route with respect to the total impedance accumulated by that route. The results of this process for the three competitive metro-rail routes are presented in Table 1 for the engineering vision.

Table 1 presents the route options defined by the respective sequences of nodes (the raster cell code identifiers together with the names of the metro-rail stations and their corresponding utility values), the length of these metro-rail lines (computed by the distance calculation module of ILWIS), the mean spatial utility measures and the total impedance of the routes.

The results in Table 1 demonstrate that there is no route option that would entirely dominate over the other options. Observe, for example, that if a route is shorter than another, then this fact not necessarily means that it represents a better route option. The best option, Route 3 (Green Line), however, outperforms the other two ones both in terms of the total impedance and the length of the line. Therefore, considering the enormous construction costs of the whole metro-rail project, the implementation of the Green Line might be proposed. Perhaps the best conceivable proposal could be to lengthen the track of the Green Line to the airport.

Table 1. Effect table of the three metro-rail routes

Route 1 (Blue Line)	Route 2 (Red Line)	Route 3 (Green Line)
(463) South Railway Station. 0.75	(463) South Railway Station 0.75	(463) South Railway Station 0.75
(400) Airport 0.78	(341) Meridian Hotel 0.70	(508) Giant Mall 0.65
(355) Riverside 0.74	(349) Central Park 0.61	(295) Royal Square 0.87
(147) Bridge Square 0.55	(118) Forbes 0.31	(265) Prince Cross 0.80
(181) North Railway Station 0.83	(181) North Railway Station 0.83	(181) North Railway Station 0.83
$L_1 = 5801$ m $MSU_1 = 0.73$ $\Omega_1 = 1566.27$ -	$L_2 = 4443$ m $MSU_2 = 0.64$ $\Omega_2 = 1599.48$ -	$L_3 = 4146$ m $MSU_3 = 0.78$ $\Omega_3 = 912.12$ $\square^*$

Conclusions

The primary objective of this paper was to facilitate the route/site selection process in the presence of multiple and diverse decision criteria using GIS equipped with multi-criteria decision support capability to incorporate the preferences of the committees making the decision. The analytical capabilities and the computational functionality of a knowledge-based, intelligent GIS promote to produce policy relevant information to decision makers in the design, evaluation and implementation of such spatial decision making processes. The suitability analysis module provides an incremental or flexible evaluation scheme applied to relevant map layers, which helps to define the desired suitability conditions by examining each composite map. In generating solutions, the user can interact with the responses of the system through the user interface. Although different stakeholders usually have different priorities to highest level objectives, however, using this approach provides a considerable help in reaching a satisfactory compromise ranking of the objectives for the conflicting interests. The deployment of such an integrated GIS/MCDM approach in urban transportation projects was demonstrated through the planning of a metro-rail network system.

References

[1] Anselin, L., Dodson, R.F., Hudak, S.: Linking GIS and Spatial Data Analysis in Practice. Geogr. Syst. 1, 3–23 (1993)
[2] Eldrandaly, K., Eldin, N., Sui, D.: A COM-based Spatial Decision Support System for Industrial Site Selection. J. Geogr. Inf. Decis. Anal. 7, 72–92 (2003)
[3] Goodchild, M.F., Haining, R., Wise, S.: Integrating GIS and Spatial Data Analysis: Problems and Possibilities. Int. J. Geogr. Inf. Syst. 6, 407–423 (1992)
[4] van Herwijnen, M.: Spatial Decision Support for Environmental Management. Vrije Universiteit. Amsterdam (1999)
[5] ILWIS 3.4 (2008), http://52north.org/index (Accessed, 22 August 2008)
[6] ILWIS User's Guide (2004), http://www.itc.nl/ (Accessed, 14 July 2008)
[7] Jankowski, P.: Integrating Geographical Information Systems and Multiple Criteria Decision-Making Methods. Int. J. Geogr. Inf. Syst. 9, 251–273 (1995)
[8] Jun, C.: Design of an Intelligent Geographic Information System for Multi-Criteria Site Analysis. URISA J. 12, 5–17 (2000)
[9] Keeney, R.L.: Value-focused Thinking. Harvard University Press, London (1992)
[10] Keeney, R.L.: Siting Energy Facilities. Academic Press, London (1980)
[11] Keshkamat, S.S.: Formulation & Evaluation of Transport Planning Alternatives Using Spatial Multi Criteria Assessment and Network Analysis. MSc thesis. International Institute for Geo-Information Science and Earth Observation. Enschede (2007)
[12] Keshkamat, S.S., Looijen, J.M., Zuidgest, M.H.P.: The Formulation and Evaluation of Transport Route Planning Alternatives: a Spatial Decision System for the Via Baltica Project. Poland. J. Transp. Geogr. 17, 54–64 (2009)
[13] Malczewski, J.: GIS and Multicriteria Decision Analysis. Wiley, Chichester (1999)
[14] Nyerges, T.L.: Coupling GIS and Spatial Analytical Models. In: Breshanan, P., Corwin, E., Cowen, D. (eds.) Proceedings of 5th International Symposium on Spatial Data Handling, University of South Carolina, pp. 534–543 (1992)
[15] Saaty, T.L.: The Analytic Hierarchy Process. McGraw-Hill, New York (1980)
[16] Sharifi, M.A., Boerboom, L., Shamsudin, K.B., et al.: Spatial Multi-Criteria Decision Analysis. In: Integrated Planning for Public Transport and Land Use Development Study in Klang Valley, Malaysia. ISPRS Technical Comission II Symposium, Vienna, pp. 85–91 (2006)
[17] Sharifi, M.A., Boerboom, L., Shamsudin, K.B.: Evaluating Rail Network Options Using Multi-Criteria Decision Analysis. Case Study Klang Valley, Malaysia. In: Application of Planning and Decision Support Systems, International Islamic University, Malaysia, pp. 21–60.(2004)
[18] Sharifi, M.A., Retsios, V.: Site Selection for Waste Disposal through Spatial Multiple Criteria Decision Analysis. J. Telecommun. Inform. Technol. 3, 1–11 (2004)
[19] Transportation Case Studies in GIS: Portland Metro, Oregon, GIS Database for Urban Transportation Planning, FHWA-PD-98-065, No. 2 (1998)

Author Index